PREPARE YOUR STUDENTS FOR TOMORROW WITH *CORD BIOLOGY: SCIENCE IN CONTEXT* TODAY

If there's one thing you know you can do for your students it's teach them biology. With **CORD Biology: Science in Context** you can do even more. You can help your students achieve two goals: build a solid foundation in biology and learn skills that will help them in their future. That's because **CORD Biology: Science in Context** presents biology content and science skills in the context of careers.

Unlike any other text on the market, **CORD Biology: Science in Context** helps students recognize the relevance of biology to their lives and in future career choices. Every student can benefit from **CORD Biology: Science in Context.** Here's why:

➡ *Interactive, contextual approach* makes concepts accessible to all students.

➡ *Learning concepts within the context of potential careers* and real-world problems helps students apply skills and concepts to new situations.

CORD BIOLOGY: SCIENCE IN CONTEXT REACHES OUT TO STUDENTS AND PREPARES THEM FOR LEARNING

CORD Biology: Science in Context puts biology in perspective for every student. The book links science concepts to careers in science and personal experience.

CORD Biology: Science in Context

➡ Demonstrates the real-life relevance of science skills and concepts

➡ Makes biology accessible to all students

➡ Meets the need of every student through a variety of coordinated materials—textbook, laboratory manual, logbook, and video

"Why Should I Learn This?"
Chapter opening scenarios provide a reason why each chapter's topics are relevant.

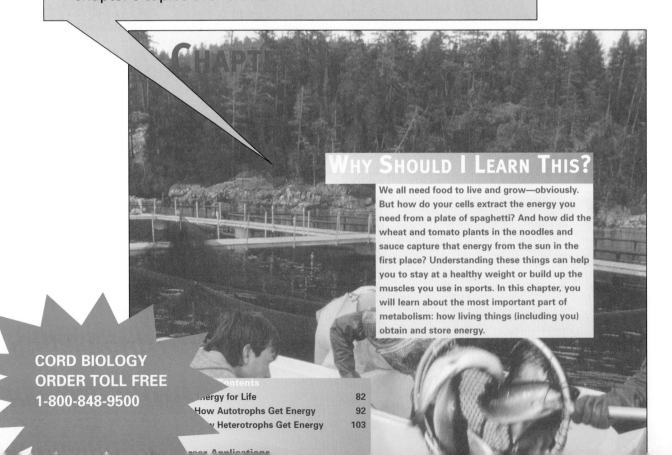

WHY SHOULD I LEARN THIS?

We all need food to live and grow—obviously. But how do your cells extract the energy you need from a plate of spaghetti? And how did the wheat and tomato plants in the noodles and sauce capture that energy from the sun in the first place? Understanding these things can help you to stay at a healthy weight or build up the muscles you use in sports. In this chapter, you will learn about the most important part of metabolism: how living things (including you) obtain and store energy.

CELL METABOLISM

WHAT WILL I LEARN?

1. how living things use chemical reactions to obtain and store energy
2. how plants convert solar energy to usable chemical energy
3. how living things break down sugars and other substances to obtain energy
4. oxygen's role in extracting energy from foods

"Fish gotta swim, and birds gotta fly," go the words to an old song. Swimming and flying take energy. Where does that energy come from? How do fish and birds or any organism put it to work? How do they store and retrieve energy? People who raise fish for market know that fish get their energy from food and oxygen. Any farmer knows that you can't grow a crop without plenty of sunlight or raise livestock without providing lots of energy in the form of feed.

In this chapter you will learn how organisms collect energy from sunlight, from minerals, and from each other and store it as chemical energy. You will also learn about chemical reactions that take place in nearly all organisms, making chemical energy available for any purpose, including swimming, flying, and growing.

CAREER FEATURES INSPIRE STUDENTS TO DO THEIR BEST AND PREPARE FOR THE FUTURE

CORD Biology: Science in Context relates what students learn in your classroom to what they experience personally and what they may encounter professionally. This contextual approach also incorporates strategies for diverse learning styles so that every student considers how science will impact his or her career.

Career Applications and **Learning Links** relate academics to careers through interviews with professionals, relevant research, intriguing debate, and Internet activities.

LEARNING LINK HOME AWAY FROM HOME

People who care for animals in captivity must provide conditions that allow the animals to maintain homeostasis. Special problems arise when animals are moved from their native environment. Zoos in the northern United States, for example, may have animals such as camels, which are native to hot desert areas. Find out from a zookeeper or an animal breeder how the requirements of such animals are met.

CAREER APPLICATIONS

The applications that follow are like the ones you will encounter in many workplaces. Use the biology you learned in this chapter to complete the activities. Share your work with the class.

AGRICULTURE & AGRIBUSINESS

Greenhouse Plants Get Extra CO_2

Talk to a commercial greenhouse operator or search for information on the Internet or elsewhere about growing plants in an atmosphere that is supplemented with CO_2. What effects does the extra CO_2 have on plants? When is CO_2 supplementation used? How feasible is it for widespread cultivation? Is there an optimum level of supplementation? Write up your findings in a brief report in your logbook and be prepared to share it with the class.

BUSINESS & MARKETING

The Tour de Plants

Ask a local travel agency about fall leaf tour packages. What does the package include? What transportation is provided? Are meals provided? How many people does the agency typically book in a year for leaf tours? How do they decide when and where the tours will go? What is the closest location? Why do people go on fall color tours? Design a brochure that will entice leaf peepers.

LESSON 5.4 MUTATIONS

The term "mutation" might make you think of monsters, but most mutations cause only minor changes in an enzyme or other protein. Still, even slight changes can be bad news for the organism.

DENTAL HYGIENIST

CAREER PROFILE

Cindy A. is a dental hygienist who works in a general dentistry office. In her daily work she checks the condition of patients' teeth and looks for possible cavities, although the dentist makes all diagnoses. She cleans and polishes teeth and applies sealant to some of the younger patients' teeth. Cindy also keeps patients' charts and takes X rays of their teeth.

When asked whether she worries about the safety of X rays, Cindy said, "While it is true that dental X rays are a form of radiation that can cause mutations, dental X rays are generally considered no more dangerous than walking in the sun. This is because the dose, or amount of radiation a patient is exposed to, is small. In our office patients receive X rays only once every two

This dental hygienist is preparing to take an X ray of a patient's teeth for a dentist to examine.

FUEL FARMS

BIOLOGY IN CONTEXT

Melvin Calvin is remembered for his work describing the reactions of photosynthesis, but photosynthesis was not Calvin's only interest. Calvin spent a good part of his later career studying plants called *Euphorbia,* or gopher plant. This plant received worldwide attention because it produces oils that are very similar to crude petroleum.

Calvin thought that the plant would adapt to growing on land that is too dry for other crops. He envisioned petroleum plantations where the plants would produce the equivalent of many barrels of crude oil each year. Calvin even showed that the sap could be poured directly into the fuel tank of a diesel-powered car or truck without any processing or purification.

In the early 1980s the University of Arizona tested Calvin's ideas about growing oil. *Euphorbia* plants were gathered from around the world and planted. The first crop was a total failure. The scientists determined that *Euphorbia* would make a better winter crop than a summer crop. The plants did not adapt well to dry soils and had to be irrigated. Also, the plants were susceptible to soil-borne diseases that are common in the Southwest. The final blow came when the crop

Euphorbia plants produce oils that can be used as fuel.

CORD Biology Builds Skills for Future Success

From math tips and vocabulary notes to laboratory work and the history of science, ***CORD Biology: Science in Context*** helps build the skills that will last students a lifetime. Students develop problem-solving skills and knowledge and use cooperative learning and data collection to complete projects throughout the text and laboratory manual.

> **Word Banks,** found in the margins of the text throughout each lesson, develop understanding of science terms.

> **Changing Ideas** feature historical perspectives on scientists' perceptions of issues and compare them to current theories.

WORD BANK

auto- = same, self, self-acting

-clave = cleave, split, break into pieces

An autoclave is an apparatus that creates heat and pressure to break up cells.

the use of an auto (Figure 7.5). An makes pressuri at temperature the boiling point to kill micr People w pressure canning use pr surized steam to

Figure 7.5: *Autoclaves use extremely high temperature to kill microorganisms that contaminate laboratory equipment.*

CHANGING IDEAS

DISCOVERY OF DISEASE AGENTS

In the late 1600s, Anton van Leeuwenhoek first saw microorganisms with his homemade microscope. But a century before that, the Italian physician Girolamo Fracastoro argued that many diseases are caused by organisms that are too small to see. When microscopes finally showed bacteria and protozoa on boiled extracts of hay and meat, scientists began to explain how they got there. One theory, called the theory of spontaneous generation, proposed that the organisms arose mysteriously out of the organic matter in the extracts.

Louis Pasteur later proved that microorganisms entered the extracts from the surrounding air and reproduced there. He used the apparatus shown in Figure 7.22, putting nutrient solutions in the flasks. Airborne microorganisms were trapped along the glass necks, and nothing grew in the nutrient solutions. When Pasteur broke off the long necks close to the flasks, the nutrient solutions soon turned cloudy with growing microorganisms.

A **acillus.** it is

INVESTIGATION 5B   THE AMES TEST

The best way to determine whether a substance is likely to cause mutations is to expose live cells or tissues to the substance and observe the effects. A laboratory test called an Ames test is used to screen for chemicals that may cause mutations. The test exposes a bacterial culture to the test substance and looks for changes in the bacterial phenotype that are known to accompany mutations. In this investigation you will test a number of household substances to determine whether they are mutagens.

DO MUTATIONS REPRODUCE?

Even without exposure to the mutagens mentioned above, DNA would still mutate. Some mutations occur spontaneously as a result

MATH TIPS

To find the probability of two events occurring when they are independent of (not related to) one another, multiply the likelihood of the first event by the likelihood of the second event. For example, if you have a 50% chance of getting an A on your next biology exam and only a 10% chance of getting an A on your next history exam, then the probability of getting As on both exams is 50% × 10% = 0.50 × 0.10 × 100% = 5%.

Use this same formula to determine the probabilities in a genetic cross when the genes for a trait assort independently. If the likelihood of a plant's producing offspring with white flowers is 50% and the likelihood of its producing a dwarf is 25%, then the probability of a white-flowered dwarf is 50% × 25% = 0.50 × 0.250 × 100% = 12.5%.

Remember that the alleles for dwarf and for wrinkled seeds are both recessive. Add blocks with the same phenotypes. The result of the Punnett square analysis is a probability (see the Math Tips):

- Nine of every sixteen offspring (or $\frac{9}{16}$) are tall with smooth seeds,

- Three of every sixteen offspring (or $\frac{3}{16}$) are tall with wrinkled seeds,

- Three of every sixteen offspring (or $\frac{3}{16}$) are dwarf with smooth seeds,

- One of every sixteen offspring (or $\frac{1}{16}$) is dwarf with wrinkled seeds.

In genetics, probabilities are often expressed as a ratio. In this case the ratio of phenotypes in the F_2 is 9:3:3:1.

Look at Figure 4.14 again and consider the ratio of phenotypes for each trait separately. There is a 12:4, or 3:1, ratio of tall plants to dwarf plants, just as you would expect in a monohybrid cross of heterozygous tall plants. The ratio of plants with smooth seeds to those with wrinkled seeds is also 3:1, like a monohybrid cross. So you could think of the ratio of phenotypes for the dihybrid cross in terms of two monohybrid crosses. Multiply the probability of finding a plant with a tall or dwarf stem by the probability of finding a plant with smooth or wrinkled seeds to get the proportion of phenotypes:

- Tall, smooth-seeded plants: $\frac{3}{4} \times \frac{3}{4} = \frac{9}{16}$

- Tall, wrinkled-seeded plants: $\frac{3}{4} \times \frac{1}{4} = \frac{3}{16}$

- Dwarf, smooth-seeded plants: $\frac{1}{4} \times \frac{3}{4} = \frac{3}{16}$

ACTIVITIES AND ASSESSMENTS ENSURE THAT STUDENTS ARE WORKING IN THE RIGHT DIRECTION

Activities and assessments throughout the text stress the relevance of biology and help students measure their progress and understanding.

Think and Discuss provokes students to consider the relevance of concepts learned in each chapter.

Hands-on Activities clarify understanding and reinforce what students have learned.

Think and Discuss

15 Why is it important for scient[...] share their findings with other scientists? Use the developm[...] the cell theory as an example.

16 When you sweat, active trans[...] carries positive sodium ions out of the cells of your sweat gla[...] Negative chloride ions and water molecules then pass [...] through membrane channels. People with the disease cyst[...] fibrosis lack working chloride channels. How would you ex[...]ect this to affect their ability to sweat normally? Explain your answer.

ACTIVITY 2-10   CELL SIZE AND CELL DIVISION

See Activity 2-10 in the TRB.

MATH TIPS

Graphing your results can help you to see whether the size of your pieces has more effect on their volume or their surface area. Set up a graph on a piece of graph paper. Label the *x*-axis "Length (cm)" and the *y*-axis (on the left) "Surface Area (cm^2)." Plot your surface area data on this graph and draw a smooth curve through all the points. Then label a second *y*-axis (on the right) "Volume (cm^3)" and plot the volume data on the same graph in a different color. Draw a smooth curve through all those points in the same color. Which curve is steeper? What does that mean?

An organism faces serious consequences if its cells don't divide normally. But why do cells have to divide at all? Is there some advantage to small cells? In this activity, you will learn how size affects objects of various shapes and develop a hypothesis about why cell division is important.

Set up a table in your logbook with three columns and four rows. Label the columns "Length of Pieces," "Surface Area (cm^2)," and "Volume (cm^3)." Use your ruler and the appropriate equations from Figure 2.21 to determine the volume and surface area of your food. Record these numbers in the first row of your table, along with the length of your food. Cut your food in half crosswise, as in Figure 2.21. Determine the surface area and volume of one piece, and complete the second row of your table. Repeat the process, cutting one piece in half twice more and

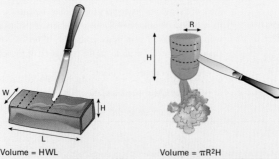

Volume = HWL
Surface area = 2(HW + HL + WL)

Volume = $\pi R^2 H$
Surface area = $2\pi(R^2 + RH)$

Figure 2.21: *How to cut and calculate volume and surface area*

Lessons Assessments are an opportunity for students to communicate what they've learned through writing and in-class discussion.

LESSON 2.2 ASSESSMENT

1. Cell membranes that are composed mostly of lipid separate the watery cell contents from the bath water.

2. Small or nonpolar molecules diffuse through the cell membrane; larger or more polar particles enter or leave via facilitated diffusion or active transport; still larger objects enter by endocytosis and leave by exocytosis.

3. The membrane has a fat-soluble region consisting of two layers of lipid molecules with their hydrophobic tails pointing inward. Protein

1 Your body consists mostly of water. Why don't you dissolve in the bathtub?

2 Compare three ways in which substances enter and leave cells. Consider the size and type of particles that pass into or out of cells and how they do so.

3 What makes up a cell membrane and how are these components organized?

4 Foods that are preserved by pickling in salt water or candying in syrup often shrink and become wrinkled. Explain how these changes come about.

Chapter Assessments give you the tools to further measure students' understanding through skill-, context-, career-, and lab-based problems.

CHAPTER 4 ASSESSMENT

Concept Review

See TRB, Chapter 4 for answers.

1 Explain how a karyotype shows genetic problems.

2 You are a genetic counselor. A married couple has come to your office to determine the possibility that they might have a child with sickle cell anemia. The father is a carrier of the sickle cell trait (Ss) but the mother is not a carrier nor does she have the disease (SS). What is the likelihood of this couple having a child with sickle cell anemia?

3 Describe how chromosomes of the same pair can be recognized on a karyotype.

4 Discuss why faulty instructions in an organism's DNA blueprint result in problems with its proteins.

In the Workplace questions encourage students to consider how they'll use what they've learned in possible future careers.

In the Workplace

19 How might an artificial membrane, similar to the dialysis tubing you used in lab, be used to purify a city's water supply? What kinds of impurities could it remove?

20 What is one important chemical property of a detergent that a manufacturer must always consider in designing a new cleaning product?

21 A substance called a surfactant is often applied as a coating to the lung lining of newborns who have breathing difficulties. Should surfactants be polar or nonpolar substances? Explain the reason for your answer.

COURSE COMPONENTS

The Total Package Gives Your Students the Big Picture

CORD Biology helps you to facilitate successful, relevant learning in the classroom. The entire package reinforces the contextual-based approach of the text. *CORD Biology* components include:

Teacher's Resource Book enriches the learning environment with additional activities, assessment materials, blackline masters, and laboratory notes.

(0-538-68169-1)

Student Edition helps students to learn biology concepts through real-life and career-related examples.
(0-538-68166-7)

Teacher's Annotated Edition makes instruction a cinch with margin notes on activities, labs, and assessment tools.

(0-538-68167-5)

CORD BIOLOGY
ORDER TOLL FREE
1-800-848-9500

Laboratory Manual
helps students complete
investigations in the
Student Edition.
(0-538-68168-3)

Video launches
each unit of study.
(0-538-68328-7)

**Computerized
Test Bank** includes
questions for assessment
(Macintosh and Windows).
(0-538-68327-9)

Logbook
provides a convenient
resource for students to
take notes and make
observations as they
would in the working
world. (0-538-68326-0)

CORD BIOLOGY: SCIENCE IN CONTEXT

Table of Contents

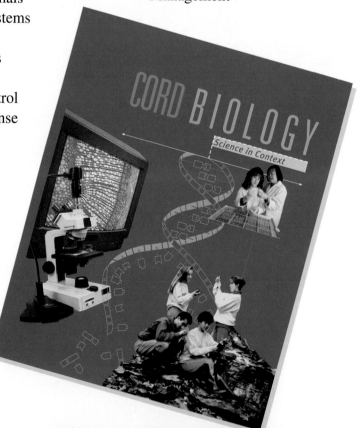

CALL TO ACTION

Order Today and Prepare Your Students for Tomorrow

Call or visit our Web site
www.globefearon.com

CORD BIOLOGY

Science in Context

Developed by the
Center for Occupational Research and Development
Waco, Texas

Upper Saddle River, New Jersey
www.globefearon.com

CORD Staff
Director: Stephen Jones, Ph.D.
Program Manager: John Souders, Jr., Ph.D.
Educational Specialist: Bonnie Rinard
Laboratory Specialist: John Reiher, Ph.D.

Editorial and Production Services: PC&F, Inc.
Design: Zender + Associates, Inc.
Photo Research: Alix Parson
Front cover image: Researchers Examine Sprouts, Ed Young/Corbis
　　　　　　　　　Videoscope photograph courtesy of Video Labs, Minneapolis, MN
　　　　　　　　　Research team photograph courtesy of LaMotte Chemicals, Chestertown, MD
Back cover image: ©1998, Corbis Images

Globe Fearon Educational Publisher
Upper Saddle River, New Jersey
www.globefearon.com

Consultants and Reviewers

Science Consultants

Larry Bond

Director of Cross Roads Educational Consortium
Fairfield, VA

James Cockerill

Science Education Consultant
Baltimore, MD

Molly Greenberg

Consultant
Columbus, OH

Mary Ann Figuly

Science Teacher
Schlagle High School
Kansas City, KS

Douglas Mandt

Safety Specialist
Edgewood, WA

Tim Ritter

Science Educator
Rancho Cucamonga High School
Rancho Cucamonga, CA

Stephen W. Taber

Biology Department
St. Edward's University
Austin, TX

Reviewers

Mary Colvard

Science Teacher
Cobleskill-Richmondville High School
Cobleskill, NY

Susan Gannaway

Professor of Education
North Georgia College and State University
Dahlonega, GA

Janis Lariviere

Science Teacher
West Lake High School
The Learning Center
Austin, TX

Lisa Merritt

Educational Consultant
ACKLAM Incorporated
Homosassa, FL

Lisa Jones-Rath

Science Teacher
Birmingham College Prep Magnet, LAUSD
Van Nuys, CA

TABLE OF CONTENTS

CORD Biology is a result of a 10-year national collaboration led by CORD, a nonprofit organization dedicated to increasing student achievement through the use of contextual and other active learning strategies. This initiative, involving literally thousands of educators and employers from across the country, sought to raise student achievement in mathematics and science by integrating both proven instructional techniques and pioneering innovation. The resulting curriculum materials opened the doors to new ways of learning for millions of students by teaching academic content in context to provide meaning and relevance for today's learners.

In developing this curriculum, the CORD staff has incorporated all that we have learned in the last 20 years about contextual learning and exemplary teaching. Our goal has been to make *CORD Biology* a highly effective and motivating vehicle through which students can learn the science skills they need to be successful in future science courses and in their future careers. Rich resources are provided through innovative laboratory activities, career-oriented problems, and research activities that enable teachers to finally answer the questions "Why do I have to learn this?" and "How will I use this?"

CORD is committed to playing a leading role in educational reform through curriculum development. We believe that learning is most effective when concepts are presented within the context of their use, and when students understand, experience, and practice the application of concepts to solve problems. We hope that *CORD Biology* provides a highly effective approach for bringing successful biology achievement to many students.

Finally, I would like to say a word to you, the teacher. Your work with our youth is absolutely critical, not only for your students, but also for our entire nation. We recognize that this textbook will be effective only to the extent that you can make it your textbook—personalizing it, customizing it, enhancing it from your teaching experience and special expertise. Please let us hear from you as you journey through the exciting world of *CORD Biology*. Tell us what you like and what works and—far more important—how we can make it better.

I wish you the greatest success in your pursuit of educational excellence.

Daniel M. Hull

Daniel M. Hull
President and Chief Executive Officer

DMH/jm

CONTEXTUAL LEARNING STRATEGIES

Contextual learning strategies have proven to be extremely useful for teachers who wish to improve learning for students with multiple intelligences and different learning styles. When teachers combine contextual learning with cooperative learning, they can greatly facilitate their students' efforts in connecting new concepts to existing knowledge and to the world outside the classroom.

DIVERSITY IN LEARNING STYLES

In a sense, there is nothing new about contextual learning. There have always been teachers who intuitively understand how to teach concepts so that all learners can grasp them—through example, illustration, and hands-on application. These teachers seemed to understand, even before the idea was published by Howard Gardner, that human capacity for learning is much broader than the traditional measurements of intelligence (verbal and analytical) would indicate. Gardner, a Professor of Education at Harvard University, argues that individuals have as many as seven forms of intelligence: linguistic, logical/mathematical, musical, spatial, kinesthetic, interpersonal, and intrapersonal. Gardner also observes that everyone has some measure of each of the seven intelligences and that specific strengths and combinations of intelligences vary for each individual.

Gardner's theory of multiple intelligences identifies a need to address diverse learning styles in the classroom. Learning theorist David Kolb reinforces this need. Kolb observes that learners tend to perceive information either abstractly (by conceptualizing and thinking) or concretely (by experiencing and feeling), and then they process that information either actively (by experimenting and doing) or reflectively (by observing and watching). These four learning styles are typically set on an axis as a way of understanding the entire realm of students' tendencies for learning.

Kolb's construction, like Gardner's, clearly indicates that most students do not fit neatly into one category or the other. Almost all students can learn by, and benefit from, all four experiences (thinking, feeling, doing, and watching). And no one type of learning is superior to another; all contribute to the process of effective learning. Nevertheless, most students will show a preference for one or two particular kinds of learning, and this preference will indicate the

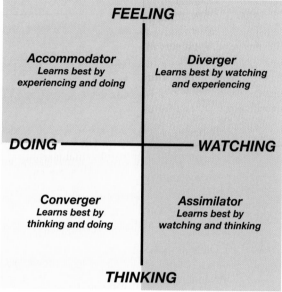

individual's primary learning style. In addition, Kolb's studies indicate that only a small percentage of students have a strong ability to learn by perceiving abstractly and processing reflectively—precisely the learning style that is rewarded in the commonly used lecture method of teaching. The majority of students tend to perceive and process information concretely and actively. Most people, in other words, are extroverted learners. They learn best through interpersonal communication, group learning, sharing, mutual support, team processes, and positive reinforcement.

MAKING CONNECTIONS

Despite individual differences in learning styles and intelligences, all learning requires and strives for connectedness. Isolated bits of information normally are not processed and retained by the mind for meaningful usage unless connections are made and points of reference or relationships are established between what is known and what is not known.

Traditionally, students have been expected to make these connections on their own. However, growing numbers of teachers are discovering that most students' interest and achievement in math, science, and language improve dramatically when they are directly assisted in making these connections. Today, more than ever before, teachers must facilitate their students' efforts in connecting new information, or knowledge, to experiences they have had or other knowledge they have already mastered. Students' involvement in their schoolwork increases significantly when they are taught why they are learning the concepts and how those concepts can be used outside the classroom, especially in the workplace. Furthermore, most students learn much more efficiently when they are allowed to work cooperatively with other students in groups or teams and to learn from one another. Collaborative learning greatly facilitates making connections.

A curriculum using contextual learning strategies and demonstrating the connections and usefulness of the curriculum should require the average student to develop a stronger academic foundation, a higher caliber of work skills, and a better understanding of how academic concepts relate to his or her environment outside the classroom and to the workplace. This is a higher level of learning that is not usually taught even to the above-average student.

CONTEXTUAL LEARNING

The premise of contextual learning theory is that learning occurs only when students process new information or knowledge in such a way that it makes sense to them in their frame of reference—their own inner world of memory, experience, and response. This approach to learning and teaching assumes that the mind naturally seeks meaning in context, that is, in the learner's own environment.

In a contextual learning environment, students discover meaningful connections between abstract ideas and practical applications in the context of their world. The students internalize concepts through the process of discovering, reinforcing, and interrelating the ideas and applications.

In *CORD Biology,* each lesson of each chapter begins with an *Activity* and a *Biology in Context* feature or *Career Profile.* Right from the beginning, students know at least one application of the lesson material, and they have the opportunity to experience some aspect of it through a hands-on activity. For example, Chapter 2, Lesson 2.2, which discusses the cell

membrane, begins with an *Activity* in which students observe a mixture of oil and water and what happens when they add a few drops of detergent and shake the mixture. They then read a *Biology in Context* about how liposomes are used in cosmetics or as a vehicle for transporting drugs across cell membranes. A text reading describes the structure of the cell membrane and a discussion of what it means for a membrane to be selectively permeable. This is followed by an *Investigation* in which students test the permeability of dialysis tubing.

This example demonstrates that contextual curricula and instruction encourage many forms of learning. These include:

- *Relating:* learning in the context of life experiences.
- *Transferring:* learning in the context of existing knowledge—using and building upon what a student already knows.
- *Applying:* learning in the context of how the knowledge or information can be used.
- *Experiencing:* learning in the context of exploration, discovery, and invention.
- *Cooperating:* learning in the context of sharing, responding, and communicating with other learners.

Exercising the different contexts in which students learn will broaden their abilities to make connections, enjoy discovery, and use knowledge. These are abilities they will need throughout their lives and careers.

RESOURCES FOR FURTHER READING

Caine, Renate, and Geoffrey. *Making Connections: Teaching and the Human Brain.* Association for Supervision and Curriculum Development, 1991.

Gardner, Howard. *Frames of Mind: The Theory of Multiple Intelligences.* Basic Books, 1983.

Hull, Daniel H. *Opening Minds, Opening Doors: The Rebirth of American Education.* CORD Communications, 1993.

Kolb, David. *Experiential Learning.* Prentice-Hall, 1984.

Secretary's Commission on Achieving Necessary Skills. *What Work Requires of Schools: A SCANS Report for America 2000.* U.S. Department of Labor, U.S. Government Printing Office, 1992.

USING *CORD BIOLOGY* EFFECTIVELY

The *CORD Biology* curriculum has several components and numerous features that work together to enhance learning by students who have various learning styles. The student text includes science readings and illustrations as well as special features that are unique to *CORD Biology*. The special features include *Biology in Context, Career Profile, Changing Ideas, Word Bank, Math Tip, Learning Link, Activities, Investigations* (the procedures are detailed for students in the Investigations Manual), and *Career Applications*. Following is a description of the various features, their purpose, and strategies for using them successfully.

CONTENT FEATURES OF THE STUDENT TEXT

CHAPTER OPENER SPREAD

Each chapter begins with a two-page spread that includes a section *Why Should I Learn This?*, an abbreviated Table of Contents, and *What Will I Learn? Why Should I Learn This?* is a student-friendly introduction to the chapter that gives examples of when the student might use the information or tells how others use the information in the chapter. After listing the specific objectives of the chapter, *What Will I Learn?* includes a more specific example of what is going to be covered in the chapter.

BIOLOGY IN CONTEXT

The *Biology in Context* feature generally appears at the beginning of each lesson and illustrates how some aspect of the biology content of the lesson is important in some situation. Topics of these pieces may be taken from the workplace, personal issues, historic events, or current research. They are based on actual interviews or derived from news sources. Let students know that new content may be introduced through these scenarios and so they are important for the understanding of the material—not just interesting trivia pieces.

CAREER PROFILE

Every *Career Profile* is based on an actual interview with the person being profiled. This feature describes aspects of the person's job related to the content of the chapter and describes the educational requirements and/or certifications that are required for students aspiring to that career. New biology content may be introduced in this feature.

CHANGING IDEAS

The intent of the *Changing Ideas* feature is to illustrate that biology is not a static science. It highlights changes in scientific thought and the people who have contributed to those changes. The nationalities of the scientists are given to show that science is a global endeavor and scientists build on ideas of others. In some instances it describes how improved technology contributes to better understanding of biological processes.

WORD BANK

Throughout the text you will find *Word Bank* features in the margins of the text. The *Word Bank* breaks new words into their prefixes, suffixes, and combining forms. Its main purpose is not to define the words—that is done in the text and the glossary. Rather, by showing how a word is formed, the *Word Bank* enables students to reason out the meanings of new scientific terms when they come across them for the first time.

MATH TIPS

A *Math Tip* is included wherever an explanation of some math concept might facilitate the understanding of the biology being covered.

ACTIVE LEARNING FEATURES OF THE STUDENT TEXT

LEARNING LINK

A *Learning Link* gives students directed help on how to find out information from a variety of print, electronic, and community resources. Some *Learning Links* have students research a historic event, some a current event. Others deal with topics on which knowledge is rapidly expanding and changing and the students must discover the latest information and current views. Still others have students research social issues and take and support a side in an ongoing debate. Generally, the work is to be done individually, but some activities have students work in pairs. The results of the research are to be recorded in the student's logbook and reported back to class. You might wish to divide the class into groups and let each group work on a different *Learning Link*. Each group can share its research with the class, making more efficient use of resources and time.

ACTIVITIES

The *Activities* involve students in short, hands-on experiences using the tools and techniques of science. Students are required to observe, measure, calculate, analyze, compare, and classify. Through the *Activities*, students build skills that they can use in future science classes as well as in the workplace. Questions at the end of each *Activity* lead the students to think and reason about their observations and data.

The *Activities* are designed to be done in small groups. A laboratory facility is the preferred setting for most, although many can be done in the classroom. The Teacher's Resource Book (TRB) provides background information and complete materials and equipment lists. Safety issues are discussed as well as any parts of the *Activity* that might be problematic. The TRB lists any advance preparations that are required. Although time and equipment limits might prevent the use of all of the *Activities,* you should try to do as many as possible. In lessons with more than one *Activity,* you might want to have part of the class complete one *Activity* while the other part completes a different *Activity.* Each group can show the other what it did and share the results.

Safety is an important aspect of any science activity. Several resources are available to you in emphasizing safety to your students. In *Appendix A* you will find a list of safety rules. Also, in the Laboratory Manual is a short lesson on safety that includes a safety quiz and safety contract. In addition, the teacher's notes in the ATE or TRB for each *Activity* and *Investigation* will include a summary of the safety issues related to that particular experience.

INVESTIGATIONS

Investigations are in depth laboratory-based experiences that often involve procedures or modification of procedures used by scientists and technicians in a variety of fields. They involve data analysis, and may involve students in higher-order skills such as organizing data, formulating hypotheses, designing experiments, predicting outcomes, and evaluating results. *Investigations* tend to be lengthier and more equipment intensive than *Activities* and should be conducted in a laboratory setting. Some *Investigations* are completed over a period of several days.

Before completing any *Investigations,* the teacher should review the scientific method with students. *Appendix C* summarizes the key elements of the scientific method. In keeping with the contextual approach of the book, the scientific method is presented in the context of a scientist researching the effects of caffeinated soda.

Each chapter of *CORD Biology* has two recommended *Investigations* that are introduced in the text. The full procedures for the *Investigations* are in the Laboratory Manual. In addition, the Laboratory Manual includes alternative investigation ideas for some chapters. The TRB describes all teacher preparation required for the *Investigations* as well as materials and equipment lists. Answers to the discussion questions are also included in the Teacher's Resource Book.

CAREER APPLICATIONS

Career Applications are activities at the end of every chapter that involve students in discovering more about various career options and how the biology in the chapter is used in a number of fields. Specific fields include Agriculture and Agribusiness, Business and Marketing, Family and Consumer Science, Health Careers, and Industrial Technology. These activities demonstrate that although a biology degree might not be required for a particular job, knowledge of biology is important in a wide range of careers.

Career Applications, like *Learning Links,* call for the students to use a variety of print based, electronic, or community resources to find information. In addition, the student is generally asked to produce some product that might be a written or oral report, a model, a poster, a video, a brochure, a skit, or other format so that different learning styles are addressed. Suggestions for sharing the information beyond the class environment are part of many of the *Career Applications.*

Career Applications may be done by individual students or small groups. It is not intended that every student complete every *Career Application.* The variety is provided so that students can be allowed to explore areas that interest them. Students will be exposed to the information from other *Career Applications* as the students completing those projects present them to the class.

ANCILLARY MATERIALS

LOGBOOK

In research institutions, business, and industry, keeping written records is an important skill. The *Logbook* is the place where students record everything they do in *CORD Biology. Appendix B* gives practical tips for setting up and maintaining the logbook. Specially labeled *Logbooks* can be purchased as part of this program, although any notebook can be used for this purpose. Bound notebooks are preferred over spirals or notebooks from which pages can be readily removed.

LABORATORY MANUAL

The Laboratory Manual is a nonconsumable manual that includes all of the information the students will need to complete all 48 of the *Investigations* from *CORD Biology* and the *Alternative Investigations* that have been suggested. For each *Investigation* there is a contextual introduction, a statement of purpose, objectives, a list of skills used, materials and equipment needed for one station, safety notes, step-by-step procedures, cleanup instructions, and a set of conclusion questions. The Laboratory Manual also includes a safety lesson. Teacher notes for the *Investigations* are found in the Teacher's Resource Book.

TEACHER'S RESOURCE BOOK

The Teacher's Resource Book is an essential part of the *CORD Biology* curriculum. It contains chapter-by-chapter objectives, assessment rubrics, complete listing of materials and equipment needed to complete all of the *Activities* and *Investigations,* a list of live specimens needed, teacher preparations required for *Activities* and *Investigations,* solutions to *Investigations* conclusion questions, blackline masters, solutions to Chapter Assessments in the student text, performance assessments, hard copy of assessment questions from the computerized test bank, and complete student and teacher procedures for Additional *Activities.*

COMPUTER GENERATED ASSESSMENT SOFTWARE

To facilitate the preparation of chapter tests, a computer-generated test bank of multiple-choice questions is provided. The test bank includes four diskettes: two in Windows format and two in Macintosh format, packaged together. A hard copy of all of the questions in this text bank is included in the TRB.

STRATEGIES FOR USING THE ACTIVE LEARNING FEATURES

COOPERATIVE LEARNING

CORD Biology is designed to foster small group interaction among students. However, merely placing students in groups and assigning a task will not ensure completion of the work and learning by all group members. Years of study completed by David Johnson and Roger Johnson have determined that teachers can overcome the limitations that are commonly associated with small-group work by using an approach called cooperative learning. By their definition, cooperative learning is a small-group process of learning that maximizes the learning of all group members by sharing resources, providing mutual support, and celebrating group success.

The idea of students collaborating on a task is not new. The method has come into and gone out of vogue many times since the early 1900s. In the past 80 years, educational researchers have spent a great amount of time and energy comparing the impact of cooperative learning with that of other instructional methods that emphasize competition and individualization. Several conclusions have evolved. Cooperative learning increases student achievement, retention, perspective, intrinsic motivation, and on-task behavior.

The cooperative learning model also has social and economic value. The skills other than academic skills that students gain complement changes in the workplace. Modern management focuses on worker involvement through participatory teams. Our workforce relies on highly developed social, interpersonal skills. With a cooperative learning style, educators can help to develop and refine these skills in their students to better prepare them for the modern workplace.

As was pointed out earlier in the description of the text features, most of the active learning can be done in groups. You should also look for ways in which students can work in groups to learn or review the text content.

In the early part of the year, you should assign students to groups and assign roles to each member of the group. You will need to be watchful at first and help students to build the skills they need to work together. As the year progresses, students can be given more flexibility in choosing groups and dividing the work among team members. For further resources and support in using cooperative learning, you can contact The Cooperative Learning Center at the University of Minnesota, Minneapolis, MN 55455.

USING COMMUNITY RESOURCES EFFICIENTLY

The community can be a treasure trove of information resources. Who can provide better information on careers than people working in those careers? Community services and social service organizations often have educational materials on a variety of topics related to the content you teach in class. Most of them will be happy to provide you with a speaker to talk to your class if they are given enough advance notice. Parents might be willing to talk about their careers and/or provide you with a point of contact at local businesses and industries. You also might find leads for information by thumbing through the yellow pages and/or business listings in your local telephone directory. As you or your students contact various resources throughout the community, build an information file on each resource to include types of interaction to which they are amenable, information they can provide, specific contact person(s), and best time to call.

Many of the *Learning Links* and *Career Applications* suggest that students contact people in the community. If every student in each of your classes were to contact each resource individually, you would soon strain your relationship with those resources. When an activity calls for using a specific community resource, assign the task to one individual or group of individuals. Before they interview the source, they should have a list of questions compiled from others in the class. After the interview they can share that information with their class and even with other classes. Copies of interviews can be kept on file so that future classes will have the information. You might have students update the information every couple of years.

The TRB has a blackline master with information that will guide students in setting up and conducting interviews with people in business and industry. Be sure to review this material with your students before having them contact people from the community for information. Stress the importance of being polite. You might wish to have students practice role-playing interviews in class so that they will be more confident when they interview adults whom they probably have never met before. Encourage them to ask questions to clarify any information they receive that they don't understand.

ASSESSMENT OPTIONS

Numerous opportunities are available for ongoing formal and informal assessment. There are lesson assessments at the end of each lesson that are primarily of a contextual nature. There is a chapter assessment section at the end of each chapter that includes Concept Review questions that relate to the facts and Think and Discuss and In the Workplace questions that require students to apply information to a given problem. Also included in the chapter assessments are questions related to the *Investigations*.

The TRB has suggested performance assessments to go with each chapter. It also contains a printed version of the test items that appear in the Computer-Generated Assessment Software package.

Tests should not be the only means of evaluating student achievement. Entries in the *Logbook, Investigations, Activities, Learning Links,* and *Career Applications* projects should all be assessed. Rubrics can be used in assessing student achievement in these areas. You might wish to have students assist in the development of generic rubrics for specific types of products—for example, an oral report to the class, a poster, or a video. This will give students a better understanding of expectations for a given assignment. Even if students do not participate in the development of the rubrics, they should know ahead of time what criteria will be assessed for a specific project.

To create a rubric for a given assignment, first list the attributes of the student's performance on that assignment that you want to assess. The attributes could be things like information presented, presentation, research skills, and so forth. For each attribute, list characteristics of quality and quantity related to that attribute. For example, for information presented, you might list accuracy and comprehensiveness, and for presentation you might list organization, format, neatness, and so forth. Then for each characteristic, ask yourself what you would consider exemplary work. Then ask yourself what beginner work would look like from someone who has not yet grasped the content or who failed to complete the assignment. Then describe competent work and work that has minor flaws. You can then assign ratings to each of the characteristics, such as a 4 for exemplary work, 1 for beginning work, 3 for competent work, and 2 for work with minor flaws. Alternatively, you can assign point values for each attribute and have them sum to 100.

CORD BIOLOGY

Science in Context

Developed by the
Center for Occupational Research and Development
Waco, Texas

Upper Saddle River, New Jersey
www.globefearon.com

CORD Staff
Director: Stephen Jones, Ph.D.
Program Manager: John Souders, Jr., Ph.D.
Educational Specialist: Bonnie Rinard
Laboratory Specialist: John Reiher, Ph.D.

Editorial and Production Services: PC&F, Inc.
Design: Zender + Associates, Inc.
Photo Research: Alix Parson
Front cover image: Researchers Examine Sprouts, Ed Young/Corbis
             Videoscope photograph courtesy of Video Labs, Minneapolis, MN
             Research team photograph courtesy of LaMotte Chemicals, Chestertown, MD
Back cover image: ©1998, Corbis Images

Globe Fearon Educational Publisher
Upper Saddle River, New Jersey
www.globefearon.com

CONSULTANTS AND REVIEWERS

Science Consultants

LARRY BOND

Director of Cross Roads
    Educational Consortium
Fairfield, VA

JAMES COCKERILL

Science Education Consultant
Baltimore, MD

MOLLY GREENBERG

Consultant
Columbus, OH

MARY ANN FIGULY

Science Teacher
Schlagle High School
Kansas City, KS

DOUGLAS MANDT

Safety Specialist
Edgewood, WA

TIM RITTER

Science Educator
Rancho Cucamonga High
    School
Rancho Cucamonga, CA

STEPHEN W. TABER

Biology Department
St. Edward's University
Austin, TX

Reviewers

MARY COLVARD

Science Teacher
Cobleskill-Richmondville High
    School
Cobleskill, NY

SUSAN GANNAWAY

Professor of Education
North Georgia College and
    State University
Dahlonega, GA

JANIS LARIVIERE

Science Teacher
West Lake High School
The Learning Center
Austin, TX

LISA MERRITT

Educational Consultant
ACKLAM Incorporated
Homosassa, FL

LISA JONES-RATH

Science Teacher
Birmingham College Prep
    Magnet, LAUSD
Van Nuys, CA

UNIT 1
CONTINUITY AND DIVERSITY OF LIFE

CHAPTER 1
Chemistry of Life   1

CHAPTER 2
Introduction to Cells   39

CHAPTER 3
Cell Metabolism
81

CHAPTER 4
Classical Genetics   119

CHAPTER 5
Molecular Genetics 157

CHAPTER 6
Evolution and Classification 189

UNIT 2
MICROORGANISMS

CHAPTER 7
Bacteria and Viruses

CHAPTER 8

Kingdom Protista

277

CHAPTER 9

Fungi

315

UNIT 3
ANIMALS

CHAPTER 10

Introduction to Animals

347

CHAPTER 11
Musculoskeletal Systems   389

CHAPTER 13
Digestion

Chapter 14
Interaction and Control

Chapter 15
Protection and Defense

CHAPTER 16
Animal Reproduction 573

UNIT 4
PLANTS

CHAPTER 19
Plant Growth and Development
663

CHAPTER 20
Plant Life Processes
691

CHAPTER 21
Plant Reproduction
715

UNIT 5
COMMUNITIES AND ENVIRONMENTS

CHAPTER 22
Populations
737

CHAPTER 23
Communities and Ecosystems 764

INTRODUCTION TO *CORD BIOLOGY: SCIENCE IN CONTEXT*

Why are you taking a biology class? Because it is required for high school graduation? Because it is the only science that would fit in your schedule this year? Ideally, most of you are in the class because you want to learn more about how your body works and about the many living things around you. No matter why you are in the class, we hope this text will make your study of biology a rewarding one.

Discoveries are made in biology and related fields every day. *CORD Biology* includes a feature, *Changing Ideas*, that points out how our understanding of life and living organisms has changed over time. The constant change of the body of knowledge called biology means that no one textbook can teach you everything there is to know about biology. There are many ways to gain the latest information about topics. *Learning Link* and *Career Applications* guide you in using print materials, the Internet, and community resources to learn more about biology and how it is applied. Most important, you will be learning biology by doing biology through this book's and its laboratory manual's numerous activities and investigations. You will use the same science investigation skills that scientists use in discovering more about life and the world around us.

Did you notice *Science in Context,* the subtitle of this text? Do you know what contextual learning is? *Science in Context* means that, as we present the science of biology, we try to show how those ideas are important in your personal life, to people in various careers, and in various issues that are critical to society. In addition to discussions in the main text, several special features emphasize how biology is applied. Each chapter has a *Career Profile* of someone whose career work focuses

An optional opening activity, The Camping Trip, can be found as a blackline master in the TRB.

on some aspect of the biology concepts that are presented in the chapter. Each chapter also has several *Biology in Context* features that show examples of how the information you are learning is used in everyday situations.

Throughout this course you will learn about life and the biological processes that go on in each living thing. You will see how this knowledge can be used to help you live a healthy life, how it is used in a wide variety of careers, and how you can use it to make informed decisions as you are called upon as a citizen to make decisions on health care, food and nutrition, and the environment.

In Unit 1 you will learn the characteristics of living things, about cells, the smallest unit of life, about processes that provide energy for cells and to carry out life processes, how each living thing goes through a specific life cycle, how genetic information is passed from one generation to another, how we inherit traits from our parents, and a system for organizing all living things through their similarities and differences.

In Unit 2 you will explore some of the tiniest living things: bacteria, protists, and fungi. You will see them in both their beneficial roles and their harmful roles. You will see how we use them as food or in processing food, in breaking down waste, and in producing beneficial substances through genetic engineering. You will also look at how some destroy property and cause disease or even death.

In Unit 3 you will move to the animal kingdom. You will look at the parts of the body, how they function, and how we maintain our health and energy. You will even investigate behavior. As you study each body system, you will look at how human biology compares to that of other members of the animal kingdom.

In Unit 4 you will look at the plant kingdom. You will look at the great diversity among plants and explore how they grow and reproduce. You will see many ways in which we use plants: from food, to fiber, to building supplies, and to products such as drugs.

In the final unit, Unit 5, we will look at how living things interact with each other and their nonliving environments. The environment presents many challenges for your generation. Issues such as global warming, pollution, habitat destruction, and the mass extinction of species are very real threats to life as you know it. Solving these problems will require knowledge of more than any one subject. Biology provides a key to understanding these issues and tools for working toward their solutions. Even if you choose a career that is not directly related to working in the environment, you will be called on as an individual and as a citizen of the global community to make decisions that will involve these and other environmental issues.

CORD BIOLOGY: SPECIAL FEATURES

CORD Biology's special features put biology in perspective for you. The book links science concepts to careers in science and personal experience.

"Why Should I Learn This?"
Chapter opening scenarios provide a reason why each chapter's topics are relevant.

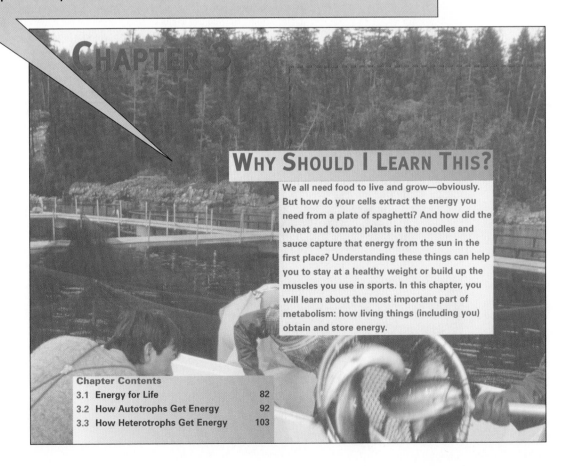

CHAPTER 3

WHY SHOULD I LEARN THIS?

We all need food to live and grow—obviously. But how do your cells extract the energy you need from a plate of spaghetti? And how did the wheat and tomato plants in the noodles and sauce capture that energy from the sun in the first place? Understanding these things can help you to stay at a healthy weight or build up the muscles you use in sports. In this chapter, you will learn about the most important part of metabolism: how living things (including you) obtain and store energy.

Chapter Contents

CELL METABOLISM

WHAT WILL I LEARN?

1. how living things use chemical reactions to obtain and store energy
2. how plants convert solar energy to usable chemical energy
3. how living things break down sugars and other substances to obtain energy
4. oxygen's role in extracting energy from foods

- -

"Fish gotta swim, and birds gotta fly," go the words to an old song. Swimming and flying take energy. Where does that energy come from? How do fish and birds or any organism put it to work? How do they store and retrieve energy? People who raise fish for market know that fish get their energy from food and oxygen. Any farmer knows that you can't grow a crop without plenty of sunlight or raise livestock without providing lots of energy in the form of feed.

In this chapter you will learn how organisms collect energy from sunlight, from minerals, and from each other and store it as chemical energy. You will also learn about chemical reactions that take place in nearly all organisms, making chemical energy available for any purpose, including swimming, flying, and growing.

CHAPTER 3 Cell Metabolism **81**

CORD Biology relates what you will learn in your classroom to what you may experience personally and what you may encounter professionally. This contextual approach also incorporates strategies for diverse learning styles.

Career Applications and **Learning Links** relate academics to careers through interviews with professionals, relevant research, intriguing debate, and Internet activities.

LEARNING LINK HOME AWAY FROM HOME

People who care for animals in captivity must provide conditions that allow the animals to maintain homeostasis. Special problems arise when animals are moved from their native environment. Zoos in the northern United States, for example, may have animals such as camels, which are native to hot desert areas. Find out from a zookeeper or an animal breeder how the requirements of such animals are met.

CAREER APPLICATIONS

The applications that follow are like the ones you will encounter in many workplaces. Use the biology you learned in this chapter to complete the activities. Share your work with the class.

AGRICULTURE & AGRIBUSINESS

Greenhouse Plants Get Extra CO_2

Talk to a commercial greenhouse operator or search for information on the Internet or elsewhere about growing plants in an atmosphere that is supplemented with CO_2. What effects does the extra CO_2 have on plants? When is CO_2 supplementation used? How feasible is it for widespread cultivation? Is there an optimum level of supplementation? Write up your findings in a brief report in your logbook and be prepared to share it with the class.

BUSINESS & MARKETING

The Tour de Plants

Ask a local travel agency about fall leaf tour packages. What does the package include? What transportation is provided? Are meals provided? How many people does the agency typically book in a year for leaf tours? How do they decide when and where the tours will go? What is the closest location? Why do people go on fall color tours? Design a brochure that will entice leaf peepers. Put together a weekend package that includes at least two types

Career Profiles introduce you to careers that use the skills and concepts you are learning.

LESSON 5.4 MUTATIONS

The term "mutation" might make you think of monsters, but most mutations cause only minor changes in an enzyme or other protein. Still, even slight changes can be bad news for the organism.

DENTAL HYGIENIST

CAREER PROFILE

Cindy A. is a dental hygienist who works in a general dentistry office. In her daily work she checks the condition of patients' teeth and looks for possible cavities, although the dentist makes all

diagnoses. She cleans and polishes teeth and applies sealant to some of the younger patients' teeth. Cindy also keeps patients' charts and takes X rays of their teeth.

When asked whether she worries about the safety of X rays, Cindy said, "While it is true that dental X rays are a form of radiation that can cause mutations, dental X rays are generally considered no more dangerous than walking in the sun. This is because the dose, or amount of radiation a patient is exposed to, is small. In our office patients receive X rays only once every two

This dental hygienist is preparing to take an X ray of a patient's teeth for a dentist to examine.

Biology in Context segments show how science concepts are applied in real-life situations.

FUEL FARMS

BIOLOGY IN CONTEXT

Melvin Calvin is remembered for his work describing the reactions of photosynthesis, but photosynthesis was not Calvin's only interest. Calvin spent a good part of his later career studying plants called *Euphorbia,* or gopher plant. This plant received worldwide attention because it produces oils that are very similar to crude petroleum.

Calvin thought that the plant would adapt to growing on land that is too dry for other crops. He envisioned petroleum plantations where the plants would produce the equivalent of many barrels of crude oil each year. Calvin even showed that the sap could be poured directly into the fuel tank of a diesel-powered car or truck without any processing or purification.

In the early 1980s the University of Arizona tested Calvin's ideas about growing oil. *Euphorbia* plants were gathered from around the world and planted. The first crop was a total failure. The scientists determined that

Euphorbia plants produce oils that can be used as fuel.

From math tips and vocabulary notes to laboratory work and the history of science, ***CORD Biology*** helps build the skills that will last you a lifetime. You will develop problem-solving skills and knowledge and use cooperative learning and data collection to complete projects throughout the text and laboratory manual.

Word Banks, found in the margins of the text throughout each lesson, develop understanding of science terms.

Changing Ideas feature historical perspectives on scientists' perceptions of issues and compare them to current theories.

WORD BANK

auto- = same, self, self-acting

-clave = cleave, split, break into pieces

An autoclave is an apparatus that creates heat and pressure to break up cells.

the use of an au (Figure 7.5). An makes pressur at temperatur boiling poin to kill mic ganisms. People w se a pressur oker in cannin vegetables also use p ssurized steam to kill microorganisms.

Figure 7.5: *Autoclaves use extremely high temperature to kill microorganisms that contaminate laboratory equipment.*

DISCOVERY OF DISEASE AGENTS

CHANGING IDEAS

In the late 1600s, Anton van Leeuwenhoek first saw microorganisms with his homemade microscope. But a century before that, the Italian physician Girolamo Fracastoro argued that many diseases are caused by organisms that are too small to see. When microscopes finally showed bacteria and protozoa on boiled extracts of hay and meat, scientists began to explain how they got there. One theory, called the theory of spontaneous generation, proposed that the organisms arose mysteriously out of the organic matter in the extracts.

Louis Pasteur later proved that microorganisms entered the extracts from the surrounding air and reproduced there. He used the apparatus shown in Figure 7.22, putting nutrient solutions in the flasks. Airborne microorganisms were trapped along the glass necks, and nothing grew in the nutrient solutions. When Pasteur broke off the long necks close to the flasks, the nutrient solutions soon turned cloudy with growing microorganisms.

A

acillus.

it is

In the late nineteenth century, German physician Robert Koch (Figure 7.23) established an undeniable connection between microorganisms and disease. Koch injected healthy laboratory mice with fluid from cattle with a disease called anthrax. The mice got sick. Koch also grew disease-

← Nutrient solution →
Microorganism trapped in neck of vessel

Figure 7.22: *Louis Pasteur's swan neck flasks that were used in his spontaneous generation experiments*

Laboratory Investigations add dimension to the classroom as you:
- See biology come to life through hands-on experiments
- Practice science methodology to reinforce learning
- Collect data for problem solving
- Build skills as team members

INVESTIGATION 5B — THE AMES TEST

The best way to determine whether a substance is likely to cause mutations is to expose live cells or tissues to the substance and observe the effects. A laboratory test called an Ames test is used to screen for chemicals that may cause mutations. The test exposes a bacterial culture to the test substance and looks for changes in the bacterial phenotype that are known to accompany mutations. In this investigation you will test a number of household substances to determine whether they are mutagens.

DO MUTATIONS REPRODUCE?

Even without exposure to the mutagens mentioned above, DNA would still mutate. Some mutations occur spontaneously as a result

Math Tips review and reinforce math concepts and offer strategies for completing calculations.

MATH TIPS

To find the probability of two events occurring when they are independent of (not related to) one another, multiply the likelihood of the first event by the likelihood of the second event. For example, if you have a 50% chance of getting an A on your next biology exam and only a 10% chance of getting an A on your next history exam, then the probability of getting As on both exams is 50% × 10% = 0.50 × 0.10 × 100% = 5%.

Use this same formula to determine the probabilities in a genetic cross when the genes for a trait assort independently. If the likelihood of a plant's producing offspring with white flowers is 50% and the likelihood of its producing a dwarf is 25%, then the probability of a white-flowered dwarf is 50% × 25% = 0.50 × 0.250 × 100% = 12.5%.

Remember that the alleles for dwarf and for wrinkled seeds are both recessive. Add blocks with the same phenotypes. The result of the Punnett square analysis is a probability (see the Math Tips):

- Nine of every sixteen offspring (or $\frac{9}{16}$) are tall with smooth seeds,

- Three of every sixteen offspring (or $\frac{3}{16}$) are tall with wrinkled seeds,

- Three of every sixteen offspring (or $\frac{3}{16}$) are dwarf with smooth seeds,

- One of every sixteen offspring (or $\frac{1}{16}$) is dwarf with wrinkled seeds.

In genetics, probabilities are often expressed as a ratio. In this case the ratio of phenotypes in the F_2 is 9:3:3:1.

Look at Figure 4.14 again and consider the ratio of phenotypes for each trait separately. There is a 12:4, or 3:1, ratio of tall plants to dwarf plants, just as you would expect in a monohybrid cross of heterozygous tall plants. The ratio of plants with smooth seeds to those with wrinkled seeds is also 3:1, like a monohybrid cross. So you could think of the ratio of phenotypes for the dihybrid cross in terms of two monohybrid crosses. Multiply the probability of finding a plant with a tall or dwarf stem by the probability of finding a plant with smooth or wrinkled seeds to get the proportion of phenotypes:

- Tall, smooth-seeded plants: $\frac{3}{4} \times \frac{3}{4} = \frac{9}{16}$
- Tall, wrinkled-seeded plants: $\frac{3}{4} \times \frac{1}{4} = \frac{3}{16}$
- Dwarf, smooth-seeded plants: $\frac{1}{4} \times \frac{3}{4} = \frac{3}{16}$

Activities throughout the text stress the relevance of biology and help you measure your progress and understanding.

Think and Discuss provokes you to consider the relevance of concepts learned in each chapter.

Hands-on Activities clarify understanding and reinforce what you have learned.

Think and Discuss

15 Why is it important for scien[...] share their findings with other scientists? Use the develop[...] the cell theory as an example.

16 When you sweat, active tran[...] carries positive sodium ions out of the cells of your sweat gl[...] Negative chloride ions and water molecules then pass [...] rough membrane channels. People with the disease cys[...] brosis lack working chloride channels. How would you e[...]ct this to affect their ability to sweat normally? Explain yo[...]swer.

17 Which organelle is most lik[...] to be the site of a defect that causes muscle weakness? [...]plain the reasons for your answer.

18 Special cells called transfer[...] ells are found on the internal surfaces of seeds. They br[...]g sugars and other nutrients into immature seeds as they d[...]velop on the plant. The walls of transfer cells often have ri[...]ges that force the cell membrane into

ACTIVITY 2-10   CELL SIZE AND CELL DIVISION

An organism faces serious consequences if its cells don't divide normally. But why do cells have to divide at all? Is there some advantage to small cells? In this activity, you will learn how size affects objects of various shapes and develop a hypothesis about why cell division is important.

Set up a table in your logbook with three columns and four rows. Label the columns "Length of Pieces," "Surface Area (cm^2)," and "Volume (cm^3)." Use your ruler and the appropriate equations from Figure 2.21 to determine the volume and surface area of your food. Record these numbers in the first row of your table, along with the length of your food. Cut your food in half crosswise, as in Figure 2.21. Determine the surface area and volume of one piece, and complete the second row of your table. Repeat the process, cutting one piece in half twice more and

MATH TIPS

Graphing your results can help you to see whether the size of your pieces has more effect on their volume or their surface area. Set up a graph on a piece of graph paper. Label the x-axis "Length (cm)" and the y-axis (on the left) "Surface Area (cm^2)." Plot your surface area data on this graph and draw a smooth curve through all the points. Then label a second y-axis (on the right) "Volume (cm^3)" and plot the volume data on the same graph in a different color. Draw a smooth curve through all those points in the same color. Which curve is steeper? What does that mean?

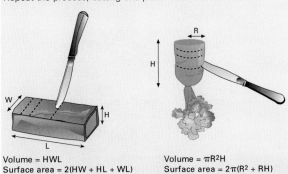

Volume = HWL
Surface area = 2(HW + HL + WL)

Volume = $\pi R^2 H$
Surface area = $2\pi(R^2 + RH)$

Figure 2.21: *How to cut and calculate volume and surface area*

CHAPTER 1

WHY SHOULD I LEARN THIS?

In the future you might train for a job that requires raising, caring for, or transporting organisms. You will need to know what these organisms require to stay alive and healthy. Or you might work in a job controlling organisms that cause problems for humans. In either case you'll need to know something about the characteristics that living organisms have in common—in other words, what it means to be alive.

CHEMISTRY OF LIFE

WHAT WILL I LEARN?

1. to distinguish between living and nonliving things
2. to identify what organisms need to stay alive and how they meet these needs
3. to describe a life cycle and relate its stages to development, growth, and reproduction
4. to observe the cycle of growth, development, and reproduction in several kinds of organisms
5. to explain how an organism's responses to changes in its environment help it survive
6. to distinguish between responses made by individual organisms and those made over time by a species

Growing plants is sometimes a challenge, but shipping them all across the country can be even harder. A grower who knows what plants need can assure customers that their orders will arrive in good condition. When a customer receives a shipment, the plants must be put in an environment that helps them to resume growth.

In this chapter you'll learn how to tell what is alive and what is not—the differences between living and nonliving things. You'll look at certain characteristics to decide whether something is alive, and you will find out what living things need to stay alive.

LESSON 1.1 WHAT IS LIFE?

How can you tell if something is alive? Some people might say it's alive if it moves or if it changes shape or grows. These statements try to identify life by physical characteristics, which

aren't good criteria. Some nonliving things can move, change shape, and grow (Figure 1.1).

On the other hand, not every living thing is always moving, changing shape, or growing. You've seen plants that look lifeless—no leaves, no new growth. Suddenly after a rain, the plant is opening buds, sprouting leaves, and sending up new shoots.

Let's try to identify life by its chemistry: Perhaps something is alive if it's made mostly of carbon. But other things can be mostly carbon, such as graphite tennis rackets, coal, and diamonds, which aren't alive.

What about energy? Doesn't something that's alive take in and use energy? True, but so do calculators, car engines, and wristwatches. Maybe we need to do a little more investigation to identify what life is.

Figure 1.1: *Glaciers, stalactites, and crystals all move, change shape, or grow, yet are not alive.*

ACTIVITY 1-1   THE MEANING OF LIFE

Before starting Activity 1-1, review the Scientific Method (Appendix C), Laboratory Safety Guidelines (Appendix A), and Keeping Your Logbook (Appendix B).

On the basis of your observations of living things, you no doubt have your own criteria for what is alive and what isn't. If everyone can agree on these criteria, you might have a useful definition of *life.*

Draw a line down the middle of a page in your logbook, and label the two columns "Living" and "Nonliving." In the first column, list the characteristics that all living things share. In the second, list the characteristics that all nonliving things share. Compare your lists with a partner's and try to agree on which should be in each column. Share your conclusions with the class and help to make a master list to be used in later activities in this chapter.

CHARACTERISTICS OF LIFE

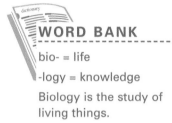

WORD BANK

bio- = life

-logy = knowledge

Biology is the study of living things.

People who study biology, called biologists, use three characteristics to determine that something is alive:

- Living things expend energy to maintain the internal conditions necessary for survival.

- Living things grow, develop, and reproduce in a repeating cycle of change.

- Living things respond to changes in the environment in ways that increase their chances of surviving.

Compare these three criteria with the ones you agreed on in Activity 1-1. The rest of this lesson will fill in a few details.

SEARCH FOR LIFE

BIOLOGY IN CONTEXT

Viking lander on Mars

Do we really know enough about life to program a space probe that could detect life on another planet? In the 1970s the National Aeronautic and Space Administration (NASA) sent two probes to Mars, Earth's nearest planetary neighbor. The probes sent back video images of the Martian surface and analyzed samples of Martian soil. Experiments checked for something that might chemically resemble life on Earth or that carried out lifelike processes, such as reproduction, growth, or the ability to break down sugars. After analyzing a lot of data, NASA scientists concluded that the probes had not detected life. This was no surprise to scientists who thought that Mars is far too inhospitable for life. Temperatures reach –130°C (–200°F), there's virtually no oxygen in the atmosphere, and all of the water on Mars is thought to be solid ice.

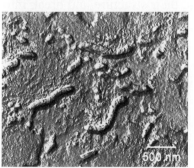

Meteorite from Mars with close-up of possible fossils

Scientists knew that Mars had many large volcanoes, which might generate enough heat to melt the ice and support life. NASA has plans to send more probes to Mars to explore more thoroughly for evidence of life. But if life is, or once was, present on Mars, will we be able to recognize it?

Searching for life on planets that are less habitable than Earth might seem like a hopeless task. Yet many hostile environments on Earth—such as the polar ice caps—support some form of life. Even the friendliest environments experience temperature changes, wind, flooding, or drought. Yet life goes on.

How do living things maintain the proper internal conditions for survival? A few organisms can wall themselves off from their environment if conditions aren't just right. A water bear (Figure 1.2) is a tiny organism that secretes a shell over itself when its environment becomes unfavorable. It can remain completely inactive for long periods. Such organisms pay a price for their isolation; as long as they're inactive, they can't grow, reproduce, or get away from something that's trying to eat them.

Figure 1.2: *Micrograph of a water bear*

On the other hand, an organism that stays active must constantly exchange materials and energy with its environment and convert them into the forms it needs. Substances in food provide the raw materials for building and maintaining an organism's highly organized structures. An organism uses its energy supply to convert raw materials into more complex forms, which are used for growth, maintaining internal conditions, and reproduction.

Living creatures' ability to stay internally organized seems to defy the natural tendency of things to lose their structure and break into simpler forms. For example, nonliving things, such as rocks and dead wood, tend to break down over time. A living tree, however, produces new wood as it maintains itself (Figure 1.3).

Figure 1.3: *A living tree collects the energy of sunlight and raw materials from the air and soil to maintain its own complex structure, but once a tree dies, it begins to break down into simpler substances.*

INVESTIGATION 1A   IS IT ALIVE?

WORD BANK

- -

micro- = small

-scope = to look at

A microscope helps you look at small things.

Before beginning Investigation 1A, allow students to examine a diverse collection of living, nonliving, and formerly living things. Ask them to determine the category in which each specimen belongs, and discuss the reasons for these choices with the class. As students refine their criteria for life, ask them to apply their tentative criteria to some unfamiliar things that they will observe in the investigation.

See Investigation 1A in the TRB (Teacher's Resource Book).

Most living things are too small to see, so it's hard to tell whether they're alive. In this investigation, you'll use a microscope to observe some very small objects. You'll learn how to estimate the size of what you see in the microscope, and you'll have a chance to figure out what's alive and what's not.

ACTIVITY 1-2   LIFE AS A CYCLE OF EVENTS

Provide groups with different sets of living materials, such as brine shrimp, aquarium fish, mealworms, earthworms, painted-lady butterflies, fruitflies, and/or various seed plants. Include as many stages of the life cycle as possible (eggs or seeds, larvae, hatchlings, seedlings, and adults). Answers to the discussion questions will depend on the materials provided. The conclusion should be that all organisms grow, develop, and reproduce.

Living things often change throughout their lives. You'll look at one type of organism at different stages of development. The sequence of these different stages is called a life cycle, which is repeated in each new organism.

With your group, look at the organisms and describe and draw them in your logbook. Make a diagram showing the different life stages the organism goes through. Try to answer the following questions. Be prepared to present your conclusions to the class and compare them to the findings of other groups.

- Describe how the organism that you examined began its life.

- On the basis of your observations, describe any changes in size or form that the organism undergoes during its life.

- How might the organism get the materials and energy it needs to grow? How could you test your ideas?

- Describe how the organism's abilities change as it enters each new life stage.

- How does the organism reproduce?

Tadpole

Figure 1.4: *This tadpole looks nothing like the frog that produced it, but it is a new organism. It will grow and develop until it resembles its parents.*

During its life cycle, every organism develops, grows, and reproduces. Organisms take in substances and use energy as they develop and grow. Development and growth are accompanied by changes in structure or activities. For example, a tadpole like the one in Figure 1.4 grows legs, stops eating plants, and starts eating insects as it develops into a frog. When an organism has completed most or all of its development, it is ready to reproduce. This reproduction results in new organisms that resemble others of its kind, or **species.** These new organisms will experience the same stages of the life cycle.

The survival of living things depends on their interaction with the environment. There are three kinds of interaction:

1. The environment has a direct effect on the organism. The body temperature of a beetle increases after it has moved from shade into sunlight.
2. Organisms **adapt,** or adjust to changes in the environment. We humans cool off by sweating or by avoiding direct sunlight; shivering warms us up when we are cold.
3. Species adapt by changing slowly over many generations. A species' adaptation depends on how well individuals with certain characteristics can survive and reproduce. For example, some organisms have camouflage, that is, colors that help them

Figure 1.5: *The Arctic fox (left) is adapted to snowy northern environments while its relative the gray fox (right) is better suited for its more southern habitat.*

to blend in with their environment. These organisms may have an advantage over organisms whose colors make them more visible. Organisms with camouflage are less likely to be seen (and eaten). This advantage increases their chances of surviving and reproducing. Eventually, their descendants far outnumber organisms of the same species that lack camouflage. This type of change has produced white Arctic animals such as polar bears, arctic foxes, and snowy owls, which can hide easily in the snow and ice of that wintry environment. By contrast, brown bears, gray foxes, and brown owls are common in temperate zone forests, where they blend in with dark trees and foliage (Figure 1.5).

ASTROBIOLOGIST

CAREER PROFILE

Dr. Gerald S. is responsible for the future of biology at NASA. As director of biology programs he plans NASA's work in astrobiology for the next 20 years. Astrobiology is the study of life from the beginning of the universe through our present search for life in space. During Jerry's career at NASA he has always worked to find life in space. He started at NASA in 1961 as a senior scientist designing experiments to look for life on Mars. He moved up through the ranks until he was the project scientist for the Viking mission to Mars. A project scientist oversees all the experiments on a specific mission.

How can you get a job like Jerry's? Jerry replied that if you want to design and work on your own experiments, you will need a Ph.D. in a biology-related field. However, you can work as a technician with a bachelor's or master's degree. Technicians and interns work under the direction of a research scientist. Courses that Jerry found especially useful were biochemistry, microbiology, genetics, cell physiology, and pharmacology.

Jerry says that his job is to try to find the answer to one of life's great questions: Is life a cosmic imperative? That is, will life occur wherever you find the right conditions, no matter what? Needless to say, he doesn't expect to find the answer next week.

LESSON 1.1 ASSESSMENT

1. They're alive if they are consuming food, producing wastes, and increasing in size or number.

2. Rocks do not grow or reproduce by producing new smaller rocks; they mechanically break down into smaller rocks and soil particles. They do not adapt to their environments.

3. The similar shapes and different sizes suggest an organized structure and the ability to grow. Perhaps the triangles change color as they grow. Examining their interiors and looking at thin slices with a microscope might reveal a complex structure that would be evidence that the objects are or once were alive.

4. Since modern conditions are much different from the conditions in which life first formed, modern-day bacteria aren't likely to be the same as early ones.

1 Bacteria are living things that are too small to see without a microscope. Cheese makers, other people in the food industry, and people who manufacture medicines grow bacteria that help to make these products. It is important to keep the bacteria alive and healthy. Describe the characteristics by which you would know that these bacteria were alive if you had to maintain them. Be specific.

2 Explain why you wouldn't consider a rock to be a living thing that grows and reproduces very slowly.

3 Suppose that a space mission returns with some triangular objects from another planet. They all have the same shape but different sizes and colors. They do not move or seem to respond in any way to changes in the environment or to being handled. Decide whether the objects represent living (or once-living) things or whether they are inanimate. Give evidence for your choice. If you need more evidence before you decide, how could you get it? What type of evidence would you need?

4 Some scientists look for bacteria in environments that they think resemble those in which the first organisms came to life. Would you expect them to find living bacteria that are the same as the first living things? Give a reason for your answer.

LESSON 1.2 MAINTAINING A STRUCTURE

In Investigation 1A you saw that living things are a lot more organized than their surroundings are. To maintain this organization, organisms must get two things from their environment: the materials they need to build and repair their structures and the energy to do this and other work. **Work** means using energy to bring about some change. For organisms, work includes movement, exchanging substances with the environment, reproducing, maintaining structures, and sometimes thinking and communicating.

PUTTING ORGANISMS TO WORK IN AN OIL WELL

BIOLOGY IN CONTEXT

Some types of bacteria live underground, feeding on substances in rock. Oil companies have started feeding these bacteria, at least the ones that live in oil deposits. The food is molasses, injected

directly into an oil well. The result: The bacteria reproduce rapidly and release carbon dioxide gas as a waste. This gas builds up pressure in an oil deposit and makes the oil flow more rapidly into the well. This technique has solved the problem of oil wells that seem to dry up before even half their oil is recovered.

Going a step further, scientists are developing special bacteria that can be injected into an oil deposit and will feed directly off the oil. Like the other bacteria, they would produce carbon dioxide that would help to pressurize the well. Then drilling crews could stop pouring molasses into holes in the ground.

ACTIVITY 1-3 WHICH FOODS CONTAIN THE MOST ENERGY?

Ensure that students work safely with fire. See Activity 1-3 in the TRB for information on the foods to provide students.

Figure 1.6 represents sources of energy in the environment available to organisms. Some organisms, especially plants, get their energy directly from the sun. The source of energy is light, a form of **radiant energy.** Some bacteria get **chemical energy** from inorganic substances, which are substances not made up primarily of carbon. Animals obtain chemical energy by eating other organisms and breaking down their organic compounds, substances

A

B

C

made up primarily of carbon. We humans get energy from substances in our foods, which were once parts of living organisms. In this activity you will judge which foods contain the most energy.

Working with a partner, set up a Bunsen burner in a fume hood or near an open window. **Put on safety goggles and observe all the safety precautions for working with fire: Tie back long hair and loose clothing and roll up long sleeves.** Light the burner. Using tongs or long metal forceps, hold a small piece of each of the foods provided in the flame. Keep each sample in the flame until it stops burning.

After removing each sample from the flame, place the hot tongs or forceps on a ceramic square and discard the burned sample in the trash. Record in your logbook your observations of how each sample burns. When you're finished, clean up your area and return the equipment. Discuss the following questions with your partner:

- Was there any evidence that energy was released when the foods burned? If so, where did that energy come from?

- Which kinds of foods burned most easily and intensely? (You may need to check out the packaging for a list of ingredients to answer this question.) Are these the same foods that released the most energy when they burned?

- What is the similarity between fire and the way your body breaks down foods to obtain energy?

Figure 1.6: *Sources of energy. (A) Radiant energy. (B) Compost heap. (C) Spider eating a beetle.* **D** *(D) Grasshopper eating a sunflower. (E) Crops are a food source for animals.*
Energy is released as heat and light when foods burn. Fatty foods release the most energy, followed by carbohydrates. Proteins release the least energy of the three major food constituents when burned. Our bodies release energy from food by a process similar to fire, though less rapid and dramatic.

E

BASIC CHEMISTRY

The carbon and water that are released when you burn foods are essential raw materials for life. Organisms must constantly take in raw materials from their environment and use them to build and repair structures. This process can be complicated, but we'll start by looking at the basic raw materials.

Every substance is composed of one or more chemical **elements,** represented by a symbol. Here are some elements you may already know: carbon (symbol C), hydrogen (H), helium (He), oxygen (O), sulfur (S), chlorine (Cl). An element is the basic substance of matter; it is composed of fundamental units called **atoms.** (Appendix D shows a table of all the elements.)

LEARNING LINK PERIODIC TABLE

On the wall of most laboratories you will find a periodic table of the elements. Besides the chemical symbols of the elements, the table displays other helpful information about the elements. Visit an Internet site or use computer software to learn more about what information the periodic table provides.

Since it is impossible, even with an expensive microscope, to observe an atom in detail, scientists have devised models of what they think an atom looks like. Figure 1.7 is a model of the hydrogen atom. In its center is a **nucleus,** which contains a very small particle, a **proton.** Outside the nucleus is an even smaller particle, an **electron.** Atoms of the same element contain the same number of protons. For instance, a hydrogen atom has one proton; a carbon atom has six protons. Each proton in an atom has a positive electrical charge, and each electron has a negative electrical charge. The difference in charge means that protons tend to attract electrons. The attraction between protons and electrons is what holds an atom together. Most atoms have no overall charge because the number of protons in an atom usually equals the number of electrons.

Sometimes, two atoms share electrons. When this happens, a **chemical bond** (Figure 1.8) forms between the two atoms. Energy stored in the chemical bond can be released when the bond breaks

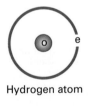

Hydrogen atom

Key
o Proton
e Electron

Figure 1.7: *Model of a hydrogen atom*

Oxygen
molecule

Figure 1.8: *Diagram of a chemical bond*

and a new bond forms. Chemical energy is a source of energy for all organisms.

Two or more atoms joined by chemical bonding form a **molecule.** When two oxygen atoms form a bond, for example, a molecule of oxygen gas (O_2) results (Figure 1.8). If atoms of different elements form a molecule, the molecule is a **compound.** When one atom of carbon and one atom of oxygen are bonded, for example, the result is the compound carbon monoxide.

Atoms are rearranged to form new compounds when organisms carry out a **chemical reaction.** In a chemical reaction, substances called **reactants** interact and form new substances called **products** (Figure 1.9). An organism must carry out hundreds of different chemical reactions to maintain its complex structure. The sum total of all the chemical reactions carried out by an organism is called **metabolism.**

$$AB + C \longrightarrow A + BC$$

reactants   products

OR: (using actual chemical symbols)

$$2C + O_2 \longrightarrow 2CO$$

2 carbon   oxygen   two molecules of
atoms   molecule   carbon monoxide

REACTANTS   PRODUCT

Figure 1.9: *A diagram of a chemical reaction 2 C (carbon) + O_2 (oxygen) = 2 CO (carbon monoxide)*

Most organisms need about 20 elements to carry out their chemical reactions. Six elements are particularly important: carbon (C), hydrogen (H), nitrogen (N), oxygen (O), phosphorus (P), and sulfur (S). These six elements make up about 95% of the body weight of most organisms. Organisms get these elements from either organic or inorganic substances. Figure 1.10 shows some examples of organic and inorganic compounds that organisms use. Can you tell which molecules are organic and which are inorganic?

The ammonia and hydrochloric acid are inorganic compounds; the rest are organic.

Each line between the elements and above nitrogen represents a pair of electrons.

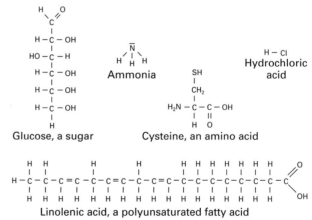

Glucose, a sugar

Ammonia

Cysteine, an amino acid

Hydrochloric acid

Linolenic acid, a polyunsaturated fatty acid

Figure 1.10: *Examples of organic and inorganic compounds used by various organisms*

BIOLOGICAL POLYMERS

WORD BANK
----

poly- = many

mono- = one

-mer = part

A polymer has many subunits. A monomer is the subunit on which a polymer is based.

Some of the compounds that organisms make are large molecules called **polymers.** Polymers have long chains of repeating subunits called **monomers.** You may be familiar with some polymers that are made by chemists, compounds such as nylon, silicone, and Teflon (Figure 1.11). Plastics contain many different types of polymers. The properties of a polymer, whether biological or manufactured, depend on the types and numbers of monomers it contains.

Figure 1.11: *Manmade (synthetic) polymers are found in many products: plastic drink bottles, telephones, and CDs.*

Monomers

Polymer

Organism

Figure 1.12: *Living organisms assemble monomers into polymers, including proteins, complex carbohydrates, and nucleic acids.*

Organisms make three important groups of polymers: proteins, complex carbohydrates, and nucleic acids (Figure 1.12). Each of the polymer types has one or several roles in an organism. The role that a biological polymer plays depends on its structure. Proteins, complex carbohydrates, and nucleic acids have different structures.

A **protein** is a polymer composed of monomers called amino acids. Proteins participate in a variety of chemical reactions or serve as structural materials. The type and number of amino acids give a protein a unique shape. The role of the protein depends on this unique shape.

A **complex carbohydrate** is a polymer composed of monomers called simple sugars. Complex carbohydrates such as cellulose are part of the structure of organisms, giving them strength or protection. Organisms use some complex carbohydrates as sources of energy: examples are starch made by plants and glycogen made by animals. The energy in complex carbohydrates is stored in the chemical bonds between the sugar units. When these bonds break and new bonds form, energy is released. Some of the released energy is used to form molecules of a substance called **adenosine triphosphate,** or **ATP** (Figure 1.13). An organism can use the energy stored in ATP molecules to move, reproduce, build polymers, or perform other activities.

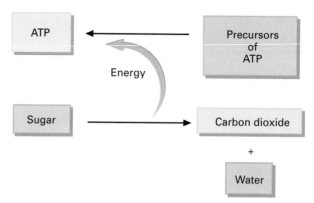

Figure 1.13: *When sugar is broken down to form carbon dioxide and water, some of the energy released is used to form ATP.*

A **nucleic acid** is a polymer composed of monomers called nucleotides. Nucleic acids store chemical information in the sequence of their monomers. Organisms use this chemical information to direct their chemical reactions.

INVESTIGATION 1B   BIOLOGICAL POLYMERS

See Investigation 1B in the TRB.

In this investigation, you will compare the behavior of two kinds of polymers (proteins and complex carbohydrates) and a third substance (fat). As you learn how these substances differ, think about their properties and behavior and what role each might play in the lives of the organisms that produce them.

REACTION TIME

Under normal temperature and pressure conditions, chemical reactions may occur so slowly that we're often unaware of them. For example, oxygen in the atmosphere combines with iron to form iron oxide, which you know as rust (Figure 1.14). You can tell when something has rusted, but rusting takes place too slowly for you to watch its progress.

One thing that distinguishes living organisms from nonliving things is that in living organisms, chemical reactions occur at breakneck speed. Chemical reactions must occur very rapidly if organisms are to maintain constant internal conditions.

Figure 1.14: *The rust on these screws was formed by iron reacting with oxygen in the atmosphere.*

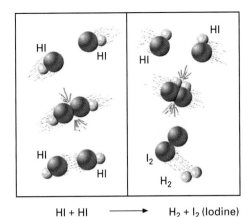

Figure 1.15: *Reactant molecules must collide in just the right way for a reaction to take place.*

$$HI + HI \longrightarrow H_2 + I_2 \text{ (Iodine)}$$

Let's suppose that a chemical production technician wanted to speed up the chemical reaction that produces nylon. He or she might increase the temperature at which the reaction occurs, increase the pressure, or increase the concentration of the reactant molecules. Any of these three changes will make reactant molecules collide more frequently. In some of these collisions the reactant molecules collide at the proper angle (Figure 1.15), converting the reactants to one or more products.

Organisms, however, can't tolerate changes in temperature, pressure, and their fluid concentrations. Instead, they make proteins called **enzymes** that speed up, or **catalyze,** their chemical reactions. Each type of enzyme has a location, the **active site,** that attracts a single type of reactant molecule (Figure 1.16). For instance, when starch binds to its enzyme's active site, it is in the correct position to react and release sugar. The active site of the enzyme is then free to receive another starch molecule, which is converted to sugars. This happens in your mouth when you chew a cracker or piece of bread. The starchy taste is replaced by the sweet taste of the sugars being released. Try it. (Be sure not to swallow the cracker too quickly.)

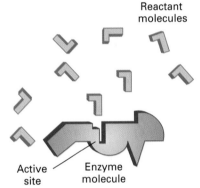

Figure 1.16: *Enzymes remain inactive until the desired molecule binds to the active site.*

LEARNING LINK ENZYME DEFICIENCIES

Animals use enzymes to help digest food. We humans use an enzyme called lactase to help us digest lactose, the sugar found in breast milk and in dairy products. Some people stop producing lactase as adults and so have difficulty digesting some of their favorite foods. Fortunately, replacement enzymes are available for deficiencies of many digestive enzymes. Using the Internet or library resources or a visit to a pharmacy, make a list of products that help people or other animals with enzyme deficiencies.

An important feature of enzymes is that the active site of one type of enzyme attracts only one type of reactant. Therefore an organism must make a different type of enzyme for each chemical reaction that it carries out.

ACTIVITY 1-4   ENZYMES ARE PICKY ABOUT WHAT THEY ATTRACT

Make up various sugar solutions by dissolving 2–3 g of sugar in 100 mL of distilled water. Use powdered glucose, sucrose, fructose, lactose, or maltose or have students test beverages that contain only one type of sugar: bottled fruit juices (fructose), powdered fruit drinks (sucrose), milk (lactose), sports drinks (glucose, but check the label to be sure it is the only sugar). Tape over the word "glucose" on the roll of glucose test strips, so that students do not know beforehand the specificity of the test strips.

A sugar is a type of carbohydrate. Organisms use a number of different sugars as energy sources. Most animals use glucose as the most readily available energy, but young mammals are weaned on the lactose sugar in their mother's milk, which splits into two simple sugar units that the blood can absorb. Sucrose is a common form of sugar consumed in a typical American diet. It is processed from sugar beets and sugar cane into the white granular sugar that you pour from a table shaker. The breakdown of various carbohydrates during digestion produces simple sugars such as fructose, galactose, and glucose. Sugars other than glucose are converted, each by a different type of enzyme, to glucose, which is then released into the blood.

In this activity you will test an enzyme that has been fixed on a strip of paper to see for which sugar it is specific. Wear gloves or use forceps during this activity so that you don't contaminate the test strips with your fingers.

The test strip changes color only when it is placed in a glucose solution or a beverage containing glucose. Therefore the enzyme on the test strip catalyzes the reaction of glucose but does not cause other sugars to react. The test strip changes color because the enzyme on it converts glucose to a compound that reflects blue or green light.

Get a sugar solution or a sugar-containing beverage. Tear off a 4-cm–long piece of test strip. Use a stirring rod to wet the test strip with one of the sugar solutions. Wait about 30 seconds for a reaction. A reaction changes the color of the strip to blue or green. Record your observations in your logbook.

Test the remaining sugar solutions or beverages, using a new test strip each time. Record these observations in your logbook. What does each observation indicate about the enzyme on the test strip?

WHY ORGANISMS NEED WATER

Exchanging substances with the environment always involves water in some way. Aquatic organisms such as *Euglena*, brine shrimp, and fish directly exchange substances between their body water and the surrounding water. The roots of plants are in contact with water in the soil. The oxygen that an animal inhales first passes across a thin layer of water in the lining of the lung. What makes water so important to an organism that is exchanging substances with its environment?

CODE PINK—BLUE BABY

Helen, a neonatal intensive-care nurse, reports to the hospital delivery room for a Code Pink. Code Pink means that a baby is being born who could have some medical problems. The neonatal team also includes a neonatologist (a doctor who specializes in the care of newborns) and a respiratory therapist. The Code Pink was called by the delivering obstetrician, who suspects that the baby may have problems breathing because it is being born prematurely.

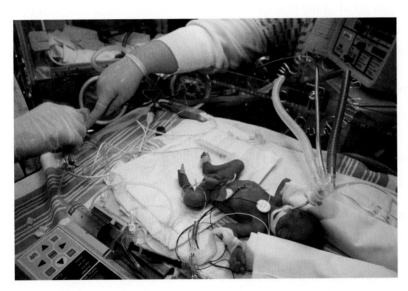

Some premature babies have trouble breathing because their lungs are not fully developed. They may need a breathing machine to help them breathe. The lungs of premature infants often do not produce enough surfactant. Surfactant, a kind of "soapy" liquid, keeps droplets of water from forming in the tiny air sacs that make up the lung. Water droplets can make the fragile air sacs in a premature baby collapse, leading to suffocation.

Surfactant treatments are given to premature babies in steps. The baby is taken off the breathing machine for a short time while the surfactant is delivered through a tube placed into the breathing tube. The baby then goes back on the breathing machine in a position that helps the surfactant to spread through as much of the lung as possible. This procedure is repeated four times with the baby in different positions until the entire dose coats as much of the lung as possible.

H₂O

Figure 1.17: *Model of a water molecule*

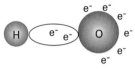

Figure 1.18: *The bond formed between the hydrogen and oxygen atoms is a polar bond. The electrons that form polar bonds are unequally shared between two atoms.*

Look at the model of a water molecule in Figure 1.17. Water has two atoms of hydrogen, each bonded to one atom of oxygen. Each bond in a water molecule is a pair of electrons (Figure 1.18). Each atom contributes one electron to the bond. This type of bond, in which different atoms provide electrons, is a covalent bond. A water molecule is held together by two covalent bonds.

The properties of water are a result of a special feature of its covalent bonds. Ignore the electrons of water for a minute and think about protons. The oxygen atom in the covalent bond has eight protons in its nucleus. The hydrogen atom has one proton. The larger positive charge of the oxygen nucleus attracts the electrons in the covalent bond more strongly than does the smaller charge of the hydrogen nucleus.

Remember that electrons have a negative charge. Because the oxygen atom in the bond attracts the electrons more strongly, it has a slightly negative charge. Likewise, the hydrogen atom has a slightly positive charge. When electrons are attracted unequally between two atoms, the bond between the atoms is called a **polar bond.** A molecule that has a polar bond is a **polar molecule.**

The fact that water is a polar molecule gives it important properties. Water can associate with other polar molecules, such as ions, salts, sugars, and proteins. In water, polar substances dissolve, forming a solution. A solution is a mixture of a substance called a **solute** dissolved in another substance, called a **solvent.** In organisms, water is the solvent for anything that is in solution. Because water is a solvent for polar substances, an organism can easily exchange polar substances with its environment. This exchange often relies on the behavior of substances in a solution.

ACTIVITY 1-5   HOW A DYE BEHAVES IN WATER

Like all particles, molecules of a solute constantly collide with other molecules in the solution. What is the result of all these collisions? They eventually distribute a solute evenly throughout a solution.

Put a drop of food coloring in a beaker of water. Observe the beaker every few minutes for the rest of the class period. Record your observations in your logbook. Explain the behavior of the dye in terms of molecular collisions.

The dye slowly spreads out as its molecules collide with the water molecules and with each another.

DIFFUSION

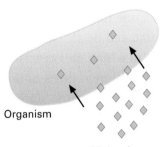

Organism

Molecules

Figure 1.19: *Diffusion occurs in the direction of the arrows until a concentration difference no longer exists.*

In Activity 1-5 you saw molecules of a dye move from a region of high concentration (the drop of food coloring) to a region of lower concentration (the water in the rest of the beaker). The movement of dye molecules that you observed is **diffusion.** Diffusion occurs only when the concentration of a substance is different in two regions. Once the dye is evenly distributed, diffusion stops, even though the dye molecules remain in constant motion (see Figure 1.19).

Organisms use diffusion when they exchange materials with their environment. For example, you take in oxygen because it has a higher concentration in the air you inhale than it has in your blood. You will learn more in Chapter 2 about how this happens.

LESSON 1.2 ASSESSMENT

1. An animal could not make ammonia or cysteine if it was deficient in nitrogen. It could not make cysteine if it was deficient in sulfur.

2. The fatty acid molecule contains the most carbon-hydrogen bonds.

3. The organisms lack an enzyme that is needed to break down and release energy from sucrose, so they starve.

4. The vinegar and spices flavor the pickles by diffusion.

1 Refer to Figure 1.10. Which of the compounds would an animal be unable to make if its diet lacked sulfur? If it lacked nitrogen?

2 A compound is a great source of energy to organisms if it contains many chemical bonds between carbon and hydrogen atoms. Select a compound in Figure 1.10 that would best serve as a source of energy.

3 You place a small organism in a solution of glucose and water. After a few days you notice that it appears to be healthy and is reproducing rapidly. One day when you are feeding the organisms, you misread a label on a stock bottle and give them sucrose instead of glucose. After a few days the organisms stop reproducing. Shortly afterward, they all die. Explain this result.

4 Pickles are made from cucumbers that are soaked in vinegar and spices. Explain how the vinegar and spices flavor the entire pickle.

LESSON 1.3 THE CHANGING ORGANISM

Pest exterminators are trained to find and identify insect pests at any stage of their life cycle.

Each species goes through a life cycle that includes development, growth, reproduction, and death. Some species have a complex life cycle in which the organism takes on different forms, usually at different times of the year. Pest exterminators are familiar with the life stages of organisms such as fleas, termites, ants, and bees. Knowing the life cycle of these pests allows the exterminator to provide effective treatment.

ACTIVITY 1-6   LIFE CYCLES: TAKE 2

Let's look for signs of growth, development, and reproduction in the organisms you observed in Investigation 1A. Take a look at the same organism to see whether it has entered a new stage of development or growth. You'll need a microscope to see hatched brine shrimp and other tiny organisms. Consider the following questions and update your logbook:

- What evidence do you find that organisms have grown or entered a new life stage?

- Estimate the size of organisms, using the technique you learned in Investigation 1B.

- How have the size and shape of organisms or their parts changed since your last observations?

BIOLOGY IN CONTEXT

MISSION: FLEA CONTROL

Getting rid of fleas on your dog or cat used to mean having your pet dipped or sprayed with powerful chemicals several times a year. With a better understanding of the life cycle of fleas, a new two-pronged attack can now be used: Kill the adult fleas, and interrupt the flea life cycle in its early stages.

Fleas lay eggs in your pet's fur. The eggs then fall off and hatch in the animal's bedding, its kennel, and, in the case of an indoor pet, the carpet, woodwork, and furnishings. The immature flea, called

If your pet scratches frequently, it may have fleas.

a larva, goes through a series of stages before becoming an adult that must have animal blood to survive. Adult fleas might remain on your pet for 3–5 months, enough time for a female flea to lay about 2000 eggs.

Unfortunately, the chemicals that kill adult fleas don't work on the larvae. Instead, insect growth regulators such as methoprene or fenoxycarb are applied to a pet's living area. These compounds prevent the eggs and larvae from developing into adults by interfering with growth hormones. Because fewer fleas become blood-sucking adults, you don't need to treat your pet for fleas as often.

STEERING THE LIFE CYCLE

Mealworms, also known as darkling beetles, have a complex life cycle (Figure 1.20A and B) that is familiar to people who raise them as pet food or fish bait. A female beetle lays eggs in some grain. The eggs hatch to produce larvae called mealworms. A mealworm spends its first days feeding on the grain. As it feeds, it sheds its "skin" every few days (an event called a **molt**) and grows larger in the short time before its new "skin" hardens.

After a certain number of molts, the mealworm surrounds itself with a hard outer case and becomes inactive for about seven days in a stage called a **pupa.** During this week the pupa undergoes major changes in form. When these changes are complete, a dark-brown beetle emerges. In a few days the beetle finds a mate. Not long afterward, the mated female will lay eggs in some grain, and the cycle continues.

Although most organisms do not have a life cycle as complex as a mealworm's, each goes through a series of predictable changes. A tadpole develops into a frog; a chestnut grows into a tree. How do these changes occur so reliably in each individual of a species?

A   B

Figure 1.20: *Larval mealworm (A) and adult darkling beetle (B)*

PREFORMATION

For thousands of years, animal breeders have known that if you mate a cow and a bull, they'll produce a calf—not a kitten. But early biologists couldn't show why "like begets like." One explanation, first advanced in the 1600s, was the idea of **preformation.** According to this notion, each organism contains within it all its future descendants in miniature form. Scientists of the time had little evidence for preformation, even though some, using crude microscopes, claimed to see miniature offspring inside human sperm and unfertilized eggs. For lack of a better explanation for reproduction, the idea of preformation persisted well into the late 1700s.

With the introduction of finer microscopes, Caspar Wolff, a German biologist, became the first to observe development in a chicken's egg. He watched the organ systems of the chick proceed through an orderly formation. Clearly, the chick had not preexisted in the egg, as the proponents of preformation thought.

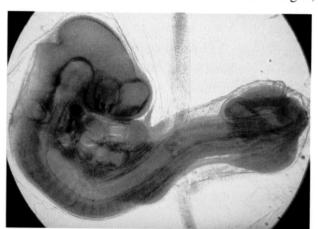

Early development of a chick as seen under a microscope

The preformationists were right in thinking that eggs and sperm contain something that dictates what offspring—and *their* offspring—will look like. It wasn't until the 1940s that researchers could determine the substances present in eggs and sperm that allow like to beget like.

BLUEPRINT FOR AN ORGANISM

Every organism contains a set of instructions that serves as a blueprint for its development, its growth pattern, and its physical characteristics and behaviors. This blueprint is coded by the order of the monomers of a nucleic acid polymer called deoxyribonucleic acid, or **DNA.** The expression of this DNA blueprint can be influenced by factors in the environment, such as nutrition, the weather, and experiences with other organisms. For example, identical human twins have the same DNA blueprint, which directs them to develop in the same way, yet identical twins often differ in characteristics such as weight or intelligence. They may eat different foods and have different learning experiences, for example, which influence the expression of their DNA.

DNA forms a continuous link among all individuals of a species—past, present, and future. Each organism inherits its DNA from its parents (or, in some species, from a single parent), which inherited their DNA from their own parents. As a result, each organism plays out the life cycle of its species in a similar way.

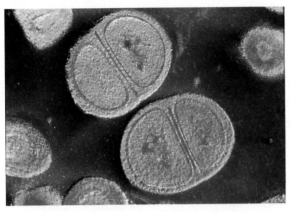

Figure 1.21: *When a bacterium divides, each new bacterium gets an identical copy of DNA.*

The structure of DNA provides a very accurate way for an organism to pass on DNA to its offspring. A crime lab technician who works with DNA samples as criminal evidence knows that DNA can be copied in a process called **replication.** The technician can artificially replicate a tiny quantity of DNA from a crime scene until there is enough DNA to analyze. When an organism grows in size or reproduces, it also replicates its DNA. For instance, before a bacterium divides to produce two new bacteria, it makes a copy of its DNA (Figure 1.21). Each new bacterium contains an identical copy of DNA. Each has inherited the ability to carry out the same biochemical reactions as the original bacterium.

LESSON 1.3 ASSESSMENT

1. The term "life cycle" implies that the stages repeat in each generation of organisms.

2. In the case of fleas, adult fleas may be of different ages and lay eggs at different times.

3. The two organisms have inherited the same DNA blueprint.

4. The two plants may be influenced by different environmental conditions, such as amount of sunlight or the nutrients in the soil.

1 Why are the stages in the life of an organism referred to as a cycle?

2 How can two or more different life stages of a species be present at the same time?

3 Explain why two organisms that result from the division of a single organism are identical.

4 Strawberry plants that develop directly from the stem of a parent plant have DNA that is identical to the parent's DNA. However, two plants that develop in this way may produce strawberries that differ in size. Explain why this is possible.

LESSON 1.4 RESPONDING TO ENVIRONMENTAL CHANGES

Every organism experiences constant changes in its environment. Think of the changes in temperature you might have experienced just since you woke up this morning. Maybe you got out of a warm bed, took a hot shower, stood in the chilly air at the bus stop without your jacket, and sat down in a too-warm classroom. Isn't it amazing that your internal body temperature never changed?

Your body cannot respond to environmental changes without first receiving some kind of signal, or stimulus, from the environment. You receive environmental stimuli primarily through your sense organs: eyes, ears, temperature and pain receptors, and—to a lesser extent—your senses of touch, taste, and smell. The sense organ sends a signal to your nervous system, often to the brain itself, which reacts by acting on your muscles or glands. In this way you make adjustments that help you to maintain a steady internal state. The maintenance of a steady internal state is **homeostasis,** another characteristic of living things.

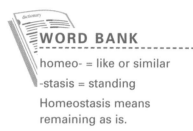

WORD BANK

homeo- = like or similar

-stasis = standing

Homeostasis means remaining as is.

ACTIVITY 1-7   ADJUSTING TO THE ENVIRONMENT

Each group of two to four students will need a container (an aquarium, large glass bowl, or cardboard box) in which half the area is covered (e.g., with black paper) and half is brightly lit. If possible, provide each group with a bright lamp and with a different species. Some species that move in response to light levels include brine shrimp, water fleas, and *Euglena* (all of which move toward light) and *Planaria,* earthworms, mealworms, and most insects (all of which tend to avoid bright light). Students can record the position of a single organism or the number of individuals (from a group of 5–10) in each area at 30-second intervals. For *Euglena,* only the general direction of movement can be observed with the microscope. Its light response can be better observed in a vial placed under a lamp. *Euglena* will collect at the surface.

Have you ever heard the comment "He doesn't have enough sense to get out of the rain"? Heading for cover is one way to cope with an unfavorable environment. Do all organisms have ways of protecting themselves from difficult environments? In this activity, you'll see one way in which organisms respond to differences between environments.

Working with a group, *gently* transfer an individual or group of organisms to a container, placing them at the border between the dark and light areas. Every 10 seconds, record in your logbook whether an organism is in the light part or the dark part of the container. (If you have several individuals, record how many are in each area every 30 seconds.) Record the organism's positions for 3 minutes. Observe the organism closely, and make notes about its movement. When you have collected all the data, answer the following questions:

Some benefits of phototaxis are acquiring light energy, finding food species that like light, finding mates, hiding from predators, and avoiding dehydration. Sessile organisms could protect themselves from extremes of light and temperature by means of reflective or absorptive pigments; insulating layers of fur, feathers, or fat; and internal thermal regulation.

- Did the organism tend to remain in the light or in the dark?

- Did the organism move directly toward the light or dark, or did it wander about a lot? Did it seem to be more active in light or in dark?

- How do the responses the organism makes to light intensity benefit it?

- How could organisms that don't move adjust to factors such as light intensity and temperature?

THE DUAL LIFE OF SALMON

BIOLOGY IN CONTEXT

Eric K. is a fisheries research team leader at the Alaska Biological Science Center in Anchorage. The team consists of biologists, technicians, and volunteers, each requiring different levels of training. Many of the projects at the Center focus on salmon, a fish on which the economy of Alaska depends heavily. Salmon have a life cycle that includes living both in the ocean and in freshwater rivers.

Here's how Eric explains the salmon life cycle: "Salmon lay eggs in freshwater streams. The hatchlings migrate to the ocean, where they spend one to four years feeding and growing rapidly. When they have become mature adults, they swim back to reproduce in the same streams where they were hatched."

But how does a fish that has spent most of its life in the sea adjust to fresh water? Wouldn't river water be too diluted for a salmon? Eric explains, "Salmon exchange water, salts and other substances with their environment across their gills and linings of their mouth. When living in salt water, a salmon, or any other fish, tends to absorb salt and lose water. In a river, the situation is the opposite; the salmon tends to lose salt and gain water."

You may be thinking that salmon are constantly shrinking and bloating during their life cycle. But, as Eric tells us, salmon are actually well adapted to both sets of conditions. "Salmon living in salt water drink a lot of water but produce only small amounts of urine. They get rid of excess salts through their gills and in their feces. This allows them to keep

Adult salmon swim upstream to spawn in the area where they were hatched.

water inside their tissues and salts out. When they enter fresh water, salmon begin making a very dilute urine, which helps eliminate excess water coming in through their gills and body linings. Their kidneys must work harder to keep substances they need from escaping in the urine."

Most species have well-established limits on what environmental conditions they can tolerate. But salmon have adapted to a dual life.

MAINTAINING HOMEOSTASIS

To survive, organisms must maintain homeostasis. In the absence of homeostasis, enzymes, which are very sensitive to changes in solutions, would stop working. Without enzymes, all life processes stop. To protect the structure of enzymes, DNA, and the other polymers you have learned about, an organism must adjust constantly to changes in the external environment.

LEARNING LINK HOME AWAY FROM HOME

People who care for animals in captivity must provide conditions that allow the animals to maintain homeostasis. Special problems arise when animals are moved from their native environment. Zoos in the northern United States, for example, may have animals such as camels, which are native to hot desert areas. Find out from a zookeeper or an animal breeder how the requirements of such animals are met.

Maintaining homeostasis is part of what it means to be alive, but each species has limits beyond which it can't maintain homeostasis. For some species the environment may be too hot or too cold, the water may be too acid or too salty, or the air may have too little oxygen. A major breakdown in homeostasis for large numbers of species results from acid rain.

An important factor of homeostasis for an aquatic species, such as a fish, is the concentration of ions in the water. An **ion** is an atom that has gained or lost one or more of its electrons and therefore has an electrical charge. Two ions that greatly affect the ability of organisms to exchange substances with their environment are the hydrogen ion (H^+) and the hydroxide ion (OH^-). A hydrogen ion is formed when a hydrogen atom loses its only electron. This leaves an ion that is just a proton and gives it an electrical charge of +1. Equal numbers of H^+ and OH^- ions can combine to form water as shown in Figure 1.22.

$$H^+ + OH^- \longrightarrow H_2O$$

Figure 1.22: *Chemical equation for the formation of water*

Organisms run into trouble when the concentration of one of these ions is higher than the concentration of the other. For instance, a body of water—or any solution—that has many H^+ ions and few OH^- ions is acidic. In fact, an **acid** is defined as a substance that raises the concentration of H^+ ions of a solution. By contrast, a body of water (or a solution) that has more OH^- ions is alkaline, or basic. A **base** is a substance that, when added to a solution, raises the solution's concentration of OH^- ions. In general, organisms can't survive conditions that are too acidic or too alkaline.

For the last 20–30 years, biologists and water quality technicians have been concerned about lakes and streams whose acidity has increased so dramatically that there is virtually no life left (Figure 1.23). The increased acidity comes from rain that is acidic because of certain substances in polluted air.

Figure 1.23: *Many lakes in the eastern US have become 100 times more acid over the last 20–30 years, with devastating consequences for life.*

To evaluate the acidity of a body of water (or solution), biologists measure a property of the water called pH. The **pH** of a solution is a measure of the concentration of hydrogen ions (H^+) present in the solution. Chemists have devised a scale of pH values (Figure 1.24). On this scale the most acidic values are at the top and the most alkaline values are at the bottom. The midpoint of the scale, the value of 7, is called **neutral pH.**

A pH value is the negative of the logarithm of the H^+ ion concentration. A solution with an H^+ ion concentration of 10^{-3} moles per liter has a pH of 3. (A mole of a substance contains about 6×10^{23} molecules or atoms of the substance.)

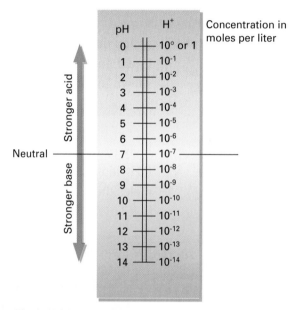

Figure 1.24: *pH scale*

ACTIVITY 1-8   CONTROLLING PH

Acids and bases don't affect the solutions found in living organisms in the same way that they affect a body of water. In this activity you'll compare the effect that acids and bases have on protein (found in organisms) with their effect on a body of water.

CAUTION: You will use acid and base solutions in this activity. Do not get acid or base on your skin, eyes, or clothing. Wear goggles, gloves, and a lab apron. Do not drink the solutions. If you accidentally get some solution on you, rinse it off thoroughly with plenty of cold water.

soap and hot water after the experiment to avoid salmonella poisoning from the egg whites. Alternatives to egg white include homogenized mixtures of equal parts water and raw potato, chicken, fish, meat, or spinach. pH test strips or meters may be used.

Put 25 mL of stream or pond water in a 50-mL beaker, and put 25 mL of egg white in another 50-mL beaker. Label each beaker according to its contents. Measure the pH of each sample using pH paper. Record the pH values in your logbook. Use a dropper pipette (medicine dropper) to add 5 drops of diluted acid to each sample. Stir gently with a stirring rod to mix. Record the new pH of each sample in your logbook. Continue to add acid to each sample, swirling and recording the new pH after every 5 drops, until you have added 35 drops of acid to each sample.

Rinse out the beakers in the sink and repeat the procedure, using diluted base this time instead of acid. When you have finished, pour your solutions down the drain. Clean up the equipment and return it to the proper storage area. **Wash your hands well with soap and hot water before leaving the lab.** Use your data to answer the following questions:

Water changes its pH easily. Organisms produce proteins such as egg white and other substances that resist changes in pH. These proteins can absorb or release H⁺ ions in response to changes in the surrounding solution. The resulting pH tends to remain nearly the same until the protein is overwhelmed by a very large change in the concentration of acid or base.

- Was the pond water or the egg white better able to resist changes in pH?

- Explain how proteins help organisms to maintain proper concentrations of H^+ ions.

ADAPTATION AND EVOLUTION

Most environmental change is more predictable than the changes caused by acid rain. Seasonal changes take place at roughly the same time each year; changes from day to day are usually not extreme. Many organisms can prepare for such predictable changes as a result of the long experience of their species. For instance, as days shorten each autumn, Arctic mammals, such as foxes and polar bears (Figure 1.25A), prepare for winter by growing a coat of insulating winter fur. Arctic squirrels store up body fat that they will use during hibernation (Figure 1.25B). Many types of trees retain water by dropping their leaves before water in the soil freezes and becomes unavailable to the roots. Leaves are ordinarily a major source of water loss for plants.

A

B

Figure 1.25: *The thick coats of the (A) polar bear and the (B) arctic squirrel will keep them warm in the winter but will be shed in the summer.*

Each of these examples involves the **adaptation** of a species to its environment, which occurs gradually over a long period of time. Biologists use the term "evolve" to mean change in a species over time. The ability to evolve is yet another characteristic of living organisms.

The **evolution** of characteristics that make organisms better suited, or adapted, to survive is most often the result of a process called **natural selection.** Here's how natural selection works. Say that an inherited characteristic, such as the pointed bill of a bird, improves the survival of birds that possess it. A pointed bill might be better at picking up highly nutritious seeds buried in grass. Birds with pointed bills live longer than birds having blunt bills and therefore produce more offspring. The offspring also have pointed bills, since bill shape is inherited. In the next generation of birds, birds with pointed bills will be more numerous. Another way of saying this is that the characteristic of pointed bill has spread. The species is undergoing evolution through this characteristic.

Have students guess possible adaptations of the species in Figure 1.26. Examples below are: The hyena's large ears aid in cooling; the shape and color of the praying mantis serve as camouflage; needles on the plant help to defend against grazing animals and retain water; the deformed red blood cells help to provide immunity to the parasite that causes malaria.

The word "adaptation" not only describes the process by which a species responds to changes in its environment. It also refers to any biologically inherited characteristic that helps an individual to survive in its environment. It may be something as simple as a bill shape that allows a bird to pick up and eat certain seeds or something as complex as a series of behaviors that allow a bird to locate seeds in the first place. Adaptations are the expression of an organism's DNA blueprint.

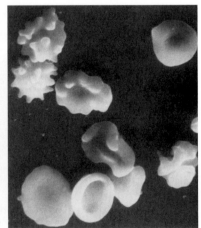

Figure 1.26: *These organisms show interesting adaptations. What adaptations can you identify?*

LESSON 1.4 ASSESSMENT

1. The pH may be dropping because of acid rain. You could test this by recording the pH of rainfall in the area. Other possible answers might include acidic runoff from surrounding area or from leaking landfills and a testing method should be recommended.

2. The technician uses shivering, which generates heat by muscle contraction.

3. Changes in the environment, such as decrease in height of grasses, might make it easier for birds with blunt bills to collect more seeds than birds with pointed bills. Then birds with blunt bills would have improved survival rates and would leave more offspring.

4. A species is unlikely to adapt to such drastic and sudden changes because none of its members is likely to have characteristics (such as enzymes that work at high temperatures) that would allow it to survive the change.

1 A factory on the bank of the lake had been releasing acidic wastewater to the lake. The factory has been shut down. Even after the plant closed the pH of the lake continued to drop over a period of years. How might you explain the falling pH? How could you test your explanation?

2 Give an example of a biological adaptation that a utility repair technician uses to maintain body temperature in freezing weather.

3 Using the case of the birds with pointed bills, show how the process of evolution could be reversed.

4 Aquatic organisms are very sensitive to changes in the concentration of oxygen in the water. Oxygen, like other gases, does not dissolve as well in warm water as in cool water. In some countries, manufacturing plants release water that was used to cool engines and turbines into nearby streams and lakes. This water, often close to 100°C, can substantially raise the temperature of a small lake. Do you think species of aquatic organisms can adapt to this change? Give a reason for your answer.

CAREER APPLICATIONS

The applications that follow are like the ones you will encounter in many workplaces. Use the biology you learned in this chapter to complete the activities. Share your work with the class.

AGRICULTURE & AGRIBUSINESS

Interrupting the Life Cycle

Some examples are releasing sterile male screwworms so that eggs are not fertilized and thus would not mature; using a insect growth regulator, hexaflumuron, that keeps termites from molting, thus halting their growth; or disrupting the natural development of adult cockroaches and specifically altering their reproductive organs, preventing mating.

Talk to an agricultural extension agent, an exterminator, or a knowledgeable pesticide salesperson. What products or techniques are available that interrupt the life cycle of a pest rather than just killing the adults? Which pests are targeted by this product or technique?

BUSINESS & MARKETING

Profitable Adaptations

We humans create products that protect us from harsh environments; as a result we can travel and live virtually anywhere on Earth. Look for ads in newspapers, magazines, and catalogs that display such products. Make a poster showing how these products help us to adjust to the environment. When you present your poster to the class, discuss whether the products would be successful in other environments.

FAMILY & CONSUMER SCIENCE

Enzyme Power

Enzymes are important polymers in living systems. Research some uses of enzymes, using the Internet or other reference sources. Report on ways in which enzymes are used in food processing, cleaning supplies, and textile production. Use keywords such as "enzymes," "enzyme technology," "detergent," and "textiles" in your search. The Biotechnology Industry Organization (BIO) Web site has some good resources.

HEALTH CAREERS

How to Feed a Microorganism

Talk to someone who works in a hospital or private microbiology lab. Ask for a description of the various kinds of media used to culture different types of microorganisms. How are the media formulated to meet the nutritional requirements of the microorganisms that grow on them? Find out the differences among growth media, selective media, and differential media. How does the lab determine which antibiotics might be effective against specific microorganisms?

Tender Treatment of Stings

Meat tenderizer includes enzymes that break down protein, including the protein venom of a jellyfish.

Jellyfish such as sea nettles infest coastal waters at certain times of the year. Many beachgoers know from experience that these creatures can inflict very painful stings. A common first-aid treatment is to apply a paste of meat tenderizer to the sting. Find out why this treatment is recommended.

INDUSTRIAL TECHNOLOGY

Living and Nonliving Problems

Talk to someone who works with condensers, cooling towers, water treatment facilities, or similar equipment. Ask about fouling and scaling. Which of these conditions involve living organisms? Which of these conditions involve nonliving materials? How is each of these conditions prevented? How are they treated? Which condition is the most problematic at your informant's location?

Enzymes for Industry

Use the Internet to search for industrial uses of enzymes, and make a list of the broad categories of industries that use enzymes. Find a local business that represents one of the categories. Contact the company and interview someone to find out whether the company uses enzymes and, if so, how. As a class, make a chart of local companies that use enzymes, the reasons they use them, and the enzymes they use.

CHAPTER 1 SUMMARY

- Something is alive if it maintains its own complex, organized structure; passes through a cycle of growth, development, and reproduction; and responds to changes in its environment in ways that help it to survive.

- Organisms collect raw materials from their environments and use them to build and maintain their highly organized structures. To do this, organisms must have a source of energy, either light energy or chemical energy.

- The raw materials that organisms collect are composed of one or more elements. Elements are composed of fundamental units called atoms. Organisms absorb raw materials and rearrange the atoms to build polymers such as proteins, complex carbohydrates, and nucleic acids.

- An organism's ability to maintain its complex structure depends on its ability to carry out hundreds of different chemical reactions, each involving a specific enzyme.

- Water is a polar molecule: this feature allows water to dissolve other substances that are polar. Water is ideally suited to be a medium for chemical reactions within living things.

- One important way in which organisms exchange materials with their environment is diffusion, the net movement of molecules from a region of high concentration to a region of lower concentration. The result of diffusion is that substances in a solution become evenly distributed throughout the solution.

- When members of a species reproduce, the offspring are of the same species. This happens because parents pass to their offspring a set of instructions, or DNA blueprint, that is characteristic of the species. Along with the environment, DNA directs the organism's growth and development as well as its activities.

- An organism must adjust constantly to changes in its external environment to maintain homeostasis, a constant internal state.

- A species changes over time to become better suited to survive in its environment. This change is an example of evolution, which may occur through the process of natural selection.

■ Natural selection occurs when organisms with characteristics best suited for survival in a particular environment leave more descendants than other organisms, thus increasing the frequency of these characteristics within the species. These characteristics, passed from parent to offspring, are called adaptations.

CHAPTER 1 ASSESSMENT

Concept Review

See TRB Chapter 1 for answers.

1 Burning a potato chip causes a chemical change in the potato chip. So does eating a potato chip. Explain how the chemical change is different in the two examples.

2 Some nonliving things, such as rivers and mountains, can grow and develop. Compare and contrast the growth and development of a nonliving river and of a living organism such as a fish.

3 Predict which would respond more quickly to environmental change, individual organisms or species? Give a reason for your prediction.

4 Describe the structure of a hydrogen atom.

5 How might diffusion help an organism to exchange substances with its environment?

6 How do oxygen and hydrogen atoms bond to form a molecule of water?

7 Using a pearl necklace as a model, describe what a polymer is.

8 Describe three types of polymers and their importance to living things.

9 What six elements make up 95% of your body weight?

10 Two colorless solutions are poured into the same beaker. After 15 minutes the resulting solution has turned blue. This experiment is repeated, but charcoal is added to the beaker. The solution turns blue almost immediately after the charcoal is added. What did the charcoal do?

11 How are enzymes specific?

12 What process ensures that an organism can provide a copy of its DNA to each of its offspring?

13 In what form does an organism trap the energy released from the "burning" of sugar?

14 Discuss, in terms of chemical bonding, why a molecule of water is polar.

15 Why can organisms of the same species carry out the same chemical reactions?

16 Explain how an organism maintains homeostasis.

17 A sea lion's flippers help it to glide swiftly through the ocean as it searches for food or escapes a predator. Explain how flippers are an adaptation.

18 A 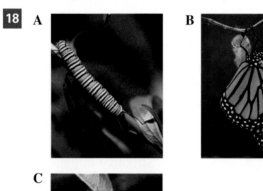 B

C

Put the photographs in the correct sequence. Explain briefly what changes in development occur from one stage to the next.

Think and Discuss

19 Discuss the relationship between the disproved idea of preformation and the passing on of DNA from parent to offspring.

20 Supermarkets and restaurants try to offer fruits and vegetables that are fresh. Is fresh produce alive? How could you be sure?

21 Could a machine that "reproduces" by making duplicates of itself be considered alive? Give the reasons for your answer.

In the Workplace

22 Explain why knowledge of a pest's life cycle is important to an exterminator.

23 Design a simple test to be sure that a cheese maker is adding live bacteria to flavor her cheese.

24 You work in an industry that uses several chemical processes. One of the steps is very slow. You have been asked to research ways to shorten this step. What might you do?

Investigations

The following questions relate to Investigation 1A:

25 How were you able to distinguish living from non-living organisms?

26 Give two examples of living and non-living organisms used in your investigation.

The following questions relate to Investigation 1B:

27 A seed is the beginning of a new plant. Why do many seeds contain a much higher percentage of fat than most other plant parts?

28 When people have a very high fever, their body can become so warm that their brain is damaged. Would you expect this damage to occur mainly to brain carbohydrates or to brain proteins? Give reasons for your answer.

CHAPTER 2

WHY SHOULD I LEARN THIS?

Cells, the basic units of life, are studied by a variety of professionals. Medical technicians count cells in blood samples. Crime lab technicians collect evidence such as hair, skin, or blood from crime scenes and test them to determine whose cells they contain. Hospital inspection teams swab objects to find out if they're contaminated with cells of organisms that might cause infections. In this chapter, you will learn about the different kinds of cells and how they function—information you might use as you make decisions about health and hygiene.

INTRODUCTION TO CELLS

WHAT WILL I LEARN?

1. how cells make up every living thing
2. how cells take in food and get rid of wastes
3. the similarities and differences among cells of animals, plants, fungi, and bacteria
4. the important parts of cells and their functions
5. how cells reproduce and how they control their reproduction

---

Why aren't we constructed like spiders or cacti? After all, every organism is made of cells. But cells come in many types. Spiders have cells that make venom to paralyze their prey. A cactus has cells that absorb energy from sunlight. Animals have cells that use electricity to communicate with one another. Differences among cells create differences among organisms. These differences allow a doctor to use antibiotics to kill disease-causing bacteria without hurting the patient. They make it possible for farmers to use pesticides to control mold and insects without harming their crops. For that matter, these differences are found even within a single organism, for example, a human. Without differences among our cells, we would be shapeless blobs, without bone, hair, brain, and other distinguishing characteristics.

In this chapter you'll learn about the structure of cells, how they control what enters and leaves them, and how they reproduce. You'll see differences among cells from different organisms and find out how medicine, agriculture, and other fields take advantage of these differences.

LESSON 2.1 BASIC UNITS OF LIFE

Your body demonstrates its great complexity when you perform highly coordinated activities, adjust to small changes in your surroundings, and express emotions. Creatures that you might consider primitive also have complex organization, just as you do. A basic unit of structure that all organisms share may become clear as you examine them under a microscope.

ACTIVITY 2-1   OBSERVING ORGANIZATION ON A SMALL SCALE

Suggested organisms for the activity are *Elodea, Fucus, Volvox, Spirogyra, Paramecium*, and small animals such as rotifers. Have students examine both live protists in pond water or pure cultures and prepared slides.

Students should discern the shape, size, and color of organisms. They should recognize cells as fundamental units with internal structures.

Organisms of all sizes have structure and organization. You'll start your study of structures by looking at some small organisms. Put a drop of pond water onto a slide and add a coverslip, or use a prepared slide. Place the slide on the stage of your microscope. Check it out under low power first, then under the high-power lens. Look at a number of different slides and draw what you see in your logbook. Compare your observations with a partner's, and then answer the following questions in your logbook:

- What features do all organisms share?
- What special structures do you see in some organisms but not in others?
- How does each organism seem to be organized?

GROWING CELLS FOR THE BAKERY

BIOLOGY IN CONTEXT

You might be surprised to learn how many jobs require a knowledge of cells. In the lab of a baker's yeast manufacturer, a technician is looking at yeast under a microscope. She is busy clicking a counter. She is checking the maturity of the yeast to see whether it is ready to be harvested for packaging. Mature yeast cells do not have buds, which are formed when the cells are reproducing. (Cells are the basic unit of life. Yeast is an organism with only one cell, so each cell is a separate organism. You will learn more about cell reproduction later in this chapter.) The manufacturer wants to package the yeast when at least 95% of the cells have completed budding. This helps to maximize the product's shelf life.

A technician examines yeast cells under a microscope.

Across the lab, another technician is working to solve a problem in a different batch of yeast. Under the normal growing conditions at the plant, yeast can double in number every 2–3 hours. This means that a drop of yeast cells can grow to over 300,000 pounds in a week. The problem batch of yeast isn't growing correctly and is taking more than twice as long to double. The technician is trying to find out what percentage of the yeast cells are alive. He has put a drop of methylene blue onto a slide with a drop of the yeast cells. If the yeast cells are alive, the cell wall around each cell remains colorless. If they are dead, the walls turn dark blue. If he finds an abnormal number of dead cells, the technician will begin to look for causes.

CELL THEORY

People who examine cells under a microscope as a routine part of their job are constantly aware that cells are the basic units of an organism. Their importance is reflected in the cell theory. (A **theory** is an explanation that has been supported by many observations.) The cell theory can be stated as follows:

1. Cells are the basic units that make up all organisms.
2. Cells are the smallest units of life; they can carry out all of the activities that constitute life.
3. Every cell comes from a pre-existing cell.

In fact, every cell in an enormous redwood tree came from a single cell. Just as you did.

Each cell of an organism carries out the basic chemical reactions of life. The chemical information that directs these reactions is the same in all cells of an organism. So it is possible to study certain characteristics of organisms by looking at a small sample of their cells. This fact makes it possible to check the sex of Olympic athletes by examining a few of their blood cells. This is done to ensure that everyone competing in the women's events is really a woman and every competitor in the men's events is a man.

The cell theory also links together different species (Figure 2.1). All cells share certain features: They all have some of the same parts and perform many of the same chemical reactions. Together with the fact that cells arise only from pre-existing cells,

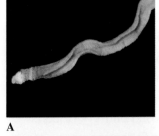

A

B   C

Figure 2.1: *All these organisms are composed of cells. (A) An acorn worm. (B) A shelf fungus. (C) A crab.*

this is evidence that all cells (and therefore all organisms) must have descended from a single ancestor cell. So a frog and, say, a duck, must share an ancient ancestor. This cell lived long before there were frogs and ducks. Of course, frog cells and duck cells are also different in some ways. They carry different chemical instructions, so they perform some different chemical reactions.

GROWING THE CELL THEORY

The cell theory was brought to us by the microscope. The first microscopes had one lens, like a magnifying glass. They were toys for the rich, used to look closely at flowers or insects. No one thought that these toys would teach us anything new.

But in 1665, Robert Hooke, an English scientist, described how he used the microscope to discover a honeycomb structure in thin sections of cork (Figure 2.2). He wanted to find out what in the structure of cork made it lightweight, able to float on water, and firm yet compressible. He named the air-filled spaces of the cork "cells," because they reminded him of the small rooms of a monastery. Hooke did not see any of the internal structures of a cell, because cork has only dead, hollow cells. Nearly a decade later, Anton von Leeuwenhoek, a Dutch scientist, provided the first description of living cells, seen in water through a microscope.

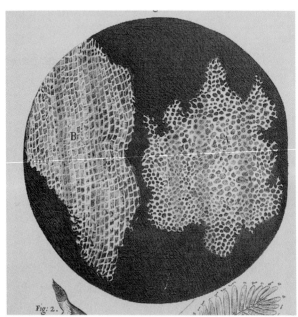

Figure 2.2: *Robert Hooke's first view of cells*

The next major breakthrough in our knowledge of cells was the discovery of the cell nucleus by the Scottish botanist Robert Brown in 1831. Improved microscopes allowed Brown and others to see more detail in cells. Matthias Schleiden, a German botanist (plant biologist), used a microscope to study many types of plants. In 1838 he published his conclusion that all plants are composed entirely of cells or products of cells. The following year, Theodor Schwann, a German zoologist (animal biologist), published his conclusion that animals are also composed of cells. In 1858, another German, Rudolf Virchow, added to cell theory the idea that all cells come from other living cells.

ACTIVITY 2-2   BUDDING YEAST CELLS

Add packaged yeast to a 10% solution of sucrose in several 250-mL beakers. Allow yeast to incubate at room temperature for 24 hours. Provide students with dropper bottles of 1% Neutral Red stain solution to use as a stain for their wet mounts. Other materials needed are microscopes, slides, coverslips, and paper towels.

You may need to explain the use of the word "culture."

Now that you have learned something about cells, you can test the cell theory by examining some living yeast cells, just as technicians do at companies that produce yeast for bakers and brewers.

Get some yeast culture in a 250-mL beaker. Use a Pasteur pipette to put one drop of the culture in the middle of a microscope slide. Put a coverslip over the drop. **Beware of sharp edges on the slide and coverslip.** You have made a wet mount. At one edge of the coverslip, add a drop of Neutral Red stain. Touch a paper towel to the opposite edge of the coverslip (see Figure 2.3) to pull the stain under the coverslip.

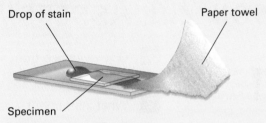

Drop of stain   Paper towel

Specimen

Figure 2.3: *Use a paper towel to draw the stain under the coverslip.*

Use your microscope to observe the slide at low power, then at high power. Draw and label what you see at high power. The circular structures are yeast cells. Focus your microscope on a yeast cell that appears to have a lobe or extension: This is a bud.

Answer the following questions in your logbook:

Yeasts are single-celled and reproduce by budding. The bud is at first a small replica of the parent cell. Budding demonstrates that new yeast cells come from existing yeast cells, as the cell theory predicts. Students may be able to detect a nucleus within the bud.

- On the basis of your observations, would you consider yeast a single-celled organism or a multicellular (many-celled) organism?

- Do buds stay attached to the cell from which they develop, or do they become independent of that cell? Explain why you think this happens. What role do you think bud formation might play in the life of yeast?

- How does the bud compare to the parent cell from which it developed?

- Explain how budding fits into the cell theory. Do your observations provide evidence for or against any part of the theory? Explain your answer.

Because every cell in a plant or animal contains the same hereditary information, it is sometimes possible to grow a large organism from a single body cell. Carrot plants were produced in this way before 1950, and in the late 1990s, scientists announced that they had cloned (grown from single cells) sheep and calves. Using the Internet and resources available at your library, investigate cloning. Report to the class on how it is done, what its limitations are, and what social and medical issues are involved in its application to humans.

LESSON 2.1 ASSESSMENT

1. Prepared slides of organisms show cellular structure; budding demonstrates that yeast cells come from preexisting cells.

2. Microscope.

3. Cells of the same organism contain the same chemical information and carry out the same basic chemical reactions.

4. Some bacterial cells would be visible at 400×. Typical cells of plants, animals, and fungi are visible at 40–100×.

1 Give evidence from your observations so far that support the cell theory.

2 What instrument was crucial to the development of the cell theory?

3 How are the cells of a multicellular organism similar to one another?

4 At what magnification can you begin to tell that organisms are made up of cells?

LESSON 2.2 WHERE CELLS LIVE

To remain alive and healthy, you must take in substances such as food and oxygen and eliminate others as wastes. Each of your cells must also control which substances it absorbs from its surroundings and which it releases, to maintain homeostasis. Cells must constantly work with or against diffusion, which tends to upset their internal environment. The **cell membrane,** a thin boundary that separates a cell's contents from its surroundings, controls the exchange of substances, allowing some in or out, but blocking others. People who have the disease called cystic fibrosis have problems with the function of the cell membrane. This prevents their cells from producing normal sweat and mucus, leading to serious medical problems. In this lesson, you will learn how membranes control the movement of substances in and out of cells.

ACTIVITY 2-3   FORMATION OF DROPLETS IN DETERGENT

Oil is not soluble in water. When you shake an oil-water mixture, oil droplets form briefly. The detergent molecules surround the oil droplets; the polar regions of the detergent molecules interact with water (a polar molecule), and the nonpolar regions interact with droplets of oil, which are nonpolar. These interactions keep the oil suspended as tiny droplets. Reflection of light from these droplets makes the mixture cloudy. Because water and polar substances that dissolve easily in water do not dissolve easily in oil, a layer of oil can act as a barrier between two solutions. In the same way, a layer of oil surrounding a cell can act as a barrier between the cell and its environment.

The cells that you observed in Activity 2-2 did not break apart and dissolve in the solution. Investigate why the cell contents didn't leak out by following this procedure.

Put a few drops of vegetable oil in a test tube. Add 15–20 mL of water. Stopper the tube and shake it. Observe the behavior of the oil and water. Add 1 mL (20 drops) of liquid detergent to the test tube. Restopper the tube and shake it. Now discuss the following questions with your partner and record your answers in your logbook:

- What made the water and oil separate?

- How did the detergent affect the oil and water? What do you think caused the change?

- How could a layer of oil prevent two solutions in water from mixing? How could this separation help cells to keep their internal solutions stable and different from the surrounding environment?

MEMBRANES FOR HEALTHY SKIN

It's hard to watch television or read a magazine without seeing ads for a multitude of cosmetic products. This lotion will moisturize your skin, that cream prevents aging and wrinkles, and another one can provide vitamins and minerals to your skin. Some of these products work by laying down a protective barrier on top of your skin, but other products contain substances that must be taken up by the skin cells to be effective. Cosmetic manufacturers now say that they have created new products to ensure the delivery of these substances to the cells of skin. These new products are called liposome concentrates.

Liposomes can be thought of as small versions of cells in which a nutrient, or other organic compound such as a drug or antibiotic, is temporarily trapped inside a membrane. They are made by mixing substances called phospholipids in a solution of water and other compounds and agitating them. Unlike some lipids, one end of each phospholipid molecule is polar, and the other is not. The phospholipids arrange themselves in two layers, forming a tiny sphere around the compound. The polar ends of the phospholipid molecules face out of the membrane, toward the water. When you apply a cosmetic that contains liposomes to your skin, the membrane of the liposome begins to merge with the membranes of cells. In the process, the liposomes release their payload of nutrients into your cells.

Liposomes are also being used for more important purposes than attractive skin. They can be designed to deliver anticancer drugs to tumor cells and antibiotics to bacterial cells. When the drug is kept inside a membrane until it reaches the target cells, it has few of the side effects of drugs that travel free in the bloodstream.

THE CELL MEMBRANE

A cell interacts with its watery surroundings much as oil droplets interact with water when detergent is added to the mixture. All cells are surrounded by a thin cell membrane of lipids and proteins that separates the fluid contents of the cells (the **cytoplasm**) from the environment outside. Because organisms must obtain food and other substances from their environments and must excrete wastes, cell membranes must be **selectively permeable:** They must allow some substances through but not others. The structure of the membrane allows the cell to maintain very fine control over the entrance and exit of substances. Two important factors affect the ability of molecules to pass through cell membranes: the size of the molecules and how polar they are.

Figure 2.4 shows a typical cell membrane. It consists of two layers of lipid molecules whose ends have different properties. The outward-pointing, polar heads of the molecules are **hydrophilic** (attracted to water). The inward-pointing, nonpolar tails of the lipid molecules are **hydrophobic** (repelled by water). So the molecules line up with their tails pointing away from water. Ions and polar molecules do not diffuse easily through the lipid layers of cell membranes. Fat-soluble, nonpolar molecules, such as oxygen and carbon dioxide, diffuse quickly through the lipid layers and enter or leave cells by passing through their membranes (see Figure 2.5). Small molecules, such as water or sugars, also pass through membranes more easily than large molecules, such as proteins.

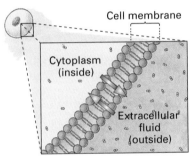
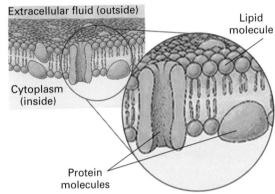

Figure 2.4: *Lipid molecules in a cell membrane line up with their polar ends (red) out and nonpolar ends in away from water. Note proteins embedded in the double lipid layer.*

Figure 2.5: *Oxygen molecules (black) diffuse into cells through their lipid membranes.*

INVESTIGATION 2A   SELECTIVELY PERMEABLE MEMBRANES

See Investigation 2A in the TRB.

Kidney dialysis machine

An important job of your kidneys is keeping your blood from getting too concentrated or diluted. People who have serious kidney disease may need a device called a dialysis machine to help them with this job. An artificial membrane, called dialysis tubing, is used in these machines. You can use dialysis tubing to learn about diffusion and the permeability of membranes. The tubing's pores allow some molecules, but not all, to pass through the tube wall. Try the available solutions and figure out which kinds of substances can pass through the tubing. Does the tubing behave like a cell membrane?

ACTIVITY 2-4   DIFFUSION THROUGH A MEMBRANE

SAFETY NOTE

Obtain a fresh red onion. Refrigerate until the start of the Activity. Prepare a 10% solution of sodium chloride in water by dissolving 10 g of table salt in 90 mL of water. Dispense in small dropper bottles for each lab station.

Cells mounted in water retain their normal appearance. In salt solution the cell membranes gradually pull away from the cell wall, leaving a boundary area between the cell membrane and cell wall in which there is no red pigment.

The cells are losing water because of the difference in concentration of solutes inside and outside the cell. As the cytoplasm of a cell shrinks, its cell membrane pulls away from the more rigid cell wall.

The cell is unable to maintain homeostasis. If the loss of water is too great, the contents of the cell become too concentrated to carry out biochemical reactions.

Students should suggest trying to add water to the slide to get the cells to take up water.

Sometimes a cell finds itself in conditions so extreme that it cannot maintain homeostasis. In this activity you will have an opportunity to observe what happens to cells when they can't adjust to their surroundings.

1. Cut a small slice from a red onion. Peel the red layer from the surface of the slice. Put this layer onto a clean microscope slide and add a drop of water. Gently lower a coverslip over the drop and put the slide on the stage of your microscope. **Beware of sharp edges on the slide and coverslip.**

2. Observe first at low power to locate the section of red onion cells. Then observe at high power and record your observations in your logbook.

3. Repeat steps 1 and 2 but apply a drop of 10% salt solution to the slide instead of water. Compare your observations to your observations of the first slide.

4. Place the microscope pointer over a large cell in the second mount and observe every 2 minutes for the next 5–10 minutes. Continue recording your observations in your logbook. Then answer the following questions:

   • What differences did you see in onion cells on the two slides? Describe these differences in terms of the cell structures you have learned about so far.

   • How can you explain the response of the cells placed in salt solution?

   • How do you think the response of the cells affects their ability to survive?

   • Explain how you would treat the cells to help them recover.

TRANSPORTATION ON A SMALL SCALE

Some substances diffuse through cell membranes when they are more concentrated on one side of the membrane than on the other. The substance moves from the side where it is more concentrated to the side where it is less concentrated. For example, wastes that are produced in a cell, and therefore are in high concentration there, will diffuse out of the cell. If the concentration on both sides of the membrane becomes equal, equal numbers of molecules pass through the membrane in both directions, and the concentration on both sides remains the same.

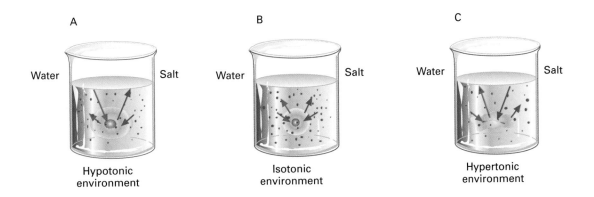

A   B   C

Water   Salt   Water   Salt   Water   Salt

Hypotonic   Isotonic   Hypertonic
environment   environment   environment

Water ⟶

Salt ⟶

Figure 2.6: *In a hypotonic solution (A), cells swell as they absorb more water than they lose. In a hypertonic solution (C), cells collapse as they lose more water than they absorb.*

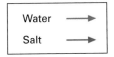

WORD BANK

hypo- = less than normal

-tonic = tension

A cell swells in a hypotonic (more dilute) solution.

hyper- = more than normal

A cell shrinks in a hypertonic (more concentrated) solution.

iso- = equal

A cell stays the same size in a solution that is isotonic (same concentration) as its cytoplasm.

turg- = swelling

Turgor keeps cells swollen out against their walls.

When a cell, such as the onion cells you examined in Activity 2-4, is in a concentrated salt solution, water is more concentrated inside the cells than outside, so it diffuses out into the solution (Figure 2.6). Diffusion of water through a membrane is called **osmosis.** Water moves by osmosis from regions of high water concentration to regions of lower concentration. Another way to look at this is that if the salt or other dissolved substances are more concentrated on one side of a membrane, water will move to that side until the concentrations become equal.

Osmosis can cause problems for cells. Water and the substances in it are apt to have different concentrations in a cell's cytoplasm and the surrounding fluid. In the cells of a potato plant, for example, sugars, salts, and other dissolved substances are more concentrated than they are in the water in the surrounding soil. Since the soil water is "more watery," or more dilute than the plant's cytoplasm, the plant's cells tend to absorb excess water and swell in this **hypotonic** environment. If the cells cannot stop this process, they could burst. You may recognize this as a problem for homeostasis, a concept you learned about in Chapter 1.

Plants are well adapted to a hypotonic environment. Their cells are surrounded by tough **cell walls.** As the cells absorb water and swell, the pressure of their cytoplasm pushes the cell membrane against the cell wall. The wall keeps the cell from bursting, and the pressure, called **turgor,** helps the plant to keep its shape. Farmers and gardeners soon learn to recognize when their plants have lost turgor. A wilted plant is a danger sign that more water is needed right away.

Organisms that live in the ocean, however, have cytoplasm that is less concentrated than the surrounding water. A whale, for example, lives surrounded by salt water, an environment that is **hypertonic** to its cytoplasm. As a result, the whale's cells tend to lose water to the sea and to absorb salt. To avoid this problem, sailors learn to store a supply of drinking water in their lifeboats. A stranded sailor may endanger his health more by drinking seawater than by not drinking at all.

Many animals have adapted to hypertonic and hypotonic environments by surrounding their cells with fluids such as blood, which is about as concentrated as the cytoplasm of the animal's cells. These cells live in an **isotonic** environment. Under normal conditions they do not gain or lose water by osmosis, and homeostasis is maintained.

In Activity 2-4, you saw onion cells shrink away from their walls as they lost water in a hypertonic environment. But how did water pass so quickly through the cell membranes, which are mostly lipids? Look again at the cell membrane in Figure 2.4. Protein molecules embedded in the membrane act as channels or transporters. They allow specific substances (such as water) to diffuse more easily through the membrane in either direction. The cell uses no energy to make this happen, aside from the energy needed to make the proteins. These proteins allow diffusion to occur more quickly, a process called **facilitated diffusion** (Figure 2.7).

Other membrane proteins pump specific substances into or out of cells. Because these proteins use the cell's energy, they can work against diffusion, actively transporting substances from regions of low concentration to regions of high concentration (Figure 2.7). Facilitated diffusion, which can only speed up diffusion, cannot do this. Some of these **active transport** proteins pump food substances into cells. Others pump wastes and toxic substances out of cells. Active transport enables organisms to survive in hypertonic or hypotonic environments by pumping salts, for example, into or out of cells. This prevents cells from bursting in fresh water or collapsing in sea water. Active transport of excess salts and water by cells in your kidneys keeps your blood isotonic. People with kidney disease may require the help of an "artificial kidney," or dialysis machine, that can perform this function for them.

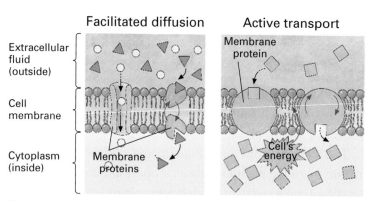

Facilitated diffusion | Active transport

Extracellular fluid (outside)

Cell membrane

Cytoplasm (inside)

Membrane proteins

Membrane protein

Cell's energy

Figure 2.7: *Cells can speed up transport with membrane proteins that allow facilitated diffusion (left) or use their own energy to oppose diffusion with active transport (right).*

ACTIVITY 2-5   LET THEM EAT YEAST

Some organisms take in or release particles that are too big to fit through channels in their cell membranes. For example, *Paramecium,* a one-celled organism, consumes yeast and other one-celled organisms. How does a paramecium consume its prey without using a mouth or membrane channels?

Put a drop of paramecium culture on a slide. Add a drop of red-stained yeast suspension and a few cotton fibers. Put a coverslip over the drop on the slide. Look at it under low power to find active paramecia. Switch to high power, focusing on one paramecium. Look for a groove in the paramecium where red yeast cells are collecting.

What is happening to yeast cells in and around the groove? Observe as many paramecia as you can. Record your observations in notes and drawings in your logbook.

Some yeast cells are rejected by the paramecia, but others are ingested and collect at the "bottom" of the preoral groove. As these ingested cells accumulate in the groove, a food vacuole forms around them.

TRANSPORTING LARGER PARTICLES

As you saw in Activity 2-5, some things, such as cells or other kinds of large food particles, leave or enter cells in bubbles of membrane rather than passing through the cell membrane. For example, your body contains many defensive cells, such as white blood cells. When these cells encounter a disease-causing bacterium or other foreign object, they wrap part of themselves around it, the way you would wrap your arms around someone to hug them. But where the tips of the arms meet, they fuse. The enclosed part of the cell membrane pinches off inside the cell, forming a closed bubble, or **vesicle,** with the bacterial cell inside (Figure 2.8A). The whole process is called **endocytosis.** The white blood cell then breaks down the bacterium. This is an important part of your defense against disease, and one that all health care workers should understand.

Endocytosis

Food particle

Vesicle

Figure 2.8A: *Some cells take in large food particles by endocytosis.*

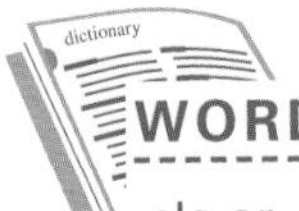

A paramecium reverses endocytosis to ship out wastes. It encloses the waste in a vesicle inside the cell (Figure 2.8B). When the membrane of this vesicle contacts the cell membrane, it joins with it and opens outside the cell, leaving the wastes outside the cell. This elimination process is called **exocytosis.**

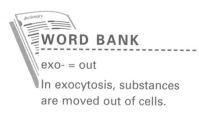
Exocytosis

Waste materials

1

2

3

Figure 2.8B: *Some cells release wastes, new cell wall material, and other substances by exocytosis.*

LESSON 2.2 ASSESSMENT

1. Cell membranes that are composed mostly of lipid separate the watery cell contents from the bath water.

2. Small or nonpolar molecules diffuse through the cell membrane; larger or more polar particles enter or leave via facilitated diffusion or active transport; still larger objects enter by endocytosis and leave by exocytosis.

3. The membrane has a fat-soluble region consisting of two layers of lipid molecules with their hydrophobic tails pointing inward. Protein molecules are embedded in the lipid layers.

4. Water diffuses out of the cells in the food and into the salt water or syrup. As the cells lose water, they shrink or collapse, causing the food to shrivel.

1 Your body consists mostly of water. Why don't you dissolve in the bathtub?

2 Compare three ways in which substances enter and leave cells. Consider the size and type of particles that pass into or out of cells and how they do so.

3 What makes up a cell membrane and how are these components organized?

4 Foods that are preserved by pickling in salt water or candying in syrup often shrink and become wrinkled. Explain how these changes come about.

LESSON 2.3 TWO KINDS OF CELLS

WORD BANK

anti- = against

-biotic = living

An antibiotic is a chemical that acts against a living thing.

Have you ever taken penicillin or put antiseptic on a cut? These antibiotics kill disease-causing organisms. Why don't they kill you, too? Because your cells are different from those of disease-causing bacteria. People in many occupations use chemicals that are meant to kill only certain organisms. When a veterinarian prescribes an antibiotic, a farmer selects a weed killer, or a food service worker sets out rat poison, each should ask, "What do I want to control? Will this substance injure crops? wildlife? people?" The answers depend on the kinds of cells that make up these organisms. In this lesson you will learn about the differences among various types of cells.

ACTIVITY 2-6   IDENTIFYING CELLS

Provide prepared slides or fresh material and stains for freehand sections. These should include several leaf and animal tissue sections with prominent nuclei, plant cells with chloroplasts and vacuoles, fungi with prominent nuclei (and walls, if possible), single-celled and colonial bacteria, and cyanobacteria.

Suppose you worked at a water treatment plant. You might examine a sample of water and find it swarming with organisms. How could you identify them? Do plant and animal cells have different parts? Review your notes and drawings of cells from Activity 2-1. Then use your microscope to examine slides of various kinds of cells. In your logbook, record a description (including cell size) and draw and label each type of cell you observe.

Working with a small group, compare your observations. Try to find a few patterns that you could use to identify what kind of organism a sample of cells came from. The following questions can help you find these patterns:

All cells except bacteria have visible internal parts. Nuclei and other internal parts that students cannot identify occur in all nonbacterial cells. Plant and fungal cells have external cell walls, and plant cells have chloroplasts. Plants and animals are made of many cell types. Other organisms consist largely of nearly identical-appearing cells and, except for fungi, are usually unicellular or colonial. Smaller cells (bacteria) lack visible internal parts and complex multicellular organization.

- Which kinds of organisms have cells with internal parts? Do larger or smaller cells have more internal parts?

- Which parts do you see in animal cells? plant cells? fungus cells?

- Which kinds of organisms are made of only single cells or groups of identical cells? Which are made of many kinds of cells?

MEDICINES FROM DIRT

BIOLOGY IN CONTEXT

About 100 years ago, tuberculosis (TB) was responsible for millions of human deaths each year. TB bacteria infect the lungs and may spread to other parts of the body. People with TB are easily tired, have a fever, and lose a lot of weight. They often cough up blood from their damaged lungs.

The 1940s saw an intense search for a way to cure TB. Selman Waksman, working at Rutgers University in New Jersey, found that some soil bacteria produce a substance, streptomycin, that interferes with the ability of other bacteria to make the proteins they need. The bacteria that cause TB are affected. Waksman and his colleagues wondered whether streptomycin could be used as an antibiotic to treat people with TB. What if the antibiotic interfered with patients' ability to make their own proteins?

Further research showed that streptomycin attacks small particles called **ribosomes** which are the sites in cells where protein molecules are synthesized. Because bacterial ribosomes are different from ribosomes of other organisms, streptomycin doesn't interfere with protein synthesis in human cells.

When TB patients received streptomycin, it stopped the infection from spreading and produced only minor side effects. Encouraged by this result, Waksman continued to identify new antibiotics produced by soil bacteria. Today, medical researchers continue to look for new treatment strategies based on differences between human cells and the cells of disease-causing organisms.

WORD BANK

biotic = living things

An antibiotic is a substance that acts against organisms that cause disease, especially bacteria.

BACTERIAL CELLS

Selman Waksman's antibiotics were extremely useful in fighting TB because of fundamental differences between cells of bacteria and cells of other organisms. Because of these differences, bacteria are classified as **prokaryotes.** Prokaryotes have simple structures (Figure 2.9), without the many internal parts of other cells. Most

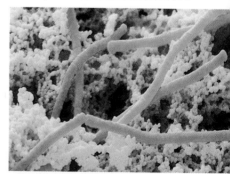

Figure 2.9: *Rod-shaped bacteria in soil (color added)*

prokaryotic cells are 1–5 microns in length. They may form clumps, pairs, or chains, but usually each bacterial cell is a separate organism.

Most bacteria have a stiff cell wall surrounding the cell membrane (Figure 2.10A). The cell wall gives the cell shape and some protection. The turgor pressure of the cytoplasm usually keeps the cell membrane pressed against the cell wall. Many bacteria are covered with sticky substances that help them cling to surfaces like your hands and teeth. Washing your hands with soap and hot water and brushing your teeth help to remove potentially harmful bacteria. This is why health care and food service workers must wash their hands so often.

WORD BANK

chromo- = color

-some = body

Chromosomes ("colored bodies") were named for the fact that they become visible when stained with colored dyes.

A ribosome is a body that contains ribose sugar, one of the components of DNA.

Powerful electron microscopes have revealed many protein-making ribosomes scattered throughout the bacterial cytoplasm (Figure 2.10B). Because the types and amounts of protein a cell makes control nearly everything it does, ribosomes are crucial to the cell's survival. Instructions for manufacturing the cell's proteins are encoded in the chemical structure of the **chromosome,** a long, tightly coiled ring of DNA attached to the cell membrane. The part of a prokaryotic cell that contains the chromosome is the nuclear region, or **nucleoid.** All cells contain ribosomes and chromosomes, although the chromosomes of prokaryotes are much simpler than those of other cells.

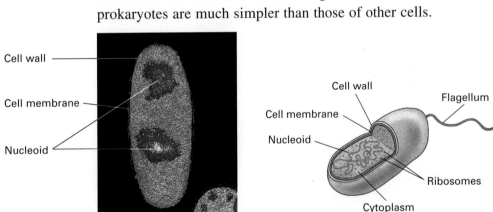

Cell wall
Cell membrane
Nucleoid

Cell wall
Cell membrane
Nucleoid
Flagellum
Ribosomes
Cytoplasm

A   B

Figure 2.10: *Typical bacterial cells. (A) Electron micrograph of typical bacterial cell (color added). (B) Parts of a typical bacterial cell.*

Some prokaryotes have other parts. Vesicles may store waste products, enzymes that would damage the cell if they escaped into the cytoplasm, or other substances. Some bacteria contain highly folded sacs of green membrane called **thylakoids** that absorb the energy of sunlight. Bacteria that contain thylakoids (Figure 2.11) can use this energy to make their own carbohydrate food just as green plants do. If you observed live bacteria with your microscope, you might have seen some of them swim by spinning a **flagellum** (plural, flagella) like a propeller. Flagella are thin protein fibers attached to the cell wall and cell membrane of some bacteria. Many of these bacteria are important in food processing, as causes of disease, and in making the nutrients in soil available to plants.

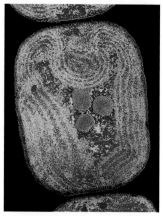

Figure 2.11: *Layers of green thylakoid membranes are visible in this bacterium (color added).*

Activity 2-7   Using Bacterial Cells in Food Processing

Provide students with a container of high-quality plain yogurt with active bacteria that has not been refrigerated for at least 4–6 hours.

The small, simple cells are prokaryotes. As the bacteria in the added yogurt multiply in the fresh milk, they consume substances in the milk and convert it to yogurt. Bacteria must be added to each new batch of milk to begin this process. In the beginning there would have been very few bacteria in the milk.

Many people think of bacteria only as causes of disease. However, most types of bacteria are helpful, not harmful. They are especially important in the production of certain processed foods, including yogurt, cheese, and other dairy products, soy sauce, and cured meats. Here's how you can see bacteria at work in yogurt.

Put a small drop of fresh yogurt on a microscope slide. **Do not eat any of the yogurt.** Mix in a drop of water and add a coverslip, and observe under low power and high power. Compare observations and discuss the following questions with your group. Record your answers in your logbook.

- Did you see anything in the yogurt that might be living cells? If so, are they likely to be prokaryotes? Why or why not?

- When dairy workers make yogurt, they always save some of each batch. They add this yogurt to fresh milk, which then slowly turns to yogurt. What do you think is happening in this process? Why is it necessary to add some finished yogurt to each new batch of milk?

- How different would your observations have been if you had examined the milk when it was just starting to turn into yogurt?

MORE COMPLEX CELLS

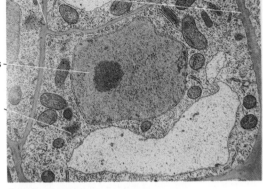

Cell walls

Nucleus

Golgi body

A

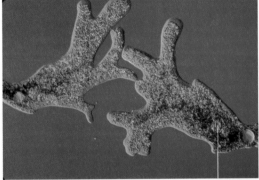

B

Nucleus

Figure 2.12: *Eukaryotic cells (color added). (A) Walled cell from a plant root. (B) Ameba, a one-celled organism without cell walls.*

As you saw in Activity 2-6, prokaryotes are smaller and simpler than other cells. The cells of plants, animals, fungi, and some organisms with only one or a few cells are larger and more complex (Figure 2.12). These more complicated cells are known as **eukaryotes.** Most are 10–50 microns in diameter, about 10 times the size of bacteria. Eukaryotic cells make up all organisms except for bacteria. Their complex structure allows them to develop into the many specialized types of cells that make up multicellular organisms such as trees, insects, and people.

If you examined stained eukaryotic cells, their most obvious difference from prokaryotes was probably a large, darkly stained round **nucleus** in the cytoplasm. A lipid membrane separates the nucleus from the surrounding cytoplasm. The nucleus contains the cell's chromosomes. Each eukaryotic cell's DNA is divided into several chromosomes, much as an encyclopedia is divided into volumes. Most eukaryotic cells have one nucleus, which may contain as few as 4 or over 100 chromosomes, depending on the type of organism. Each chromosome is a dense bundle of DNA wrapped around a series of protein "spools" like beads on a string. The chromosomes are usually crammed tightly into the nucleus, but when they separate, they look like short rods (Figure 2.13).

WORD BANK

eu- = good

pro- = simple or early form

karyos = nucleus

A eukaryote has a "good" or fully formed nucleus.

Prokaryotes have a primitive nucleus: the nucleoid.

Multicellular organisms such as apple trees (or people) begin life as a single cell. Each cell in a tree came from that single ancestral cell and carries in its nucleus a complete copy of that cell's chromosomes, with all the instructions needed to produce roots, leaves, and fruit. This makes it possible for orchard owners to grow new trees from branches cut from existing trees.

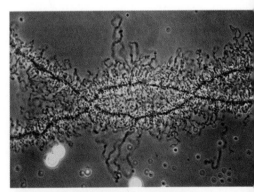

Figure 2.13: *These two chromosomes are partly unravelled, revealing "pearl chains" of DNA strands wrapped around protein "spools."*

The nucleus is one of the many specialized parts, or **organelles,** found in eukaryotic cells. As in prokaryotes, a cell wall surrounds the cells of plants, fungi, and some single-celled eukaryotes, providing mechanical support and protection and preventing the cells from bursting when in a hypotonic environment (Figures 2.14A and 2.14B). The cell walls of plants, bacteria, and fungi are each made of a different type of carbohydrate. Animal cells do not have cell walls.

You may have seen small green **chloroplasts** in the cells of plants and some one-celled eukaryotes. Like some bacteria, each of these organelles contains a thylakoid membrane that absorbs the energy of sunlight. Enzymes in chloroplasts convert this energy to the chemical energy of carbohydrates. The thylakoid is surrounded by an outer chloroplast membrane. Chloroplasts also contain a few ribosomes and a small ring of DNA, like a bacterial chromosome. Deep green leaves assure growers that their crops have plenty of chloroplasts to provide the plants with energy.

Not all eukaryotes have chloroplasts, but they all have several other membrane-bounded organelles (Figure 2.15). One is the **mitochondrion** (plural, mitochondria), where carbohydrates and other foods are broken down to yield usable chemical energy, much as a power plant burns coal to produce electric power. Like chloroplasts, mitochondria have highly folded inner membranes, a small chromosome, and ribosomes. A cell may have dozens or hundreds of mitochondria, depending on its energy needs. A defect in the mitochondria can cause people to feel weak and tired. A dangerous weight-loss pill of the 1960s (now banned) worked by preventing cells from using the energy released by their mitochondria.

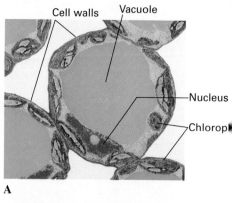

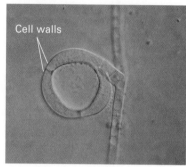

B

Figure 2.14: *Eukaryotic cells with cell walls (color added). (A) Thick-walled zinnia leaf cells. (B) Thin-walled cells of a fungus.*

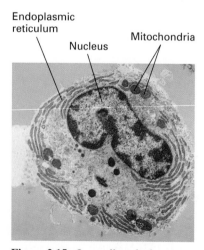

Figure 2.15: *Organelles of a human bone-forming cell (color added)*

CHANGING IDEAS

Biologists have long wondered how eukaryotic cells originated, presumably from the simpler prokaryotes. In the 1960s, biologist Lynn Margulis proposed that the ancestors of today's eukaryotes developed out of a partnership of several kinds of prokaryotes. She suggested that chloroplasts and mitochondria developed through endocytosis. If an engulfed cell was not digested by the surrounding cell, both cells could have helped each other to survive (Figure 2.16). For example, the cell inside may have used a thylakoid membrane to produce usable chemical energy, while the surrounding cell provided protection and collected nutrients. In this way, the thylakoids in the chloroplasts of today's plants may have developed from the thylakoids of ancient bacteria (Figure 2.17A and B). The cell membranes of other ancient prokaryotes may have become the inner membranes of mitochondria. The outer membranes of these organelles would then be descended from the vesicles in which these ancient bacteria were enclosed by endocytosis.

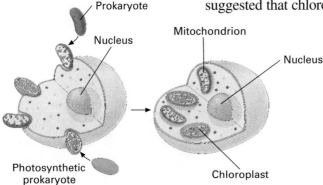

Figure 2.16: *One version of the possible origin of eukaryotic cells*

At first, many biologists thought this hypothesis was unlikely, but today it is widely accepted. Evidence that supports Margulis's hypothesis includes the double membranes of mitochondria and chloroplasts and these organelles' bacteria-like chromosomes and ribosomes. You might think of the mitochondria that provide you with energy as a vast number of bacterial "guests" distributed among all the cells of your body.

SOME OTHER ORGANELLES

WORD BANK

extra- or exo- = outside

endo- or intra- = inside

reticulo- = net

The endoplasmic reticulum looks like a net inside the cell.

Extracellular means outside the cell; intracellular means inside the cell.

A system of membranes in eukaryotic cells helps to move substances about the cell. (Prokaryotic cells must rely on diffusion to move dissolved substances through their cytoplasm.) The **endoplasmic reticulum** (ER) is a set of membrane-enclosed channels that carry substances throughout the cell (Figure 2.15). Ribosomes attach to some parts of the ER. Protein molecules that are synthesized at these ribosomes pass directly into the ER, which carries them off to their assigned places in the cell. The **Golgi body** (Figure 2.12A), a group of membranous sacs that looks like

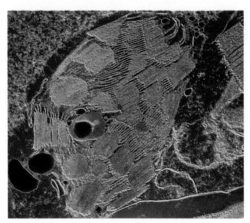

Figure 2.17A: *A closer view of a chloroplast packed with layers of thylakoids (color added). Where did chloroplasts come from?*

a stack of pancakes, collects substances from the ER and passes them through the stack, from one sac to the next. Vesicles form at the edges of the last sacs in the series and carry their contents to various parts of the cell. Some fuse with the cell membrane, discharging their contents outside the cell. The membranes of these vesicles are added to the cell membrane, allowing the cell to expand as it grows. The movement of substances from the Golgi body to the outside of the cell is especially important in cells that are building extracellular structures, such as the walls of plant and fungal cells and the networks of protein fibers that help to hold skin and bone together.

Plant cells often contain a large vesicle called a **vacuole** that may occupy up to 90% of the space inside the cell (Figure 2.14A). Active transport moves excess salts, wastes, and water from the cytoplasm into the vacuole. Smaller vesicles include **lysosomes,** which contain enzymes that help to break down proteins and other substances. Vesicles formed by endocytosis often fuse with lysosomes. The enzymes in the lysosome then break down the food or other substances in the vesicle. Defensive cells in your body use this process to destroy disease-causing bacteria that they engulf, keeping you healthy.

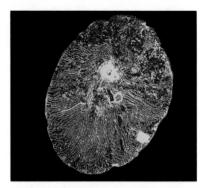

Figure 2.17B: *Could bacterial cells like this one, filled with layered thylakoids, be the ancestors of chloroplasts? (color added)*

WORD BANK

vacuo- = empty

lyso- = loss

A vacuole appears in the microscope to be empty (like a "vacuum").

Objects in lysosomes are broken down and "lost."

While looking at cells with the microscope, did you notice that many animal cells have definite shapes? Perhaps you saw bloblike cells wrap themselves around food particles and absorb them by endocytosis. How does this work? An internal framework of protein fibers in eukaryotic cells, the **cytoskeleton,** gives these cells shape and helps them to move, as your skeleton does for your body (Figure 2.18). As connections

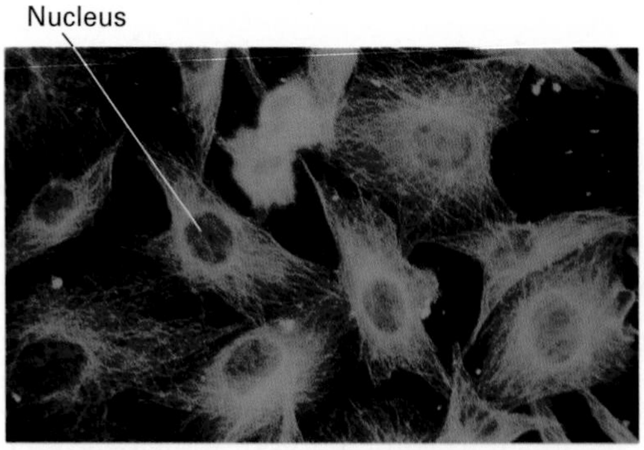

Nucleus

Figure 2.18: *The cytoskeletons of these skin cells have been stained with a fluorescent dye. Note that the cytoskeleton does not penetrate the nucleus.*

WORD BANK

Ameboid movement is named for amebas, which are one-celled eukaryotes that move in this way.

between these fibers are made and broken, the cytoskeleton changes shape. These changes allow some eukaryotic cells to creep along surfaces (Figure 2.19A) in a process called **ameboid movement.** When you get a splinter, some of your defensive cells creep along the surface of the wood in this way, engulfing and breaking down any bacterial cells that they find.

Eukaryotic cells may move in other ways. (See Figures 2.19B, 2.19C, and 2.19D.) Some one-celled eukaryotes swim with structures that look like prokaryotic flagella but are very different. Eukaryotic flagella contain cylinders of fine protein tubules anchored just inside the cell membrane in a structure that provides energy for their movement. They lash back and forth like tails, unlike prokaryotic flagella, which rotate like propellers. Sperm cells have flagella. Other eukaryotic cells may have many short fibers called **cilia** on their surfaces. Cilia beat together in rhythm, like oars. The cells that line your breathing passages have cilia that move dust particles you inhale back up toward your nose and mouth. One way in which smoking damages your lungs is by paralyzing these cilia. The smoke particles then accumulate in your lungs, eventually killing lung cells or turning them cancerous.

A pair of cylinders made of protein tubules similar to those in eukaryotic flagella are also found in the cytoplasm of animal cells and many one-celled eukaryotes. These **centrioles** may help to organize the separation of chromosomes when cells divide in two.

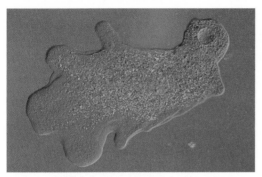

Figure 2.19A: *An ameba. Continuous reshaping of the cytoskeleton allows the cytoplasm of this one-celled eukaryote to flow into fingerlike extensions, producing ameboid movement.*

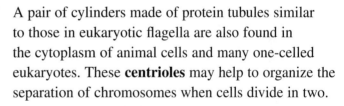

Flagellum

Figure 2.19B: *Euglena, a one-celled eukaryote, swims with a flagellum (color added). Eukaryotic flagella and cilia have an internal structure (Figure 2.19D) not found in bacteria.*

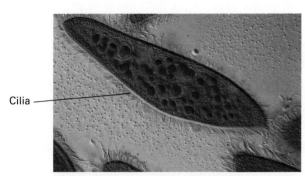

Cilia

Figure 2.19C: *Paramecium, a one-celled eukaryote that uses cilia to move (color added)*

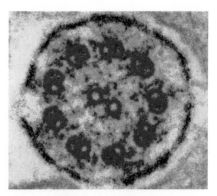

Figure 2.19D: *Cross section of a cilium of a seahorse cell (color added). The ring of protein tubules does not occur in bacterial flagella, which are solid.*

ACTIVITY 2-8   ORGANELLE HUNT

Provide specimens that are easily identified as plant tissue by their prominent chloroplasts, some free-living flagellates and ciliates, and prepared slides of animal tissue. Also provide one or two numbered samples or prepared slides of bacteria. Do not label the source. Cover the labels of prepared slides with a removable number. Provide general-purpose stains such as methylene blue or Congo Red. Protein stains such as Fast Green will make mitochondria visible.

Point out that students' answers to some of the questions will necessarily be speculative, and they should feel free to hypothesize without worrying about being correct. Encourage students to discuss conflicting observations and interpretations.

When hospital laboratory technicians examine cell samples from patients, they must consider the types and numbers of organelles that they see in deciding whether a cell is normal or cancerous or whether a nonhuman cell that they find causes disease or is normally present in the human body. As you look at cells through a microscope, draw and describe in your logbook each type you see. Label any organelles you see. Then record your answers to the following questions about each type of cell:

- Is this a eukaryotic or prokaryotic cell? What evidence supports your answer?

- Is this cell part of a larger organism, or does it live on its own? What evidence supports your answer?

- Do you think this cell can move on its own? If so, explain how it moves and what you saw that supports your answer. If not, explain what makes you think so.

- How do you think this cell obtains its food? What evidence supports this conclusion?

- How is this cell specialized or adapted? For example, if you think it is a one-celled organism, where would you expect to find it living? If it is part of a larger organism, what special tasks might it do? Explain the reasons for your ideas.

TARGETING PROKARYOTES

Some structures that look similar in eukaryotes and prokaryotes have important differences (see Table 2.1). The cell walls of prokaryotes and eukaryotes are made of different materials. Eukaryotic ribosomes are also larger than those of prokaryotes and different in structure. These differences allow medicine to attack prokaryotes that cause diseases, without harming eukaryotes such as people. Antibiotics that you may have taken prevent bacteria from building cell walls or making protein on their ribosomes. For example, penicillin prevents many kinds of bacteria from building cell walls. As bacteria grow in the presence of penicillin (Figure 2.20), their walls develop gaps.

Structure	Prokaryotes	Eukaryotes
Cell wall	Present in nearly all types	Present in plant and fungal cells and some one-celled types; absent in animal cells
Organelles	Very few	Many surrounded by membranes
Chromosomes	One circle of DNA in part of the cytoplasm	Several rod-shaped chromosomes with proteins bound to the DNA; in the nucleus, surrounded by a double layer of membrane
Ribosomes	Small	Large
Mitochondria	Absent	Present
Thylakoid membrane	In the cytoplasm of some bacteria	In chloroplasts of plant and algae cells, surrounded by a membrane
Cytoskeleton	Absent	Present
Flagellum	Present in some types; simple structure	Present in some types; complex structure

Table 2.1: Major Differences Between Prokaryotic and Eukaryotic Cells

Eventually, they encounter hypotonic conditions that cause them to swell and burst.

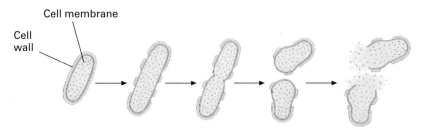

Figure 2.20: *How penicillin works. Bacteria exposed to penicillin (left) continue to grow, but their cell walls do not. Gaps in the cell wall increase as cells grow and divide (center). Without a complete cell wall, the cell membrane bursts when a hypotonic environment causes the cell to swell.*

ACTIVITY 2-9   MAKING CELL MODELS

Working in a small group, make a model of a prokaryotic cell and a model of a eukaryotic cell. You may want to bring materials from home. Think about the best way to represent each part as you choose your materials. When your models are complete, present them to the class. Explain how the materials you used represent the structures.

LESSON 2.3 ASSESSMENT

1. Eukaryotic cells can distribute materials internally more quickly than prokaryotic cells because eukaryotes have internal transport systems.

2. A safe antibiotic that kills prokaryotes is easier to find because the differences between prokaryotes and human eukaryotic cells allow the drug to kill prokaryotes but not eukaryotes. A filter would more easily remove eukaryotic cells, which are larger.

3. Ameboid movement would be the most affected, because it depends completely on changes in cell shape caused by movement of parts of the cytoskeleton. Flagella and cilia can still operate even if the cytoskeleton is defective and the cell loses its usual shape.

4. A chemical that attacks mitochondria could damage both fungi and corn plants. An antiribosomal chemical could kill bacteria, but it could also damage the ribosomes of the corn's mitochondria and chloroplasts. The most promising approach would be to find a chemical that would block the action of bacterial and fungal cell wall–synthesizing enzymes, but not those of plants.

1 Would you expect food to be distributed more quickly through a prokaryotic cell or a eukaryotic cell? Explain the reasons for your answer.

2 Campers and other people who drink water from streams can become sick when the water carries tiny organisms that can live in the human digestive system. Would it be harder to find an antibiotic that is safe for humans but kills disease-causing bacteria or one that kills disease-causing eukaryotes? Would eukaryotes or prokaryotes be easier to remove by pouring the water through a filter? Explain the reasons for your answers.

3 Which kind of eukaryotic cell movement would be most affected by a defect in the cytoskeleton? Explain the reasons for your answer.

4 Suppose you wanted to develop a chemical to protect corn plants from bacteria and fungi that kill them before they can produce edible ears. Would it be best to try to find a chemical that attacks mitochondria, cell wall formation, or ribosomes? Explain the reason for your answer.

LESSON 2.4 THE CELL CYCLE

Have you ever had a doctor check a mole or remove one from your skin? Most of your cells grow and reproduce, but sometimes a few skin cells multiply faster than normal, producing a mole. If they multiply even faster, they may become a cancer. A doctor who suspects that a mole is cancerous can remove it and have a cytotechnologist examine it with a microscope. This person has learned to distinguish healthy skin cells from cancerous ones. How do cells reproduce, and why do the normal controls on this process fail in a cancer cell? Can those controls be repaired to stop the growth of a cancer? In this lesson you will learn about the cycle of cell growth and reproduction and how it is controlled.

CAREER PROFILE

CYTOTECHNOLOGIST

Jose Y. is a cytotechnologist at a hospital lab. Cytotechnologists spend a lot of time looking at cells under a microscope. This morning, Jose is looking at cells from a Pap smear. He stains them to bring out the detail in the cells and their organelles.

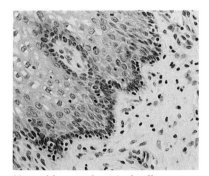

Normal layers of cervical cells

As he prepares to examine the slide under the microscope, he explains that Pap smears are cells taken from the cervix in the female reproductive tract. They are used to diagnose certain types of cancer, including cervical cancer, the second most common cancer among women. "What I see," Jose says, "are lots of cells, each with a nucleus that has an unusual shape and is enlarged." After he moves the slide around, he steps back and says, "Look at the two cells at the end of the pointer. This is not a normal condition found in such a young female. These cells are transformed. They have undergone some basic change. Compare them to these photos of normal cells. Although the cells are not working properly, they are still able to divide. That could be a big problem for the patient if the dividing cells invade the underlying tissues."

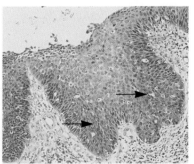

Section of a cervix with abnormal cells (arrows)

If you are interested in a career as a cytotechnologist, Jose recommends that you take as much science in high school as possible, including biology, chemistry, math, and computer science. You need to get a bachelor of science degree in biology or a related field and to attend an accredited cytotechnology program. To maintain your certification once you graduate, you will take courses and attend seminars to stay current in lab methods.

ACTIVITY 2-10 CELL SIZE AND CELL DIVISION

See Activity 2-10 in the TRB.

MATH TIPS

Graphing your results can help you to see whether the size of your pieces has more effect on their volume or their surface area. Set up a graph on a piece of graph paper. Label the *x*-axis "Length (cm)" and the *y*-axis (on the left) "Surface Area (cm^2)." Plot your surface area data on this graph and draw a smooth curve through all the points. Then label a second *y*-axis (on the right) "Volume (cm^3)" and plot the volume data on the same graph in a different color. Draw a smooth curve through all those points in the same color. Which curve is steeper? What does that mean?

Volume grows faster than surface area.

The more membrane surface area a cell has, the more rapidly it can absorb and expel substances. Since a larger cell consumes more food and produces more waste, it must have a larger surface area through which to exchange materials with its environment. Since growth increases volume more than surface area, a cell's need to exchange materials with the environment increases faster than its ability to do so.

An organism faces serious consequences if its cells don't divide normally. But why do cells have to divide at all? Is there some advantage to small cells? In this activity, you will learn how size affects objects of various shapes and develop a hypothesis about why cell division is important.

Set up a table in your logbook with three columns and four rows. Label the columns "Length of Pieces," "Surface Area (cm^2)," and "Volume (cm^3)." Use your ruler and the appropriate equations from Figure 2.21 to determine the volume and surface area of your food. Record these numbers in the first row of your table, along with the length of your food. Cut your food in half crosswise, as in Figure 2.21. Determine the surface area and volume of one piece, and complete the second row of your table. Repeat the process, cutting one piece in half twice more and

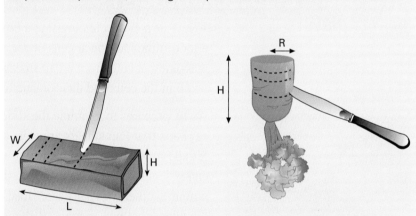

Volume = HWL
Surface area = 2(HW + HL + WL)

Volume = $\pi R^2 H$
Surface area = $2\pi(R^2 + RH)$

Figure 2.21: *How to cut and calculate volume and surface area*

recording the results after each cut. With your group, review your results and answer the following questions in your logbook:

- As an object with the shape you investigated gets larger, which increases faster: its surface area or its volume?

- How does a cell's surface area affect its ability to transport substances in or out through its cell membrane? Would a bigger cell, with more cytoplasm, need more surface area than a smaller one?

- Would it be more difficult for a large cell or for a small cell to absorb everything it needs from its environment and to get rid of wastes? Explain how your results support your answer.

If humans had only one large cell, we would be unable to exchange substances with the environment quickly enough to support life, and any injury might be fatal, since no other cells would be available to replace an injured or dead cell.

The ratio of volume to surface area increases with size, regardless of shape.

- What problems would we have if our bodies consisted of one enormous cell?

- Compare results with a group that worked with food of a different shape. Does size have the same effect on the relationship between volume and surface area of all shapes?

CELL DIVISION

Cytotechnologists aren't the only people watching cell division. Remember the lab technicians working for a yeast manufacturer. They're checking yeast cells to see that they are dividing properly. Researchers who develop growth regulators for plants check the regulator's effect on actively dividing cells of test plants. Each worker is aware that growing cells need more surface area for exchanging food, water, oxygen, and waste with the environment. If a cell didn't divide but kept growing larger, its volume would increase more than its surface area. If it got too large, it would die. Are very big cells exceptions to this rule? The yolk of a bird's egg is a single large cell, but its nucleus, cytoplasm, and organelles are concentrated in a tiny zone just inside the cell membrane on one side (Figure 2.22). The rest of the yolk consists of stored lipid and protein that nourish the developing chick after fertilization. This material does not consume food or oxygen or produce waste.

Figure 2.22: *From left to right, a quail's egg, a chicken's egg and an ostrich egg. The yolk of an ostrich egg is the largest known animal cell, but only a small part of its volume consists of cytoplasm and organelles.*

Cells remain small for other reasons, too. For example, the nucleus controls everything that happens in a cell. Molecules produced in the nucleus act as messengers that travel throughout the cytoplasm. If a cell becomes too large, part of its cytoplasm will be too far away to exchange chemical signals quickly with the nucleus. In a very large cell, the nucleus may not be able to produce enough of these chemicals to reach the entire cell (Figure 2.23).

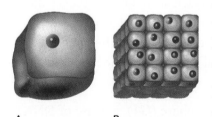

A   B

Figure 2.23: *Communication between cytoplasm and nucleus is slow and inefficient in a large cell (A). Nuclei in smaller cells (B) communicate quickly with a small volume of cytoplasm.*

Many types of cells undergo **cell division** whenever they grow to a certain size. When food and growing conditions permit, these cells continue to grow and divide. In organisms that have many cells, cell division makes the whole organism grow and replaces injured or dead cells. For example, the cells that line your stomach divide often, producing replacements for cells that are rubbed off by food or damaged by stomach acid.

THE CELL CYCLE

The cycle of cell growth and division occurs in both prokaryotes and eukaryotes. Prokaryotic cells divide by **fission** (Figure 2.24). First the chromosome duplicates, starting at the point where it attaches to the cell membrane. Continued cell growth separates the attachment points of the two DNA molecules. The cell pinches in two, producing two small cells with identical chromosomes. The two daughter cells are usually about the same size. Each can grow to the full size of the parent cell and then divide again to produce two new cells. (The products of cell division are called "daughters," but these cells are not really female.) Fission can occur as quickly as every 20 minutes. This rapid multiplication is part of the reason that you begin to feel ill so quickly after being exposed to disease-causing bacteria.

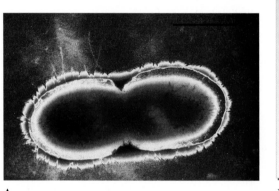
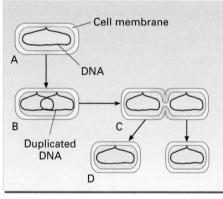

A   **B**

Figure 2.24: *Bacterial fission. (A) A bacterial cell in the process of division (color added). (B) The cell's DNA is duplicated, providing each daughter cell with a complete chromosome.*

Growth and division in eukaryotic cells goes through a series of phases called the **cell cycle** (Figure 2.25). During **mitosis,** or the **M phase,** the cell divides in two. The nuclear membrane breaks down, and the chromosomes separate into two identical groups, which become the nuclei of the daughter cells. During **interphase** the cell grows and produces additional ribosomes, mitochondria, and other organelles.

Interphase has three parts: a growth phase (**G1**), a phase of DNA synthesis (**S**), and a second growth phase (**G2**). During the S phase the DNA molecule in each chromosome is duplicated. Most eukaryotic cells pass through the S, G2, and M phases in a few hours, but the length of the G1 phase varies a lot. Skin cells and

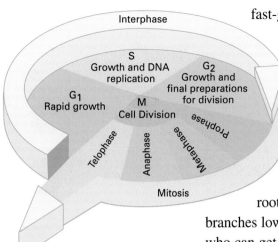

Figure 2.25: *The cell cycle. A second cell is formed in telophase (arrow) and continues with its own cycle.*

fast-growing fungi, for example, complete the cycle in less than a day. Nerve cells, however, may remain in G1 throughout an animal's life. Medical researchers are trying to stimulate nerve cells to divide and replace damaged nerves in people who are paralyzed, deaf, or blind.

Most plant cells remain in G1 once they mature to become part of a stem, leaf, or root. Only special cells in growing zones, such as the tips of stems and roots, cycle rapidly. So trees usually don't produce new branches low on their trunks—an advantage for lumber producers, who can get more boards from tall, thick, straight trunks. However, certain hormones can stimulate plant cells (or animal cells) to resume cycling. Gardeners can cut off a branch, dip it in a solution of plant hormones, and put it in soil to root.

CELL DIVISION IN EUKARYOTES

Remember that the long DNA strand in each chromosome is duplicated during the S phase of the cell cycle. During mitosis a eukaryotic cell provides each daughter cell with a complete copy of each chromosome. Other cell parts may or may not be divided equally.

Mitosis is a continuous process, in which cells pass gradually from one stage to the next. To describe the process, however, it's convenient to divide it into four stages (Figure 2.26A–D).

In **prophase,** the first stage of mitosis, the nuclear membrane breaks up and the chromosomes separate from one another. At this stage, each chromosome consists of two identical strands, called **chromatids,** each with a protein structure called a **centromere.** Each pair of chromatids is connected by their centromeres. At each end of the cell a group of fine protein tubules join at one end and point out in all directions, like a star. Animal cells make a new pair of centrioles during prophase. One old centriole and its duplicate move to each end of the cell.

In **metaphase,** enzymes use some of the cell's energy resources to line up the chromosomes across the center, or equator, of the cell. Additional protein tubules join each chromosome's centromeres to the clusters of tubules at the ends of the cell, in an arrangement called a spindle.

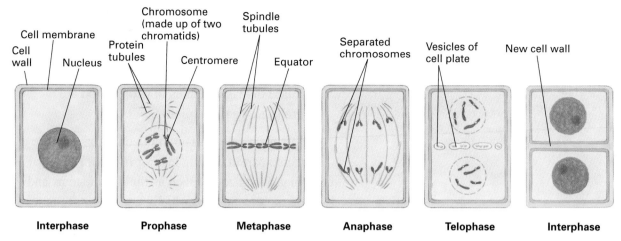

Figure 2.26A: *Stages of mitosis in a plant cell*

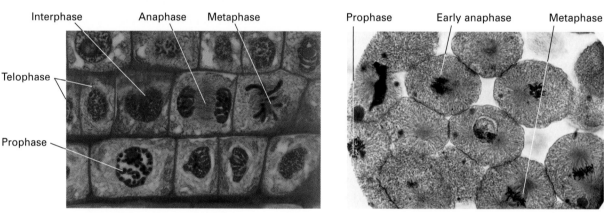

Figure 2.26B: *Dividing cells in a plant root, stained to make their chromosomes visible*

Figure 2.26C: *Cell division in stained cells of a whitefish embryo*

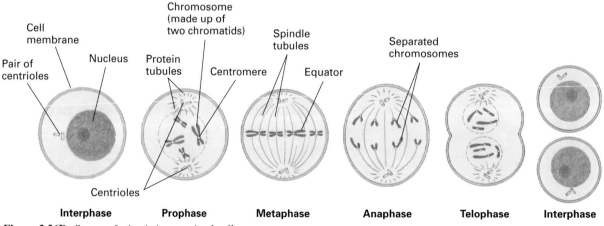

Figure 2.26D: *Stages of mitosis in an animal cell*

After the chromosomes are lined up along the equator, **anaphase** starts. Each pair of chromatids comes apart, and from now on, each chromatid is called a chromosome. During anaphase, enzymes again use the cell's energy to move the chromosomes. The spindle tubules shorten, pulling one chromatid from each pair to each end of the cell. An identical set of chromosomes clusters at each end. Cells without walls begin to narrow at the equator as they begin to divide in two.

During the last stage of mitosis, **telophase,** a new nuclear membrane forms around the cluster of chromosomes at each pole. Cells without walls finish dividing as they pinch in two at the equator. In cells with walls a **cell plate** of vesicles forms along the equator. The vesicles join with each other and the neighboring cell membrane, eventually dividing the cell in two. At the same time, enzymes assemble the material in the vesicles into a layer of cell wall on the new surfaces between the two daughter cells, so that each new cell is completely encased in its own cell wall.

ACTIVITY 2-11 STAGES OF MITOSIS

Provide students with microscopes and prepared slides of longitudinal sections of onion root tips. They will also need variously colored pipe cleaners or strings of beads to demonstrate mitosis to their partners and for the class.

You could ask students to come up with a phrase to help them remember the order of the stages, using the first letter of the name of each stage. Ask the class to vote for the best or funniest phrase.

This activity will help you make sense of the process of mitosis.

Use a microscope to look at slides of onion root tips stained to show the chromosomes. Use low power to find a region near the root tip where you can see many cells in various stages of mitosis. Center this region in your field of view, and use high power to observe the cells. Draw five to eight cells in your logbook, labeling the chromosomes and spindle of each cell (if they are present) and writing which stage of mitosis you think the cell is in. Try to find at least one example of each stage.

Check with a partner to make sure you agree on the stages. Show each other your cells and quiz each other about the stage they are in. Choose cells that are typical of a stage, not borderline examples.

When you have finished, use the materials your teacher provides to make a model of two pairs of chromosomes. Use the model to demonstrate the process of mitosis.

CONTROL OF CELL DIVISION

A landscaper trying to control weeds, a beautician struggling to keep a client's unwanted facial hair from returning, and a doctor fighting the growth of a patient's cancer are all trying to stop growth. Most growth, whether it involves a plant, hair, or cancer, depends on cell division. A cell's progress through the cell cycle depends on its production of specific proteins during each phase of the cycle.

In single-celled eukaryotes the speed of the cycle depends mostly on environmental conditions. Factors such as food supply and temperature determine how quickly the cell grows and divides. Plants and animals have many specialized cell types. Factors such as hormones, position in the body, and age of the organism have important effects on the cycling of these cells. Many mature cells in these organisms, such as nerve cells, stop dividing and stay in G1. In contrast, many cancer cells seem to cycle and divide uncontrollably. Taking advantage of this difference, many anticancer drugs kill cells during mitosis. Since most normal cells do not cycle as rapidly as cancer cells, they are more likely to survive the treatment. Another approach to fighting cancer may some day restore normal control of the cell cycle in cancer cells.

INVESTIGATION 2B   GROWTH AND THE CELL CYCLE

See Investigation 2B in the TRB.

How do organisms and tissues grow? Both cell division and cell enlargement are part of growth, but which is more important? When a doctor tries to stop the growth of a cancer, or a poultry farmer tries to manage chickens so that they will grow to market size as quickly as possible, should they be more concerned with the rate of cell division or the rate of cell enlargement? In this Investigation, you will use growing roots to study how growth occurs.

LESSON 2.4 ASSESSMENT

1. Extensions of the cell membrane provide these large cells with extra surface area, and extra nuclei control and provide instructions to their large volume of cytoplasm.

2. Protein tubules are needed to pull the chromosomes apart during mitosis. Since colchicine prevents their formation, the chromosomes cannot separate. Both copies remain in the same cell, doubling the chromosome number as the cell grows larger.

3. The poison would prevent synthesis of protein tubules and other proteins needed to progress through the cell cycle.

4. If a pair of chromatids fails to separate during anaphase, both chromatids will wind up in the same daughter cell. That cell will then have two copies of one chromosome plus a copy of its partner. The other daughter cell will receive one chromatid from the normally divided chromosome and no copy of its partner.

1 Your muscles are made of very long cells. These cells have folds in the cell membrane that reach deep into the cell and many nuclei distributed along their length. How do these folds and extra nuclei help muscle cells to survive and function?

2 Colchicine is a chemical that can cause plants to produce giant fruits. It prevents cells from forming protein tubules. Cells treated with colchicine are often larger than normal and contain twice as many chromosomes as normal cells. Explain how colchicine might cause these changes.

3 Some chemicals poison cells by preventing protein synthesis. How would such a chemical affect the cell cycle?

4 Abnormal cells may contain one or three copies of one chromosome, instead of the normal two copies. What could go wrong during cell division to produce this result? Give reasons for your answer.

CAREER APPLICATIONS

The applications that follow are like the ones you will encounter in many workplaces. Use the biology you learned in this chapter to complete the activities. Share your work with the class.

AGRICULTURE & AGRIBUSINESS

Who Uses Microscopes in Agriculture?

Using the *Dictionary of Occupational Titles,* the *Occupational Outlook Handbook,* O*NET, or other occupational resources, identify some agricultural jobs that require knowledge of cell structure and mitosis. Compare your list with those of other students, and put together one list of all the occupations identified by your class.

BUSINESS & MARKETING

Maintaining Healthy-Looking Plants

Turgor in plant cells determines whether the plant looks strong and healthy or weak and wilted. Interview an owner of a produce market, a flower shop, or a nursery to find out how produce or plants are kept looking their best until

they are sold. Tell the class what you found out. Be ready to explain how methods of maintaining turgor work on a cellular level.

Ad Campaigns for Cells

In this chapter you read about many different aspects of cells and how cells are important in various occupations. Work in small groups. Create an imaginary company that will deal with cells in some way. Create an advertising campaign that will feature a cartoon character cell that you can use in marketing your product or service. Make a presentation to the class describing what product or service your company provides and introduce your advertising campaign.

FAMILY & CONSUMER SCIENCE

Food Preparation Makes Use of Cell Knowledge

Humans' food comes from plants and animals. Research some food preparation techniques. For example, why do people (1) soak carrots and other raw vegetables in cold water, (2) soak shredded potatoes in salt water before making hash browns, and (3) cook cheaper cuts of meat in broth or stews rather than broiling or roasting them. Explain how the techniques you investigated depend on the structure of cells. Someone who works in food services or your local home extension agent might help you figure this out.

Skin Care Products

Talk to a cosmetologist or dermatologist to research three categories of skin care products—exfoliants, cleansers, and moisturizers. Determine how each affects skin cells. Create a poster that would explain to the general public what each type of product does. Find out what other steps people can take to maintain healthy skin. Present your poster to the class in a five minute presentation. Include any additional information you found out about maintaining healthy skin in your presentation.

HEALTH CAREERS

Lab Tests

Compare the tasks performed by these technicians who work with human cell structure: microbiology technologist, cytotechnologist, medical technologist, and histotechnologist. Research and describe the most common tests these technicians use to diagnose disease or cell damage in human tissues.

Clues to Cancer

Using a microscope and prepared slides, look at several cancer cells and several normal cells of the same type (or examine a set of photomicrographs). Focus on and sketch at least two cancer cells. Try to identify and label different cell structures in your drawing. Label and describe each type of abnormality in these cancer cells. Find out what types of cancer are produced by these cells and how cancer specialists (oncologists) treat the disease.

Importance of Surface Area in Industrial Processes

Ask an HVAC (heating, ventilation, and air conditioning) technician about the importance of fins and heat sinks in heating and cooling. Someone at a local chemical or food-processing company could tell you about the importance of surface area in chemical processing and the rate at which solids dissolve in liquids. Ask a representative of a coal-burning power utility about fluidizing coal (breaking it into small pieces) before burning it.

Correlating Cell Discoveries with Advances in Technology

Create a timeline of major discoveries about cell structure and function. Put on the same timeline the major developments in imaging technology. Analyze how advances in imaging technology (equipment and staining techniques) have helped us to expand our knowledge of cells.

Local Microscopists

Contact one or more local businesses that you think may employ research assistants or technicians who use microscopes as a regular part of their workday. Some likely candidates are workers at a water treatment plant, developers of personal-care products (such as shampoo, shaving cream, and body lotion), and quality-control workers in the semiconductor industry. Get the workers' job titles. Ask them how they qualified for their jobs and how they use microscopes. What kind of microscopes and what magnifications do they typically use?

CHAPTER 2   SUMMARY

■ The cell theory states that (1) the smallest, simplest living thing is a single cell; (2) cells make up all living things; and (3) cells come only from existing cells. In multicellular organisms that develop from a single cell, every cell inherits the same chemically encoded information in its chromosomes.

■ All cells are surrounded by a selectively permeable membrane composed of a double layer of lipid molecules with proteins embedded in it. Small, lipid-soluble molecules diffuse easily through cell membranes. Larger or more polar molecules pass into and out of cells more slowly.

- Proteins in the cell membrane serve as channels for facilitated diffusion and pumps for active transport of specific substances into or out of cells.

- Cells use endocytosis to absorb larger particles such as food and exocytosis to deliver particles to the outside environment.

- The simple cells of bacteria have cell membranes and usually cell walls. Their cytoplasm contains a single chromosome and numerous ribosomes. Some have flagella.

- Eukaryotic cells have numerous organelles, including several types surrounded by membranes that divide the cell into compartments. One organelle, the nucleus, contains the chromosomes. Cell walls surround the cells of plants, fungi, and some single-celled eukaryotes.

- Cells reproduce by making a copy of their DNA and then dividing in two. Division ensures adequate surface area for each cell and limits the volume of cytoplasm each nucleus must serve. It also provides a source of new cells for growth and healing of multicellular organisms.

- Prokaryotic cells divide by fission. Eukaryotic cells have a cell cycle consisting of a division phase (M, mitosis), a growth phase (G1), a DNA synthesis phase (S), and a second growth phase (G2). During mitosis, duplicate chromatids separate, and each daughter cell receives one copy of every chromosome.

- The rate of cell growth and division is affected by many factors, including food supply, temperature, and other environmental conditions. In multicellular organisms, age, hormones, and position in the organism also affect growth and division. These controls are lost or defective in cancer cells.

CHAPTER 2 ASSESSMENT

Concept Review

See TRB, Chapter 2 for answers.

1 Would a poison that prevents cells from using their energy reserves interfere with facilitated diffusion? Explain the reason for your answer.

2 All living things are composed of _____.

3 In your own words, explain the main points of the cell theory.

4 In what fundamental way are your cells identical to those of every other organism?

5 Describe the two processes that transport substances through selectively permeable membranes without the aid of cells' energy reserves.

6 How does a cell membrane separate two water-filled compartments?

7 Describe two transport processes that are assisted by proteins of the cell membrane.

8 Describe how some cells are able to ingest large particles, such as food.

9 Explain why the Golgi body, ER, and cell membrane can together be considered one system.

10 Which organelle might you study to learn how an animal cell changes shape as it develops? Explain the reason for your answer.

11 Which organelle would you expect to change in appearance when leaves turn from green to red? Explain the reasons for your answer.

12 If you wanted to collect a lot of DNA from growing cells, would it be better to extract their DNA when they are in G1 phase or in G2 phase? Explain the reason for your answer.

13 Does each of your cells contain a copy of every chromosome you inherited from each of your parents, or do some of your cells have more of your mother's chromosomes and others have more of your father's?

14 Give at least three reasons that cell division is important to the survival of living things.

Think and Discuss

15 Why is it important for scientists to share their findings with other scientists? Use the development of the cell theory as an example.

16 When you sweat, active transport carries positive sodium ions out of the cells of your sweat glands. Negative chloride ions and water molecules then pass out through membrane channels. People with the disease cystic fibrosis lack working chloride channels. How would you expect this to affect their ability to sweat normally? Explain your answer.

17 Which organelle is most likely to be the site of a defect that causes muscle weakness? Explain the reasons for your answer.

18 Special cells called transfer cells are found on the internal surfaces of seeds. They bring sugars and other nutrients into immature seeds as they develop on the plant. The walls of transfer cells often have ridges that force the cell membrane into a wavy pattern of hills and valleys. How can this help transfer cells to absorb nutrients?

In the Workplace

19 How might an artificial membrane, similar to the dialysis tubing you used in lab, be used to purify a city's water supply? What kinds of impurities could it remove?

20 What is one important chemical property of a detergent that a manufacturer must always consider in designing a new cleaning product?

21 A substance called a surfactant is often applied as a coating to the lung lining of newborns who have breathing difficulties. Should surfactants be polar or nonpolar substances? Explain the reason for your answer.

22 Suppose a poisonous substance were chemically bonded to tiny plastic beads so that cells could absorb it only by endocytosis. Could these poison beads be useful for protecting apple trees from bacteria that cause fruit to rot? Could they help to protect the trees from insects that eat the fruit and leaves? If they were used, what problems might they cause? Explain the reasons for your answers.

23 Various methods are used to test chemicals for the possibility that they will cause cancer. Would you have more confidence in a test that measures the ability of a substance to move cells from the G1 phase to the S phase or from the G2 phase to the M phase? Explain the reasons for your answer.

Investigations

Items 24 and 25 relate to Investigation 2A: Selectively Permeable Membranes.

24 How could a real cell take in grains of starch that could not pass through the membrane of your model cell?

25 What are two ways in which the controlled, selective permeability of cell membranes can protect cells from toxic substances?

Items 26 and 27 relate to Investigation 2B: Growth and the Cell Cycle.

26 Do your conclusions in Investigation 2B suggest that trees grow tall mainly by producing many cells or by producing very large cells?

27 Suppose you found that certain plant hormones are more concentrated at the tip of a growing root than farther up the root. How could that help to explain your observations in Investigation 2B?

CHAPTER 3

WHY SHOULD I LEARN THIS?

We all need food to live and grow—obviously.
But how do your cells extract the energy you
need from a plate of spaghetti? And how did the
wheat and tomato plants in the noodles and
sauce capture that energy from the sun in the
first place? Understanding these things can help
you to stay at a healthy weight or build up the
muscles you use in sports. In this chapter, you
will learn about the most important part of
metabolism: how living things (including you)
obtain and store energy.

CELL METABOLISM

WHAT WILL I LEARN?

1. how living things use chemical reactions to obtain and store energy
2. how plants convert solar energy to usable chemical energy
3. how living things break down sugars and other substances to obtain energy
4. oxygen's role in extracting energy from foods

"Fish gotta swim, and birds gotta fly," go the words to an old song. Swimming and flying take energy. Where does that energy come from? How do fish and birds or any organism put it to work? How do they store and retrieve energy? People who raise fish for market know that fish get their energy from food and oxygen. Any farmer knows that you can't grow a crop without plenty of sunlight or raise livestock without providing lots of energy in the form of feed.

In this chapter you will learn how organisms collect energy from sunlight, from minerals, and from each other and store it as chemical energy. You will also learn about chemical reactions that take place in nearly all organisms, making chemical energy available for any purpose, including swimming, flying, and growing.

LESSON 3.1 ENERGY FOR LIFE

Lots of people talk all the time about fat, calories, and sugar: "How many calories are in this sandwich?" "I really burn off calories in aerobics class." What exactly is a calorie? Does your body really burn them the way a car burns gasoline? What makes an exercise aerobic? Air, food, and exercise all seem to be connected, but how? Do calories work the same way for worms, plants, and bacteria as they do for people? Plants certainly don't eat or exercise, so how do they stay healthy and grow? In this lesson, you'll find answers to some of these questions and learn how all living things make or use the food they need.

ACTIVITY 3-1   THE COLOR OF LIFE

See TRB for instructions on preparing the solutions needed for this activity. If necessary, students can use test tubes in place of petri dishes. Have them hold a piece of paper behind their test tube racks. Before students begin the activity, explain that you will be preparing a similar experiment with a plant for them to observe.

Explain what you are doing as you turn on the projector and add a sprig of *Elodea* or *Anacharis* to two dishes. Place one dish with a plant in a dark place such as a drawer. Explain how carbon dioxide makes solutions in water acidic. Ask students to record predictions in their logbooks of what will happen to the colors of the three solutions.

A change should be apparent within 5–20 minutes. After discussing the change in the two illuminated solutions with the class, ask them to predict the color of the darkened dish. Then return it to the projector for comparison and discussion.

Students may be misled by the turbidity and lighter shade of plates 2 and 3 caused by light scattering by the suspended yeast cells. Explain that they should concentrate on hue (yellow, green, or blue), not on the richness or brightness of the color.

All living things need a source of energy. Most of what organisms do to obtain, store, and use energy involves chemical changes. In this activity you will detect these changes with a dye that changes color in response to one kind of chemical reaction: a change in pH. Bromthymol blue is a dye that is yellow in acids, blue in bases, and green near neutrality (pH 6.0–7.6). See how living organisms affect the pH of their environment, and try to explain your observations.

Working with a group, label the covers of three petri dishes: "1. Sugar," "2. Yeast," "3. Sugar + yeast." Use a pipette to add 2 mL of bromthymol blue solution to each dish. Use a clean pipette to add 1 mL of sugar solution to dishes 1 and 3. Use a clean pipette to add 10 mL of lukewarm water to dish 3 and 11 mL to dish 1. Cover and swirl the dishes gently to mix their contents.

Set all three dishes on a piece of clean white paper. Add 1 mL of yeast suspension to dishes 2 and 3, replace their covers, and gently swirl to mix.

Record the color of the solution in each dish in your logbook. After discussing what you think will happen to the color of each solution with your group, write your predictions in your logbook.

Swirl each of your dishes gently every 10 minutes. Over the course of the class period, observe your own dishes, those of other groups, and those on the overhead projector. Record any changes you see in your logbook. Discuss the following questions with your group and record your answers in your logbook:

Yeast cells obtain energy by breaking down sugar to CO_2, which acidifies the environment.

Plants obtain energy from light and store it in the form of carbon compounds that they synthesize from CO_2. By absorbing CO_2 from the water, they removed an acid, raising the pH of the water.

Yeast cells and plants affect the pH of the environment in opposite ways.

- How did the yeast cells affect their environment? Explain the effect of adding sugar to the yeast.

- How did the plants affect their environment? Explain the importance of light in the way that plants affect their environment.

- Did the yeast cells and plants affect their environments in the same way? In opposite ways?

FISH FARMS AND AQUARIUMS

BIOLOGY IN CONTEXT

In Activity 3-1 you saw that plants extract carbon dioxide (CO_2) from their environment, and other organisms, such as yeast cells, produce it. At the same time, plants produce oxygen gas. As you probably know, not only yeast cells, but many other organisms, including people and other animals, consume oxygen.

Land animals can get oxygen from the atmosphere, but how do fish and other aquatic organisms get the oxygen they need? The water they're in contains dissolved oxygen gas (O_2). Most fish grow and thrive in water that contains between 5 and 12 parts per million (ppm) of oxygen. At 3–4 ppm oxygen, many fish stop feeding. At 2 ppm oxygen, fish surface and gasp for air. At 1 ppm, fish die.

Fish, corals, and other water-dwelling animals absorb oxygen dissolved in the water around them.

Oxygen level is probably the most critical condition for businesses that raise fish for food or home aquariums or for restocking lakes and rivers. Healthy oxygen levels are hard to maintain in a commercial fish farm or an aquarium. The fish use up the oxygen quickly. The bacteria in a fish pond or aquarium also use oxygen in breaking down ammonia waste from the fish. In an established aquarium the fish use about 80% of the oxygen, and the bacteria use about 20%. If the organisms in the aquarium use up oxygen faster than it can enter the water from the atmosphere, they will die.

Fish farmers and aquarium keepers try hard to increase the rate at which oxygen dissolves in the water. They use

The wide, floating leaves of water lilies absorb CO_2 from the air and release O_2.

agitators and pumps to bubble air through the water and may grow aquatic plants that produce oxygen. In large aquariums or the ponds and tanks of a fish farm, pure oxygen may be pumped into the system so that more oxygen dissolves in the water.

If you have plants in your aquarium, another gas, CO_2, will also be important to you. CO_2 from the atmosphere dissolves in water, and most organisms, including fish, produce CO_2 as a waste product. In the presence of light, plants use CO_2 to make sugar. Under normal home lighting, aquarium plants get enough CO_2 to produce food and grow. Under plant-growth lights, however, this process speeds up, and you may need to bubble extra CO_2 into the aquarium. Extra CO_2 is required to get the really lush plant growth seen in aquarium books and magazines.

LIFE: PUTTING CHEMICAL ENERGY TO WORK

In the ponds of a commercial fish farm, pumps or paddle wheels spray water into the air to help it absorb oxygen and release CO_2.

The water at a commercial fish farm can be so packed with fish that the pond seems like one big living, growing mass. If all those fish are to grow, reproduce, and remain healthy, they must have a source of energy. What exactly is energy? **Energy** is the ability to do work. It occurs in several forms: Chemical energy can change one substance into another; mechanical energy causes movement; thermal energy increases temperatures; radiant energy is seen as light; and electrical energy is essential to brain functions.

Most of the work that living things do involves chemical energy. The energy boost that you may feel after eating comes from chemical energy. In your body, enzymes catalyze the breaking of bonds between atoms that make up the molecules of food. This reduces the molecules to unstable, high-energy fragments. As these fragments of molecules form simpler, more stable molecules of CO_2 and other waste substances, they release much of their chemical energy. Some of this chemical energy becomes thermal energy, which keeps your body warm. Living cells retain the rest of the chemical energy by using it to form other compounds. Some of these compounds, such as fats and sugars, can be stored until they are needed. Others, such as proteins, serve as structural supports, enzymes, and other cell components.

Cells also perform work by converting stored chemical energy to other energy forms. When a fish sees an obstacle ahead, its eye converts radiant light energy into chemical energy. The fish's nerve

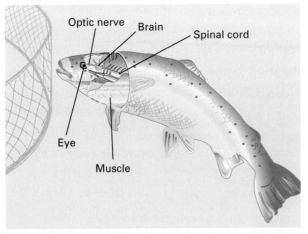

Figure 3.1: *Seeing an obstacle and swimming away from it depend on energy conversions in nerve and muscle cells.*

cells convert this chemical energy into electrical energy, and a series of electrical pulses pass through the fish's nerves to its muscles, signaling them to make the fish swim in a different direction. The muscle cells, in turn, convert some of their own stored chemical energy to mechanical energy when they contract, producing swimming movements (see Figure 3.1).

OXIDATION

Let's return to the crucial oxygen supply. If the oxygen supply fails in a fish farm, huge numbers of fish can die within hours. Why do fish and most other living things need oxygen?

When oxygen combines with other substances, electrons tend to be strongly attracted to the oxygen atoms. The tendency of oxygen to steal electrons from other substances causes many chemical reactions. A substance that loses electrons is **oxidized.** For example, many foods oxidize when they are exposed to the oxygen in air: Cut fruits turn brown, and butter and other fats become rancid. As various substances in these foods oxidize, the electron acceptor (usually O_2 from the air) may combine with them to produce new compounds.

A substance such as O_2 that accepts electrons is **reduced.** Oxidation and reduction always occur together. When electrons move from one atom or molecule to another, the electron donor becomes oxidized, and the electron acceptor becomes reduced (see Figure 3.2).

Organisms can obtain energy from chemical reactions that involve oxidation and reduction. Oxygen and other easily reduced substances attract

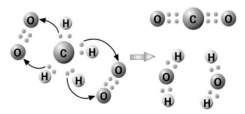

Figure 3.2: *When CH_4 (methane, or natural gas) is oxidized by O_2, it loses the 4 hydrogen atoms and their electrons to form water.* $CH_4 + 2O_2 \rightarrow CO_2 + 2H_2O$

electrons, much as Earth's gravity attracts falling objects. A falling object loses energy. As water falling over a dam loses energy, for example, some of that energy can be captured by a waterwheel and used to produce electrical energy (see Figure 3.3). In a similar way, electrons lose energy when they are transferred to oxygen.

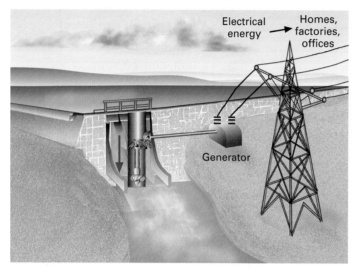

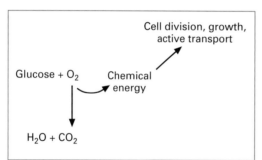

Figure 3.3: *Just as a generator converts the energy released by falling water to electrical energy for people to use, living cells convert the chemical energy released by oxidized foods to forms they can use.*

Cells capture some of this energy when oxygen oxidizes reduced food substances such as sugars and fats. They store the energy lost by the transferred electrons as chemical energy. But how do they do this?

You could think of a fish farm as a gigantic biological generator. Just as huge quantities of water pour through the power station at a dam, generating great quantities of electrical power, the fish in a pond oxidize the food they consume, producing chemical energy that they store in their bodies. Parts of some organisms accumulate so much chemical energy, especially in the form of highly reduced fats, that they can burn. You may have seen vegetable oil or fatty meat burst into flame when it was too close to a stove burner. But a fire oxidizes the entire organism, releasing much of its stored energy all at once. This rapid oxidation is no more practical than cooking all the fish in a pond at once for a single meal would be. Organisms need to release chemical energy slowly, without wasting it or overheating. Oxidation is a powerful energy source, but it must be controlled to be useful.

Most chemical reactions in organisms need only small amounts of energy, because enzymes reduce the amount of energy needed to start a reaction. In Activity 3-1, you saw that yeast cells, like most living things, obtain energy by breaking down sugars to CO_2. This process also produces water. But the amount of energy released when molecules of sugar are completely oxidized to CO_2 and water is far more than most enzyme-catalyzed reactions need. Breaking down a sugar molecule each time a chemical reaction needs energy would be as wasteful as paying for a $1 item with a $100 bill and getting no change back.

Cells save most of the energy released when they oxidize glucose or other fuel molecules. This energy is converted to smaller units that cells use one at a time, just as you might get change for a $10 bill to use in a vending machine. ATP is ideal for this purpose (see Figure 1.13, p. 14).

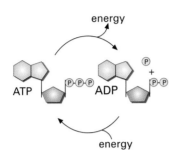

Figure 3.4: *Energy is released when ATP loses a phosphate group to become ADP. Enzymes can add a third phosphate group to ADP, storing energy in the form of ATP.*

When a molecule of glucose, the most commonly oxidized sugar, breaks down completely to CO_2 and water, it releases enough energy to produce about 94 molecules of ATP. (In reality, cells are not 100% efficient, and a large part of this energy is lost as heat.) Energy stored in the chemical bonds of ATP can be used efficiently, since the relatively small amount of energy released when a molecule of ATP breaks down is just enough to power many cellular reactions (see Figure 3.4).

Some enzymes aid reactions that absorb or release energy. In living organisms this energy is usually stored in the form of ATP. Many reactions, each assisted by a different enzyme, occur when organisms synthesize or break down the many substances that they contain. For example, organisms that oxidize glucose completely to CO_2 and water produce a set of enzymes to carry out each of the 20 steps in this process. The product of each reaction is the starting material for the next reaction. A series of linked chemical reactions in a living organism provides a **biochemical pathway** from one substance to another. Organisms have pathways that synthesize every needed substance that they can't get from their environment. Organisms also have pathways that break down wastes to safe forms that organisms recycle as raw materials or return to the environment.

INVESTIGATION 3A   THE ENERGY VALUE OF FOODS

See Investigation 3A in the TRB.

In Chapter 1 you saw that some foods release enough energy to produce a flame as they oxidize. If you could measure this energy, you could estimate how much energy you get by eating different foods. Do you think the more fattening foods release more energy when they oxidize by burning? Is there a way to measure this energy? In this investigation you will answer these questions.

Copy the following list of organisms into your logbook:

apple tree	human
cow	mold
dog	mushroom
earthworm	tapeworm

Take the list with you as you research each organism in the library, on the Internet, or in other reference materials. In your logbook, answer the following questions about each organism:

- How does the organism obtain carbon atoms to build new cell materials?

- How does the organism obtain energy?

- What other substances, such as water, oxygen, and CO_2, does the organism need?

- What waste materials, such as water, oxygen, and CO_2, does the organism produce?

Divide the list of organisms into two or more groups that get energy and carbon in similar ways and that need and produce many of the same substances. List the members of each group in your logbook. Write a few sentences explaining how each group might depend on the other groups.

ENERGY IN THE ENVIRONMENT

This beefsteak fungus is a heterotroph that consumes the nutrients in dead trees.

As organisms take in nutrients and release wastes, they change their environments. In fact, every organism can be considered part of the environment of every other organism. For example, animals, fungi, and many one-celled organisms get their nutrients from other living things. These nutrients include carbon compounds such as sugars, proteins, and lipids that serve as a source of energy when they are oxidized. Most animals obtain these substances by eating other organisms. Many fungi and bacteria and some animals depend on dead plants and animals or animal wastes as a source of these reduced carbon compounds. All organisms that

The lion's mane jellyfish is a heterotroph.

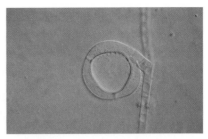

These fungus cells are heterotrophs.

WORD BANK

hetero- = other or different

auto- = self

-troph = nourishment

Heterotrophs are nourished by other living things. Autotrophs synthesize their own nutrients.

obtain energy by oxidizing carbon compounds produced by other organisms are called **heterotrophs.**

Other organisms collect energy directly from the nonliving environment. Plants and some bacteria and one-celled eukaryotes obtain energy from sunlight. Other bacteria gain energy by oxidizing minerals, such as sulfur, iron, or copper. All of these organisms are **autotrophs.** They do not consume carbon compounds produced by other living things. Instead, they reduce the CO_2 in the air or water in which they live to produce

Like all plants, a cactus is autotrophic.

CO_2 in water combines with water molecules to form carbonic acid (H_2CO_3), which dissociates to form H^+ and carbonate (HCO_3^-) ions.

their own carbohydrates, lipids, and amino acids. As you can see in Figure 3.5, in plants and some other autotrophs this process also produces oxygen, which returns to the atmosphere.

As you may have figured out from the Learning Link on page 88, all heterotrophs depend on autotrophs for their food supply. Your next meal may include plant products, such as potatoes, and animal products, such as meat or cheese. But even if you ate only tuna or hamburgers, your meal would still depend on all the plants eaten by animals.

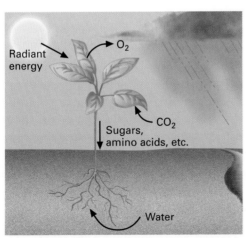

Radiant energy

O_2

CO_2

Sugars, amino acids, etc.

Water

Figure 3.5: *Plants use the energy of sunlight and the hydrogen atoms (H) and electrons of water molecules (H$_2$O) to reduce CO$_2$, forming sugars. Oxygen atoms (O) from water are released as oxygen gas (O$_2$).*

The fact that heterotrophs depend on autotrophs for food is important to farmers and forest and wildlife managers. In a forest, for example, mice may eat seeds, and insects may eat leaves. The mice become food for wolves, and birds devour the insects. Hawks may prey on these birds. The hawks and wolves may never eat plants directly, but the whole series of organisms, the **food chain,** depends on plants for carbon compounds and energy (Figure 3.6). And that chain includes insects, worms, bacteria, and other small heterotrophs that feed on dead plants, dead animals, and animal wastes. Wildlife managers try to ensure that the environment provides the resources needed to support a diverse and healthy food chain. Farmers and foresters often try to control species that prey on crops or commercially useful trees and maximize the production of these valuable plants.

Autotrophs also depend on heterotrophs. Heterotrophs break down carbon compounds produced by plants and return the carbon to

Figure 3.6: *Food chains. Algae feed shrimp and other small sea animals that are eaten in turn by fish. Grass provides energy to cattle and other farm animals that feed humans.*

Figure 3.7: *The carbon and oxygen cycles. Plants and other autotrophs absorb CO_2 and produce O_2. Carbon passes from autotrophs to animals and other heterotrophs, which return CO_2 to the atmosphere. Most autotrophs and heterotrophs consume oxygen.*

the atmosphere as CO_2. Plants and other autotrophs capture and reuse this carbon. Earthworms, fungi, and other heterotrophs act as decomposers, breaking down dead materials and releasing nutrients such as nitrogen, phosphorus, and sulfur to soil and water, where they can be absorbed by the roots of plants and by other autotrophs. Figure 3.7 shows the larger processes in this carbon cycle.

Alongside the carbon cycle is the oxygen cycle. Oxygen produced by autotrophs is consumed by both autotrophs and heterotrophs as they oxidize carbohydrates and other carbon compounds to produce ATP. The carbon and oxygen cycles will be described in more detail in Chapter 24.

LESSON 3.1 ASSESSMENT

1. Human muscles convert chemical energy to mechanical energy; their eyes convert radiant energy to chemical and electric energy; they convert chemical energy to thermal energy in generating body heat.

2. Animals need carbon compounds to use as raw materials for growth and repair and to oxidize to generate usable chemical energy (ATP).

3. CO_2 was reduced in the dish containing a plant that received light. Sugar was oxidized in the dish containing sugar and yeast.

4. The statement is correct: Both autotrophs and heterotrophs oxidize substances such as sugars to liberate usable chemical energy. Oxygen is reduced during this process. Autotrophs also use an energy source such as sunlight to reduce CO_2. This process also oxidizes water.

1 List three ways in which your body does work by converting one form of energy to another.

2 What are two reasons that kennel owners and cattle producers need to feed their animals carbon compounds such as carbohydrates?

3 In which petri dishes in Activity 3-1 was a carbon compound oxidized? In which dishes was a carbon compound reduced?

4 Comment on the following statement: Heterotrophs are mainly oxidizers, but autotrophs are both oxidizers and reducers. Is the statement correct? Explain what these organisms oxidize and reduce and how this supports your comments.

LESSON 3.2 HOW AUTOTROPHS GET ENERGY

Cotton and linen are plant fibers. Their production depends on photosynthesis.

Just about everything you eat—bread made from wheat flour, milk or meat from a corn-eating cow, a salad—came from plants. Your clothes are made of fibers from cotton plants, wool from grass-eating sheep, and polyester made from petroleum that formed from plants in ancient swamps. Besides us consumers, farmers, textile workers, and oil refinery workers are a few of the people whose livelihood depends on **photosynthesis,** the use of light energy by organisms to produce carbon compounds. Just as people use plants as raw materials to produce many useful things, plants get raw materials from the environment. With water, oxygen, CO_2, and the energy of sunlight they produce sugars and other substances that provide chemical energy. How do plants do that? Why do they produce oxygen in the process? You will learn the answers to these questions in this lesson.

ACTIVITY 3-2   OXYGEN IN PHOTOSYNTHESIS

Provide students with 0.25% sodium bicarbonate solution and boiled aquarium water. Show students how to measure dissolved oxygen with a reagent capsule or dissolved oxygen probe.

Plants use the energy of sunlight to convert CO_2 and water to sugar, but when you look at plants, they don't seem to be doing much of anything. In this activity you will see a result of photosynthesis and figure out what's happening inside the plants.

Working with two other students, number two large beakers 1 and 2, and fill each with boiled aquarium water. For every 100 mL of water in each beaker, add 2 mL of sodium bicarbonate solution. Following your teacher's directions, take a sample from each beaker and test it for the presence of dissolved oxygen. Record the results of this test in your logbook.

While one or two group members are testing for dissolved oxygen, another member should put three or four sprigs of a water plant such as *Hydrilla* or *Elodea* (also called *Anacharis*) in one of the beakers. Put a glass funnel upside down over the plants. Fill a small test tube with the solution in the beaker and put it upside down over the stem of the funnel. Don't let any air into the test tube. Set up a funnel and test tube in the second beaker in the same way but without plants. Set a bright lamp next to your two beakers, at an equal distance from both beakers, and turn it on.

On the next two days, look closely at your setups and describe in your logbook anything that has accumulated in the test tubes or funnels. After two days, take a sample of the water under each funnel and test it for dissolved oxygen as before. Record your results in your logbook. Discuss the following questions:

- How did the concentration of dissolved oxygen in each beaker change?

- What effect did the plants have on the dissolved oxygen level?

- Did the plants produce or consume oxygen?

- Look again at the chemical formula for glucose on p. 12. Compare the ratio of carbon to oxygen atoms in glucose, water, and CO_2. How can you explain the fact that photosynthesis produces oxygen gas (O_2)?

Dissolved oxygen increased in the beaker containing a plant and remained unchanged in the other beaker.

The plants increased the dissolved oxygen level.

Photosynthesizing plants produce O_2.

CO_2 and water contain more oxygen atoms than carbon atoms, but glucose contains equal numbers of carbon and oxygen atoms. The excess oxygen is released.

PHOTOSYNTHESIS AS A TOURIST ATTRACTION

BIOLOGY IN CONTEXT

Fiery, vibrant, brilliant, yellow, orange, purple, scarlet, maroon, gold, bronze, burgundy—these are a few of the adjectives people use to describe autumn leaves. Each fall, "leaf peepers" spend billions of dollars traveling to see the colorful leaves. In some states these people make fall the most profitable time of year for tourism. Many other people enjoy the colors for free in their neighborhoods, on school grounds, at parks, and along highways.

Why do leaves change color? Green leaves contain the pigment **chlorophyll.** Chlorophyll is a major component of chloroplasts, the organelles where photosynthesis occurs. Chlorophyll is not the only pigment present in a leaf. Red and orange carotenes and yellow xanthophylls help the chlorophyll to absorb light energy. During the growing season the chlorophyll masks these other pigments. As the days grow shorter, trees don't produce enough chlorophyll to replace what they break down each day. The other pigments become visible, turning leaves yellow and orange. Warm, dry summers and early autumn rains allow leaves to survive and photosynthesize longer. Rainy weather in late fall reduces photosynthesis, so leaves break down their pigments more quickly, producing rather drab colors. Some oak leaves also produce tannin, a brown, indigestible substance that protects them from leaf-eating insects.

WORD BANK

pigment

Any colored substance that does not dissolve in water.

As winter approaches and photosynthesis declines, a sealing layer forms inside the leaf bases so that evaporation from the leaves doesn't dehydrate the trees. Sugars trapped in the leaves are converted to anthocyanin, a pigment that gives the leaves red and purple tones. Warm, sunny fall days with nighttime temperatures below 7°C but above freezing allow photosynthesis to produce these sugars during the day and reduce their breakdown at night, leading to a spectacular accumulation of red and purple anthocyanins. Early frosts, on the other hand, bring photosynthesis to an end. With less sugar available, the leaves produce less anthocyanin, and the red color is softer. Combined with windy conditions, early frosts may produce early leaf drop.

Spectacular fall foliage such as this attracts many tourists.

The display of autumn color is as predictable as the weather. When autumn arrives, the people who profit from tourism watch the weather forecasts closely, hoping that travelers will soon be coming to watch the leaves change colors. The U.S. Forest Service, many state agencies, and travel agencies provide hotlines that tell you the best places and times to see the leaves.

THE GREEN MACHINE

People have known at least since the time of Aristotle (about 2300 years ago) that plants get their energy from sunlight, but only in the twentieth century have people had the tools to figure out how plants do this. Some people have hoped to increase the world's food supply by increasing the rate of photosynthesis. To do this, you must understand how photosynthesis works. In Activity 3-2 you learned an important piece of this puzzle: Photosynthesis produces oxygen. But where does this oxygen come from? From CO_2? From water? And why do plants release any oxygen at all?

The answers lie in the two stages of photosynthesis. During the first stage, called the **light reactions,** chlorophyll, embedded in

See TRB Chapter 3 for a photosynthesis demonstration that you may perform at this point.

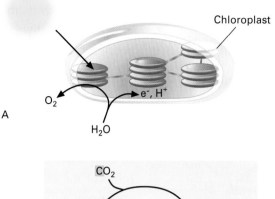

A

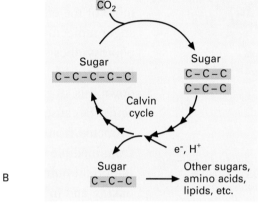

B

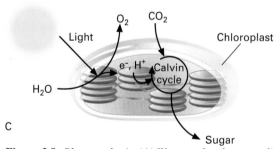

C

Figure 3.8: *Photosynthesis. (A) Water molecules are split in the light reactions. (B) CO$_2$ is used to produce sugar in the Calvin cycle. (C) In daylight, the light reactions provide energy to drive the Calvin cycle.*

the thylakoid membranes of the chloroplasts, absorbs light energy (see Figure 3.8A). This energy splits water molecules into highly oxidized oxygen and highly reduced hydrogen. Pairs of oxygen atoms form O$_2$. This oxygen is released as a gas. You found it dissolved in the water in Activity 3-2. Light energy absorbed by chlorophyll is transferred to the electrons of the hydrogen atoms. This energy and the reducing power of the electrons drive the second stage of photosynthesis, called the **Calvin cycle.**

In the Calvin cycle, an enzyme combines a molecule of CO$_2$ with a molecule of a five-carbon sugar to produce two molecules of three-carbon sugar (Figure 3.8B). Other enzymes then catalyze reactions in which the energy and reducing power accumulated by the light reactions convert this simple sugar to a more reduced form with greater bond energy. Some of this reduced sugar replenishes the supply of the five-carbon sugar, allowing the process to continue. The rest of the reduced sugar travels from the chloroplasts into the cytoplasm, where the cell may use it, or it may feed other parts of the plant, such as the roots. Sugar produced in photosynthesis is also the raw material for the synthesis of many other substances, such as amino acids and lipids. The events of the Calvin cycle are also called the dark reactions, but this name is a bit misleading. In most plants, the Calvin cycle takes place in daylight, when the light reactions provide energy (Figure 3.8C), although the cycle itself

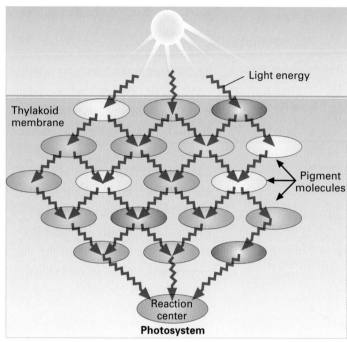

Figure 3.9: *The light reactions in more detail. In the thylakoid membrane, groups of chlorophyll, carotene, and xanthophyll molecules absorb light energy and transfer it to a central chlorophyll molecule at the "reaction center," where the water splitting takes place. The group of pigment molecules makes up a photosystem. Figure 3.10 shows how the photosystems power the production of ATP and reduced compounds used in the Calvin cycle.*

Figures 3.9, 3.10, and 3.11 can be used as the basis of a more detailed discussion of the mechanism of photosynthesis.

does not require light. The structure of the thylakoid membrane and the biochemical pathways of photosynthesis are described in more detail in Figures 3.9, 3.10, and 3.11.

The Calvin cycle may have an extra contribution to make to our food supply. Many plants store extra protein in their leaves by producing much more of the enzyme that adds CO_2 to sugar molecules than they need for photosynthesis. When the plant needs to make other proteins, it can break down this stored protein into amino acids. Because the enzyme is easily extracted from leaves, protein extracts from inedible leaves could become an important nutritional supplement in baking and in other commercial food preparation.

Hopes for increasing food production by making photosynthesis more efficient have often focused on the reaction in which CO_2 combines with existing sugar molecules to form more sugar. The enzyme that catalyzes this reaction often "mistakenly" combines a five-carbon sugar molecule with a molecule

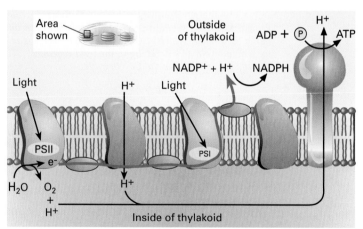

Figure 3.10: *How thylakoids work. Light energy absorbed by one kind of photosystem (PSII) is used to split water molecules. The electrons this releases pass from one protein to another in the membrane, receiving an energy boost from photosystem I (PSI). H+ ions from water reduce NADP+, converting it to the reduced form NADPH. As concentrated H+ ions pass out through the membrane, they donate energy to ATP formation.*

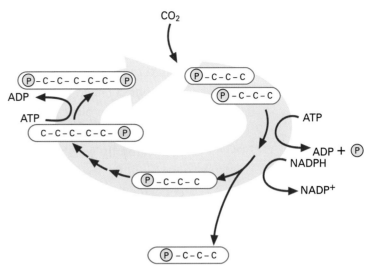

Figure 3.11: *A closer look at the Calvin cycle. ATP donates phosphate groups and energy to sugar molecules. NADPH donates electrons that reduce the carbon from CO_2. Three turns of the cycle "fix" 3 carbon atoms from CO_2 into sugar, producing one 3-carbon molecule of sugar-phosphate. A different enzyme catalyzes each step in the rearrangement of 5 3-carbon molecules into 3 5-carbon molecules.*

of O_2 instead of CO_2. When this happens, the products of the reaction are not two molecules of three-carbon sugar but one three-carbon sugar-phosphate molecule and one two-carbon molecule that breaks down to CO_2 and is lost (Figure 3.12). This process, called **photorespiration,** looks like a total waste. No one knows whether it benefits plants in any way. Some kinds of tropical grasses (including corn) and desert plants have evolved ways to decrease photorespiration. Many scientists who have studied photorespiration have given up on finding a way to decrease it artificially.

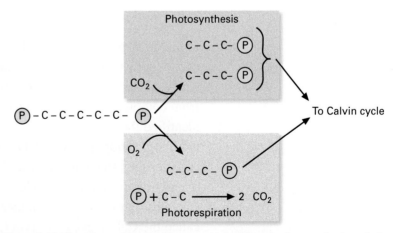

Figure 3.12: *Reactions of the 5-carbon sugar-phosphate in photosynthesis and photo respiration. In photosynthesis, addition of CO_2 produces 2 usable molecules of 3-carbon sugar-phosphate. In photorespiration, addition of CO_2 produces one usable 3-carbon molecule, and an unstable 2-carbon molecule that breaks down and is lost, draining the Calvin cycle of carbon.*

FIGURING OUT PHOTOSYNTHESIS

Early progress in understanding the chemistry of photosynthesis was very slow. In the seventeenth century, Jean Baptiste van Helmont, a Dutch physician, conducted the first quantitative experiment on plant growth. He weighed a small tree and the soil in the pot in which he planted it. He grew the plant for 5 years and then carefully weighed it and the soil again. The tree's weight increased by 164 pounds, and the soil lost only about 2 ounces. He concluded that the plant gained its weight from water. He was partly right.

No one challenged van Helmont's conclusions for about a century. In 1772, Joseph Priestley discovered that plants can change the air. He conducted a series of experiments with burning candles in jars. The candles would burn for a while and go out. If he grew a plant in the jar, the candle could be relit after a few days. He assumed correctly that a burning candle needs a particular substance in the air. Plants can generate that substance, which at the time was still unidentified.

Others contributed to our understanding of photosynthesis over the next century and a half. Jan Ingenhousz, a Dutch scientist, showed that plants can "regenerate" the air only in the presence of light. Nicholas de Saussure tied together all of the earlier findings. He concluded that plant growth in light uses CO_2 and water. In 1845, Julius Robert Mayer outlined the essential steps in photosynthesis: light absorption, conversion of light energy into chemical energy, and storage of chemical energy in plant compounds. But there was still no way to find out how plants did all this.

In the mid-1940s, Melvin Calvin adapted two new techniques to study photosynthesis: labeling molecules with radioactive isotopes and using chromatography to separate tiny amounts of compounds from complicated mixtures. Over the course of years, Calvin identified the reactions in the cell during photosynthesis. Between the first reaction involving CO_2 and the final product, a three-carbon sugar, there are at least 10 intermediate products, and the reactions between these products are catalyzed by 11 different enzymes. For his research in identifying these reactions, now called the Calvin cycle, Melvin Calvin received the 1961 Nobel Prize for Chemistry.

INVESTIGATION 3B   STARCH PRINTS

See Investigation 3B in the TRB.

Somehow plants use the energy of sunlight that they capture and store to make all the carbohydrates, lipids, proteins, and other substances that they need. Does this all happen at once, in each leaf, or are some substances formed first and others later? Do different parts of a plant make different substances? In this investigation you will see how leaves temporarily store the substances they make.

FUEL FARMS

BIOLOGY IN CONTEXT

Melvin Calvin is remembered for his work describing the reactions of photosynthesis, but photosynthesis was not Calvin's only interest. Calvin spent a good part of his later career studying plants called *Euphorbia,* or gopher plant. This plant received worldwide attention because it produces oils that are very similar to crude petroleum.

Euphorbia plants produce oils that can be used as fuel.

Calvin thought that the plant would adapt to growing on land that is too dry for other crops. He envisioned petroleum plantations where the plants would produce the equivalent of many barrels of crude oil each year. Calvin even showed that the sap could be poured directly into the fuel tank of a diesel-powered car or truck without any processing or purification.

In the early 1980s the University of Arizona tested Calvin's ideas about growing oil. *Euphorbia* plants were gathered from around the world and planted. The first crop was a total failure. The scientists determined that *Euphorbia* would make a better winter crop than a summer crop. The plants did not adapt well to dry soils and had to be irrigated. Also, the plants were susceptible to soil-borne diseases that are common in the Southwest. The final blow came when the crop was harvested. The yield of oil was only about a third of what had been predicted.

As long as petroleum is relatively cheap and available, *Euphorbia* will not be an economical substitute fuel. As our supply of fossil fuels dwindles, however, researchers may try to cultivate *Euphorbia* in temperate areas that have more rainfall.

ACTIVITY 3-3   NO FOOD, NO LIGHT . . . NO PROBLEM!

So far, you've learned about autotrophs that depend on solar energy and heterotrophs that obtain energy from food molecules that they collect. Does that exhaust the possibilities, or are there organisms that don't depend on either of these energy sources? In this activity you will observe the Winogradsky columns that you put together when you began Chapter 1 to answer this question.

Compare the columns that you grew with and without a sulfur supplement and light. Then observe closely as your teacher performs a demonstration. Record your observations in your logbook. Discuss the following questions:

- Is there any evidence that something grew in the columns when no light was available for photosynthesis?

- Does sulfur burn? What happens to things when they burn? Can sulfur be oxidized and reduced?

- How might organisms use sulfur to survive without food or light? Did adding sulfur help organisms to grow in the absence of light?

CHEMOAUTOTROPHY

Photoautotrophs, including plants and photosynthetic microorganisms, are the basis of the food chains in most environments. However, some environments are always dark and poor in oxygen, such as the bottoms of oceans and swamps, or are too hot or acidic for most plants, such as mineral hot springs. In these environments, bacteria known as **chemoautotrophs** take the place of photoautotrophs. Many of these organisms are members of the Archaea, an unusual ancient group of bacteria. Although the Archaea are not an important part of the food chain in most environments, they are the basis of most life on the ocean floor. Some biologists who study that environment have suggested that these nonphotosynthetic food chains may make up most of the living matter on Earth.

Scientists collect samples of deep-sea organisms for study.

Several types of chemoautotrophs are of practical importance. The methane producers gain energy by reducing CO_2 (produced by other organisms) to methane (CH_4, a gas), using hydrogen gas (H_2) as a reducing agent. These bacteria thrive in the digestive systems of cattle, termites, and other plant-eating animals, where they contribute to the animals' digestion. Methane-producing bacteria in the digestive systems of herds of cattle may contribute to global warming by producing large quantities of methane. Some farmers use these bacteria to produce methane for fuel from manure.

Sulfur bacteria obtain energy by oxidizing hydrogen sulfide gas (H_2S), much as plants oxidize water (Figure 3.13). The Earth releases hydrogen sulfide at hot springs, cracks in the ocean floor, and other volcanic areas. The bacteria produce solid sulfur instead of oxygen. The sulfur accumulates around the hot springs and volcanic vents in the ocean floor. These bacteria are the source of the large sulfur deposits that are mined throughout the world.

A third group of chemoautotrophs lives in soil, obtaining energy by oxidizing nitrogen compounds. These nitrifying bacteria convert ammonium ions (NH_4^+) to nitrate ions (NO_3^-). Nitrogen compounds are important plant nutrients. Most plants absorb nitrate from soil and use it as a source of nitrogen for the synthesis of amino acids. Partly because of the importance of nitrifying bacteria in plant nutrition, farmers pay close attention to soil drainage. Nitrifying bacteria depend on the oxygen in the air spaces in soil. Flooded soils can lose much of their nitrogen as ammonium ions are converted to ammonia (NH_3), which quickly evaporates.

How Autotrophs Convert CO_2 to Sugars

In general:
n = number of carbon atoms present

$$CO_2 + H_2X \longrightarrow C_nH_{2n}O_n + H_2O + 2X \text{ (sugar)}$$

In plants:
(X = oxygen)

$$CO_2 + H_2O \longrightarrow C_nH_{2n}O_n + H_2O + O_2$$

In sulfur bacteria:
(X = sulfur)

$$CO_2 + H_2S \longrightarrow C_nH_{2n}O_n + H_2O + 2S$$

Figure 3.13: *Autotrophs use hydrogen compounds (H_2X) to reduce CO_2 to sugars. In photosynthesis, the hydrogen compound is water. In some chemoautotrophic bacteria, it is hydrogen sulfide gas (H_2S).*

LEARNING LINK USING CHEMOAUTOTROPHIC BACTERIA

Use the library or the Internet to research one kind of chemoautotrophic bacteria. You may choose to concentrate on bio-gas generators that depend on methane producers, the sulfur bacteria that live at hydrothermal volcanic vents on the sea floor, or the importance of nitrifying bacteria in agriculture. Prepare a report to the class on your findings.

LESSON 3.2 ASSESSMENT

1. Photosynthesis in the green skin of the immature fruit contributes to the growth of the fruit. As the fruit matures, chlorophyll in the skin breaks down, revealing carotenes and xanthophylls. In some fruits, such as apples, sugar produced in the skin is converted to anthocyanins. As long as the fruits remain green, it is advantageous to let them receive sunlight and continue to photosynthesize. This feeds the growing fruit.

2. No, the O_2 produced in photosynthesis is from water, not from CO_2.

3. Increasing the supply of CO_2 can increase photosynthesis, but only if additional light is also available to allow the plants to take advantage of this source of carbon. Converting more carbon to sugar requires more energy in the form of light.

4. Sulfur deposits and fossil fuels are both formed by bacteria that live in oxygen-poor environments

1 Many fruits are green when they begin to form and are considered ripe when they become red or yellow. What is going on in the skins of these plant parts as they grow and develop? Will fruit growers have a better crop if they cover their fruits up or if they let the sun shine on them? Give a reason for your answer.

2 Suppose you labeled the oxygen atoms in some CO_2 by making them radioactive and fed this gas to a plant. Would you expect the O_2 the plant produced to be radioactive?

3 Greenhouse operators sometimes add CO_2 to the air in their greenhouses to stimulate photosynthesis. How would you expect this to affect the plants' need for light? Give a reason for your answer.

4 Sulfur is usually found with fossil fuels (petroleum, natural gas, and coal), which are formed by the bacterial decay of dead plant or animal material in deep, oxygen-poor water or sediments. Why do mineral prospectors often expect to find sulfur and fossil fuels together?

LESSON 3.3 HOW HETEROTROPHS GET ENERGY

All organisms obtain energy from oxidation-reduction reactions, and autotrophs make the sugars and other reduced carbon compounds that all organisms use as building materials and energy sources. But how do heterotrophs, including you, oxidize carbon compounds and retain the released energy for their own needs? Once you learn this, you'll understand a lot about weight gain and loss, dieting, and the importance of diet for athletic performance.

ACTIVITY 3-4   ENERGY NEEDS

You may need to explain that the Calorie is a unit of energy.

Exercise stimulates appetite so that the person will eat to replace energy lost during exercise. This is an example of homeostasis.

Increasing activity (but not dietary intake) uses energy stored in the form of fat. Breakdown of fat reserves leads to weight loss. Exercise also increases wear and tear on the body, especially the bones and muscles.

Extra protein would provide raw material for repair and growth.

How much energy do humans need in their diets? Look at Table 3.1. How do you explain the different numbers in the table? Do you see any patterns? Discuss your ideas with your partner, and answer the following questions:

- How would an increase in exercise affect a person's appetite? What characteristic of living things does this demonstrate?

- Why should people who need to lose weight increase their exercise?

- Why do very active people need more protein than people who get little exercise?

Activity Level	Males	Females
Sedentary	3,000	2,400
Light exercise	3,300	2,700
Moderate exercise	3,600	3,000
Vigorous exercise	4,000	3,400

Subtract 10% if dieting for weight loss.

Table 3.1 Energy Requirements For Teenagers (Calories/Day)

Jane R. (not her real name) was 16, very overweight, and very unhappy with her weight and everyone else's stares and comments. She had already tried several fad diets. After getting nowhere with them, she went to see Melanie P., a nutritionist.

Melanie counsels patients on diet and nutrition. Some people come to Melanie because they have a medical problem, such as high blood pressure, diabetes, or ulcers, that requires a special diet. Many come to her because they're worried about their weight. Her favorite patients are teenage girls who want to lose weight. Melanie explains, "When I can work with young people and help them understand the principles of achieving their weight goals while maintaining the needed nutrient levels, I can help prevent lifetimes of yo-yo dieting that is inefficient, and sometimes even harmful."

Melanie explained that to lose weight, Jane would have to use up more calories than she took in through the things she ate. The first thing they decided was that Jane should keep a diary of everything she ate for a week and when she ate it. Melanie also asked Jane to write down why she ate when she did and what she did each day.

After a week, Jane and Melanie reviewed Jane's food diary and found several trends. Her meals had no schedule. Some days she skipped breakfast and lunch and then snacked all afternoon and evening. She was taking in more calories than she needed to support her current weight, and she wasn't getting much exercise.

Melanie describes the plan that they worked out together: "By making adjustments in what Jane ate and spreading the meals and snacks out over the course of the day, we made sure that she wouldn't feel hungry. We also discussed ways she could become more active. For instance, she started riding her bicycle to school some days instead of getting a ride. She started walking her dog in the evenings. She really liked doing that because she met a boy in the park who walked his dog there, and they became good friends.

"Jane made slow but steady progress. Whenever she slipped, she learned not to feel guilty but to get back with the program. At the beginning she thought the progress was too slow and asked if she could lose weight by cutting down to half the calories that I recommended. I had to explain that if you take in too few calories, your body will conserve its energy reserves and you will not burn off fat as quickly as you would if you were eating sufficient calories. It has taken her about 18 months, but Jane has reached her goal. She's going to sign up for a bicycle camp this summer. Her self-esteem has greatly improved."

METABOLIZING SUGAR

Dieters think a lot about calories and often use artificial sweeteners instead of sugar, but they aren't the only ones who depend on the energy released when living cells break down sugars. Sugar, specifically glucose, a six-carbon sugar, is the main form of food that nearly all cells break down to produce ATP. Before other foods, such as other carbohydrates, lipids, and amino acids, can be used for energy, they must be converted to sugars or to fragments of sugar molecules. Athletes, soldiers, and other people whose work requires physical endurance often keep carbohydrate-rich snacks handy for a quick energy boost. The best snacks for this purpose are not sugary candies, but foods such as peanut butter sandwiches and trail mix that also contain other nutrients that hard-working bodies need.

The process of breaking down sugars to release energy is called **cell respiration.** Sugars are metabolized in two phases. The first phase, a biochemical pathway called **glycolysis,** occurs in nearly all cells. Each glucose molecule breaks down to two molecules of a three-carbon substance called pyruvate (Figure 3.14). The process partially oxidizes the original glucose molecule, releasing a little energy. During glycolysis the cell gains two ATP molecules from each molecule of glucose (Figure 3.15). Remember that completely oxidizing a glucose molecule releases almost as much energy as breaking down 94 molecules of ATP. Some of this energy is lost during glycolysis as heat. Much more energy can be released during the second phase of cell respiration, when pyruvate is oxidized further.

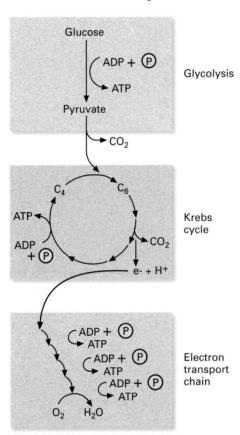

Figure 3.14: *Cell respiration is the breaking down of sugars to release energy. In the presence of oxygen, complete oxidation of glucose to CO_2 produces electrons (e^-) whose reducing power is used to produce many ATP molecules.*

WORD BANK

glyco- = sweet or sugar

-lysis = loss

Glycolysis is the loss, or breakdown, of sugar.

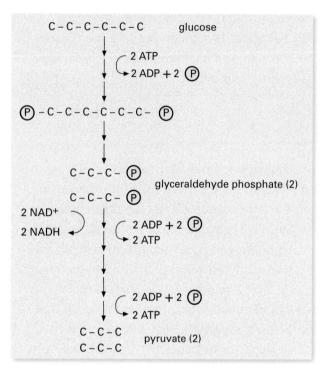

Figure 3.15: *A closer look at glycolysis. Breakdown of one molecule of glucose to 2 molecules of pyruvate consumes 2 molecules of ATP and produces 4, for a net gain of 2. NAD$^+$ is reduced as glucose is oxidized, just as NADP$^+$ is reduced during the light reactions of photosynthesis.*

When oxygen is available, much more energy can be used for ATP production during the second phase of cell respiration. In the Krebs cycle (named for its discoverer, Sir Hans Krebs) the partially oxidized pyruvate produced during glycolysis is oxidized completely to CO_2. As pyruvate is oxidized, it loses electrons (and hydrogen ions). These are transferred by a series of proteins called the electron transport chain to oxygen molecules, which are reduced to become water (H_2O). This is the main reason that you need to breathe. Your body takes in oxygen as you breathe in and gets rid of CO_2 as you breathe out. Most of the oxygen that your body consumes as you breathe is used in cell respiration, so you breathe harder during exercise, when cell respiration is very active.

Some of the energy that electrons lose as they are transferred from pyruvate to oxygen is saved in the form of ATP. In this way the reducing power saved from glucose produces useful energy. Up to 34 more ATP molecules can be produced during this process, so it's much more efficient than relying on glycolysis alone, as organisms do that can't use oxygen (Table 3.2). This is 18 times as much energy as glycolysis provides, or about 38% of the energy released in the oxidation of glucose. In practice, however, cell respiration is not always this efficient, and some energy may be lost as heat or used by the cell for other purposes without being saved as ATP.

Pathway	ATP Molecules Formed
Aerobic respiration	36
Fermentation	2

Table 3.2: Energy Saved as ATP During Fermentation and Aerobic Respiration of One Molecule of Glucose

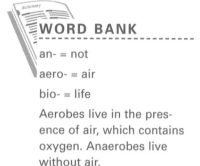

WORD BANK

an- = not

aero- = air

bio- = life

Aerobes live in the presence of air, which contains oxygen. Anaerobes live without air.

Organisms that use oxygen in this way to oxidize glucose completely are called **aerobes;** they are said to perform aerobic respiration. If you have ever felt tired and out of breath after heavy exercise, it should not be hard for you to understand that humans (and other animals) are aerobes. In aerobic eukaryotes the enzymes of the Krebs cycle and the electron transport chain occur inside the mitochondria. Glycolysis occurs in the cytoplasm, so pyruvate must be transported into the mitochondria before it is oxidized.

ACTIVITY 3-5   THE BREATH OF LIFE

Provide each student with two test tubes of bromthymol blue or phenol red indicator solution adjusted to pH 7.0–7.5 and several clean soda straws. Consider safety, practicality, and available space and facilities when choosing exercise activities for students.

What does it mean to be "out of breath"? After running or other heavy exercise you may feel that you need to breathe more deeply. After all, you do reduce more oxygen when your cells are respiring vigorously. But then why do you need more oxygen after exercising? Do your muscles work hard while you exercise and then somehow make up afterward for the extra oxygen they needed? In this activity you will investigate how exercise affects the cell respiration that keeps you going.

Step 1: Obtain a test tube of indicator solution and a clean straw. Place the straw in the tube. Note the color of the solution and record a description of it in your logbook.

Step 2: Take three deep breaths and exhale each one slowly through the straw into the solution so that your breath bubbles up through the solution. Observe and record the color of the solution in your logbook again.

Step 3: Exercise vigorously for 5 minutes as your teacher directs. Be sure to follow all safety instructions.

Repeat steps 1–3 with a fresh straw and tube of indicator solution. Compare your results with those of two other students. Discuss the following questions:

During exercise, the muscles consume energy faster than the heart and lungs can supply oxygen. The muscles temporarily use another, nonaerobic form of cell respiration to produce energy (and CO_2), until the exercise slows down or stops, and the oxygen supply again becomes sufficient.

- CO_2 dissolves in water and makes it more acidic. How did exercise affect the change in color of the indicator solution when you blew into it?

- What might be the reason that exercise had this effect?

- What does it mean to be out of breath? Did you feel that you were able to obtain all the oxygen you needed during exercise by breathing?

- Do you think that your muscles have a way to obtain energy during exercise without using oxygen? Explain the reasons for your answer.

FERMENTATION

Organisms that do not use oxygen are called anaerobes. These organisms, mostly bacteria and fungi, are common in oxygen-poor environments such as deep water and flooded soil. Anaerobes perform anaerobic respiration, also known as **fermentation.**

As you exercised in Activity 3-5, you became temporarily anaerobic. Fermentation continues the oxidation of glucose a bit beyond what glycolysis accomplishes. Some anaerobes can also switch over to aerobic respiration when oxygen is available; others are actually poisoned by oxygen. As you might imagine, fermentation does not produce nearly as much ATP as aerobic respiration, so aerobes have a great advantage wherever oxygen is present. For example, some anaerobic disease-causing bacteria live within our bodies, where our aerobic cells rapidly consume all available oxygen. Knowing this helped the chemists at one company to develop a better toothpaste. They added a small amount of hydrogen peroxide (H_2O_2) to their toothpaste. This unstable compound breaks down in the user's mouth to water and oxygen gas. The oxygen kills many of the bacteria that cause tooth decay and gum disease.

How does fermentation work? Anaerobic cells convert the pyruvate they produce during glycolysis to a stable waste product and excrete it into the environment. There are two main types of fermentation (Figure 3.16). Some anaerobes convert pyruvate to a substance called lactate, while others break it down to CO_2 and ethyl alcohol. Many bacteria that convert glucose to lactate are

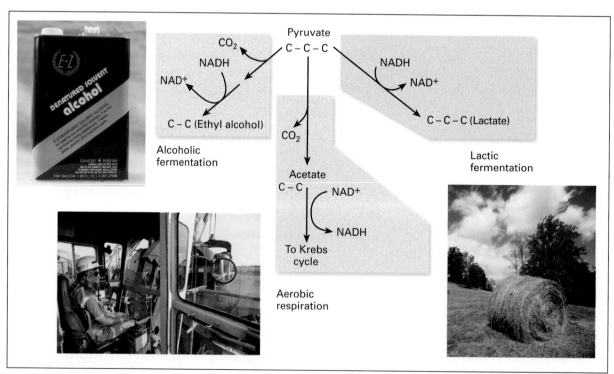

Figure 3.16: *Three possible fates for pyruvate. Anaerobes ferment pyruvate to lactate (blue) or ethyl alcohol and CO_2 (pink). In aerobic respiration, pyruvate is broken down completely to CO_2 by the Krebs cycle (green).*

Lactic acid-producing bacteria are used in the dairy industry to produce cheese (shown here) and yogurt.

used in food and feed processing, especially in the conversion of milk to cheese, yogurt, and other fermented products. These bacteria ferment the sugars in milk and excrete lactate. In combination with hydrogen ions, lactate forms lactic acid, which preserves these foods by making them too acidic for the growth of bacteria and fungi that cause spoilage. You have probably noticed the sharp, sour taste of lactic acid in many cheeses, yogurt, or sauerkraut, which is made by treating cabbage with lactate-producing bacteria. The same process is also important in agriculture. Lactic acid produced by bacteria present on grass, corn, and other animal feeds quickly makes them acidic, helping to preserve them as hay and silage. Bacteria and fungi, such as brewer's yeast that convert pyruvate to ethyl alcohol and CO_2, are used to produce industrial alcohol and alcoholic beverages. Alcoholic fermentation also occurs in the roots of crops and other plants when flooding deprives them of air. The alcohol produced can damage the roots just as it does the brain and liver cells of heavy drinkers.

Although most animals, including humans, are aerobes, many have developed ways of coping with shortages of oxygen. In general, your muscle cells rely on aerobic respiration. But what happens when you are working very hard, as you did during Activity 3-5, and your blood cannot provide enough oxygen to your rapidly respiring muscles? Your muscle cells temporarily switch to lactic acid fermentation. Eventually, when you slow down and catch your breath, the lactic acid that has built up in your muscles converts back to pyruvate, and cell respiration becomes aerobic again. Lactic acid can make your muscles so acidic that they cannot function properly. The result is a cramp. People who are out of shape often experience cramping when they first begin to exercise, but regular exercise will increase the ability of their heart and lungs to deliver oxygen to their muscles. Occasional breaks also help to prevent cramping.

You may notice in Figure 3.16 that, before pyruvate enters the Krebs cycle, it first breaks down into CO_2 and the two-carbon

Alternative Investigation 3C, Fermented Foods, may be performed here. See the *Laboratory Manual* for this investigation. See also Chapter 3 of the TRB.

compound acetate. The CO_2 is released as a waste gas, and the oxidation of acetate continues in the Krebs cycle. Some bacteria continue the process only as far as the production of acetate. Like lactate, acetate can combine with hydrogen ions to form acetic acid, also known as vinegar. These bacteria produce vinegar and can spoil wine and cider.

CAREER PROFILE
ATHLETIC TRAINER

On a hot August afternoon at football practice, Mickey S., an athletic trainer for a major university, explains the nutritional needs of athletes. Mickey is in charge of planning all pregame meals for the football players. He has a bachelor's degree in physical education and specialized training in sports nutrition. Mickey talks about how athletes maintain their energy levels. He nods toward the field where football practice is going on. "Football players eat diets that are high in carbohydrates all season long. They eat lots of food. Where you or I will sit down and eat one plate of food, these guys may eat two or three plates of the same type of food."

You may have heard of carbohydrate loading. Mickey says that this is something different. He goes on to explain that carbohydrate loading can improve performance for athletes in some sports such as bicycle road racing or marathon running, but not for football players. The body stores carbohydrates as a kind of starch called glycogen in the liver and muscles. Glycogen provides readily available energy during a race. Athletes who participate in races use up all of their glycogen by diet modifications a few days before the competition. Then three days before the event they eat a diet that includes lots of complex carbohydrates, such as pasta, potatoes, and bread. Because their carbohydrate reserves are depleted, the athletes' bodies compensate by storing more glycogen than normal. Because the glycogen depletion part of this strategy is stressful, the body needs about six months to recover before the strategy is used again. Because football

Athletic trainers manage athletes' diet and exercise before a major competition.

players need to maintain energy levels for intense practices and weekly games, the strategy doesn't work well for them.

Mickey describes what a football player eats before a game: "The meal is mostly carbohydrates, but we do give players a small piece of steak because they think they need the protein provided by meats more than they actually do. The meal is usually served about four hours prior to the game because it takes longer to digest the protein. We want the athletes to have a little something in their stomachs so they are not hungry during the competition."

LESSON 3.3 ASSESSMENT

1. The bacteria are using lactic fermentation, which produces no CO_2.

2. As these substances are broken down in cell respiration, much of the energy released is lost as heat.

3. As the oxygen supply in a poorly ventilated, crowded place is depleted, cell respiration must slow down. The result is that less energy is available in the form of ATP.

4. Admitting air allows the organisms to switch from fermentation to more efficient aerobic respiration, which produces the same amount of energy (ATP) from less sugar.

1. Suppose you are growing a batch of bacteria that produce a valuable medicine or other chemical. If the sugar disappears in the container of bacteria and nutrient solution but no CO_2 is produced, which metabolic pathway are the bacteria using? How do you know?

2. Explain how eating foods that are rich in carbohydrate and lipids can help a ski instructor to keep warm in the snow.

3. From what you have learned in this lesson, what might be one reason that train and bus passengers often become sleepy in stuffy, crowded waiting rooms?

4. Louis Pasteur was a chemist who made many important contributions to our understanding of living cells. A group of French vineyard owners hired him to find out why their wines were spoiling. Pasteur found that when air was allowed to leak into the bottles, the yeasts and bacteria inside used up much less of the sugar in the grape juice as they fermented it. How can you explain this observation?

CAREER APPLICATIONS

The applications that follow are like the ones you will encounter in many workplaces. Use the biology you learned in this chapter to complete the activities. Share your work with the class.

AGRICULTURE & AGRIBUSINESS

Greenhouse Plants Get Extra CO_2

Talk to a commercial greenhouse operator or search for information on the Internet or elsewhere about growing plants in an atmosphere that is supplemented with CO_2. What effects does the extra CO_2 have on plants? When is CO_2 supplementation used? How feasible is it for widespread cultivation? Is there an optimum level of supplementation? Write up your findings in a brief report in your logbook and be prepared to share it with the class.

BUSINESS & MARKETING

The Tour de Plants

Ask a local travel agency about fall leaf tour packages. What does the package include? What transportation is provided? Are meals provided? How many people does the agency typically book in a year for leaf tours? How do they decide when and where the tours will go? What is the closest location? Why do people go on fall color tours? Design a brochure that will entice leaf peepers. Put together a weekend package that includes at least two types of leaf tours: a walking tour and a driving tour for people who want less physical activity.

Selling Sauerkraut and Cheese

Choose a fermented food to research. You might investigate cheese, yogurt, sauerkraut, soy sauce, or some other food. Find out how the food is made and how the industry promotes its products. Devise a new marketing plan for your product that emphasizes product quality: Explain to consumers how they benefit from the way the food is produced. Try to think of reasons that some people may be reluctant to try this food. Your plan should deal with these concerns by presenting the product in a positive way.

Who Needs How Many Calories?

People's need for energy varies throughout life and at different levels of activity. Using various sources, calculate the calories that you and each member of your family need to maintain a healthy weight for their stage of life. You may also recommend changes in activity level for some people to help them meet their weight goals.

Summer Camp Nature Counselor

Design a lesson that you could use to teach nine- and ten-year-olds about photosynthesis. Talk first with an elementary school teacher to find out how much detail and which aspects of photosynthesis children that age are ready to learn. Design handouts and a hands-on activity. When you have planned your lesson, try to arrange to teach it at a local school or camp. If it can't be arranged through school, check with organizations such as your local Boy Scouts, Girl Scouts, Campfire, or Boys and Girls Clubs to try to find an appropriate group.

HEALTH
CAREERS

Emergency Oxygen

Talk to a paramedic, an EMT, an emergency room doctor, or a first-aid instructor. Find out about the ABCs of first aid. Why is maintaining oxygen levels in the body so important? What happens if the body is without oxygen? What are some typical emergencies that affect oxygen level? How are they diagnosed? How do caregivers increase oxygen levels?

Hyperbaric Oxygen Therapy

Using the Internet, medical or nursing books, or interviews with health care workers, find out about hyperbaric oxygen therapy. Write a short report in your log book that answers the following questions: For what conditions is it used? How does it work? What is the length of a typical session? How often are treatments repeated? Who is not a candidate for the treatments? What are the possible negative side effects? How often do they occur? Be prepared to present your findings to the class.

Oxygen Essential to Waste

Microorganisms use oxygen to break down organic materials for energy. People use microorganisms to break down the wastes found in sewage. Talk to someone at your local sewage treatment facility and find out how oxygen levels are maintained in sewage treatment.

Oxygen for Aquariums

Make a survey of all of the ways to aerate aquariums and aquaculture ponds. You might consult the Internet, aquaculture journals and catalogs, aquarium supply stores, or other sources. What equipment is necessary for each method? What are the advantages and disadvantages of each method?

CHAPTER 3 SUMMARY

- Nearly all activities of living things depend on chemical energy.

- Organisms store chemical energy in the form of reduced compounds, especially carbohydrates and lipids. Oxidizing these compounds releases energy.

- Chemical energy is stored in small, convenient quantities as ATP.

- Plants and many other photoautotrophs absorb solar energy and store it as chemical energy by synthesizing sugars from CO_2 and water.

- Food chains, which include heterotrophs, depend on autotrophs for energy and carbon compounds.

- Photosynthesis replenishes the atmosphere's supply of oxygen and removes the CO_2 produced by cell respiration. Autotrophs and heterotrophs depend on one another to maintain the carbon and oxygen cycles.

- During the light reactions of photosynthesis, light energy absorbed by chlorophyll is responsible for splitting water molecules. The oxygen produced is released as a gas (O_2), and the remaining hydrogen ions and electrons are used to reduce CO_2 in the Calvin cycle.

- The enzymes of the Calvin cycle combine CO_2 with existing five-carbon sugars, reduce the product, and convert it to a three-carbon sugar that can be used for any of the plant's needs.

- Chemoautotrophic bacteria are important organisms in dark, oxygen-poor environments such as the deep ocean.

- Partial oxidation of glucose during glycolysis or fermentation produces a small quantity of ATP.

- Fermentation may produce lactate or ethyl alcohol.

- Complete oxidation of glucose during aerobic respiration conserves about 38% of the energy released, in the form of ATP.

- The Krebs cycle completes the oxidation of glucose, disposes of the oxidized carbon as CO_2, and passes electrons and hydrogen ions to the electron transport chain.

- Electrons and hydrogen ions produced by the Krebs cycle pass to the electron transport chain, where they combine with oxygen, producing water. Some of the energy released in this process is conserved as ATP.

- Organisms vary in their reliance on fermentation and aerobic respiration. Some rely on only one, while others can use aerobic respiration when oxygen is available and fermentation when it is not.

CHAPTER 3 ASSESSMENT

Concept Review

See TRB, Chapter 3 for answers.

1 Does the oxygen that plants produce in photosynthesis come from water or from CO_2?

2 Do plants use oxygen? Give a reason for your answer.

3 Workers in a dairy processing plant make skim milk by removing its fat, so the milk has fewer calories. Is it possible to remove calories from a food without affecting its fat, carbohydrate, or protein content? Give a reason for your answer.

4 In photosynthesis, does CO_2 absorb solar energy?

5 Plants that are grown in the dark are often pale green or white. How do you explain this?

6 How would you expect the increasing CO_2 concentration in the atmosphere to affect the competition between a crop that has developed a way to reduce photorespiration and a weed that has not?

7 Can carbohydrate loading help to prevent muscle cramping due to lactate accumulation? Give a reason for your answer.

8 Can the Krebs cycle be seen as a system for getting rid of waste? What waste does the Krebs cycle get rid of?

9 Some exercise is called aerobic, because it involves deep breathing. Why would aerobic exercise be recommended for people who are trying to lose weight?

10 Do fermenting cells produce sugar that aerobic cells can break down? Give a reason for your answer.

11 Cell respiration requires many enzymes, and it might be simpler for cells to release all the energy of carbohydrates and lipids by oxidizing them in one step. Why don't they?

12 Which part of the pathway of cellular respiration is involved in both fermentation and aerobic respiration?

Think and Discuss

13 Facultative anaerobes often consume much less carbohydrate in the presence of oxygen than without it. Explain why this is so.

14 Why do low-fat diets for weight loss usually include some carbohydrate?

15 Leaves of sick plants are sometimes slightly fluorescent: After being exposed to light, they glow slightly. Explain why this is so. (*Hint:* Think about what normally happens to light energy absorbed by chlorophyll.)

16 Is this statement completely correct, partly correct, or false? "Plants consume CO_2 and produce O_2, while animals consume O_2 and produce CO_2." If it is not completely correct, explain what is wrong or missing from this description of gas exchange.

In the Workplace

17 You run a pet store. A business purchased a large aquarium setup from you, including the tank, filters, air pump, and a variety of fish, snails, and live plants. About a month later the receptionist comes back to your store and tells you that the plants look as though they are dying. What would you recommend?

18 Some chemical engineers hope to capture the power of photosynthesis for industry. Is it possible to produce sugar by passing sunlight through a test tube containing chlorophyll, CO_2, and water? Give a reason for your answer.

19 A fish farmer raises fish from a hatchery until they are large enough to ship to market. The fish are shipped live. A scheduled pickup of adult fish is missed because market workers are on strike. In the meantime the fish farmer receives a truckload of newly hatched fish. Having no choice, she adds them to the tank with the mature fish. She keeps watch on the tank and begins to notice that fish are gasping at the surface of the tank. What is causing the problem? What should the fish farmer do?

20 You are a mine and mineral refinery manager. Part of your job is extracting metals from rocks (ores) and purifying the metals. This is expensive and requires a lot of energy. You are looking for an organism that can help to break down metal ores. Would the organism most likely be a heterotroph, a photoautotroph, or a chemoautotroph? Would you have to provide it with light, a nutrient sugar solution, both, or neither? Give reasons for your answers.

Investigations

The following questions relate to Investigation 3A:

21 Some organisms, such as lightning beetles (fireflies), produce light. Suppose you could measure exactly how much light a lightning beetle produced. Would it make sense to say that it was converting a certain number of calories of the energy from its food to light? Give a reason for your answer.

22 Is it possible to describe the chemical energy a car's engine releases from gasoline as a certain number of Calories? Give a reason for your answer.

The following questions relate to Investigation 3B:

23 If you tested the protein and lipid content of the leaves that you used to make starch prints, would you expect to find more protein and lipid in the lit areas of the leaves than in the dark areas? Give a reason for your answer.

24 Leaves store the energy they collect in photosynthesis in several chemical forms. Which of these chemical forms would not be evident in a starch print?

CHAPTER 4

WHY SHOULD I LEARN THIS?

If you ever raise dogs, horses, or cattle or cultivate plants, you'll want to maintain desirable characteristics in the organisms you produce. When *you* reproduce, you might want to know whether there's any chance that you could pass a disorder such as cystic fibrosis to your children. Knowing about heredity and genetics will help you to understand the great variability of the human species and of other organisms.

CLASSICAL GENETICS

WHAT WILL I LEARN?

1. how a faulty DNA blueprint affects an individual
2. how biological traits are inherited
3. how to determine the chances that an offspring will display a certain characteristic
4. methods of diagnosing inherited diseases
5. how to trace the transmission of an inherited disease in a family

In this chapter you will learn the basic principles of heredity, the transmission of biological traits from parent to offspring. You'll learn why these principles are important to plant and animal breeders as they try to produce organisms that have commercially valuable characteristics, to genetic counselors who advise parents about the chance that their children will have a hereditary disease, and to public health officials as they look for ways to wipe out disease. You'll also understand how the DNA blueprint you inherited from your parents makes you resemble your relatives in some characteristics but unique in other characteristics.

LESSON 4.1 HOW CHROMOSOMES CONTROL THE CELL

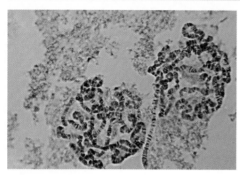

Figure 4.1: *Micrograph of chromosomes in the cell of a fruit fly*

You learned in Chapter 1 that an organism carries out its biochemical reactions according to the instructions in its DNA blueprint. Each cell has a set of these instructions. They are stored in structures called chromosomes, which in eukaryotic organisms are present in the nucleus of a cell (Figure 4.1). Because chromosomes store instructions for cell activities, they can be examined to determine whether an organism has the potential for a disorder. A **karyotype** is an orderly display of an organism's chromosomes.

ACTIVITY 4-1   LOOKING AT CHROMOSOMES

WORD BANK

karyo- = nucleus

-type = representation in terms of characteristics

A karyotype represents a cell's chromosomes.

For sample karyotypes see the TRB. The normal human karyotype contains 23 chromosome pairs. Pair 23 for females has two X chromosomes; males have one X and one Y in pair 23. Except for the XY pair in the male, chromosomes in a pair have similar length, shape, and banding patterns. The Y chromosome is much shorter than the X and is stubbier.

Some abnormal karyotypes have extra or missing chromosomes. A trisomy is the presence of an extra chromosome. An example is the form of Down syndrome in which a person has three copies of chromosome 21. Extra or lengthened chromosomes indicate that the DNA blueprint contains more genes than are needed. A monosomy represents a missing chromosome. Females with Turner syndrome have only one X chromosome. Missing or shortened chromosomes indicate that the DNA blueprint is incomplete.

People who have family histories of hereditary disease may seek genetic counseling. Such diseases are often detected in a karyotype. The person seeking counseling donates a blood sample. A lab technician treats the blood to release chromosomes from blood cells that are in mitosis (when chromosomes are fully visible with a microscope). The technician stains the chromosomes and scans them to create a digital image that a computer can analyze.

In this activity you will look at a normal karyotype from a healthy person and several karyotypes of people who have genetic defects. Use the karyotypes to answer the following questions:

- How many pairs of chromosomes are in the normal karyotype? How are chromosomes in a pair similar?

- How do the karyotypes of the normal human male and female differ?

- How does each of the abnormal karyotypes differ from the normal karyotype?

- What do the abnormalities in the chromosomes suggest has happened to the normal DNA blueprint?

DIAGNOSING HEREDITARY DISEASE

Babies with cri du chat syndrome have a high-pitched cry, low birth weight, poor muscle tone, microcephaly, and potential medical complications.

WORD BANK

syn- = with, together

-drome = to run

A syndrome is a group of signs and symptoms that occur together and characterize a particular abnormality.

Diagnosis of genetic disease is a relatively new science. Karyotyping was not developed until the early 1950s, and techniques for seeing chromosomes have been improving ever since. In 1956 the correct number of chromosomes in humans was identified. A few years later came the discovery that a form of Down syndrome involves an additional chromosome number 21. Soon after, other syndromes and disorders were correlated with abnormal chromosome numbers. A **syndrome** is a set of symptoms that, together, indicate the presence of a particular disease or abnormality.

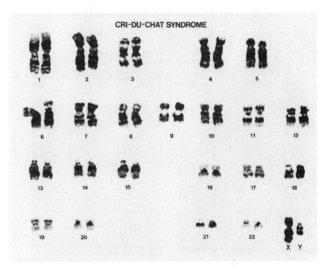

Cri du chat is characterized by a missing tip from chromosome 5. People with this disorder have language difficulties and as infants have a cry that sounds like a mewing cat.

Children with cri du chat syndrome may have poor muscle tone, feeding difficulties, delays in walking, hyperactivity, scoliosis, and significant retardation. A small number of these children are born with serious organ defects and other life-threatening medical conditions, although most individuals with the syndrome have a normal life expectancy.

With early and consistent educational intervention, as well as physical and language therapy, children with cri du chat syndrome can reach their fullest potential and lead full and meaningful lives.

A special staining technique, chromosome painting allows chromosome abnormalities associated with many diseases to be detected more rapidly and accurately. In the photograph, part of one chromosome 7 (stained red) has been exchanged with part of one chromosome 12 (stained green) yielding two abnormal (bicolored) chromosomes. There is only one normal chromosome 7 and one normal chromosome 12.

Further advances in technology led to techniques that emphasize horizontal bands on the chromosomes. The bands make it possible to identify chromosomes with added or deleted sections. Chromosome painting is a new technology based on special staining techniques.

Chromosome painting detects the exchange of material between two chromosomes.

Scientists can now analyze chromosomes at a molecular level, and more than 450 tests are available for screening for genetic disorders.

HEREDITARY PROBLEMS INVOLVE PROTEINS

A cell with a faulty DNA blueprint can't perform the proper biochemical reactions. In Williams syndrome, for example, cells that form blood vessels can't produce a protein called elastin, which gives blood vessels the ability to stretch. A number of inherited disorders involve enzymes, which are proteins. For example, an infant with PKU, or phenylketonuria, lacks an enzyme that processes an amino acid. The by-products of this amino acid interfere with proper brain development. If the disease is detected, the amino acid can be removed from the diet; if not, the infant will become mentally retarded. Diseases such as PKU led early genetic researchers to suspect that something in the chromosomes controls an organism's ability to make its proteins.

The unique characteristics of these organisms result from the unique proteins each one makes.

Remember how to build a protein? Proteins are large polymers built from monomers called amino acids. Twenty different amino acids are found in proteins. Cells assemble proteins by putting amino acid monomers into a chain. The instructions for this assembly are carried by the DNA blueprint on the chromosome(s). (Chapter 5 tells how DNA instructions are carried out to produce proteins.)

Proteins are in some way responsible for all of an organism's characteristics. For instance, a person with brown hair has a different enzyme to make hair pigment than does a person who has black hair. What allows proteins to fill such a variety of roles in an organism? It is the nearly infinite number of ways in which the different amino acids bond to form the polymers that are proteins.

INVESTIGATION 4A   PROTEIN SEPARATION

See Investigation 4A in the TRB.

Proteins are found in a wide variety of substances, and they have a wide variety of functions. In this investigation you will chemically separate two forms of proteins found in a jellyfish.

These proteins are special because they cause the jellyfish to glow. By using an ultraviolet lamp you will be able to distinguish between two fluorescent protein pigments. The two proteins that you will be using in this investigation are the green and blue fluorescent protein pigments. The green fluorescent protein pigment is the most common of the two proteins, while the blue fluorescent protein pigment is considered to be a bad copy of the fluorescent green protein pigment. Each pigment has a different chemical structure. Because of their structural differences you will be able to separate these proteins by their difference in weight.

GENETIC COUNSELOR

CAREER PROFILE

Cathy W. is a genetic counselor. She works at the University of Texas Houston Medical School. She sees several patients a day, most often pregnant women who are over 35 years old or have abnormal screening tests or ultrasounds. Other patients may have had multiple miscarriages or be from families that have genetic disorders or mental retardation. Sometimes she talks with couples who want to know, before they make the decision to have a child, the chances of having a baby with a genetic disorder.

Cathy W., pictured here, counseling a client

When patients who are expecting a child first come to her for genetic counseling, Cathy finds out as much as she can about the family history of both parents. She then describes how cells will be collected for culturing and how they will be analyzed through karyotypes or other techniques. She explains the risks that are involved in the procedure. If the patients decide to have the procedure done, Cathy schedules the tests.

Usually, it takes a couple of weeks to get the test results. After a doctor has interpreted the results, Cathy explains to the patients what the tests show. In most cases she can give her patients good news. Sometimes, though, the tests show a genetic or chromosome problem. Cathy works closely with her patients to explain what the defect means and to support their decisions.

Cathy says patients who find out that their baby won't be the normal baby they have pictured must go through a period of grief, much like those who have lost a loved one. She helps them to prepare for the birth and for life with the child. For example, if a characteristic of the genetic disorder is a **congenital** (present at birth) heart defect, Cathy may recommend that the mother give birth at a larger hospital with the facilities to handle the condition. She also may suggest that they join support groups of people who have had children with similar defects.

Genetic counselors usually have a master's degree in human genetics or genetic counseling and course work in genetics, biology, and psychology. Some have volunteered in programs that offer counseling in family planning or to victims of rape or abuse. Many counselors specialize in one area—prenatal, pediatrics, or adults. Some counselors focus on specific disorders such as cancer, cystic fibrosis, or Huntington's disease.

THE DNA BLUEPRINT IN GENES

A **gene** is a section of an organism's DNA blueprint. Each gene carries instructions for building one type of protein. In most of the organisms that you will study this year, the directions for building one type of protein are carried on genes located at corresponding positions on the maternal and paternal chromosomes (Figure 4.2). (The members of a pair of maternal and paternal chromosomes are called homologues, or homologous chromosomes.) A maternal chromosome contains genes contributed by the mother; a paternal chromosome contains genes contributed by a father. Together, an organism's maternal gene and paternal gene make up its **genotype.**

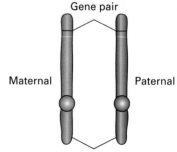

Figure 4.2: *These homologous chromosomes will combine to form the genotype of an organism. The new organism will possess traits from both chromosomes as well as traits that are unique to itself.*

You can represent a genotype by a symbol for the maternal gene and a symbol for the paternal gene. If maternal and paternal genes produce the same protein, the same symbol represents them. For example, in Hereford cattle (Figure 4.3) a

Figure 4.3: *Red and roan coat colors in Hereford cattle*

red cow or bull has two identical genes that direct the production of red pigments in the coat. Coat color is an example of a **phenotype,** a visible trait. If R_1 stands for a gene that produces the red phenotype, then the genotype of a red Hereford is R_1R_1. An organism that inherited the same form of a gene (called an **allele**) from each parent is **homozygous** for the trait. A Hereford cow with genotype R_1R_1 is homozygous. A cow with genotype R_2R_2, which produces a white coat, is also homozygous for the trait of coat color.

Different alleles produce different forms of a protein. An organism with different alleles for a trait is **heterozygous.** An example of a heterozygote is a Hereford cow or bull that has one allele that directs the production of red pigment (the R_1 allele) and another allele (R_2) does not direct the production of any pigment. This R_1R_2 genotype results in a milky red coat, called roan, which looks like cream of tomato soup.

RELATIONSHIP BETWEEN GENOTYPE AND PHENOTYPE

Sometimes a tiny difference in two alleles is responsible for a major difference in phenotype. People who have sickle-cell anemia have red blood cells shaped like a sickle or crescent (Figure 4.4). This shape is due to an abnormal protein called hemoglobin. Hemoglobin is the protein that transports oxygen in the bloodstream. Sickle cells tend to clump and move too slowly through the blood to deliver adequate oxygen to the cells. People with the disease suffer from fatigue, shortness of breath, delayed growth, bone and joint pain, stroke, and eventually damage to major organs, such as kidneys, liver, and lungs, which often results in an early death.

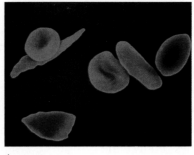

A   B

Figure 4.4: *Normal (A) and sickle-shaped (B) red blood cells (Magnification: ×2,300)*

People who have sickle cells are homozygous for an allele that directs the production of the abnormal form of hemoglobin. The genotype of a person with sickle-cell anemia is ss, while the genotype of individuals with the normal form of hemoglobin and normal red blood cells is SS. Some individuals have both alleles. They are heterozygotes with a genotype of Ss. A small portion of their blood cells is sickled, but they lead normal lives. A heterozygote whose blood behaves normally but who has a defective allele is a **carrier.** As you will see later, carriers are important in the transmission and control of certain genetic diseases.

Organisms that have the same genotype may not display the same phenotype if they live in different environments. For instance, Himalayan rabbits raised in cold climates have black tips on their noses, ears, tails, and feet. The black pigment helps them to absorb heat from the sun. But if Himalayan rabbits are raised at very warm temperatures, they have no black pigment in their fur. The alleles that direct the production of an enzyme that forms black pigment do so only at low temperatures. When rabbits grow up in the heat, the enzyme is not made and rabbits have no black pigment. Hydrangeas are a plant whose genes for flower color are affected by the acidity of the soil in which they grow (Figure 4.5). Cuttings from the same plant might produce pink or blue flowers depending on the pH of the soil.

Figure 4.5: *Hydrangea color is influenced by soil pH. Alkaline soil produces pinkish blossoms, acid soil produces bluish ones.*

As you can see from the previous examples, organisms are not strictly the products of their genes. However, a few people have denied this and have said that human traits are determined only by genes; in its extreme form, this idea of genetic determinism might be misused to exclude certain people from reproducing and passing on "undesirable" genes. Although genes play a role in determining nearly all traits, many traits, such as intelligence and complex behaviors, are greatly influenced by environment. It is best to say that genes *influence* biological characteristics, determining a range of phenotypes that an organism may express. The range is sometimes quite wide.

LESSON 4.1 ASSESSMENT

1. The woman has three genes present for each trait controlled by chromosome 21.

2. A karyotype shows only gross defects in chromosomes. Sickle-cell anemia relates to a defective allele, which cannot be distinguished in a karyotype.

3. A carrier possesses one normal allele and one defective allele. In some cases, one normal allele results in a phenotype that does not include symptoms of the disease.

4. A person usually seeks this advice when he or she is considering having children.

1 The karyotype of a young woman shows three chromosomes rather than two for chromosome 21. What does this mean in terms of the woman's genes?

2 Explain why a karyotype could not be used to determine whether a person has sickle-cell anemia.

3 Why would a carrier for a disease not show the symptoms of the disease?

4 When would it be helpful to get advice from a genetic counselor about whether you are a carrier of a genetic disease?

LESSON 4.2 HOW AN ORGANISM INHERITS ITS GENOTYPE

Organisms such as these yeast cells that reproduce asexually produce genetically identical offspring.(×430)

When an organism reproduces, it passes a copy of its DNA blueprint to its offspring. For organisms that reproduce asexually, such as bacteria and yeast, the offspring have a DNA blueprint that is identical to the parent's. In organisms that reproduce sexually, each parent contributes one half of the DNA blueprint of the offspring. So offspring that result from sexual reproduction may have different biological characteristics from those of their parents.

ACTIVITY 4-2   SAME GENES, DIFFERENT LOOK

Good subject organisms are strawberry plants, which give rise to new plants via runners; kalanchoe, which form plantlets along leaf margins; "airplane" plants, which drop plantlets from stems; and plants such as pothos, wandering jew, and English ivy, which grow from leaf or stem cuttings.

Since the plants are genetically identical, variation must be a result of differences either in the environment or in the different ages of the plants.

Consumers look for uniformity in species of ornamental plants. To obtain this, plant breeders can cultivate plants by vegetative propagation and keep environmental conditions the same for all plants.

Organisms that reproduce asexually, such as some plants, bacteria, and many molds, produce offspring that have identical DNA blueprints. So why aren't the offspring impossible to tell apart?

Examine the offspring of an organism that reproduces asexually. Choose a characteristic of the organism and write in your logbook about the variation among the offspring. Classify groups of offspring according to their variations. For some traits, such as size and weight, make measurements to describe accurately the variation.

- Offer an explanation for variation in the trait you studied.

- Why might variation in traits of a commercially valuable species be undesirable? How could breeders reduce undesirable variation?

BIOLOGY IN CONTEXT

STAMPING OUT VARIETY

The 1990s saw the first successful experiments to clone, or reproduce identical copies of, animals asexually. But for centuries, plant cultivators have used techniques to clone plants so that the desirable plants survive. These techniques continue to provide a livelihood for greenhouse and nursery workers.

Many ornamental plants and some food crops are grown through vegetative propagation. Planters cut leaves or stems of plants and simply stick them in soil to grow new plants that are exactly like the original. For example, sugar cane is replanted each year by burying pieces of stalk from last year's harvest. Sugar growers pick the best pieces of cane to bury. New plants grow from the buried stalk. The poinsettias that you see everywhere in December grow from stem cuttings that are stuck right into a pot. A plant that is produced from a cutting is genetically identical to the parent plant because it grows directly from cells of the parent.

Vegetative propagation isn't very colorful, especially compared to the flowers that many plants produce for sexual reproduction, but it does have advantages. A horticulturist can be sure plantlets that grow from cuttings will have all the desirable qualities of the parent, which is exactly what customers want. Horticulturists are not interested in variety when they already have the perfect poinsettia, onion, strawberry, or begonia.

Large greenhouses propagate plants from cuttings.

HOW AN ASEXUAL OFFSPRING GETS ITS GENOTYPE

Since an organism that is produced by asexual reproduction does not have two parents, all its chromosomes must come from one parent. Some asexual organisms, such as bacteria, have just one set of chromosomes, a condition referred to as **haploid.** Haploid cells have just a single gene to direct the production of each protein.

In a species that reproduces sexually, an organism inherits its genotype from both of its parents. For each protein it produces, an organism has two alleles: one on the maternal chromosome and one on the paternal chromosome. A cell that has pairs of chromosomes, and therefore has two genes at each location on a chromosome, is a **diploid** cell. Most cells of plants and animals except for reproductive cells are diploid. As you have already seen, the genotype of a diploid organism is indicated by a pair of symbols (R_1R_1, Ss, and the like).

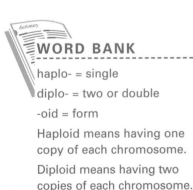

WORD BANK

haplo- = single

diplo- = two or double

-oid = form

Haploid means having one copy of each chromosome.

Diploid means having two copies of each chromosome.

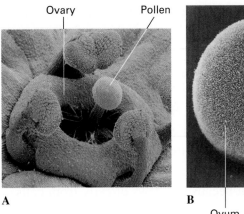

Ovary　　Pollen

A

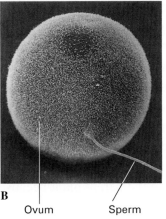

B

Ovum　　Sperm

Figure 4.6: *(A) The pollen and ovary of a flower. (B) Sperm and ovum of a hamster*

Sexually reproducing organisms produce special haploid cells, called **gametes,** for reproduction. Male gametes are called sperm or pollen; female gametes are called eggs or, more properly, ova (Figure 4.6). Haploid male and female gametes fuse during fertilization to produce a **zygote,** the first cell of the offspring, which is diploid.

MEIOSIS

Diploid cells produce diploid cells when mitosis sorts the chromosomes. But how do organisms made up of diploid cells produce haploid gametes? This happens in a special process called **meiosis.** During meiosis, homologous chromosomes move apart to opposite poles of the dividing cell. This process looks a lot like mitosis, which precedes cell division in diploid cells. Mitosis, however, maintains the diploid chromosome number in the daughter cells. Meiosis separates the chromosomes of a pair and therefore reduces the chromosome number by half (the haploid number).

ACTIVITY 4-3　MEIOSIS SIMULATION

This activity simulates meiosis as it occurs in the fruit fly, which has four pairs of chromosomes in each of its diploid cells. A set of pipe cleaners consists of two white and two colored pipe cleaners. Create four sets for each group of students. Cut pipe cleaners in each set to the same length. Each set of pipe cleaners should be cut to a different length. Use the black line masters of meiosis in the TRB to produce handouts for identifying the stages.

You will model the process of meiosis, using pipe cleaners to represent chromosomes. As you saw in Activity 4-1, chromosomes come in pairs that you can identify by their length. Before meiosis begins, each chromosome replicates and is therefore present as a pair of identical chromatids, called sister chromatids (represented by pipe cleaners of the same length and color). In this activity the color of a chromosome tells you whether it is a maternal chromosome or a paternal one.

Get 16 pipe cleaners in two colors and four lengths, a piece of string, and scissors. Form eight pairs of pipe cleaners by twisting together two of the same color and length. Put these inside the loop of string, as shown in Figure 4.7.

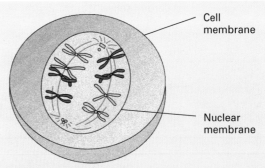

Figure 4.7: *Chromosomes in the nucleus of a fruitfly cell*

What does this model represent? The pairs of pipe cleaners represent the eight chromosomes you would see in the nucleus of a fruit fly cell under a high-powered microscope (refer to Figure 4.1). Chromosomes of different lengths contain different sets of genes, and chromosomes of the same length but of a different color contain alleles of the same set of genes. One color of pipe cleaner represents a set of maternal genes; the other color represents a set of paternal genes. Each part of the pair represents a sister chromatid. (Chromatids result from DNA replication.) Sister chromatids carry identical copies of all their genes. The string represents the membrane around the cell nucleus.

Refer to drawings of a cell undergoing meiosis. Arrange your pipe cleaners to indicate the position of the chromosomes after each stage of meiosis. Use scissors, when necessary, to cut the string. Form a new loop of string around chromosomes each time the nuclear membranes are re-formed. Draw and label the stages as you represent them. Discuss the following questions:

- In terms of their chromosomes, how do the cells formed in meiosis compare to the original cell?

- In terms of their genes, how do the cells compare?

- How else could you have positioned maternal and paternal chromosomes after they are separated in meiosis?

The cells have half the number of chromosomes of the original cell. They contain one homologue from each pair of maternal and paternal chromosomes. They also contain half the number of genes of the original cell.

Each contains some maternal genes and some paternal genes.

For each chromosome pair, maternal and paternal chromosomes move randomly to one or the other pole of the cell.

HOW MEIOSIS WORKS

Meiosis occurs in two stages (Figure 4.8). The first stage, called meiosis I, begins with prophase I, as the nuclear membrane breaks up and chromosomes appear as distinct bodies in the cell.

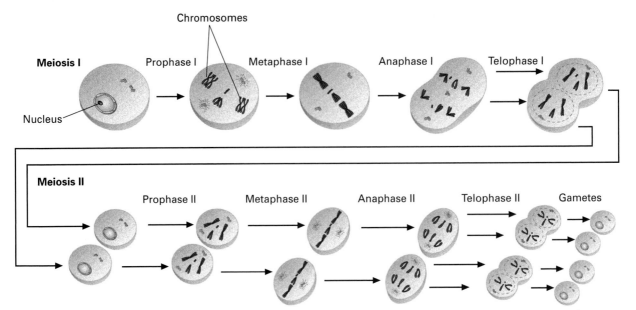

Figure 4.8: *Meiosis involves two stages, meiosis I and meiosis II. Meiosis I is quite different from mitosis. During prophase I homologous chromosomes come together and lie side-by-side. The double stranded chromosomes from each homologous pair will then move to opposite poles. At the completion of meiosis I the number of chromosomes is halved. Meiosis II is similar to mitosis in terms of mechanics, causing the single strand chromosomes to separate. After completion of this stage four haploid cells are formed.*

In prophase I, homologous chromosomes come together. At this time, each chromosome is made up of two identical chromatids. These copies were produced during DNA replication in the S phase of the cell cycle.

In metaphase I, chromosomes, still in pairs, line up along the equator of the cell, positioned along the spindle (Figure 4.8). During anaphase I, chromosomes of each homologous pair are pulled along the spindle tubules toward opposite poles of the dividing cell. In telophase I, a new nuclear membrane has formed around each haploid set of chromosomes. Each nucleus contains one chromosome of each homologous pair. Because the movement of homologous chromosomes to opposite poles during anaphase I is completely random, the haploid nuclei contain a combination of maternal and paternal chromosomes. That's the major difference between meiosis and mitosis: In mitosis, identical chromatids separate at the centromere of the chromosome and pull toward opposite poles of the dividing cell. The result of mitosis is that each daughter cell has the same number and kind of chromosomes as the parent cell.

In meiosis II, each haploid nucleus divides again in a process similar to mitosis. The sister chromatids move apart to opposite poles of the cell. Meiosis II produces four haploid cells. In males, each of these cells becomes a sperm or pollen cell. In females, one of the four cells becomes an ovum and the other three die. The number of possible combinations of chromosomes in the gametes depends on how many chromosome pairs an organism has. In humans, who have 23 pairs of chromosomes, the number of possible combinations of chromosomes in a sperm cell or ovum is 2^{23}, or 8,388,608. Since each parent can form this many genetically different gametes, each of us (except for identical twins) represents one unique genetic combination in over 70 trillion (8,388,608 $\times$ 8,388,608) possibilities.

MENDEL'S RESEARCH IN HEREDITY

Gregor Mendel

In the nineteenth century, biologists were not aware of the role of chromosomes in the cell. Plant and animal breeders didn't understand much of how inheritance worked. They used mainly trial and error to establish their varieties and couldn't explain why certain traits appeared or didn't appear when they crossed different varieties. An Austrian clergyman named Gregor Mendel was the first to discover the basic principles of inheritance. He controlled breeding in different varieties of pea plants (Figure 4.9). He removed the seeds and planted them to see what characteristics the offspring showed. He found that, for any characteristic he studied, the phenotypes occurred in certain proportions.

Mendel worked with varieties of plants that were true-breeding; that is, they always produced plants of the same phenotype (as long as environmental conditions were constant). Mendel crossed plants (from the P, or parental, generation) from a tall variety of pea with plants of a dwarf variety. All the offspring (known as the F_1, or first filial, generation) grew tall. Then he allowed the tall plants of the F_1 generation to self-fertilize. Pea plants self-fertilize when pollen from one flower on a plant lands on the female part

Figure 4.9: *Tall and dwarf pea plants*

of a flower on the same plant (see Figure 4.10). Following self-fertilization, Mendel observed that roughly three quarters of the offspring (known as the F_2, or second filial, generation) grew tall and roughly one quarter were dwarfs. Mendel was careful to use a statistical test to be sure the proportions fit a ratio of whole numbers, in this case 3:1.

Before Mendel, people had tried to explain inheritance as a blending of traits from the two parents to produce an intermediate trait. But since Mendel didn't observe offspring plants with stems midway between tall and dwarf, he ruled out the idea of blending inheritance as a general hereditary principle.

Figure 4.10: *Pea plants are capable of reproducing by self-pollination, because both the male and female reproductive organs are tightly enclosed within the flower's petals.*

Mendel explained his results by assuming that each parent had two "factors" somewhere within each cell. Each parent produced gametes by doling out randomly one or the other factor in roughly equal frequency (Figure 4.11). The factors combined when pollen containing male gametes found their way to the egg. Mendel referred to the equal frequency of factor contributions as the genetic law of **segregation.** He assumed that the factors segregated, or separated, by some random process. (He didn't know about meiosis.)

To explain the 3:1 ratio of phenotypes, Mendel assumed that one factor was **dominant** and the other was **recessive.** When an organism that is heterozygous for a trait contains one dominant and one recessive factor, the dominant factor alone shows up

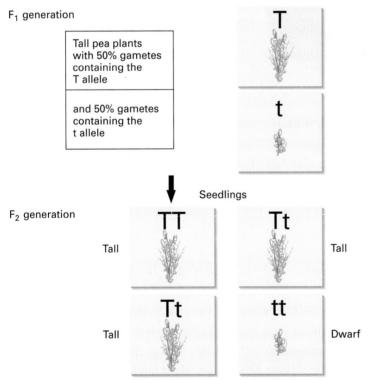

Figure 4.11: *The F_1 generation was produced by crossing tall and dwarf pea plants. The resulting offspring (F_1) were all tall but had factors from both parents. These pea plants self-fertilized producing the F_2 generation. Within the F_2 generation are both tall and dwarf plants in a ratio of 3:1.*

in the phenotype. The dominant factor masks the effect of the recessive factor. Recessive factors don't appear unless they are in an organism that is homozygous for a trait.

Further work showed that the factors that Mendel hypothesized for pea plants are present on the chromosomes of all organisms. We now call these factors genes or, more specifically, alleles.

FROM FACTORS TO GENES

When Gregor Mendel published his work on pea plants in 1866, his colleagues seemed impressed with his use of the scientific method and especially his use of statistics. But overall, his research didn't receive much attention. Scientists studying heredity weren't yet prepared to give wider significance to Mendel's work. For the next 35 years his report gathered dust in a few libraries in eastern Europe. Mendel died in 1871, probably feeling that he hadn't contributed much to the study of heredity despite the quality of his work.

Shortly after Mendel's death a number of European scientists began to turn their attention to the nucleus of the cell. In 1882, Walther Flemming developed a stain that made chromosomes and their movements visible during mitosis. One striking observation was that chromosomes nearly always occur in pairs, a fact that Flemming deduced from their similar size and shape.

This fact did not go unnoticed by two scientists working 4,000 miles apart in 1900. German biologist Theodor Boveri and U.S. graduate student Walter Sutton were intrigued by the movements of chromosomes in the testes of various insects during meiosis. It occurred to both of them that chromosomes segregate in meiosis like the factors Mendel had proposed for pea plants. Boveri and Sutton together proposed that Mendel's hypothetical factors were something on the chromosomes. Scientists started calling Mendel's factors *genes* and saying that genes were on chromosomes.

Throughout the first decades of the twentieth century, biologists around the world reported findings that were consistent with Mendel's laws and the Boveri-Sutton hypothesis. By the 1920s they began to associate genes with particular locations on chromosomes, and the new field of genetics was launched.

LESSON 4.2 ASSESSMENT

1. Given their basic understanding of genetics, students should have no reason to expect that one group would show more variation than the other. For leaf shape, all the asexual offspring will have round leaves, and all the sexual offspring will have arrow-shaped leaves.

2. Half the gametes would have one more than the haploid number of chromosomes; half the gametes would have one less than this number.

3. In metaphase. If the chromosomes are paired, the division is meiosis; if chromosomes are in no particular order, the division is mitosis.

4. The general formula is 2^n where n is the number of chromosome pairs: $2^{30} = 1.515 \times 10^{18}$ possible combinations in the gametes.

5. According to a blending theory, Mendel should have seen plants with stems that were intermediate in length between the tall and dwarf varieties in the F_1 and all succeeding generations. However, he saw only tall and dwarf plants.

1 A horticulturist is growing an ornamental plant that comes in two leaf shapes, a genetic trait. She produces offspring in two ways: by cuttings from a single plant with round leaves (asexual reproduction) and by cross-pollination (sexual reproduction) of two homozygous plants, one with round leaves and one with arrow-shaped leaves. (The arrow shape is dominant, and the round shape is recessive.) Would you expect one group of plants to show more variation in leaf shape than the other? Give a reason for your answer.

2 In meiosis in a male, one pair of chromosomes fails to segregate. How would this affect the number of chromosomes in the gametes?

3 You are observing the nucleus of a cell that is undergoing division. At what point would you be able to tell whether the type of division is mitosis or meiosis?

4 You have obtained semen (a fluid that contains sperm) from a prize Hereford bull. (The diploid chromosome number in Herefords is 60.) How many different combinations of chromosomes would you expect to find in this semen?

5 How did Mendel's results with tall and dwarf varieties of pea plants immediately rule out a blending theory of inheritance?

LESSON 4.3 PATTERNS OF INHERITANCE

Each of the traits that Mendel studied in pea plants was determined by a single gene. Single-gene inheritance in which one allele is dominant and the other is recessive leads to a 3:1 ratio of phenotypes in the F_2 generation. In other words, the probability that an F_2 offspring will show the trait controlled by the dominant gene is $\frac{3}{4}$ and the probability that it will show the recessive trait is $\frac{1}{4}$. In this lesson you will learn about techniques that breeders use to determine patterns of inheritance.

ACTIVITY 4-4   DROP THOSE HORNS

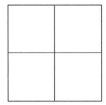

Figure 4.12: *Punnet square*

The completed Punnett squares should look like the following: Mating polled cows to the heterozygous bull leads to one half of the calves polled and one half heterozygous for horns. Using this method reduces the number of horned cows by one half in each generation.

	Cow	
	p	p
P	Pp (polled)	Pp (polled)
p	pp (horned)	pp (horned)

(Bull labels rows P and p)

	p	p
P	Pp (polled)	Pp (polled)
P	Pp (polled)	Pp (polled)

The polled (hornless) trait in cattle results from a homozygous recessive condition. Imagine that you are a cattle breeder who has a few cows that are polled and a pair of horned bulls. You would like to build a herd of polled cows and are considering mating polled cows with one of your two horned bulls. One bull has tested homozygous for horns (PP), and the other has been tested heterozygous (Pp).

In your logbook, draw a square and divide it into four equal blocks as shown in Figure 4.12. This is called a Punnett square. Beside each row, write one of the two possible genotypes for the sperm of the bull with genotype Pp. Above the columns, write the possible genotypes for the ova of a polled cow with which this bull will mate. Inside the blocks of the Punnett square, combine the alleles you have written. Each combination is a possible genotype of a calf produced by this mating. Each combination occurs with a frequency of $\frac{1}{4}$.

Draw a second Punnett square and repeat these steps for the homozygous bull. Answer the following questions:

- Which type of mating would produce a hornless herd more quickly?

- Explain what crosses you make in the F_1 generation to continue the "dehorning."

DAY BLINDNESS IN MALAMUTES

A **pedigree** is a diagram of the inheritance of a trait through a family. Pure-bred animals are sold with a pedigree so that the buyer can be sure of their genetic history. Pedigrees are also useful to veterinarians who are trying to determine the pattern of inheritance for certain disorders.

Alaskan malamute

One disorder that has been traced by pedigree is day blindness in dogs. About 40 years ago an avid breeder of Alaskan malamutes reported a litter of ten pups, three of which seemed to have problems seeing during the day but not at night. Dr. Lionel Rubin, a well-known specialist in inherited eye disorders, analyzed a series of test matings, some of which are shown in Figure 4.13. A normally sighted brother and sister from the litter of ten were mated and produced three day-blind pups in a litter of nine. Another female in the litter of ten was bred back to her father, delivering six pups, two of which were day-blind. Lumping the three litters together, Rubin tallied 17 normal and 8 day-blind pups. This ratio is reasonably close to 3:1, which would be observed if day blindness were determined by a recessive allele and both parents were carriers of that recessive allele. Other test matings that Rubin carried out were consistent with the idea that day blindness in malamutes results from a recessive allele.

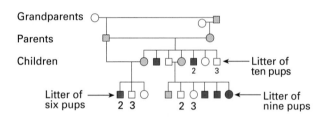

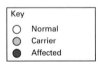

Key
- ◯ Normal
- ◔ Carrier
- ● Affected

Figure 4.13: *Pedigree for a family of Alaskan malamutes affected by day blindness*

SINGLE-GENE INHERITANCE

When just a single pair of alleles at one location on the chromosomes influences a trait, the pattern of inheritance is called single-gene inheritance. The work of Mendel dealt with single-gene inheritance, in which one allele was dominant and another allele was recessive. For some traits, both alleles of the heterozygote contribute to the phenotype. Such traits are examples of **incomplete dominance** of one allele (in the case of sickle-cell anemia) or the **codominance** of two alleles (as in the case of coat color in Hereford cattle).

Activity 4-5   Incomplete Dominance or Codominance?

See Activity 4-5 in the TRB for required advance preparation.

A pair of alleles, G and G', influence the color of soybean leaves. In this activity you will determine the pattern of inheritance for leaf color.

Plant the soybeans in two rows about 10 cm apart in a tray of soil. Set the tray in a well-lit area. Water as needed to keep the soil just moist to the touch.

Once the seeds have germinated, record the leaf color of each plant. Pool your data as a class to determine the number of plants in each color group. Construct a ratio of phenotypes and discuss the following questions:

One fourth of the seedlings should have green leaves (genotype of GG), one half should have yellow-green leaves (GG'), and one fourth should have yellow leaves (G'G').

Each plant contributes one of its two genes in equal frequency, as dictated by meiosis.

The yellow-green color is an example of codominance because both alleles are fully expressed in the heterozygote.

- What are the genotypes producing each of the phenotypes you observed?

- In deciding on genotypes, what assumptions did you make about the parent plants' contribution of genes to the offspring?

- Is the yellow-green color of soybean leaves an example of incomplete dominance or codominance?

Independent Assortment

Genetic varieties of plants and animals usually differ in more than one trait. Figure 4.14 shows the expected results when a tall variety of pea plant that produces smooth seeds was crossed with a dwarf variety of pea plant that produces wrinkled seeds. All the F_1 generation are heterozygous tall and smooth. (You know from this result that the alleles for tall stem and for smooth seeds are dominant.)

Assume that the alleles for stem length and those for seed texture are on different chromosomes. You can use a Punnett square with four rows and four columns to predict the phenotypes of offspring when the heterozygous tall and smooth plants are allowed to self-fertilize. Such a cross between organisms heterozygous for two traits is called a **dihybrid** cross. (A cross between organisms heterozygous for one trait is called a **monohybrid** cross.)

When writing the genotypes of the gametes on the top and side of the Punnett square, remember that chromosomes sort randomly in meiosis I anaphase. This random behavior is described in Mendel's **law of independent assortment:** The movement of factors of one pair does not depend on how the factors of any other pair move in meiosis. In a dihybrid cross, this means that there are four possible combinations of genotypes for both sperm and ova (Figure 4.14). Again this assumes that factors are on different chromosomes.

To analyze the Punnett square for the dihybrid cross in Figure 4.14, first combine symbols from each row and column. For instance, in the upper-left-hand block, write TTSS as the genotype of one of the offspring. Repeat this process for each block of the square. On the basis of the genotype, write the phenotype that applies to each block.

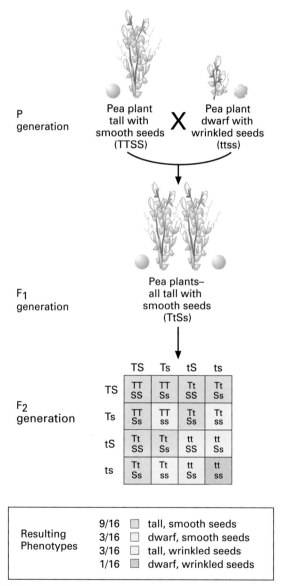

Figure 4.14: *If the genes for two traits are heterozygous, they will undergo independent assortment. This will cause new combinations of traits.*

Remember that the alleles for dwarf and for wrinkled seeds are both recessive. Add blocks with the same phenotypes. The result of the Punnett square analysis is a probability (see the Math Tips):

- Nine of every sixteen offspring (or $\frac{9}{16}$) are tall with smooth seeds,

- Three of every sixteen offspring (or $\frac{3}{16}$) are tall with wrinkled seeds,

- Three of every sixteen offspring (or $\frac{3}{16}$) are dwarf with smooth seeds,

- One of every sixteen offspring (or $\frac{1}{16}$) is dwarf with wrinkled seeds.

In genetics, probabilities are often expressed as a ratio. In this case the ratio of phenotypes in the F_2 is 9:3:3:1.

Look at Figure 4.14 again and consider the ratio of phenotypes for each trait separately. There is a 12:4, or 3:1, ratio of tall plants to dwarf plants, just as you would expect in a monohybrid cross of heterozygous tall plants. The ratio of plants with smooth seeds to those with wrinkled seeds is also 3:1, like a monohybrid cross. So you could think of the ratio of phenotypes for the dihybrid cross in terms of two monohybrid crosses. Multiply the probability of finding a plant with a tall or dwarf stem by the probability of finding a plant with smooth or wrinkled seeds to get the proportion of phenotypes:

- Tall, smooth-seeded plants: $\frac{3}{4} \times \frac{3}{4} = \frac{9}{16}$
- Tall, wrinkled-seeded plants: $\frac{3}{4} \times \frac{1}{4} = \frac{3}{16}$
- Dwarf, smooth-seeded plants: $\frac{1}{4} \times \frac{3}{4} = \frac{3}{16}$
- Dwarf, wrinkled-seeded plants: $\frac{1}{4} \times \frac{1}{4} = \frac{1}{16}$

ACTIVITY 4-6   MEIOSIS REVISITED

Students use the same materials as in Activity 4-3.

You saw in Activity 4-3 that Mendel's law of segregation is obeyed in meiosis. How is the law of independent assortment obeyed in meiosis? You'll see how if you repeat Activity 4-3, using just two pairs of chromosomes. Assign the stem length alleles (T or t) to one pair of chromosomes. Choose one color to represent the T allele and the other color to represent the t allele.

The law of segregation and the law of independent assortment are demonstrated in anaphase I of meiosis.

If alleles are on the same chromosome as stated, gametes produced by the F_1 would have either the TS genotype or the ts genotype (assuming no crossover). Without independent assortment, this produces 25% TTSS, 50% TtSs, and 25% ttss in the F_2, or a 1:2:1 ratio.

Do the same for the seed texture alleles (S or s). After you finish, answer the following questions:

- When does Mendel's law of segregation play its part in meiosis?

- When does Mendel's law of independent assortment play its part in meiosis?

- How would you expect the results of the cross to differ if both dominant alleles (T and S) were on the same homologue and both recessive alleles (t and s) were on the opposite homologue?

GENE SWAPPING

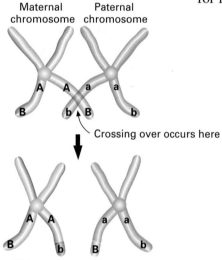

Maternal chromosome   Paternal chromosome

Crossing over occurs here

Alleles exchanged between maternal and paternal chromosomes

Figure 4.15: *During meiosis I metaphase homologous chromosomes may exchange portions of their chromosomes in a process called crossing over.*

Chromosomes contain many genes. Each human chromosome, for instance, contains several thousand genes. Genes that occur together on a chromosome are linked. Linked genes, as you have just seen, should not obey Mendel's law of independent assortment. Sometimes, however, alleles on the same chromosome appear to assort independently. Breeders note that sometimes the observed ratios of a cross do not fit the expected ratio.

Crossing over is sometimes observed during meiosis I metaphase. In crossing over alleles at the same locations on homologous chromosomes change places, or cross over, from one chromosome to another (Figure 4.15). Alleles separated by a crossover assort independently during meiosis. How often two pairs of genes on the same chromosome display crossing over depends on the distance that separates the genes. Crossovers between genes that are far apart occur more often than they do for genes that are nearby.

MANY OPTIONS

You may have the impression that each gene has only two alleles that determine a characteristic. Individual organisms carry no more than two alleles for a trait, but a population of organisms often carries many alleles.

ABO Blood Type	Form of Polysaccharide	Genotypes
A	A	I^AI^A, I^Ai
B	B	I^BI^B, I^Bi
AB	A and B	I^AI^B
O	none	ii

Table 4.1: Alleles, Genotypes, and Phenotypes for ABO Blood Groups

One example is the multiple alleles that determine the human ABO blood type. Red blood cells contain a cell membrane carbohydrate. There are two forms of the carbohydrate. Form A is produced by an allele called I^A. A person with two I^A alleles has blood type A. Form B is produced by an allele called I^B. A person with two I^B alleles has blood type B. The alleles for forms A and B are codominant, that is, a person with both alleles produces both forms of the polysaccharide. This person has blood type AB (see Table 4.1).

Some people, however, have a recessive allele called i, which does not produce a carbohydrate on the cell membrane. A person with two i alleles has blood type O. Since i is recessive, a person with genotype I^Ai has type A blood and a person with I^Bi has type B blood.

ABO blood-group testing is often presented as evidence in court cases to determine whether a certain man is (or isn't) the biological father of a child. For instance, a man with type AB blood can't be the father of a child with type O blood. Similarly, a man with type O blood can't be the father of a child with type AB blood. Use Punnett squares to convince yourself that these statements are true.

INHERITANCE OF SEX

In humans and most other animals that reproduce sexually, sex chromosomes determine maleness and femaleness. A person with two X chromosomes is female; a person with one X chromosome and one Y chromosome is male. As you saw in the karyotypes in Activity 4-1, some individuals have genotypes other than XX and XY, usually because of errors in meiosis. These unusual sex genotypes are often accompanied by disorders. In one disorder called Klinefelter's syndrome, males have an extra X chromosome. The XXY males have sparse body hair, breast development, long limbs, and some degree of mental deficiency, and they are sterile.

LEARNING LINK SEX-CHROMOSOME ABNORMALITIES

Males with XYY genotype appear to be normal except for having severe acne and being unusually tall.

Owing to mistakes in meiosis, some people inherit extra copies of sex chromosomes. In Turner's syndrome, for example, a person inherits only an X chromosome. This produces a female with poorly developed sexual characteristics. Using the library or the Internet, find out about other abnormal sex-chromosome genotypes and how they influence sexual development in an individual.

SEX-LINKED TRAITS

Like other chromosomes, sex chromosomes contain genes. Some traits are determined by an allele that is carried only on the X chromosomes (called an X-linked trait). For example, color blindness is determined by a recessive allele that is carried only on the X chromosome. Females, therefore, are carriers of a sex-linked trait such as color blindness. There is no corresponding allele on the Y chromosome. The Y chromosome is much shorter and therefore contains fewer genes than the X chromosome.

If a male inherits the allele for color blindness, he will be color-blind, even though the allele is recessive. A recessive allele is expressed if no dominant allele is present. X-linked recessive traits such as color blindness occur much more often in males than in females. A color-blind female has two copies of the allele for color blindness.

WORD BANK

hemo- = blood

-philia = tendency toward

Hemophilia is a tendency to bleed.

Sex-linked traits follow an inheritance pattern, shown in the pedigree of a family with **hemophilia** (Figure 4.16). Hemophilia is an X-linked recessive disease in which a blood-clotting factor is absent. This means that a person with severe hemophilia could die from a wound that won't stop bleeding. Notice that the allele passes from mother to son in one generation and from father to daughter in the next generation.

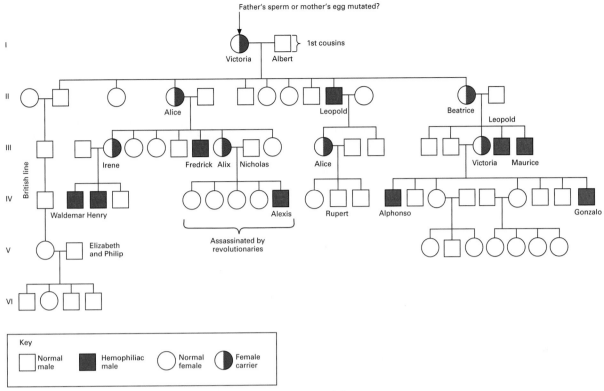

Figure 4.16: *Pedigree of European royal families affected with hemophilia*

OTHER INHERITANCE PATTERNS

Sometimes many genes, occurring on any number of chromosomes, influence one trait. This polygenic inheritance exhibits a continuous range of phenotypes, meaning that any value from one extreme to the other is possible, though some occur more frequently than others (Figure 4.17). Traits such as yield in crop plants, milk production in dairy cows, and body weight, hair color, and eye color in mammals are determined by dozens of genes. Each gene makes a small contribution to the overall phenotype, and the environment plays a much greater role than it does in single-gene traits. Polygenic inheritance is more common than single-gene inheritance.

Most often, a gene contributes to more than just one trait. These multiple effects result from interrelated biochemical reactions in an organism. One biochemical reaction catalyzed by one enzyme (the

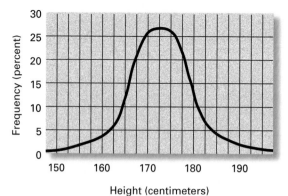

Figure 4.17: *A trait determined by many genes has a continuous number of values. A graph of these values often forms a bell-shaped curve.*

product of one gene) may be part of many biochemical pathways. A defect in this enzyme would therefore affect the products of numerous biochemical reactions.

INVESTIGATION 4B   INHERITANCE OF FACIAL TRAITS

See Investigation 4B in the TRB.

There are some human facial traits that follow the pattern of single-gene inheritance with dominant and recessive alleles. In this investigation you will have an opportunity to create a human face by randomly determining the genes that the parents contribute to a child.

What physical features do you share with your relatives? Do you have a widow's peak? Are your ear lobes attached or unattached? What traits have you inherited? Which did you inherit from your mother? Which did you inherit from your father? Do you have biological characteristics that neither of your parents has?

While genetic counselors are concerned primarily about inherited disorders, there are many other inherited traits that are quite easy to observe. The genes for these traits may be recessive, like the one for cystic fibrosis, or they may be dominant, such as the gene that determines Huntington's disease. In some cases, traits are determined by genes that are neither dominant nor recessive.

LESSON 4.3 ASSESSMENT

1. There will be nine possible phenotypes with the following probabilities of occurrence:

   $\frac{1}{4} \times \frac{1}{4} = \frac{1}{16}$

   $\frac{1}{4} \times \frac{1}{2} = \frac{1}{8}$

   $\frac{1}{4} \times \frac{1}{4} = \frac{1}{16}$

   $\frac{1}{2} \times \frac{1}{4} = \frac{1}{8}$

   $\frac{1}{2} \times \frac{1}{2} = \frac{1}{4}$

   $\frac{1}{2} \times \frac{1}{4} = \frac{1}{8}$

   $\frac{1}{4} \times \frac{1}{4} = \frac{1}{16}$

   $\frac{1}{4} \times \frac{1}{2} = \frac{1}{8}$

   $\frac{1}{4} \times \frac{1}{4} = \frac{1}{16}$

   This is a ratio of 4:2:2:2:2:1:1:1:1.

2. The child's genotype must be ii. To produce a child with type O blood, the father's genotype must be $I^A i$. The mother must also have a genotype that includes the i allele, either $I^A i$ or $I^B i$ or ii.

3. d is correct. A crossover in males is impossible, since males have only one X chromosome and therefore carry only one allele for A and one allele for B. A crossover between a heterozygous gene pair and a homozygous gene pair, such as an AABb female, would not be detectable.

4. The daughter can't be color-blind. Color blindness is caused by a recessive allele on the X chromosome. Any daughters this couple produces will inherit a normal allele for color vision from their father. Although each daughter has a 50% chance of inheriting the allele for color blindness carried by their mother, this allele is recessive, so daughters have no chance of being color-blind.

1 What ratio of phenotypes is expected in a dihybrid cross when each trait crossed separately shows a 1:2:1 ratio?

2 A man with type A blood claims to be the father of a child with type O blood. For his claim to be valid, what blood types are possible for the child's mother?

3 A crossover between genes A and B in meiosis sometimes occurs in a population of flies. Both of these genes are on the X chromosome. In which of the following flies would new combinations of genes occur in the gametes?

   a. A male with genotype AB

   b. A female with genotype AABb

   c. A male with genotype ab

   d. A female with genotype AaBb

4 A man and woman, both having normal color vision, have a color-blind son. The woman is pregnant again, this time with a daughter. What is the probability that the daughter will be color-blind?

CAREER APPLICATIONS

The applications that follow are like the ones you will encounter in many workplaces. Use the biology you learned in this chapter to complete the activities. Share your work with the class.

AGRICULTURE & AGRIBUSINESS

Genetic Variation in Animals

Select a pet or livestock breed. Using available resources such as a local veterinarian, a breeder, or an instructor of animal husbandry or genetics, breed journals, and the Internet, find out what traits define the breed. Are these traits dominant or recessive? Are they determined by a single gene or multiple genes? Are any genetic diseases or disorders of particular concern to breeders? Make a class list of findings. Discuss similarities and differences among different breeds of the same species, such as Herefords and Holsteins, quarter horses and thoroughbreds, or collies and poodles.

Endangered Species

Contact a local zoo or wildlife park or an organization that promotes the protection of endangered species, use the Internet, or find other resources related to the preservation of endangered species. Find out about breeding programs for endangered species. What genetic screening is done to select appropriate mates? What is the purpose of this screening? How successful are these breeding programs?

BUSINESS & MARKETING

Breeding Registry

Contact an organization that registers certain breed(s) of animals. Find out the purpose of animal registry. What are the restrictions for registering animals? How do these organizations preserve the breed? How are new breeds established? What services does the organization provide? How much does it cost to register an animal?

Ethical Issues

Insurance companies, employers, and others who require physical examinations may face tough decisions when they discover genetic conditions that the person having the exam doesn't know about.

Talk to someone who sells insurance or performs the physicals and ask how he or she handles this information. Can insurance or employment be denied a person with a hereditary disease?

FAMILY & CONSUMER SCIENCE

Special Diets Fight the Symptoms of Genetic Disease

Some genetic diseases cause failure to produce the enzymes that are used to break down certain compounds in a human. One such disease is phenylketonuria (PKU). Talk to a nutritionist or dietitian and find out how people born with this disorder should modify their diet. Will an infant who is born with PKU outgrow the problem? Can dietary supplements or drugs replace the missing enzymes? Ask about dietary restrictions for people with other genetic diseases.

HEALTH CAREERS

Ingenious Careers

Use the Department of Labor on-line database O*NET, the Dictionary of Occupational Titles, or other career guides and interview someone in human resources at a local hospital to make a list of medical careers that involve genetics. Consider people who do the screening, provide counseling, and provide therapy as well as those who are more indirectly involved in fields such as medical statistics. After compiling a class list, select one career to research. Interview someone in the job, contact a professional organization for people in the occupation, or use various career resources to find out what training is required, what the person does, and what working conditions are. If you interview someone on the job, find out what she or he considers most rewarding and most frustrating in the work.

INDUSTRIAL TECHNOLOGY

Industrial Uses of Cornstarch

Research industrial uses of cornstarch. Contact a supplier of industrial cornstarch and find out how corn varieties with specific alleles that affect starch synthesis are used. What corn breeds does the supplier typically purchase? How do you think they were developed?

CHAPTER 4 SUMMARY

■ A karyotype, a photographed display of a person's chromosomes, can be analyzed to detect certain inherited diseases.

■ Each gene in the DNA blueprint of an organism carries instructions for building one type of protein. Proteins are responsible for all characteristics of an organism.

■ The genotype is the combination of genes an organism has for a particular biological trait or characteristic. The phenotype is the characteristic that the organism displays for that trait.

■ In a species that reproduces asexually an organism inherits its single parent's identical DNA blueprint. In a species that reproduces sexually an organism inherits its genotype from both parents.

■ Cells that have one complete set of chromosomes are haploid; cells that have two complete sets of chromosomes are diploid.

■ In sexual species, haploid male and female gametes fuse during fertilization to produce a diploid zygote.

■ Diploid cells undergo meiosis to produce haploid gametes.

■ Gregor Mendel discovered the basic principles of inheritance using the scientific method and probabilities to analyze the results of pea plant matings. Mendel found a number of traits that result from a single gene, in which one allele is dominant and another allele is recessive.

■ Mendel's law of segregation says that maternal and paternal chromosomes (and the genes on them) separate during meiosis.

■ Mendel's law of independent assortment says that the separation of each pair of chromosomes (and the genes on them) during meiosis is independent of the movement of other pairs of chromosomes.

■ Crossing over, the exchange of alleles between maternal and paternal chromosomes, can occur in meiosis. Genes separated by a crossover assort independently in meiosis.

■ A diploid individual can have no more than two alleles for each gene pair, but a population of organisms can have numerous alleles for each gene.

- In humans and most other sexual organisms a pair of X chromosomes specifies a female; X and Y chromosomes specify a male.

- For some traits a gene is carried on the X chromosome but not on the Y. Such X-linked traits appear more frequently in males than in females.

- Most traits are influenced by many genes. These traits are examples of polygenic inheritance and, depending on the number of genes involved, may exhibit a continuous range of phenotypes.

- Most genes contribute to many traits of an organism. One reason for this is that the products of several different biochemical reactions may be controlled by the same gene, usually one which produces an enzyme common to several reactions.

CHAPTER 4 ASSESSMENT

Concept Review

See TRB, Chapter 4 for answers.

1 Explain how a karyotype shows genetic problems.

2 You are a genetic counselor. A married couple has come to your office to determine the possibility that they might have a child with sickle cell anemia. The father is a carrier of the sickle cell trait (Ss) but the mother is not a carrier nor does she have the disease (SS). What is the likelihood of this couple having a child with sickle cell anemia?

3 Describe how chromosomes of the same pair can be recognized on a karyotype.

4 Discuss why faulty instructions in an organism's DNA blueprint result in problems with its proteins.

5 A mother passes on to both her daughter and son the X-linked recessive allele for hemophilia. (Their father does not have the disease.) Will both children have hemophilia? Why?

6 In species that reproduce sexually the combination of a maternal gene and a paternal gene for a trait is the _____.

a. phenotype

b. blood type

c. genotype

d. zygote

7 Distinguish between a homozygote and a heterozygote.

8 Describe the genotype of a person who has sickle-cell anemia.

9 Distinguish between an organism's genotype and its phenotype.

10 The philosophy that places much more emphasis on the role of genes than on the role of environment in influencing biological characteristics is _____.

a. dominance

b. blending inheritance

c. crossing over

d. genetic determinism

11 In a species that reproduces asexually, how do parent and offspring compare genetically?

12 How many sets of chromosomes are present in the cells of a haploid organism? How many in the nonreproductive cells of a diploid organism?

13 By what process does a diploid organism produce its haploid gametes?

14 A zygote that results from fertilization by haploid gametes has how many sets of chromosomes?

15 Discuss the law that Mendel proposed for the separation of genetic factors before fertilization in sexually reproducing organisms.

16 In what phase of meiosis is Mendel's law of segregation realized?

17 What does the law of independent assortment mean?

18 How does independent assortment of chromosomes in meiosis affect the phenotypic variation in offspring?

19 Describe the relationship between alleles determining a trait resulting from a monohybrid trait with a phenotype rate 3:1 in the F_2 generation.

20 Calculate the ratio of phenotypes in the F_2 generation for a dihybrid cross if independent assortment takes place in the gametes of the F_1 generation. (Assume that dominant and recessive alleles operate for both traits.)

21 What is the event in meiosis that causes alleles on the same chromosome to assort independently?

22 How does crossover in one or both parents affect the expected number of phenotypes produced in their offspring?

23 How many alleles determine ABO blood groups in humans? How many different genotypes do these alleles produce? How many different phenotypes do they produce?

24 How is the karyotype of a human female distinguished from that of a human male?

25 A couple with normal color vision has a color-blind child. Do they have a boy or a girl? Explain.

26 Describe the type of variation in phenotype when a trait is determined by a large number of gene pairs.

Think and Discuss

27 How would you modify the view of the cell cycle (from Chapter 2) for cells in the testes and ovaries that give rise to gametes?

28 Sometimes the expression of a gene shuts off during an organism's life. An example is the human gene for an enzyme called lactase, which helps to digest milk sugar. People without lactase who ingest milk sugar have gastrointestinal discomfort such as cramps, diarrhea, and bloating. Does this shutdown in lactase production, which may occur in adult life, represent a change in a person's genotype, phenotype, or both? Give a reason for your answer.

29 Which would you expect to have more potential for genetic variation: a species with a small number of chromosomes, each containing a large number of genes, or a species with a large number of chromosomes, each containing a small number of genes?

30 Why is using a pedigree to determine a pattern of inheritance more reliable for animals that produce large litters?

In the Workplace

31 You are a dog breeder trying to determine whether a certain disorder that appears in the pedigree in Figure 4.18 is inherited and, if so, how.

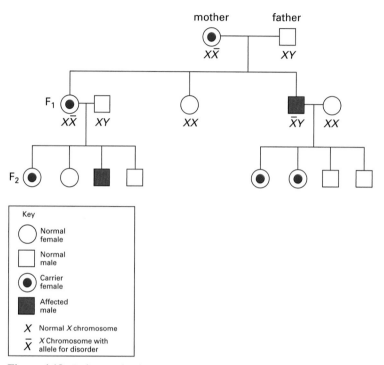

Figure 4.18: *Pedigree of a dog*

The disorder is _____.

a. dominant

b. not inherited

c. dominant, sex-linked

d. recessive, sex-linked

32 Why do scientists use mice to study the inheritance of a metabolic disorder that affects both humans and mice?

33 Suppose you were studying a fairly rare disorder caused by the failure to produce a necessary protein. You think the disorder might be inherited. What evidence would support or disprove your idea?

34 Select an occupation in which the information you learned in this chapter would be useful. Write a paragraph that describes how genetics could be used on the job and in the development of new products.

Investigations

The following questions relate to Investigation 4A:

35 Which of the following protein fractions below will travel through the column faster:
protein fraction A: molecular weight = 5400 g/mole
protein fraction B: molecular weight = 1258 g/mole
(*Hint:* your answer should be based solely on the molecular weight of each fraction)

36 What physical properties cause protein fragments to separate as they travel through a chromatography column?

The following questions relate to Investigation 4B:

37 If you were to determine the genotypes for human facial traits by flipping a coin to determine the gamete contributions of the parents, what would you assume about the genotypes of the parents? How likely is this assumption? Give a reason for your answer.

38 What does the coin-flipping method assume about the number of linked genes that determine human facial traits? How likely is this assumption? Give a reason for your answer.

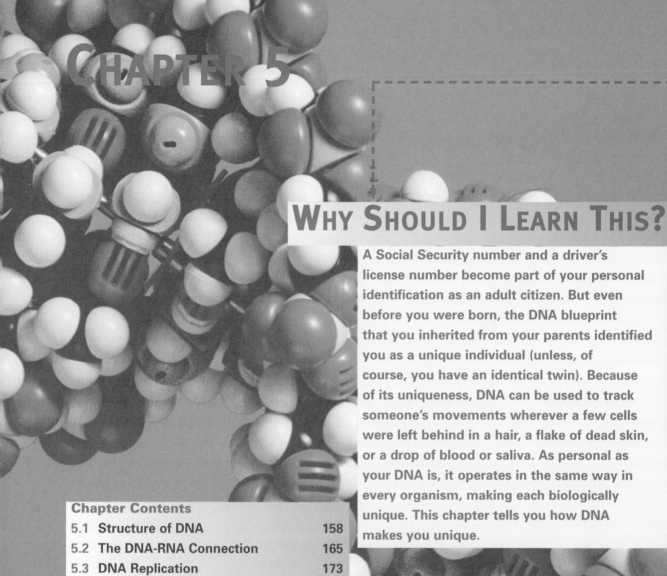

CHAPTER 5

WHY SHOULD I LEARN THIS?

A Social Security number and a driver's license number become part of your personal identification as an adult citizen. But even before you were born, the DNA blueprint that you inherited from your parents identified you as a unique individual (unless, of course, you have an identical twin). Because of its uniqueness, DNA can be used to track someone's movements wherever a few cells were left behind in a hair, a flake of dead skin, or a drop of blood or saliva. As personal as your DNA is, it operates in the same way in every organism, making each biologically unique. This chapter tells you how DNA makes you unique.

MOLECULAR GENETICS

WHAT WILL I LEARN?

1. how the chemical structure of DNA encodes information in genes
2. the role of molecules that interact with DNA to regulate a cell's activities
3. how DNA directs the synthesis of proteins
4. how changes in DNA affect an organism
5. when offspring inherit changes in DNA

The plant and animal breeders you read about in Chapter 4 were at a disadvantage because they didn't really understand what genes were or how they produced traits in organisms. In the 1950s, scientists began to learn a lot about what genes are made of and how they work. This research has opened up a whole new frontier to people in many fields. Medical personnel, for instance, use genetic knowledge to diagnose and treat disease. They can figure out whether part of a gene is missing and link it to a disease or disorder. Researchers are even looking at ways to replace faulty genes with normal ones.

LESSON 5.1 STRUCTURE OF DNA

DNA gets a lot of publicity. You've probably seen its 3D shape in magazines, in computer graphics, and on T-shirts. But DNA is more than just a colorful spiral of atoms. It is a complete set of instructions for making and running an organism, mapping the differences between species and individuals. In this lesson you learn how DNA conveys instructions.

BLUEPRINT FOR A HUMAN

BIOLOGY IN CONTEXT

A massive technological effort is under way to reveal the entire DNA blueprint of a human. It started in the late 1980s and will continue into the first decade of the twenty-first century. Teams of genetic laboratories all over the world are sequencing the more than 80,000 genes that are present on human chromosomes. **Sequencing** means determining the exact order of the DNA subunits in a gene. All of the DNA within one cell of an organism makes up a **genome,** so the project is called the Human Genome Project, or HGP for short. HGP is creating large databases of DNA sequences that researchers and product developers in every country can access.

The HGP investigated first the genomes of some simple organisms. After a decade of work, researchers had sequenced the small genomes of three species of bacteria. The first complete DNA sequence of a eukaryote, baker's yeast, was completed in 1996. Scientists will also research how its DNA blueprint, consisting of about 6000 genes, directs all the activities of a eukaryotic cell.

Knowing the details of human genes will give us new insights into human diseases and how they can be treated and cured. Genes involved in breast and prostate cancer, muscular dystrophy, kidney disease, and various blood disorders, to name but a few, have been located on particular chromosomes, and important parts of their sequences are now known. Knowing the genomes of other organisms will lead to new applications in energy, environmental protection, agriculture, and industrial processes.

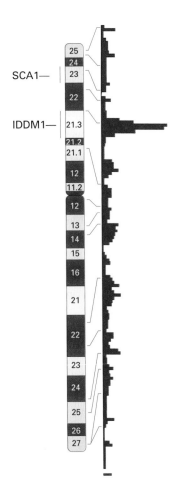

Chromosome 6

Simplified map of chromosome 6 of a human shows locations of genes that play a role in juvenile onset diabetes (IDDM1) and a degenerative problem in the brain leading to loss of muscle coordination (SCA1).

Tiny Strings of Information

It took seven years of work in more than 100 labs to sequence the relatively small genome of baker's yeast. It's fantastic that all this information, in DNA form, fits into a microscopic organism. If you could remove all the DNA from the nucleus of just one of *your* cells, the molecule would stretch from the floor to a height of 2.5 meters. You wouldn't be able to see it, though; it is less than 1/10,000th the width of the average human hair. To fit into a cell nucleus requires that a DNA molecule be tightly coiled.

DNA, like any other polymer, is a string of monomers that repeat many times in the molecule. A monomer of DNA, called a **nucleotide,** is shown in Figure 5.1. The three subunits of a nucleotide are a phosphate group, a sugar called deoxyribose, and one of four different molecules called DNA bases. A DNA base, which contains a ring that includes nitrogen atoms, is the information part of the DNA molecule, as you will learn later. The phosphate groups and sugar molecules make up the structural supports, or backbone, of the molecule. The inherited differences among organisms trace back to the differences in the long sequences of bases in their DNA molecules.

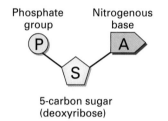

Phosphate group   Nitrogenous base

5-carbon sugar (deoxyribose)

Figure 5.1: *A DNA nucleotide formed with adenine*

Activity 5-1   Cryptography, DNA Style

In this and the next two activities the recommended DNA model kit (see TRB) allows students to build individual nucleotides, to link them in one strand, and to build a complementary strand that will bind to the first strand. The DNA bases are easily read from one or the other strand, by letter or by color. The model is three-dimensionally correct in terms of the helical structure of DNA.

Genes are sections of the DNA blueprint that determine the sequence of amino acids in protein. The chemical structure of DNA makes it a perfect molecule for carrying the cell's instructions. In this activity you will relate the structure of DNA to its ability to encode a chemical message.

Pick up a DNA model kit and identify the following parts: deoxyribose sugar subunits, phosphate subunits, and base subunits: adenine (A), thymine (T), guanine (G), and cytosine (C). Look at the monomer in Figure 5.1. Use the kit parts to build four kinds of DNA monomers. How are the four monomers alike? How do the four monomers differ? Join the four monomers into a DNA fragment, in any order, as shown in Figure 5.2.

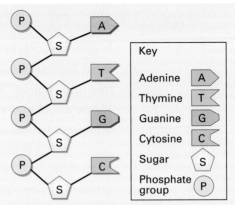

Figure 5.2: *A fragment of DNA formed from four DNA nucleotides*

Compare the order of your monomers to that of other students in the class. Answer the following in your logbook:

- How many different sequences did the class come up with?

- How could the four different DNA bases encode the 20 different amino acids that cells use to make proteins? (*Hint:* Think of each part of the code as a very small fragment of DNA, just a few bases long.)

There are 24 possible sequences.

Students should immediately see that the code for an amino acid must be more than a single nucleotide, since there are 20 amino acids and only four nucleotides. Further thinking will lead them to see that a two-nucleotide code has enough information to code for only 16 amino acids, whereas a three-nucleotide code (triplet codon) will code for all 20 amino acids.

Use the Math Tip in Activity 5-1 of the TRB to assist students in this activity.

INFORMATION STORAGE

In Activity 5-1 you made a model of a very small piece of DNA. You can see that a very long molecule of DNA could store an amazing number of chemical instructions. These instructions come in the form of genes. Genes can be represented as strings of letters, such as AGCTTACCGATAAGTAC, each letter standing for a DNA base. Surprisingly, in eukaryotic organisms, less than 5% of the total amount of DNA on chromosomes stores instructions for making proteins. We don't know yet what the other 95% does, although some of it might help to determine when genes are active.

ACTIVITY 5-2 DNA IN 3-D

The setup is the same as for Activity 5-1. At the end of this activity, save the class model for Activity 5-5.

DNA stores chemical instructions in a cell, transmits instructions between cells when cells divide, transmits instructions to offspring, and directs protein manufacture. It will be easier to understand how DNA carries out these roles if you know more about its structure.

Assemble a set of four different nucleotides from your kit, as you did in Activity 5-1. Form a double strand of DNA by matching nucleotides from the ones remaining in the kit to those in the strand you just assembled. Figure 5.3 shows you how to match nucleotides. Bases A and T can combine, and bases G and C can combine. These matches are the only possible combinations of base pairs in a DNA molecule.

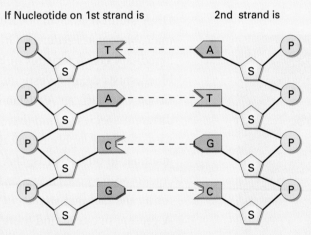

Figure 5.3: *Complementary strands of DNA*

Read the four base pairs of your double-stranded DNA fragment from top to bottom. The two strands of the DNA molecule are **complementary.** A base on one strand always pairs with a particular base on the other strand.

Add your DNA fragment to fragments built by other students to form a longer fragment. Answer the following in your logbook:

DNA is a double helix.

The strands spiral together and are joined by pairs of complementary bases.

• What is the overall shape of DNA? Be as specific as possible.

• Describe the relationship between the two strands of a DNA molecule.

WHAT'S SO SPECIAL ABOUT DNA?

In 1869, Johann Friedrich Miescher, a German biochemist, found an acid in a cell nucleus. This substance became known as nuclein, or nucleic acid, because of its location. The most abundant source of nucleic acid is DNA. Biochemists later analyzed DNA and found that it has subunits that they named nucleotides. Each nucleotide consists of a sugar called deoxyribose, a phosphate group, and one of four ring-shaped compounds called nitrogenous bases. Scientists began to hypothesize about how the subunits of DNA were put together. The most popular idea up until about 1930 was that DNA was a long polymer of "tetranucleotides," that is, a set of the four bases repeated again and again. Thinking that DNA was a rather boring molecule, biochemists preferred to explore proteins, which they knew had great variety.

DNA didn't reenter the biochemical limelight until a British microbiologist, Frederick Griffith, showed that it is responsible for hereditary characteristics of bacteria that cause pneumonia. So biochemists again investigated DNA's structure. In the late 1940s, Erwin Chargaff, an Austrian biochemist, analyzing DNA from different species, found that the proportions of adenine and thymine were always roughly equal; the same was true of cytosine and guanine. In other words, %A = %T and %G = %C. (These equations are now known as Chargaff's rule. They are accounted for by the fact that strands of DNA are complementary. For every A present in one strand there must be a T present in the other strand and vice versa.)

New techniques for looking at molecules led to solid evidence of DNA's shape. Rosalind Franklin, a biophysicist working in England in the early 1950s (Figure 5.4), took X-ray photographs of the crystal structure of DNA (Figure 5.5). These photographs clearly show that the DNA molecule is a helix. They also show that DNA, for its great length, is incredibly thin, only 2 nanometers wide. (A nanometer is a metric unit of length equal to 10^{-9} meters.)

WORD BANK

-ous = possessing, having

A nitrogenous base contains the element nitrogen.

Figure 5.4: *Rosalind Franklin*

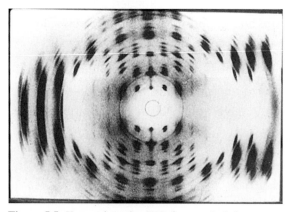

Figure 5.5: *X rays show that DNA has a spiral shape.*

In 1953, James Watson, an American biologist, and Francis Crick, a British physicist, put together all the clues (Figure 5.6). They made a model of cardboard and metal plates that fit all the known facts about DNA. Their model showed DNA as a double helix, whose two strands are made up of a backbone containing sugar and phosphate groups and of bases that are joined in pairs to hold the two strands together. The pairing of A with T and G with C accounts for Chargaff's rule.

Figure 5.6: *Francis Crick and James Watson*

INVESTIGATION 5A DNA EXTRACTION

See Investigation 5A in the TRB.

Up to this point, you have seen only molecular models of DNA. Can people actually see DNA? If so, what does it look like? In this investigation you will remove and observe DNA from cells. You will also study some of the physical and chemical properties of DNA.

LEARNING LINK DNA EVIDENCE IN COURT CASES

Alternative Investigation 5C, DNA Fingerprinting in a Criminal Case, may be performed here. See the Lab Manual for this investigation.

The DNA sequences in our chromosomes distinguish each individual (except for identical twins) from all others. Each person's DNA is unique and can be used as a "fingerprint." If law-enforcement personnel collect DNA from evidence at crime scenes, they can compare it to the DNA of suspects in the crime. DNA fingerprinting may help to place a suspect at the scene of a crime or free someone from suspicion. Find an article in a newspaper or magazine or on the Internet that reports a crime in which DNA fingerprinting helped to convict or exonerate a suspect. Explain how the DNA evidence was used in the case. If you had been on the jury, how would the DNA evidence have influenced your decision?

LESSON 5.1 ASSESSMENT

1. The amount of DNA in gametes is one half the amount in any other cell.

2. Since A always pairs with T, the percentage of thymine in the complementary strand is 24%.

3. 2

4. 8

1 In Chapter 4 you learned that the process of meiosis halves the chromosome number of a cell. On the basis of this, tell how the amount of DNA in a gamete compares with the amount of DNA in any other cell of the same organism.

2 In a certain chromosome the nitrogenous base adenine is 24% of one strand of the DNA helix. What does this tell you about the base content of the complementary strand?

3 Your instructor has asked you to make several DNA models that consist of only four base pairs from a DNA model kit that contains five adenines, six thymines, five guanines, and eight cytosines. How many DNA models can you make? (Hint: Once a DNA model is made, you cannot remove bases from the first model to make additional models.)

4 How many base pairs would be present if you link together all of the models that were made in Question 3?

LESSON 5.2 THE DNA-RNA CONNECTION

If you were building one small room in a large palace of several hundred rooms, you would need to consult the architect's blueprint for the palace. You wouldn't need a blueprint for the whole palace, so the architect might copy only the part that shows the room you need to build. This situation resembles what happens inside a cell when it prepares to make a particular protein. Making one protein doesn't require the entire DNA blueprint stored in the cell, only the gene or perhaps several genes that have the chemical information for that protein.

In this lesson you will learn how a cell uses a bit of the DNA blueprint to make a protein. The process uses ribonucleic acid (RNA). RNA is composed of nucleotides like those of DNA (see Figure 5.7). The differences are that RNA has ribose (a sugar) instead of deoxyribose and uracil (U) in place of thymine. RNA can be a single strand of nucleotides.

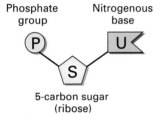

Phosphate group   Nitrogenous base

5-carbon sugar (ribose)

Figure 5.7: *An mRNA nucleotide formed with uracil*

ACTIVITY 5-3   TRANSCRIPTION: A SIMPLE BUT ELEGANT CODE

Students continue using the kit from Activities 5-1 and 5-2. Give students the sequence TTTCAAAAACCAGC AACACTTAATCGA to transcribe. Although the message for making the protein is said to be contained in the sense strand of DNA, it is the complementary strand, or antisense strand, that serves as the template for transcription. Have students save the product of this activity to use in Activity 5-4.

In this activity you will learn how the DNA blueprint of a cell is copied, or transcribed, into a molecule of RNA. This step creates a pattern for a protein molecule.

The DNA sequence TTTCAAAAACCAGCAACACTTAATCGA is part of a gene for the protein called lysozyme. Match RNA nucleotides to the DNA nucleotides for this sequence, using the rules in Figure 5.8. After you have paired RNA and DNA (which begins the

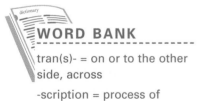

WORD BANK

tran(s)- = on or to the other side, across

-scription = process of writing a script

In transcription, the DNA base sequence is processed as an RNA base sequence.

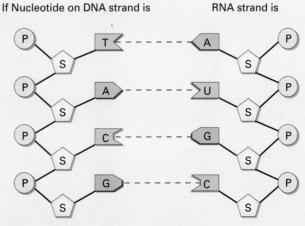

If Nucleotide on DNA strand is   RNA strand is

Figure 5.8: *Base pairing for transcription*

The messenger RNA sequence is AAA/GUU/UUU/GGU/CGU/UGU/GAA/UUA/GCU.

The RNA and DNA contain the same number of nucleotides; the nucleotides are complementary: A pairs with U; G pairs with C.

The RNA is just one small part of a DNA molecule, containing only the information in a single gene. A DNA molecule (or all of the DNA on one chromosome) might contain millions of nucleotides, depending on the species.

Transcription takes place in the nucleus of a eukaryotic cell and in the cytoplasm of a prokaryotic cell.

process called transcription), break the bonds between the DNA and RNA nucleotides and read the sequence of RNA bases. Record them in your logbook, and answer the following questions:

- How is the RNA sequence you transcribed similar to the DNA sequence?

- How does the RNA molecule compare to an entire DNA molecule?

- Where does the process of transcription take place in a eukaryotic cell? In a prokaryotic cell?

MESSENGER RNA

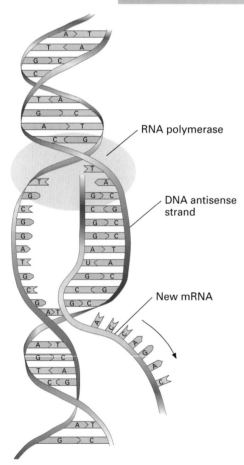

RNA polymerase

DNA antisense strand

New mRNA

Figure 5.9: *mRNA getting transcribed from DNA*

DNA serves as a template for the making of RNA, an event called **transcription.** (You demonstrated this in Activity 5-3.) This RNA carries DNA instructions to another part of the cell. (That's why it's called **messenger RNA,** or mRNA.) Usually, it travels to a ribosome to direct the manufacture of a protein.

Transcription occurs when the cell needs a certain protein. Here's an outline of the setup work that starts with DNA and leads to the protein factory, or ribosome:

The strands of a DNA molecule are unzipped by an enzyme breaking the bonds between base pairs (see Figure 5.9). This exposes the DNA bases on the antisense strand. This strand is a pattern for the complementary bases that make up the new mRNA. (You followed these rules in Activity 5-3.) The enzyme RNA polymerase joins the RNA nucleotides into a single strand of mRNA. After all the nucleotides in the gene have been transcribed, the mRNA separates from the DNA.

The new mRNA gets edited before leaving the nucleus. In one type of editing, enzymes remove sequences that do not contain coding for proteins, called **introns.** The remaining sequences that code for proteins, called **exons,** are joined to produce the final mRNA (Figure 5.10). At last, the mRNA is ready to bind to a ribosome.

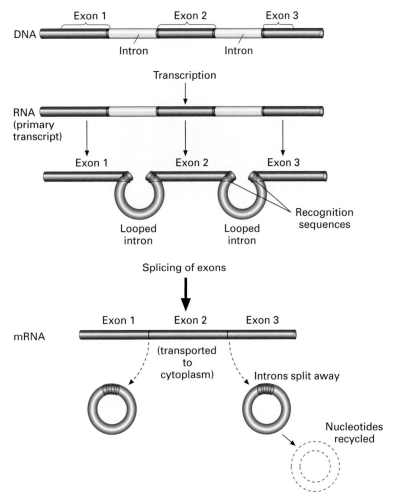

Figure 5.10: *mRNA gets a cut and paste job after transcription; introns are removed and exons are joined together.*

ANTISENSE MAKES GOOD SENSE

Scientists, as well as specialists in skin and hair products, have known for a long time about an enzyme that converts the male hormone **testosterone** to a superactive form. Superactive testosterone can cause skin cells to secrete too much of an oily substance called sebum. Sebum sometimes clogs skin pores and leads to acne. It can even make hair fall out in pattern baldness (Figure 5.11). Though more common in men, these problems can also affect women, who have less testosterone than men do.

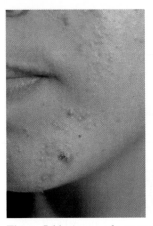

Figure 5.11: *Acne and pattern baldness are two conditions that may respond to antisense RNA technology.*

Understanding how mRNA moves within the cell is one route to controlling this enzyme that is involved in acne and baldness. When skin cells express the gene for the enzyme, its mRNA is present in them. Researchers have synthesized a piece of RNA that is complementary to the mRNA for the enzyme. They deliver this artificial RNA, called antisense RNA, to skin cells. There, it may bind to mRNA so that the message isn't translated into the enzyme. Antisense RNA is being tested in a cream that acne patients can rub on their skin.

Antisense RNA technology may also have potential as a treatment for cancer, leukemia, viral infections, and a number of other diseases triggered by the expression of genes.

IN THE RIBOSOMAL FACTORY

At the ribosomes the mRNA code directs the assembly of amino acids into protein, a process called **translation.** How does the ribosome use an mRNA message to construct a protein?

Besides mRNA, translation requires other types of RNA. Transfer RNA, or tRNA (Figure 5.12), transports amino acids in the cytoplasm to the ribosome. Each of the 20 amino acids recognizes one of 20 individual tRNA molecules by a unique nucleotide sequence on that tRNA. Figure 5.12 shows where an amino acid binds to tRNA. Another location on tRNA has a nucleotide sequence that is complementary to the RNA message on the ribosome.

The other form of RNA is ribosomal RNA, or rRNA. Several dozen different proteins and rRNA molecules form the complex structure of a ribosome. Once the mRNA reaches its proper position on the ribosome, tRNA nucleotides begin to pair up with mRNA

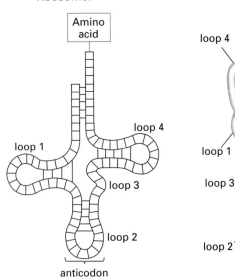

Figure 5.12: *A molecule of transfer RNA*

nucleotides. As you may have figured out in Activity 5-1, the mRNA transcribed from DNA contains its message in triplets of nucleotides. A sequence of three consecutive bases, such as CCA or UGC, specifies one amino acid of the protein to be made. Chemical bonds form between the mRNA triplet and a complementary three-base sequence on the tRNA called an **anticodon.** This bonding positions the first amino acid of the protein (Figure 5.13).

WORD BANK

anti- = opposite

-codon = triplet of nucleotide bases that identifies a particular amino acid

An anticodon is a triplet of nucleotide bases in tRNA that identifies an amino acid and binds to a complementary codon in mRNA.

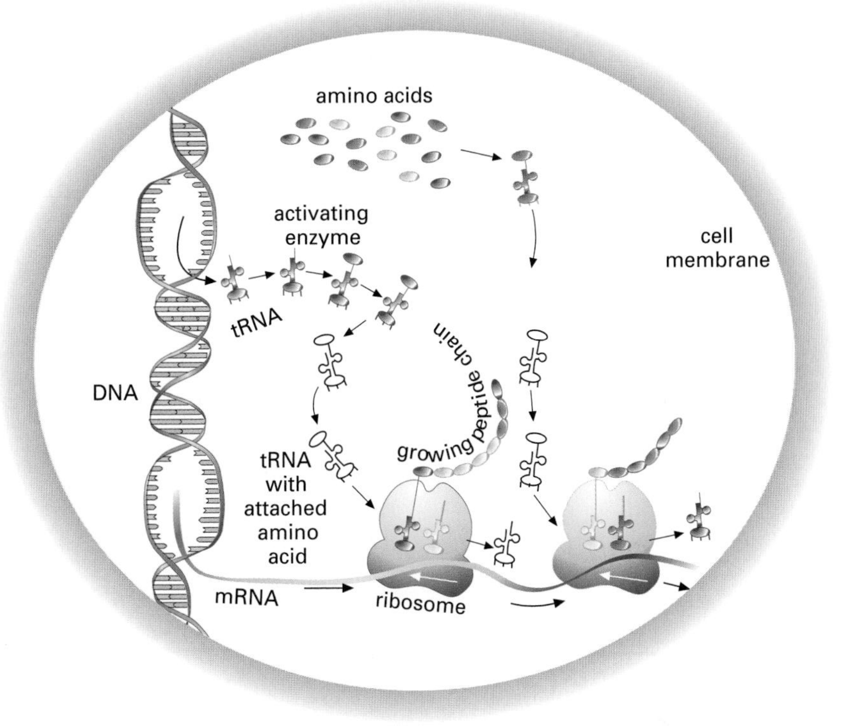

Figure 5.13: *Translation of mRNA's message into protein*

The ribosome moves along the mRNA, bringing the next codon into the proper position (refer to Figure 5.13). This codon specifies the second amino acid of the protein. When tRNA transports the second amino acid to the ribosome, a **peptide bond** forms between the first two amino acids. (A peptide bond is a bond between two amino acids.)

Translation of a single protein continues until the ribosome has moved along the entire mRNA molecule and all of the amino acids of the protein are in proper sequence. The last codon of the mRNA, called a stop codon, signals that the protein is complete. While still on the assembly line, the protein starts to fold into the shape that fits it for a specific role in the cell: transport molecule, membrane channel, enzyme, or whatever.

Students should be aware that multiple copies of the same protein are translated at many ribosomes at the same time.

ACTIVITY 5-4   MIMICKING A RIBOSOME

This activity uses the DNA and mRNA sequences from Activity 5-3.

Now that you understand some of the details of protein synthesis, try your hand at building a protein from mRNA code.

Write in your logbook the mRNA sequence that you transcribed from DNA in Activity 5-3. Write an amino acid below each codon of the mRNA sequence. Table 5.1 shows which mRNA codons specify which amino acids.

First Nucleotide	Second Nucleotide				Third Nucleotide
	U	C	A	G	
U	Phe	Ser	Tyr	Cys	U
	Phe	Ser	Tyr	Cys	C
	Leu	Ser	Stop	Stop	A
	Leu	Ser	Stop	Trp	G
C	Leu	Pro	His	Arg	U
	Leu	Pro	His	Arg	C
	Leu	Pro	Gin	Arg	A
	Leu	Pro	Gin	Arg	G
A	Ile	Thr	Asn	Ser	U
	Ile	Thr	Asn	Ser	C
	Ile	Thr	Lys	Arg	A
	Met	Thr	Lys	Arg	G
G	Val	Ala	Asp	Gly	U
	Val	Ala	Asp	Gly	C
	Val	Ala	Glu	Gly	A
	Val	Ala	Glu	Gly	G

Full names of amino acids can be found in the TRB.

Table 5.1: The Genetic Code (codon assignments in messenger RNA)

After you have translated every codon of the mRNA code, have your teacher check your amino acid sequence. You have just constructed the first part of the sequence for lysozyme. Lysozyme is an enzyme in saliva, sweat, and tears that helps to destroy bacteria. If the complete lysozyme molecule has 129 amino acids, how long is the edited mRNA? Don't count the start and stop codons.

The first nine amino acids in lysozyme are lysine, valine, phenylalanine, glycine, arginine, cysteine, glutamic acid, leucine, and alanine. The edited mRNA has 387 nucleotides.

AVOIDING OVERPRODUCTION

How does a cell control transcription and translation so that it makes just the right amounts of the proteins it needs at the proper time?

It was first shown in prokaryotes that transcription is often turned on by the presence of a reactant in a biochemical pathway. The genes that get transcribed are those that produce enzymes in the biochemical pathway. Transcription is often shut down when the products of the biochemical pathway appear in the cell. Since the DNA of every cell, except for gametes, is the same, this mechanism keeps a cell from producing an unneeded protein. For instance, white blood cells have no need to produce the protein insulin, which controls the glucose level in the blood.

There are a number of ways that cells can control protein production. Listed here are some of the more important ones.

A

- Before transcription of mRNA can begin, RNA polymerase must bind at a special region on the DNA called a promoter sequence. A cell can control transcription by the binding and releasing of special regulatory proteins that affect the binding of the RNA polymerase.

- When mRNA attaches to a ribosome, an initiation sequence on the mRNA must be read in order for translation to begin. One major method by which a cell controls its protein production is the binding and releasing of regulatory proteins at this sequence.

- Special proteins may bind to mRNA and prevent its attachment to the ribosome; this means the protein will not be produced.

- Cell enzymes break down proteins and mRNA that are no longer needed.

B

Figure 5.14: *Dr. Barbara McClintock's (A) interest in corn kernels that developed spotty or speckled appearance led to her discovery of jumping genes. (B) The movement of a jumping gene from one chromosome to another turns the color genes on and off resulting in the streaked kernels shown here.*

Other control elements for protein production include DNA sequences called jumping genes. Barbara McClintock, an American biologist, was the first to discover genes that move around on an organism's chromosomes, turning on and off the transcription of adjacent genes (Figure 5.14).

LESSON 5.2 ASSESSMENT

1. A DNA molecule has many genes, which together may contain as many as 100 million base pairs. An mRNA molecule is no longer than the gene from which it is transcribed, usually about several hundred base pairs.

2. The DNA code is being put into another form in mRNA. It is being transcribed as one might transcribe a typed message into handwritten form.

3. Protein is translated from the instructions on mRNA just as a code might be translated into an English message or words of one language might be translated into another language.

4. Only two, the amino acids coded for by UAU and AUA.

5. RNA contains uracil and not thymine. DNA contains thymine and not uracil. Therefore the nucleotide sequence that contains U is the RNA fragment, and the sequence that contains T is the DNA fragment.

1 Compare the length of mRNA molecules to the length of DNA molecules.

2 Why is "transcribe" an appropriate term to describe the copying of DNA to mRNA?

3 Why is "translation" an appropriate term to describe how a protein is synthesized from coded instructions?

4 How many different amino acids will be produced from mRNA that contains only two of the possible four bases in an alternating sequence such as UAUAUAUAUAUA . . . ?

5 A technician was asked to determine the nucleotide sequence of DNA and RNA fragments from a tissue sample. After determining the sequence, the technician realized that he had forgotten to label the fragment samples. How could he tell the DNA and RNA fragments apart if all he had were the sequences ATCGAAATGGC and UGCCGUUGGCA?

LESSON 5.3 DNA REPLICATION

A computer operator can copy files from a computer's memory onto a floppy disk. A cell, although it is thousands of times smaller than a computer chip, also can make copies. As you know, a cell copies its DNA each time it goes through the cell cycle. It passes on these copies of DNA to its daughter cells when it divides and to its offspring when it reproduces. If the new cells (or offspring) are to have the same characteristics as the original cell (or parent), DNA must be copied very accurately. In this lesson you will learn how this special copying process, called DNA replication, takes place within cells.

COPYING DNA

BIOLOGY IN CONTEXT

Alternative Investigation 5D, Polymerase Chain Reaction, may be performed here. See the Lab Manual for this investigation.

If you have the right lab equipment, you can copy DNA much as a cell does. But why bother, when cells make accurate copies all by themselves? There are some good reasons. DNA is used more and more often as evidence in criminal cases and paternity lawsuits. The amount of DNA that is collected may be so small that it is hard to analyze. This problem has been solved by a technique called polymerase chain reaction (PCR). This technology has been called a copying machine for DNA, creating thousands of copies of DNA fragments collected from clothing, skin, and other objects. PCR is a test-tube version of DNA replication.

Making copies of DNA isn't as easy as using a copying machine. You start with a test tube containing the four nucleotide bases, an enzyme called DNA polymerase, a short piece of DNA that starts the copying (called a primer), and the DNA that you want to copy. You heat the tube to 90°C, which separates the DNA strands in the double helix (Figure 5.15). After 30 seconds you cool the vial to 55°C so that the primers bind to the DNA strands. This process takes about 20 seconds. Then you raise the temperature to 75°C, and DNA polymerase begins adding nucleotides to the primer until it composes a complete complementary copy of the DNA. The temperature for each step is automatically controlled by a computerized temperature controller called a thermocycler, which runs the entire PCR sequence.

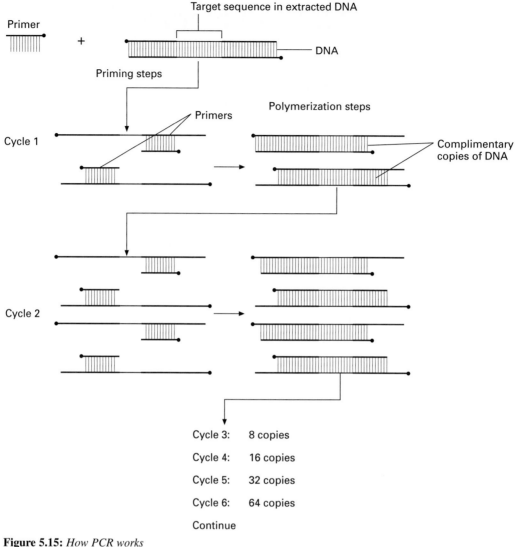

Figure 5.15: *How PCR works*

ACTIVITY 5-5   DOUBLING THE DOUBLE HELIX

This activity uses the class DNA model from Activity 5-2. Provide more nucleotides, the same number and type as those in the class model.

At some point, most cells divide into two daughter cells, which have identical DNA. That means that they have to replicate their DNA before they divide. In replication, one DNA molecule produces two molecules.

Working with one or two other students, get the class DNA model from Activity 5-2. Write down the base sequences from top to bottom on both strands of the double helix that the class

Students simulated the breaking of bonds between base pairs and the forming of new bonds between base pairs of one old strand to form a second strand of a new double helix. Each replicated helix becomes one of a pair of sister chromatids, so each daughter cell gets half of the mother cell's "old" DNA and half of the "new" DNA.

made. Break the connections between the strands. Using the base-pairing rules for DNA (refer to Figure 5.3), add DNA nucleotides to each single strand to create a double strand.

Compare each double strand to the sequence of base pairs that you wrote down from the original model. Answer the following in your logbook:

- Explain the steps that you followed in this activity.

- Does each double strand represent an identical copy of DNA?

- Of the DNA that each daughter cell receives following cell division, how much is "old" DNA and how much is "new" DNA?

- Relate the result of DNA replication to the chromatids in mitosis (described in Chapter 2).

FAIR SHARES

Think about your work in Activity 5-5 and compare it to the different chemical reactions that take place in a cell. Before DNA replication begins, bonds between complementary base pairs must break (Figure 5.16). The double helix must unzip into single strands to allow new DNA nucleotides to bond with complementary nucleotides in the exposed strands. The enzyme DNA polymerase catalyzes this reaction. Each exposed strand acts as a template for replication. In this way, two double helices are created, identical to each other and to the original double helix.

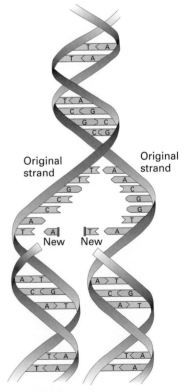

Figure 5.16: *How DNA replicates in the cell*

LESSON 5.3 ASSESSMENT

For help with Problems 1 and 2, have students refer to page 68 in Chapter 2.

1. DNA molecules are present on each of the cell's 46 chromosomes, and each DNA molecule replicates once. Therefore 46 DNA molecules replicate.

2. The new cells will be half as radioactive, because each replicated DNA molecule consists of one strand of tagged nucleotides and one strand of untagged nucleotides.

3. Both PCR and DNA replication require DNA, DNA polymerase, and the four nucleotides. The DNA double helix must break into two strands before either reaction can begin. Both processes reproduce the complementary strand on the template strand in the same way.

1 In a diploid human cell, how many molecules of DNA replicate during S phase?

2 A lab technician radioactively tags the DNA in a cell culture that has just entered the G1 phase of the cell cycle. She supplies the culture with untagged DNA nucleotides. She returns after the cells have taken up the untagged nucleotides, replicated their DNA, and gone through one cell division. Compare the radioactivity of the new cells to that of the mother cells. (Radioactive tagging is a method used by scientists to track the actions of a molecule as it reacts with other molecules. The radioactive tag allows the technician to identify portions of the tagged molecule within other cells or other molecules.)

3 How is PCR similar to DNA replication in a cell?

LESSON 5.4 MUTATIONS

The term "mutation" might make you think of monsters, but most mutations cause only minor changes in an enzyme or other protein. Still, even slight changes can be bad news for the organism.

DENTAL HYGIENIST

Cindy A. is a dental hygienist who works in a general dentistry office. In her daily work she checks the condition of patients' teeth and looks for possible cavities, although the dentist makes all diagnoses. She cleans and polishes teeth and applies sealant to some of the younger patients' teeth. Cindy also keeps patients' charts and takes X rays of their teeth.

This dental hygienist is preparing to take an X ray of a patient's teeth for a dentist to examine.

When asked whether she worries about the safety of X rays, Cindy said, "While it is true that dental X rays are a form of radiation that can cause mutations, dental X rays are generally considered no more dangerous than walking in the sun. This is because the dose, or amount of radiation a patient is exposed to, is small. In our office patients receive X rays only once every two years. To further reduce the exposure, we cover the patient with a lead apron, which can stop harmful radiation. I stay behind the wall because I do several X rays throughout the day, and the effects of repeated exposure are cumulative. I don't have to wear a badge that monitors exposure, though, because the equipment we use produces such low levels of radiation and it is well maintained to prevent radiation leakage. On the other hand, X-ray technicians who work in a hospital lab and take diagnostic X rays all day long must wear badges that monitor cumulative exposure. They use higher levels of radiation and spend more time conducting X rays, increasing their dose."

To prepare for her career, Cindy earned a bachelor's degree in dental hygiene, but she says that there are also two-year associate degree programs in dental hygiene. All dental hygienists must be licensed.

Can DNA Get Messed Up?

WORD BANK

muta- = change

-gen = generate

A mutagen generates a change in DNA structure (a mutation).

DNA can reliably store an enormous number of chemical instructions, but the DNA structure is vulnerable to changes. Changes in the coded instructions carried by DNA are called **mutations.**

Our environment contains a number of **mutagens** that can cause DNA to mutate. For instance, ultraviolet light can alter base pairs in the DNA in your skin cells which can lead to skin cancer. Radiation such as X rays and gamma rays penetrate your skin and can pass through an internal organ or tissue. There they can break the sugar-phosphate backbone of DNA. When this happens, whole stretches of DNA will not be correctly transcribed. Certain chemicals act as mutagens when they modify DNA bases and change the message encoded in DNA (Figure 5.17). Most known mutagens can also trigger tumor formation.

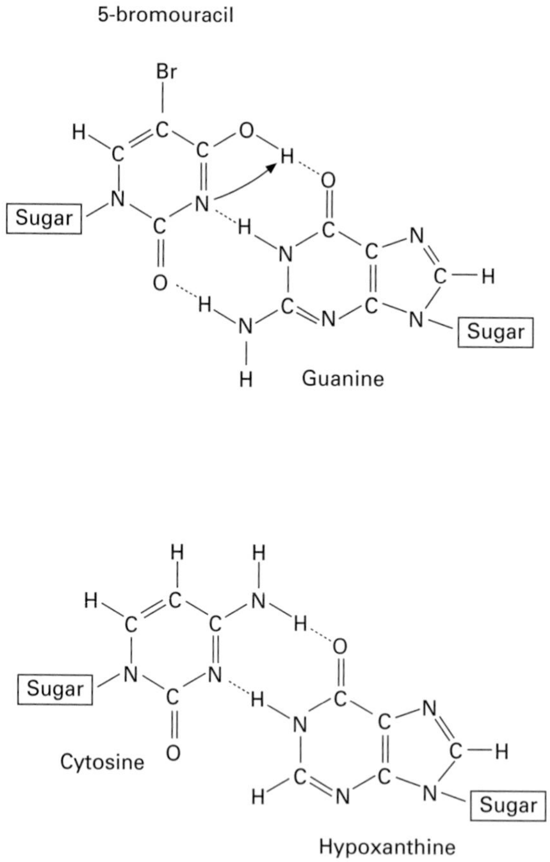

Figure 5.17: *Some chemicals can cause mutations by taking the place of nucleotides and changing the sequence. For example, 5-bromouracil can replace cytosine, and hypoxanthine can replace guanine.*

INVESTIGATION 5B   THE AMES TEST

See Investigation 5B in the TRB.

The best way to determine whether a substance is likely to cause mutations is to expose live cells or tissues to the substance and observe the effects. A laboratory test called an Ames test is used to screen for chemicals that may cause mutations. The test exposes a bacterial culture to the test substance and looks for changes in the bacterial phenotype that are known to accompany mutations. In this investigation you will test a number of household substances to determine whether they are mutagens.

DO MUTATIONS REPRODUCE?

Even without exposure to the mutagens mentioned above, DNA would still mutate. Some mutations occur spontaneously as a result of mistakes by enzymes. For instance, DNA polymerase might pair C (instead of T) with A during replication. The spontaneous rate of mutation for humans is very low; the chance that one of your gametes contains a mutation for a particular gene is smaller than one chance in ten thousand. That doesn't seem like a lot to worry about, but once a mutation occurs, it can spread. Because DNA replication produces exact copies of DNA, mutations are copied and passed on to other cells and even to offspring. To spread to the next generation, the mutation must be in a gamete. You couldn't pass on to your offspring a mutation that occurred only in a skin cell.

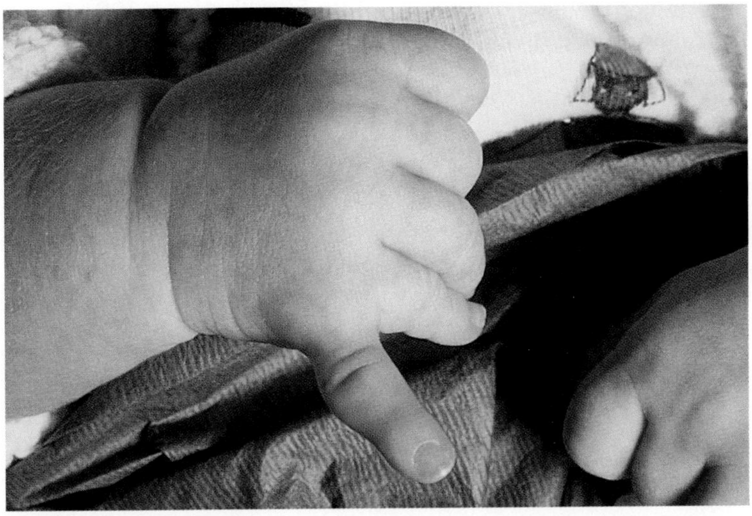

A cell with mutated DNA may have an incorrect instruction for one or more of its biochemical reactions. A recessive mutation has no effect on a cell if the cell still has a normal gene to direct protein synthesis. However, some mutations, such as having more than five fingers, or polydactylism (Figure 5.18), are dominant. They produce a protein that interferes with the normal activity of the cell.

Figure 5.18: *Polydactylism is a dominant mutation.*

Many mutations produce no effect whatsoever. One reason is that cells contain enzymes that recognize and repair mutations soon after they occur. Another reason is that the DNA code has some slack or redundancy. You can see in Table 5.1 that most amino acids are specified by more than one codon, sometimes as many as six. So a mutation might occur that does not affect the sequence of amino acids in a protein. Some mutations produce such a slight change in a protein that the protein's function is not affected.

ACTIVITY 5-6   ARE MUTATIONS ALWAYS A PROBLEM?

Students use the sequence of DNA transcribed in Activity 5-3.

Why don't mutations always affect the proteins in a cell? You'll use a DNA model to explore this question.

Write down the DNA sequence of bases from the class model from Activity 5-2 or the sequence from Activity 5-3. Change the first base in the sequence to a C (or to a G if it's already a C). Using the base-pairing rules for transcription (Figure 5.8) and the mRNA codons for translation (Table 5.1), show how the protein that is produced changes because of the mutation.

Again, write down the original sequence of DNA bases in your model. Change the third base in the sequence to a C (or G). Transcribe and translate the mutated sequence to show how this mutation affects the protein. Finally, add a C between the second and third triplets of the DNA sequences. Transcribe and translate the mutated sequence and demonstrate how it affects the protein. Answer the following in your logbook:

Some codons change the second or third base and still specify the same amino acid.

If the gene on the homologous chromosome is not mutated, the organism may still be able to make sufficient quantities of protein. As long as the incorrect protein does not interfere with other biochemical processes, the organism's phenotype may be unaffected.

- Does a mutation always change the order of amino acids within a protein?

- Explain why a mutation might not affect the phenotype of an organism (or create a monster), even when it changes the order of amino acids in a protein.

WHEN TO WORRY ABOUT MUTAGENS

A point mutation substitutes one base for another in a DNA sequence. As you saw in Activity 5-6, point mutations do not always cause problems. Other types of mutations usually have serious consequences. Suppose a base were deleted from or added to a DNA sequence (Figure 5.19). Every triplet after the addition or deletion would change. This type of mutation is called a frameshift; it throws off the translation sequence of the mRNA at the ribosome.

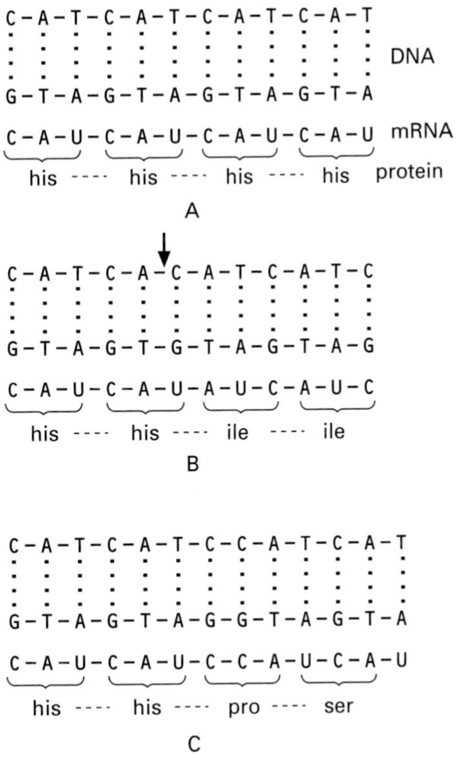

Figure 5.19: *Effect of a deletion or insertion in DNA sequence. (A) original sequence, (B) deletion, (C) insertion.*

LESSON 5.4 ASSESSMENT

1. The following 32 mRNA codons could withstand any change in the third base and still specify the intended amino acid: CUU, CUA, CUG, CUC (all leu); CCU, CCA, CCG, CCC (all pro); CGA, CGU, CGC, CGG (all arg); GUA, GUU, GUC, GUG (all val); GCA, GCU, GCC, GCG (all ala); GGA, GGU, GGC, GGG (all gly); ACA, ACU, ACC, ACG (all thr); UCA, UCU, UCC, UCG (all ser).

2. Mutations spread as a result of DNA replication and cell division. DNA replication copies the mutation, and cell division passes the mutation to the daughter cells.

3. During replication and transcription.

1 Using Table 5.1, determine how many of the 64 mRNA codons would still translate correctly after a mutation in their third base.

2 Explain how a mutation occurring in one cell of an organism can spread to other cells of the organism.

3 DNA is susceptible to mutation whenever the DNA helix is unzipped, exposing DNA nucleotides. During what two processes is DNA likely to mutate?

CAREER APPLICATIONS

The applications that follow are like the ones you will encounter in many workplaces. Use the biology you learned in this chapter to complete the activities. Share your work with the class.

AGRICULTURE & AGRIBUSINESS

Mapping Genomes

For this activity, work in pairs. Using the Internet and other resources such as a local agriculture extension service, college, or university, select a species of plant or animal whose genome is being sequenced. Try not to duplicate species within the class. Find out the following: Who is doing the work? What are the goals of the project? How long has it been under way? What percentage of the genome has been mapped? How is that information being used? Record your information in your logbook and share it with the class. Make a class data table of all species.

DNA Paw Prints

Investigate ways in which DNA technologies are used to improve breeding programs for pets, livestock, and exotic animals. Use the Internet; talk to local experts such as a veterinarian or an animal breeder (particularly breeders of exotic birds, reptiles, ostriches, emus, or zoo animals); or refer to breed journals. How are DNA technologies used for determining sex of some birds and reptiles, selecting breeding pairs, identifying individuals, tracing pedigrees, recovering stolen animals, or registering animals? What are the benefits? What are the costs? How are the techniques similar to those used in mapping a species' genome? How are they different? Write a short paper on your findings.

FAMILY & CONSUMER SCIENCE

Preventing Birth Defects

Call your local chapter of the March of Dimes or use the Internet to contact the organization. Find out the major causes of birth defects. What percentage of birth defects are caused by genetic mutations? Are the causes of the mutations known? How might they be prevented?

Carcinogens

Use the Internet or contact the National Toxicology Program to get a copy of a recent "Report on Carcinogens." Select from the lists of known carcinogens a chemical that you or your family might come into contact with in the course of normal events. What is the chemical used for? When is it considered carcinogenic? What precautions should be taken in using or coming into contact with the chemical? Make a short presentation of your findings to the class.

HEALTH CAREERS

Sensible Antisense

Use the Internet or other resources to research an antisense drug using both "antisense" and "drug" as keywords. What disease or condition does the drug target? How is the antisense drug introduced? What enzyme does it block? Has the drug been approved by the FDA? If not, at what stage of development is it? Make a class table of all drugs researched.

X-ray Careers

In groups, choose a health career that uses X rays for diagnostic or therapeutic purposes. Make sure each group picks a different career. Contact people in that career and find out what their job title is. How did they prepare for their career? What certifications or licenses do they have? How do they protect themselves and others from harmful effects of radiation? Each group should write up a career profile for a class career notebook.

DNA Technologies for Genetics Research

In Chapter 4 you learned about testing people for a genetic defect. Use the Internet or other medical or genetic reference materials or interview a cytogeneticist to get descriptions of various techniques that identify gene abnormalities. Some common methods are fluorescence in situ hybridization (FISH), Southern blot technique, and restriction fragment length polymorphisms (RFLP). Using what you know about how DNA replicates, describe how the methods work.

INDUSTRIAL TECHNOLOGY

Contact a forensics lab to find out more about DNA fingerprinting. Ask your local law-enforcement agency what lab it uses. What samples are suitable for DNA fingerprinting? When the lab has only a small sample, how does it get enough DNA for the test? What is the procedure for constructing a DNA fingerprint? Is there a central file of DNA fingerprints? Are there standardized enzymes for creating human DNA fingerprints, or does each lab choose its own?

Industrial Uses of Mutagens

Use the Internet to research mutagens, or call or write government agencies, such as the National Institute of Occupational Safety and Health (NIOSH), the Centers for Disease Control (CDC), and the Food and Drug Administration (FDA), that maintain information on known mutagens. Select a mutagen to research. What are the industrial uses of the mutagen you selected? At what concentrations does it become a problem? How can someone prevent harmful exposures to the mutagen at home or at work? Contact a local business or industry that might use the mutagen you selected or some other mutagen. How do they protect their employees?

X-ray Technology

Using texts, encyclopedias of science and technology, or Internet resources, research X-ray technology. Draw a diagram of how the technology works. Don't get bogged down in the calculations required; just explain the process. How is X-ray crystallography used? The Lawrence Livermore National Laboratory has information on its Web site about crystallography.

CHAPTER 5   SUMMARY

■ All of the DNA within one cell of an organism makes up a genome. The genomes of various species, including humans, are being sequenced. DNA sequencing determines the exact order of the subunits of DNA.

■ DNA is a polymer made up of monomers called nucleotides. A DNA nucleotide has a phosphate group, a deoxyribose sugar, and one of four different nitrogenous bases: adenine, thymine, cytosine, and guanine.

■ The two strands of a DNA helix are complementary; a base on one strand always pairs with a particular base on the other strand. A pairs with T, and C pairs with G.

■ The code in the sequence of bases in DNA is transcribed to a messenger (mRNA), which carries the transcribed code to the ribosome. There the code, composed of codons that have three bases, is translated into a sequence of amino acids that will form a protein.

■ Transcription occurs whenever the cell needs a certain protein. Protein production is controlled by a cell in a number of ways, most importantly by proteins that bind to DNA and mRNA.

■ DNA replicates in the S phase of the cell cycle. From one DNA molecule, replication produces two DNA molecules with identical base sequences.

■ Polymerase chain reaction (PCR) copies a fragment of DNA in the presence of the enzyme DNA polymerase. In a laboratory vessel it rapidly produces copies of DNA for various applications, including DNA fingerprinting.

■ Changes in the base sequence of DNA due to ionizing radiation and certain chemical reagents are mutations. The agents that cause mutations are mutagens, many of which are also involved in tumor formation.

■ Mutations are copied during DNA replication and passed on to daughter cells during cell division. A mutation in a gamete can be passed on to an organism's offspring.

CHAPTER 5 ASSESSMENT

Concept Review

See TRB, Chapter 5 for answers.

1 Law-enforcement agencies use several techniques to distinguish tissues from different individuals. Predict what problems a technician might have with samples collected at a crime scene. How could the technician eliminate the problems?

2 Give the sequence for the DNA strand from which the mRNA fragment UUUACAUGAUCCCCGC was transcribed.

3 What are some differences and similarities among the four monomers of a DNA molecule?

4 List the three forms of RNA that can be transcribed from DNA.

5 Give the complementary strand for the DNA sequence ATAAACGGGGCT.

6 Define PCR and identify the enzyme that this process uses.

7 What roles does DNA play in protein synthesis?

8 A sequence of DNA contains 100 base pairs. How many nucleotides are in this sequence?

9 The diploid number of chromosomes in a species is 8. How many chromosomes are contained in its genome?

10 An RNA strand contains uracil. What DNA base must be present in the DNA strand that produced the RNA?

11 Using Table 5.1, determine the amino acid sequence from the mRNA fragment AAAUCGCCCGCGUUUUAA.

12 Give the amino acid sequence for Problem 11 if a point mutation removed the first adenine (A).

13 Would the amino acid sequence in Problem 11 change if the third adenine (A) were replaced with guanine (G)? If so, what would be the new sequence?

14 Describe the role of tRNA in translation.

15 List the four components that PCR needs to copy DNA.

16 Give two biological roles of the DNA molecule.

17 Describe how one DNA molecule becomes two DNA molecules.

18 Describe the Watson and Crick model of DNA.

Think and Discuss

19 Errors sometimes occur during base pairing on a DNA template. Which would have more serious consequences, an error during DNA replication or an error during transcription? Explain.

20 Explain why point mutations usually have less serious consequences than other mutations.

21 What is a typical pattern for a group of codons that specify the same amino acid? (*Hint:* See Table 5.1.)

22 Look at the following example. Why is the coding strand of a DNA molecule almost identical to the RNA that the DNA molecule transcribes? How does the RNA strand differ from the DNA coding strand?

Example of DNA Strands

Coding   ATTTCGCGCGTAGAT
Antisense   TAAAGCGCGCATCTA
RNA strand AUUUCGCGCGUAGAU

In the Workplace

23 Why would it be useful for all DNA labs to use the same set of enzymes for making DNA fingerprints?

24 You are a dog breeder. One of your female dogs is accidentally bred to two different males. How can you determine which puppies were fathered by which dog so that the puppies can be registered?

25 You are working in a lab mapping a genome. On the basis of a protein's composition you have come up with a possible DNA sequence of a gene that might direct the building of that protein. How could you determine which chromosome carries the gene?

Investigations

The following question relates to Investigation 5A:

26 Outline a general procedure for isolating chromosomal DNA from cells.

The following questions relate to Investigation 5B:

27 You are testing a new chemical synthesized by Richmond Labs. After an Ames test, petri dishes with the new chemical have over twice as many bacterial colonies as the negative control petri dishes. In addition, the positive control dishes have twice as many colonies as the negative control dishes. On the basis of these observations, would you recommend that Richmond Labs be allowed to market this chemical as a new food preservative? Why or why not?

28 What would you conclude if the new chemical in Problem 27 produced less than twice the number of colonies in the negative control dish, and the bacterial lawn was absent?

CHAPTER 6

WHY SHOULD I LEARN THIS?

If you worked outdoors as a surveyor, you would need to watch out for ticks or mosquitoes that carry diseases. In the Northeast you had better know the difference between a dog tick and a deer tick, which might give you Lyme disease. Out West, look out for Rocky Mountain wood ticks, which can transmit a painful disease. Why are some species plentiful in the forests of the Northeast and others in forested mountains of the West? How did there get to be over 2 million species, anyway? In this chapter you will learn about how species develop and how they are classified.

Chapter Contents

EVOLUTION AND CLASSIFICATION

WHAT WILL I LEARN?

1. how species change and adapt to their environment
2. how new species arise
3. evidence and ideas about how life began
4. what various types of evidence suggest about the history of life
5. how to classify species

If you wonder why life has so many forms, each successful in its own way, you're not alone. Since the 1700s, philosophers and scientists have worked on theories about the development and classification of species, gathering evidence from living and fossil organisms. Some scientists have extended their research to look for how life first took shape. In this chapter you'll learn theories about the origin and diversity of life and the evidence for them. You'll also find out about many occupations in medicine, agriculture, and produce marketing that depend on using these theories.

LESSON 6.1 VARIATION IN POPULATIONS

How many kinds of dogs can you name? How many kinds of beans? Where did this diversity come from? Plant and animal breeders, depending on variation in a species, can choose the genes that will improve the plants and animals that give us food, clothing, wood, medicine, and companionship. A cattle breeder might select certain phenotypes to improve milk or meat production, to increase heat tolerance, or to add resistance to diseases. In this lesson you will learn about the genetic diversity within species and how the environment causes it to change.

ACTIVITY 6-1   OBSERVING GENETIC VARIATION

Purchase *Arabidopsis* or other seeds that have been subjected to various levels of ionizing radiation.

In science fiction, mutant potatoes produced by radiation wouldn't just be blight resistant. A huge dose of radiation would change them into bomb-resistant carnivores that eat human flesh! But that's fiction. In this activity you will see how radiation really affects a plant.

Plant an equal number of seeds of each radiation dose in a separate pot, and record this number in your logbook, along with the date. Keep the pots in a warm, sunny spot, and water them frequently. Keep the soil moist but not soggy.

Observe the *Arabidopsis* plants as they develop, and record your observations in your logbook every few days. Each group of seeds has been exposed to a different level of radiation. Record the number of surviving plants of each group, and describe their growth, including height, number of leaves and flowers, and any other information that seems important. After you have watched *Arabidopsis* plants grow for a few weeks, the following questions will help you to analyze your data. Record your answers in your logbook.

Increasing doses of radiation produce increasing numbers of mutations, eventually leading to reduced survival.

Natural radiation produces some of the genetic change that is responsible for the great variety among living things.

- Does radiation seem to change the way organisms develop? Report any changes.

- A small amount of radiation is always present. Would you expect to see at least a few altered individuals in a field of thousands of plants that had not been X-rayed?

- Could radiation cause some of the natural genetic variety that you see in living things?

OREGON REDS AND YUKON GOLDS

You don't have to visit a national park to see variety among organisms. Just visit the produce section of a supermarket and look at the potatoes. You're likely to find large brown russets, yellow-fleshed Yukon Golds, and smooth, red-skinned Oregon Reds.

The original potatoes grew in the Andes Mountains of Peru and were about the size of a small plum. During the sixteenth century, Spanish explorers introduced the food to Spain. From there it spread throughout Europe. During the seventeenth and eighteenth centuries, traders took potatoes to other parts of the world.

Because potatoes are cheap, easy to grow and store, nutritious, and filling, they became a staple food for poorer members of society. Unfortunately, the original Peruvian potatoes were often killed by late blight, a disease caused by a fungus. In the late 1840s, late blight devastated potato crops throughout Ireland, which led to a terrible famine. The natural genetic variation in the potato population produced some plants that resisted the fungus. From those plants, growers developed new blight-resistant potato varieties. By the early 1900s, potato varieties numbered in the thousands. Today, more than 5,000 varieties of potatoes exist.

SEARCHING FOR POTATOES IN A STUBBLE FIELD.

The Irish potato famine caused widespread poverty and starvation.

SOURCES OF VARIATION

An organism resembles its parents because it inherits their traits. A variation in an inherited trait makes one organism different from others of the same species. Fur texture in dogs is a trait that shows variation. Each puppy inherits genes that determine the texture phenotype (softness or coarseness) of its fur. An important source of variation is the genetic recombination that occurs during sexual reproduction. During meiosis and fertilization the genes of two parents recombine to produce a unique individual. That's why one litter can produce puppies with coarse fur and with soft fur. Sexual recombination of genes can produce new traits and thus variation.

Recombination also occurs during meiosis when pairs of chromosomes exchange matching pieces. This crossing over creates new combinations of alleles. For example, dark hair and brown eyes often occur together in humans, though different genes control the traits. Blond hair and blue eyes are another common combination of traits. But crossing over may cause the children of a person with blond hair and blue eyes and a person with dark hair and brown eyes to have blond hair and brown eyes. Most phenotypes result from the activity of many genes, so sexual recombination and crossing over can result in many unexpected new traits.

Another source of variation is mutation. Most mutations are random and can benefit or harm the individual that carries them. Some silent, or neutral, mutations have little or no apparent effect. The first blight-resistant potato was probably a variation due to a mutation. Small numbers of mutations happen all the time, often when DNA replicates incorrectly. Radiation, certain chemicals, and other stresses can make mutations much more frequent in any cell, including gametes. People who might someday have children should avoid exposure to high levels of radiation and mutagenic chemicals, so that they won't pass on harmful mutations.

These young trees are part of a population that includes all of the lodgepole pines in the area.

Migration can also increase or reduce variation in a population. A **population** is a group of individuals of the same species that live in one area and breed with others in the group. For example, the lodgepole pines of Yellowstone National Park are one population. Those in Yosemite are another population of the same species. In recent years the population of the endangered Florida panther has become so small that there is very little genetic variation among the animals. To try to save the species, wildlife managers have brought in mountain lions from Texas to breed with the panthers and increase the variation in their population. However this experiment turns out, the results will be important for tourists, park rangers, game wardens, and other wildlife workers, hunters, and Florida's commercial and sport fishing industries.

GENES IN A POPULATION

All the genes in a population at any one time make up the population's **gene pool.** Members of the next generation of a population will draw their genes from that gene pool.

The amount of individual variation in a population depends on the gene pool. Sexual recombination and crossing over alone cannot alter the gene pool. Unless other forces change the frequency of dominant and recessive alleles in a gene pool, it is constant from generation to generation. This constancy is the basis for the Hardy-Weinberg law, named for two scientists who studied the frequency of alleles in populations in the early 1900s.

Hardy-Weinberg Law

Allele frequencies in a population's gene pool remain the same from one generation to the next unless they are affected by

- mutations,

- migration into or out of the population,

- effects of genotype on survival,

- effects of genotype on mating choices, or

- effects of genotype on successful reproduction.

The Hardy-Weinberg law holds true only for large populations in which individuals mate randomly and in which nothing changes the proportions of alleles. This rarely happens in nature. In addition to the factors listed in the box above, the frequencies of alleles in small populations can easily be changed by random events such as natural disasters that kill large numbers of individuals.

The Hardy-Weinberg law is important in a number of occupations. For example, suppose a public health worker has been spraying an insecticide for several years in an area to control mosquitoes that carry a disease. After a few years of killing most of the mosquitoes in the area, the worker finds that some survive and that these resistant mosquitoes carry an allele for resistance to the insecticide. The worker can use the Hardy-Weinberg law to calculate the allele's frequency in the mosquito gene pool. An increase in that frequency is a warning to stop using the same insecticide and find another way to control the mosquitoes. This kind of study can also monitor resistance to antibiotics in disease-causing microorganisms, to insecticides in agricultural pests, and to weed killers in invasive plants.

INVESTIGATION 6A   CHANGING GENE POOL FREQUENCIES

See Investigation 6A in the TRB.

What happens to a species when individuals that carry certain alleles are more likely than others to survive? How does the population change? In this investigation you will use tokens to represent alleles as you simulate changes in a ragweed population in response to an herbicide.

LESSON 6.1 ASSESSMENT

1. Mutation usually increases variation by creating new alleles.

2. Migration can increase or reduce variation, depending on whether the migrating individuals carry rare or common alleles into or out of the population. Therefore blocking migration keeps it from changing the gene pool, bringing the population closer to Hardy-Weinberg equilibrium.

3. Variation increases the chances that a population will include individuals that can succeed in new or changed environments, so it is especially valuable in rapidly changing environments.

4. Environmental changes such as an increase in salt concentration often reduce variation by favoring the survival of part of the population, such as the more salt-tolerant fish, over the rest.

1 Does mutation increase or decrease the amount of variation in a population? Give a reason for your answer.

2 Suppose a new highway is built. Turtles can no longer migrate from one side of the road to the other. How would you expect this to affect the amount of genetic variation in the turtle population? Would it bring the turtles closer to the Hardy-Weinberg conditions or farther away? Explain your answer.

3 What advantages could variation give a population? Is variation more likely to be an advantage in an environment that experiences a great deal of change, such as a seashore, or in a relatively unchanging place, such as the ocean floor?

4 As salt that is used to melt ice in the winter washes off of roads, it often winds up in streams and lakes. How would a fisheries worker expect this to affect variation in the fish population? Explain the reasons for your answer.

LESSON 6.2 ADAPTING TO THE ENVIRONMENT

More than two million species live on Earth. Add up all the variation within each species, and you have a fantastically huge number of unique individuals. It's a good thing they don't all try to live in the same place. Consider the cat family, for example. People have carried domestic cats nearly everywhere on Earth, but relatives of cats have narrower ranges: The lion lives only in the grasslands of Africa, the snow leopard in central Asia, and other

members of the cat family in the rain forests of Latin America. Each species survives best in one environment. In this lesson you will learn how competition for resources and changes in gene pools combine to adapt species to their environments.

ACTIVITY 6-2 THE RIGHT TOOLS FOR THE JOB—SURVIVAL

Each student needs safety goggles, and each pair of students needs regular pliers, needle-nose pliers, uncooked rice, unshelled pecans or almonds, and a stopwatch or timer with a second hand.

The needle-nose pliers are better for picking up small grains.

The regular pliers grip large objects such as nuts better.

In general, a beak like the regular pliers would give a bird a better chance of survival in the forest

Few animals use tools, but their adapted body parts, such as beaks or tails, may function like tools. How do animals develop the right tools for survival? Scatter 30 grains of rice on a flat table or desk. Use needle-nose pliers to pick up as many grains of rice as you can in 20 seconds. Record your results in your logbook. Return the rice to the table. Use regular pliers to pick up as many grains as possible in 20 seconds. Record your results. Repeat the same procedures, using nuts instead of rice. Record your results.

Put on your goggles and try to use each type of pliers to break open a nut. Which type of pliers works better? Answer the following questions in your logbook:

- Which tool—the regular pliers or the needle-nose pliers—made it easier to pick up the rice? Why?

- Which tool made it easier to pick up and open the nuts? Why?

because it makes it easier for the bird to pick up and crack the larger nuts. The "needle-nose" beak would be better for a bird living in a grassland where the food consists of small seeds.

- Suppose the two pliers were two beaks. If the birds lived in a forest where the main source of food was large, hard nuts, which beak would give the bird a better chance of survival? In what kind of environment would "needle-nose" birds have a better chance of surviving?

THE ENGLISH PEPPERED MOTH

The English peppered moth has coloring that helps it escape predators. The peppered moth comes in two varieties. One is splotchy gray, and the other is dark (Figure 6.1). This moth spends most of its time resting on trees. Before the industrialization of England the light variety was more common. It blended in with the lichens growing on the trees, and predators had a difficult time finding it. (Lichens are mixtures of algae and fungi that grow on trees and rocks.) Dark moths were easily seen and eaten by birds, so few of them survived long enough to reproduce. In the mid-1800s the dark peppered moth was extremely rare.

Figure 6.1: *The English peppered moth may be light gray or dark.*

As England became more industrialized, air pollution killed off most of the lichens, uncovering the trees' dark bark. Eventually, predators could easily see the light peppered moth, and the dark moth was better camouflaged. As the pollution worsened, the dark moth became common, and the splotchy gray moth became rare.

THE STRUGGLE FOR SURVIVAL

In 1798 an English economist named Thomas Malthus wrote *An Essay on the Principle of Population.* He pointed out that people have the potential to produce far more offspring than the resources in the environment can possibly support. Malthus predicted that the human population would eventually cover the Earth, producing mass starvation.

Malthus's argument seems oversimplified today. Improvements in medicine, sanitation, and agriculture have greatly increased the ability of large human populations to survive. However, his main insight—that humans and other species have an unlimited potential for population growth—still holds true. Most organisms have the potential to produce many more offspring during their lifetime than the environment can support, but not all offspring survive and reproduce (Figure 6.2). The rest become another creature's food, starve, die from disease, or for some other reason do not reproduce. For example, a codfish might lay 1000 eggs. Of these, 900 might be eaten by other fish. Perhaps 20 of the 100 that hatch survive to adulthood. Of the survivors, maybe two female fish will mate and lay eggs. (These numbers are bad news for cod fishers, who depend on fish reproduction to replace their catch.)

Figure 6.2: *Sea turtles produce huge numbers of offspring, but only a few of these newly hatched babies will survive. Individuals with a genetic advantage are more likely to survive long enough to reproduce.*

But which two females will survive? Remember the variation that is common in most populations. Some individuals compete more successfully than other members of the population for available nutrients or space, or they may escape predators more easily. If only two of the 1000 original cod eggs produce female fish that survive long enough to reproduce, they will probably be two of the healthiest, fastest-growing, or best-camouflaged fish of the 1000.

NATURAL SELECTION

Organisms that are better adapted to the environment are likely to produce more offspring. In turn, these offspring will inherit their parents' alleles, so they are also likely to be well adapted to survive and reproduce in their environment. This process, in which better-adapted individuals survive longer and produce more offspring, is called **natural selection.** Natural selection affects inherited variation. Variations caused by accidents, such as injuries and the effects of weather or nutrition, affect survival and reproduction, but they are not inherited.

When the environment changes, what happens to organisms that were well adjusted to the old conditions? Look at the peppered moth. When conditions changed and darkened the trees, the light gray moths became more visible to their predators, and the number

Figure 6.3: *Charles Robert Darwin (1809–1882), author of* On the Origin of Species *and* The Voyage of The Beagle

of light moths in the population decreased. Changes in the environment can drive an allele—or even a whole species—into extinction. An explanation of natural selection was first published in 1858 by Charles Darwin (Figure 6.3) and his cousin Alfred Wallace. Darwin had collected and studied samples of a huge number of plant and animal species while exploring the world on a ship called the *Beagle*. He noted the tremendous diversity of organisms everywhere he went. He also noted that, although some organisms in different parts of the world closely resembled each other, each was uniquely adapted to its particular environment. Natural selection was his explanation for the biological diversity and similarity of living organisms.

Darwin and Wallace were familiar with artificial selection, used by farmers and other breeders to generate new varieties of plants and livestock (Figure 6.4). Breeders choose desirable traits in existing organisms and then breed only the organisms that have those traits. For example, artificial selection was used to breed blight-resistant potatoes. Farmers used the potatoes that resisted blight to create new varieties. Darwin suggested that nature does the same. His key idea was that organisms with beneficial variations tend to be more fit to survive in a particular environment and to produce more offspring than those with less fit variations. "Fitness" does not always mean strength, speed, or large size. It simply means the ability to survive long enough to produce offspring that can themselves reproduce.

Figure 6.4: *Artificial selection has steadily increased farm productivity for decades, while natural selection has been too slow to change wild trees within recorded human history.*

Natural selection changes a population much more slowly than artificial selection does. Breeders can often make dramatic improvements in a few years. If you look at a seed or bulb catalog, you will see the great variety of flower shapes and colors that a plant breeder can produce in a few years. In natural selection there is no breeder. Only natural variation and competition determine which individuals are fittest in that environment. Genetic variation may affect an individual's fitness only slightly. Thousands of generations may pass before the combined effects of many small genetic differences produce a noticeable change in a population. The pronghorn antelope that roam Wyoming today, for example, appear to be identical to their fossilized ancestors that lived millions of years ago.

Natural selection can change the frequencies of alleles in a population's gene pool. As individuals that carry beneficial alleles produce more offspring than other members of the population, a greater proportion of each successive generation will carry those alleles. Natural selection can also work against particular traits, so that a trait becomes less frequent, as you saw in the case of the peppered moth.

Natural selection does not always change a population's average phenotype, however. If the most typical individuals in a population are the most likely to survive and pass on their characteristics, natural selection may eliminate alleles that produce unusual or extreme phenotypes. Some species of frogs, insects, and plants seem hardly to have changed at all in millions of years. Natural selection may also maintain variety in a population; a trait that is harmful in one situation may be helpful in another. The great variety of wild dogs and grasses, for example, has helped these species to survive. This variety also enables breeders to develop new types that are suitable for a wide range of uses and environments.

Variety among dogs, as shown here, helps the species to survive.

LESSON 6.2 ASSESSMENT

1. A thinner, sharper beak might help the bird to pick small insects out of the tree bark. The brown feathers might help it to hide from predators. The advantages of these traits will make birds that have them more likely to survive and reproduce. This could increase the frequencies of the responsible alleles.

2. Tall trees shade their neighbors. Competition for light favors taller trees, so they are more likely to survive and reproduce.

3. No, traits that are acquired through exercise are not inherited.

4. Natural selection is much slower than artificial selection and may not produce a phenotype that is useful to farmers.

1 You're a park ranger studying variation in a bird that eats insects found in tree bark. Some birds have thin, sharp beaks; others have shorter, wider beaks. Some have brown feathers, and others have red and white feathers. Which of these variations, if any, would you expect to increase in the population in the future? How could this happen?

2 Describe how competition for light among trees in a forest could cause trees to become taller over many generations.

3 If you practice body-building and develop huge, strong muscles, will this cause you to have stronger children? Give a reason for your answer.

4 Why do farmers buy disease-resistant livestock and seeds produced by breeders instead of relying on natural selection to provide them with disease-resistant varieties?

LESSON 6.3 HOW SPECIES CHANGE OVER TIME

By now you can imagine how natural selection could have produced plants with broad, flat leaves to collect sunlight; predators with big, sharp teeth and claws; or turtles and clams with tough, protective shells. But how do new species come into being? How can so many species survive in the same environment? Why doesn't competition drive out or kill off all but one species—the most fit? Natural selection is only one part of the slow process of change in species. In this lesson you will learn more about how species are thought to change and some of the evidence that led to these ideas.

ACTIVITY 6-3   RANDOM SELECTION?

Provide each pair of students with 50 mixed red and white tokens in a paper cup. Use poker chips, bits of colored paper, or dry beans as tokens.

Natural selection is not the only way gene pools change. In this activity you will see a population change in a way that does not favor one genotype over another.

Put 50 tokens on the table. Each token represents a member of the same fish species. Color is an inherited characteristic among the fish in this pond. Red tokens represent red fish, and white tokens represent white fish. Record in your logbook the numbers and percentages of red and white fish in the population.

Put all 50 tokens in the cup. Imagine that a drought shrinks the pond and kills most of the fish. Without looking, take 10 tokens out of the cup and place them on the table to represent the survivors. Record the number and percentage of red and white fish among the survivors in your logbook, and answer the following questions:

The drought changed the frequencies of the alleles for red and white color.

Large populations are less susceptible than small ones to change from random events.

The effect of the drought on the fishes' gene pool was random.

Other such events might include fires, floods, and volcanic eruptions.

- How did the drought affect the fishes' gene pool?

- Suppose you had started with 500 fish. Do you think the drought would have had more or less effect on a larger population? Explain the reasons for your answer.

- Did competition or differences in the fitness of red and white fish have any effect on this change?

- What other natural events might have random effects on gene pools?

If natural selection leads to the elimination of harmful alleles, why do we still have deadly genetic diseases such as cystic fibrosis and sickle-cell anemia? Advocates of Darwinian medicine look for answers to questions like this.

Many of these mortal diseases are caused by recessive alleles. People who inherit a recessive allele from both parents may be sick or even die, but heterozygotes, who inherit a single copy of the allele, often have an advantage over the rest of the population. The allele for sickle-cell anemia, for example, is most common in the part of Africa where malaria is also found. Malaria is caused by a protist. The symptoms include high fevers, chills, and profuse sweating and may linger for years or cause death. People who have only one copy of the sickle-cell allele tend to be resistant to malaria, and if they contract malaria, they have a much higher survival rate. While having two copies of the allele is deadly, having one copy is a good thing when you live in an area where malaria is prevalent.

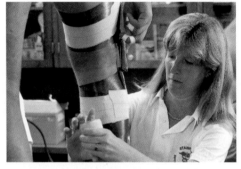

An athletic trainer bandages an injured leg. Knee injuries are a common problem for athletes.

Besides explaining the continued existence of alleles that cause disease, the Darwinian approach to medicine suggests possible explanations for other common diseases and conditions. For example, flat feet and backache may be common because humans haven't been walking upright long enough for our skeletons and muscles to adapt fully to the extra weight that our upright posture puts on our backs and feet. The human skeleton and muscles are adaptable to all sorts of tasks—at a price. Sprains, dislocations, torn ligaments, and hernias from athletic and workplace injuries are common. In contrast, dogs, members of a less adaptable species, seldom suffer sprains or hernias. Advocates of the Darwinian approach to medicine have also suggested that our big brains (and skulls) may be why giving birth is often more difficult for women than it is for females of other species.

What good is understanding the evolution of ailments if it doesn't offer a treatment or cure? In some cases, proponents of the Darwinian approach to medicine say, what we consider a problem is a defensive adaptation against a more harmful problem, like the sickle-cell protection against malaria. While many doctors question the usefulness of this approach, others believe that understanding the evolution of disease and response can lead to strategies for better health care.

EVOLUTION WITHOUT SELECTION

Populations can change when events such as natural disasters alter their gene pools in random ways. This type of change is called **genetic drift;** it may or may not affect the fitness of the population. Activity 6-3 showed that genetic drift is especially likely in small populations, because they are more sensitive to small random influences.

LEARNING LINK ISLAND SPECIES

An extreme form of genetic drift is the **founder effect,** which occurs when an isolated population forms from a few individuals. Random variation often causes the gene pool of this small group to differ from that of the original, large population. As a result, populations that developed from the few individuals who landed on remote islands, such as Hawaii and the Galapagos, are often quite different from their relatives on the mainland.

Use the library, the Internet, or other resources to learn about the animals of the Galapagos and Hawaiian Islands and their relatives in North and South America. Use pictures and the results of your research to explain to the class how the founder effect can help to explain the differences between an island species and its mainland relatives.

NONRANDOM MATING

Sometimes individuals choose or require mates of a particular phenotype. When mating is not random, the Hardy-Weinberg law is violated, and allele frequencies may change. For example, frogs that breed at various times of the year can mate only with individuals that have the same breeding season. This nonrandom mating can divide a species' gene pool among groups that breed separately. Eventually, the various breeding groups may become separate species (Figure 6.5).

Nonrandom mating and natural selection can change species gradually by slowly changing allele frequencies. Fossil evidence suggests that most species do not change very much between the time they first appear and the time they die out.

Figure 6.5: *Nonrandom mating reduces the exchange of alleles among wood frog populations. The stripes on this frog are different from those of its neighbors that breed at different times.*

LEARNING LINK COMPATIBLE MATES—FOR PLANTS!

Nonrandom mating occurs in plants, too. Some valuable vegetable and fruit crops cannot fertilize their own flowers. To produce seeds and edible fruits, they must exchange pollen with plants that carry a different "compatibility" allele. Use the library, the Internet, or an interview with an agricultural extension agent to learn about "self-incompatibility" and how it promotes genetic diversity. Find out which important crops have self-incompatibility genes. Report to the class on how these genes work and their importance to agricultural breeders and producers.

ACTIVITY 6-4   SEPARATING POPULATIONS

You've learned how species can change. But how does a new species begin? This activity looks at changes that push populations apart to become separate species.

Suppose a population of rabbits lives near a wide, deep river and the foxes there prey on them. Most of the rabbits are large and brown, but a few are smaller or of other colors. One day, an earthquake changes the course of the river. All of the foxes end up on one side of the river with about half of the rabbits. The rabbit populations on the two sides of the river can no longer interact in any way.

Imagine coming to the area 5,000 years later to observe the descendants of the original rabbits. As a class, discuss the following questions. Record your answers in your logbook.

One population is preyed on by foxes, and the other is not.

The groups of rabbits would no longer be exactly like each other or their ancestors. Without foxes a larger rabbit population and the resulting increase in competition for food and nesting sites might have selected for smaller body size, while mutation and genetic drift might have reduced the frequency of the protective brown fur trait. Less change would be expected among the rabbits that were still exposed to foxes.

If the populations do not interbreed well, they may become separate species. They may not be able to interbreed at all in another 5,000 years.

- How are the environments of the two rabbit populations different?

- Would both populations of rabbits be like their ancestors, or would they have changed? Describe possible changes in either population.

- Suppose you try to mate rabbits from the two sides of the river. You find that matings between rabbits from the two sides of the river produce fewer and less healthy offspring than matings between rabbits from the same side. What could cause this? Are the two populations still one species? How will the situation change in another 5,000 years?

HOW NEW SPECIES FORM

When a population is divided by geographic barriers such as rivers or mountains, or even highways or fences, the two populations can't interbreed. Climate barriers, such as different temperature zones on a mountain, can also isolate part of a population. Because the environment differs from place to place, natural selection may cause separate populations of a species to adapt in differing ways. Over time, the populations may become very different.

If the populations change so much that their members can no longer breed with each other, they become separate species. The process of forming a new species is called **speciation.** Species may grow very different from their ancestors and from other species that evolved from the same ancestor (Figure 6.6).

Speciation requires a **reproductive barrier** that prevents populations from mating with each other. Geographic barriers such as oceans and deserts are one kind of reproductive barrier. For populations that live in the same area, other barriers can promote speciation or maintain separate species. For example, horses and donkeys remain separate species because matings between them produce only mules, which cannot reproduce. Differences in mating times and behaviors, flowering seasons, or the types of insects that carry pollen from one flower to another keep mating within a species. Closely related species of dragonflies are mechanically incompatible: Their sex organs do not fit together.

Figure 6.6: *The Kaibab and Abert squirrels are closely related species but cannot reproduce together. The Grand Canyon keeps them apart.*

Natural selection favors speciation when members of a species compete intensely for the same food or other resources. When Darwin visited the Galapagos Islands, he found many small birds called finches. He was surprised to discover more than a dozen types of finches (Figure 6.7). Some had short, powerful beaks, which were good for cracking hard nuts and seeds. Others had long, narrow beaks and ate small seeds. The finches' beaks seemed to be adaptations that improved their chances of surviving in different island environments.

Figure 6.7: *Two or more species of finch may live in the same area. Their specializations allow them to avoid competing with each other for food.*

Two or more closely related species can survive in the same area if they do not need to compete for the same resources. In the case of the Galapagos finches, random mutations apparently produced birds that differed in preference for food, nesting sites, or other resources. Like a manufacturer that avoids competing with big auto makers by producing specialized vehicles such as golf carts or ambulances, the mutants survived by avoiding competition with the rest of the population. While these finches nested on the ground or hunted for seeds, most others nested in trees or ate nuts. Each type of finch tended to meet and mate with others of the same type. Eventually, the two types may have become so different that they were unable to interbreed. Once their gene pools divided in this way, they became separate species. This splitting of species, or **adaptive radiation,** benefits the whole population by reducing competition.

EXTINCTION OF SPECIES

Over long periods of time, some species become extinct, or permanently disappear, as a result of competition, climate change, or other factors. Extinction can also result from sudden events, such as floods, volcanic eruptions, or human interference, that quickly change the environment, possibly destroying food or nesting sites. For example, the dinosaurs may have died out because of a combination of a cooler climate and competition from smaller, faster-moving mammals and birds that could tolerate the cooler climate, run faster than the dinosaurs, and eat the dinosaurs' eggs and food.

The most common reason for extinction today is human intervention in habitats. We can destroy an environment or change it so quickly that natural selection can't keep up. A bird called the dodo once lived on Mauritius, an island in the Indian Ocean (Figure 6.8). There were no predators on the island, and over time the species lost its ability to fly. When humans colonized the island, they easily captured the flightless bird, as did their cats and dogs. In a few years the dodo disappeared forever.

Figure 6.8: *Dodo birds are now extinct.*

THE EVOLVING THEORY OF EVOLUTION

CHANGING IDEAS

Darwin's and Wallace's ideas about natural selection and evolution were revolutionary when they were first proposed. Before that, people thought that species do not change. It was widely believed that the familiar types of plants and animals had remained the same since they first existed.

During the 1700s and 1800s many people became interested in fossils, the preserved or mineralized remains or traces of organisms that lived long ago. The first person to propose a way in which living species could have developed from the ancient ones we know only as fossils was the French biologist Jean Baptiste de Lamarck. In the late 1700s he noticed that when he arranged fossils according to age, they showed patterns of gradual change in size and shape. Lamarck suggested that changes in organisms, acquired through the use or disuse of body structures, are passed on to their offspring. He suggested that animals' necks can grow longer from stretching to reach leaves high up on trees. The animal's offspring would then be born with longer necks. After many generations, Lamarck wrote, this process led to today's giraffe.

But observations do not support the idea that characteristics acquired by one generation can be inherited by the next. Lamarck's theory lost support and was rejected.

Darwin proposed natural selection as an alternative to the inheritance of acquired characteristics. Later discoveries in genetics, such as Mendel's work and the discovery of DNA, helped to explain how favorable traits are inherited. Experiments and debates continue as

This may be a good time to remind students of how the word "theory" is used in science: to describe a hypothesis that has been supported by numerous repeated observations. Many students take "theory" to be synonymous with "opinion." Explain that in science it is important to distinguish between theory (or interpretation) and observation, but this is not the same as the distinction between fact and opinion.

scientists try to understand better exactly how and at what rate evolution occurs and how much competition, environmental change, and other factors contribute to changes in species.

EVIDENCE FROM FOSSILS

A **B**

Figure 6.9: *Similarities and differences between fossils and living species provide evidence of the course of evolution.* Notungulata (A) was an early plant-eating mammal. The fern (B) lived about 300 million years ago.

Much of our knowledge of early life comes from fossils (Figure 6.9). Fossils usually are found in rock layers. Over time, new layers of rock form on top of old layers. Fossils that are found in older layers are older than those found in younger layers. By determining the age of various fossil specimens, paleontologists (scientists who study fossils) construct a rough timetable of life on earth.

Although the fossil record is not complete, many early life forms resemble modern species. This suggests that they may be the ancestors of species living today.

The fossil record also indicates that changes in life forms occurred over long periods of time. For example, fossil evidence shows that

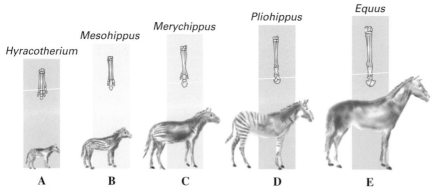

Figure 6.10: *The evolution of the horse: A. Hyracotherium (50 million years ago); B. Mesohippus (30 million years ago); C. Merychippus (20 million years ago); D. Pliohippus (5 million years ago); E. Equus, the modern horse. Each species did not just give rise to the next in a simple, linear series. There are several branches to this family tree that are not shown here. Note the reduction in the number of toes in the skeleton of the leg as shown above each animal.*

horses evolved over millions of years from animals the size of a small dog (Figure 6.10). Each species seems to have changed little during its existence, but more recent species are progressively larger and have necks and legs that are more like those of today's horses.

The fossil record also supports natural selection as an important force in changing species and promoting the development of new species from existing ones. For example, the horse evolved as it adapted to a progressively drier climate. Today's horses have hooves and long legs that enable them to move quickly on hard, dry land. Their speed and size offer protection from predators.

PALEONTOLOGIST

CAREER PROFILE

Paleontologists study the fossil remains of ancient organisms. After they dig up fossils or pry them from rocks, they spend a lot of time in the lab, comparing the specimens to others that have been identified.

Some paleontologists work for oil companies or mining companies. Their duties often include studying microscopic fossils in rock samples. Certain microfossils are often found in or near rock layers that contain oil or valuable minerals. Other paleontologists work at universities or museums or may be hired by builders or government agencies to determine whether fossils found at construction sites are important or new to science. If they are, the building project may be delayed while the fossils are removed and preserved for study. Most paleontologists have college degrees in biology, geology, or a related field.

EVIDENCE FROM BODY STRUCTURES

WORD BANK

homo- = same

ana- = again, repeated

-logo- = ratio, relationship

Homologous structures evolved from the same ancestral structure.

Analogous structures have similar functions but developed from different ancestral structures.

Comparisons of different species often reveal similarities in structure, even in structures with very different functions. What does your arm have in common with a bird's wing and a dolphin's flipper? They don't look the same or work the same, but they all have approximately the same number and types of bones and muscles, suggesting shared ancestry (Figure 6.11). Body parts that evolved from the same ancestral structure are **homologous.**

Before birth or hatching, some animals pass through stages during which they resemble ancestral species. The embryos of air-breathing vertebrates (animals with backbones), including humans, have slits in their necks that are homologous to the gills of fish.

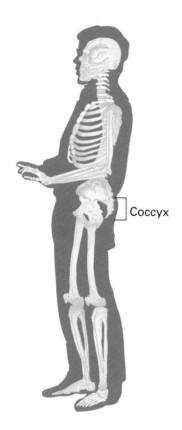

Coccyx

Sometimes, structures become alike as natural selection modifies different body parts to perform the same task. A bat's wing (Figure 6.11) is homologous to the human hand, not to our whole arm or to the wings of birds. The wings of bats, birds, and insects resemble each other, but each evolved from a different body part. Unrelated structures with the same function are **analogous.**

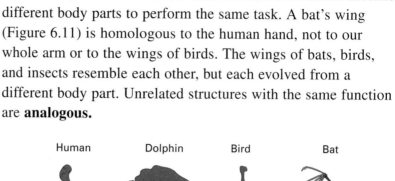

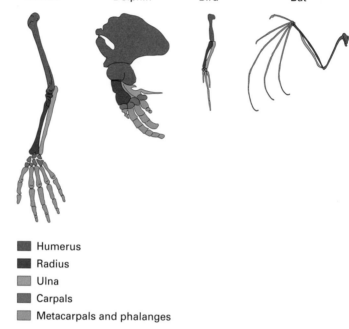

■ Humerus
■ Radius
■ Ulna
■ Carpals
■ Metacarpals and phalanges

Figure 6.11: *Homologous and analogous structures. Bones drawn in the same color are homologous. The bat's wing is analogous to the other limbs; it is not homologous to the bird's wing.*

Figure 6.12: *The human coccyx is a vestigial structure. It is homologous to the bones of the tail in other animals.*

Some body structures seem to have little or no function and are called **vestigial.** The human coccyx, or tailbone (Figure 6.12), is an example. The coccyx is homologous to the tails of other mammals. Homologous and vestigial structures are evidence that different species evolved from shared ancestors.

EVIDENCE FROM DNA

Individuals (and species) differ largely because of differences in their DNA. As species change over time, their genes also change. The longer two species have been separate, the more different their genes become.

We can use this difference to estimate how closely related two or more species are. Species with a recent common ancestor have nearly identical nucleotide sequences; species whose common ancestor was farther back in time have more sequence differences. Humans and chimpanzees, which are thought to be the closest relative of humans, have DNA that is about 99 percent identical. Human DNA and snake DNA are about 80 percent identical. Human DNA and bacterial DNA are even less alike. Using data obtained from DNA sequences, scientists produce evolutionary family trees, showing the genetic relationships among species (Figure 6.13).

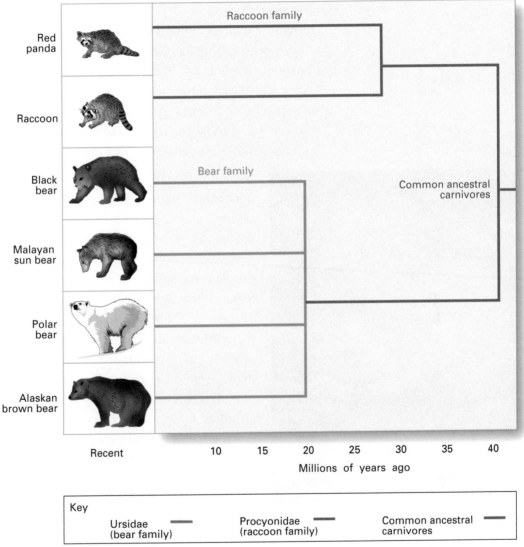

Figure 6.13: *An evolutionary family tree based on anatomical, fossil, and DNA evidence. All bears are thought to share a common ancestor; so are members of the raccoon family. A more ancient species was the ancestor of both families.*

Family trees drawn from DNA comparison are very similar to those drawn from fossil evidence and comparisons of living species. Close agreement of these three lines of evidence provides very strong support for the existence of evolution. Checking more than one line of evidence is good scientific practice and inspires confidence in a theory.

EVIDENCE FROM EXPERIMENTS AND FIELD OBSERVATIONS

Many scientists have tested parts of the theory of evolution by observing species in nature, such as the English peppered moth. Darwin's meetings with pigeon breeders showed him the power of artificial selection and helped to convince him of the possibility of natural selection. More recently, medical observations and experiments have shown that resistance to antibiotics evolves in disease-causing bacteria. Antibiotic-resistant bacteria and insects and weeds that resist pesticides usually appear a few years after introduction of a new antibiotic or a pesticide.

Experiments with swordtails have shown that mating preferences maintain the long-tail phenotype, even though it reduces survival.

Several lines of evidence support the idea that nonrandom mating changes allele frequencies. Human diseases caused by recessive alleles are more common in isolated island and mountain communities where marriages between relatives have been common. Most people carry about five recessive alleles that would be lethal if they were homozygous. Many societies discourage or forbid marriages between close relatives, partly because these marriages are more likely to produce children with defects. Plant breeders have also noticed that many crops grow poorly from seeds produced by crossing closely related plants. Male swordtail fish have large, showy tails that slow them down and make them easy prey. The long-tail allele survives, though, because female swordtails prefer to mate with large-tailed males. Experimenters removed the swords from some males and attached plastic swords to the tails of others. In the next generation, alleles carried by false-tailed males were more common in the population, showing that the females' mating preferences cause natural selection.

A current example of speciation has important practical consequences for landscapers and apple growers. A certain species of fly lays its eggs on apple and hawthorn trees. Flies that grow up from eggs laid on apples tend to lay their own eggs on apples, while those that grow up on hawthorn trees tend to lay their eggs on hawthorns. Matings between the two types are less common than they were in the past. The gene pool seems to be dividing in two. Apple growers have tried to control the flies by cutting down hawthorns near their orchards. They might want to reconsider this practice if hawthorn flies stop attacking apples.

LESSON 6.3 ASSESSMENT

1. Lack of change does not mean that selection is not occurring. In well-adapted populations, selection can reduce variation and prevent change in the typical phenotype.

2. No, speciation does not occur without a reproductive barrier. The goats are not likely to divide into two populations unless some remain in the valley and others stay in the mountains.

3. Because rats produce fewer offspring per year, mutations that produce resistance to rat poison occur less often than mutations that provide resistance in the other organisms. Rats' longer generations also slow down natural selection, compared with bacteria and insects.

4. None. Natural selection does not affect the frequency of mutation.

1 Some species have changed very little in millions of years. Does this mean that natural selection is not occurring in those species? Explain your answer.

2 Suppose a species of wild goat spends each summer grazing in the mountains and spends the winters down in a grassy valley. Would the goats be likely to divide into a mountain species and a valley species? Explain the reasons for your answer.

3 Bacteria and many insects have more generations and produce a lot more offspring in a year than rats do. Relate this difference to the fact that the same rat poisons work for many years but resistance to new antibiotics and insecticides often becomes common after a few years of use.

4 Suppose you raised and sold trees and shrubs to landscapers and home gardeners. What effect, if any, would natural selection have on the chance that a favorable mutation will occur in your tree population?

LESSON 6.4 THE ORIGIN OF LIFE

Where did the first species come from? Did all life on Earth descend from one species? Because life on Earth began so long ago, we will never be certain of the scientific answers to these questions. However, we can use fossils, rock formations, and the chemical reactions that are needed to form the basic materials of living things to distinguish likely ideas from unlikely ones about the early Earth and how life might have begun. In this lesson you will look at some theories that scientists consider the most likely.

ACTIVITY 6-5   IMAGINING BEGINNINGS

You've learned to recognize living things, and you've had some experience with the nonliving substances such as proteins and lipids that make up living cells. But can you imagine what the "almost-living" chemical systems that developed into the first living cells were like? Did they look like bacteria or viruses, or maybe like microscopic blobs of mayonnaise? Which could have come first: protein, DNA, or RNA?

Discuss the following questions with two other students. Then write in your logbook your ideas and a description of what you think "pre-cells" might have been like.

This is a speculative activity that is meant to raise issues and encourage students to think about the issues involved in the origin of life. Accept all reasonable answers, but remind students that they are only ideas. Evidence and logic will be needed to support them.

- Today's cells depend on DNA and RNA to direct protein synthesis and on proteins (enzymes) to control DNA and RNA synthesis. Some forms of RNA can also catalyze chemical reactions. Which of these three types of compounds would pre-cells have needed the most to survive and begin metabolizing? How can you imagine the other two forming later?

- Think about the functions of the cell membrane. Which probably came first, the cell membrane or metabolism? Explain the reasons for your answer.

- Were the first cells probably heterotrophs or autotrophs? Explain the reasons for your answer.

- How big do you think the first organisms were? Explain the reasons for your answer.

- In your logbook, draw and label diagrams of what you think early pre-cells and cells might have looked like. Add a paragraph describing your ideas about what each was made of, what it could do, and how it survived.

THE YOUNG EARTH

CHANGING IDEAS

How old is the Earth? Some early scientific attempts to find this out relied on measurements of the saltiness of sea water or the thickness of sea-floor sediment. But many factors affect the rate at which salt and sediment wash into the sea and the rate at which chemical reactions remove salt from sea water. By using measurements of the high temperatures in deep mines and by considering how quickly rocks cool, Lord Kelvin concluded in the late 1800s that the Earth had been cooling since it formed 20 to 40 million years ago. But he did not know that radioactive decay and friction from the movement of rock within the Earth also generate heat. Since 1905, measurements of radioactive decay in ancient rocks have led to the conclusion that the Earth is about 4 billion years old.

Figure 6.14: *What the early Earth might have looked like*

The Earth and the other planets may have formed from material pulled from the Sun by the gravity of a passing star. Hot molten rock cooled to form a crust surrounded by an atmosphere that was mostly hydrogen. Volcanoes added water vapor and other gases to the atmosphere. There was probably little, if any, oxygen gas in the atmosphere at first. The first seas formed from rains that began when water in the atmosphere condensed (Figure 6.14). Lightning, volcanoes, and intense ultraviolet radiation from the Sun bombarded the young Earth. Fossil evidence indicates that life began on this harsh and barren planet 3.5 to 3.9 billion years ago.

THE PRIMORDIAL SOUP

All organisms are built from the same organic molecules, including amino acids, fatty acids, and carbohydrates. Living cells synthesize these substances, but what created these chemical building blocks before cells existed?

In the 1920s a Russian scientist named A. I. Oparin proposed an explanation for the formation of the first organic molecules on Earth. He suggested that when inorganic molecules in the oceans were heated by the Sun and electrified by lightning, chemical reactions produced simple organic molecules such as amino acids. As the compounds accumulated, the oceans became a vast soup of these molecules. This theory is called the primordial soup theory. (*Primordial* means "first" or "primitive.") Oparin suggested that the first cells formed from these simple organic compounds.

In 1953, two American scientists, Stanley L. Miller and Harold C. Urey, tested Oparin's hypothesis in a laboratory. They made a chamber containing water vapor, methane (CH_4), hydrogen, and ammonia (NH_3), the gases that Oparin thought were present in the early atmosphere. They heated the mixture and added electric sparks to simulate lightning (Figure 6.15). After a few days, Miller found new chemicals in the chamber, including a variety of amino acids and other organic molecules. The experiment demonstrated that some of the basic chemicals of life could have formed spontaneously on the early Earth. Others have since repeated the Miller experiment with different gas mixtures, replacing the ammonia and methane, for example, with CO_2 and nitrogen gas (N_2). They obtained similar results as long as they included little or no oxygen gas. Other experiments have produced simple sugars, lipids, and adenine, one of the bases in DNA, from simple substances in the absence of living cells.

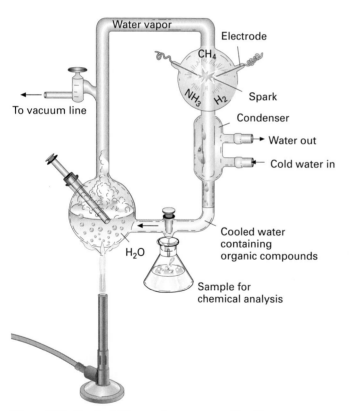

Figure 6.15: *Stanley Miller used an apparatus like this one to expose a mixture of gases to heat and electric sparks.*

FROM SIMPLE MOLECULES TO COMPLEX MOLECULES

Once amino acids began to accumulate in the oceans, they could join to form proteins. Scientists have discovered a few ways in which this process could happen. Amino acids could have washed onto shoreline rocks heated by volcanoes and the Sun. Under hot, dry conditions, amino acids link together, forming long, proteinlike polymers. Or perhaps amino acids joined spontaneously in the water or atmosphere. Similar reactions may have formed other polymers, such as complex carbohydrates, lipids, and nucleic acids.

NASA scientists have applied these ideas to their work. They study the origin of life so that they can predict where else life might exist in the universe. They have found that some clay minerals catalyze the formation of polymers of amino acids or nucleic acids. Perhaps clays helped to create these molecules on the young Earth.

FROM COMPLEX MOLECULES TO CELLS

Several hypotheses suggest how organic molecules evolved into self-replicating cells. All suggest that polymers formed from amino acids and nucleic acids combined to form structures of increasing complexity. These assemblies eventually became part of even more complex organized structures that could carry out the processes of life: the first cells.

Some RNA molecules can act as catalysts, much as enzymes do. The "RNA world" hypothesis suggests that in the primitive ancestors of the first true cells, short, self-duplicating chains of RNA acted as both catalysts (as enzymes do) and information storage (as DNA does). In other words, RNA may have done the work we commonly associate with DNA and proteins. DNA and protein synthesis would have developed later.

A second hypothesis is based on the observation that short chains of amino acids assemble into tiny droplets of proteinlike substances when mixed with cool water. These tiny drops are called microspheres. Microspheres exhibit some of the properties of life, including a selectively permeable boundary like a cell membrane. Microspheres may have had a role in the development of the first cells.

Like microspheres, other primitive cell-like structures have been created in the laboratory. Liposomes can form spontaneously in mixtures of water and certain lipids. Oparin shook up mixtures of proteins, carbohydrates, and nucleic acids to form microscopic structures that he called coacervates. Like microspheres, liposomes and coacervates are selectively permeable. Yet another possibility is that water droplets condensed on dust particles in the atmosphere. Warmed by the Sun, these floating droplets could have been sites of chemical reactions that led to life.

INVESTIGATION 6B — MAKING MICROSPHERES AND COACERVATES

See Investigation 6B in the TRB.

Microspheres and coacervates are a long way from living cells, but they can give you an idea of what "pre-cells" might have been like. In this investigation you will make microspheres and coacervates and observe some of their properties. Like the liposomes you learned about in Chapter 2, collections of enzymes enclosed in these "artificial cells" might be used to convert chemicals into useful medicines, to break down toxic substances, or to change simple materials into useful foods or fibers. What uses can you think of for them?

EARLY BIOLOGICAL EVOLUTION

Because the Earth's early atmosphere probably contained very little oxygen, the first living organisms were almost certainly heterotrophic, anaerobic bacteria. Natural selection among heterotrophs competing for food is thought to have produced autotrophs such as cyanobacteria (and eventually plants). These organisms started using carbon dioxide and water for photosynthesis, giving off oxygen gas (O_2) as a waste product. Atmospheric O_2 dramatically changed life on Earth. Until then, most organisms lived under water, where they were protected from harmful ultraviolet radiation. Ozone (O_3), formed in the upper atmosphere from O_2, screened out some radiation and made life on land safer. Aerobic respiration became possible. More complex life forms such as eukaryotes and, later, multicellular plants and animals began to evolve.

Lesson 6.4 Assessment

1. One can't be sure of a theory of the origin of life, because it occurred long ago and cannot be observed.

2. These structures cannot defend themselves, reproduce, or collect materials and energy to support their own maintenance and growth.

3. As the heterotroph population increased, food would have become scarce, and competition would have increased. Autotrophs would have been favored by natural selection because they do not compete for food.

4. This is extremely unlikely. The beads cannot maintain themselves, reproduce, or carry out any life functions. They can only catalyze one reaction.

1 How is studying the origin of life different from studying most other scientific questions?

2 What important characteristics of living cells are missing in microspheres and coacervates?

3 Describe how competition among heterotrophs for food could have led to the evolution of autotrophs.

4 The food-processing and pharmaceutical industries sometimes use coacervatelike "artificial cells" as catalysts. A solution flows through a layer of gel beads that contain an enzyme. Substances in the solution undergo chemical reactions as they flow through the beads. How likely is it that these "cells" would develop into real living cells? Explain the reasons for your answer.

LESSON 6.5 FIVE KINGDOMS

You've discovered how great wild berries taste, and you want to collect some to sell to restaurants. But you need to be sure you get the edible ones. A mistake could kill someone. Common names aren't safe to use, because a name such as "bush berry" can mean different species in different places. You'll have to classify them with scientific names, only one per species. When you're sure, you can give them to others without endangering anyone's life. This lesson explores the usefulness of classification in cooking, gardening, and disease control.

ACTIVITY 6-6 CLASSIFYING FAMILIAR OBJECTS

Provide a collection of related and unrelated familiar objects, such as garden tools, construction tools, lab equipment, art supplies, office supplies, sports equipment, and articles of clothing. You may wish to have students write a classification system for objects in the lab.

You use a classification system when you put your clean clothes or groceries away, when you arrange books on a library shelf, and when you organize art supplies or sports equipment. In this activity you take a close look at a skill you already have.

With a partner, set up a system for classifying familiar objects according to their function. Tools, for example, might be classified in these categories: for measurement, for the garden, for a science lab, and for artists. State the criteria you use in your logbook, and answer the following questions:

A hierarchy of subcategories increases precision, and each item can be classified uniquely. Responses will vary, depending on the objects that students classify.

- How could you further classify items within each category?

- If an item seemed to fit in more than one category, how did you decide where to classify it?

- What advantages are there to subdividing each category?

USING CLASSIFICATION TO HELP STOP EPIDEMICS

BIOLOGY IN CONTEXT

Some diseases caused by bacteria or other microscopic organisms are transmitted by ticks, fleas, and other blood-sucking animals. Mosquitoes transmit malaria, yellow fever, and encephalitis. Ticks transmit Rocky Mountain spotted fever, Lyme disease, and another kind of encephalitis. Most disease-causing organisms infect only a few species. Epidemiologists have to recognize these species if they want to control the spread of disease. Epidemiologists are

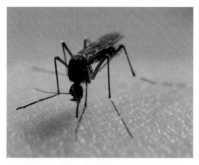

Different species of mosquitoes carry different species of parasites.

specialists who determine what causes a disease, how it spreads, where it is, and how to control it.

Two species of mosquitoes can carry malaria. One species breeds in marshes. The other prefers to breed along river banks. A public health team that failed to recognize the river-breeding species would waste their efforts on the marsh.

In the United States, five or six species of ticks carry Lyme disease bacteria. However, only two of them normally feed on people: the deer tick in the Northeast and the Pacific black-legged tick in the Northwest.

In epidemiology, classification is critical for identifying what is making people sick, controlling carriers that transmit the disease to humans or animals, and developing treatment strategies.

CLASSIFYING ORGANISMS BY KINGDOMS

As recently as the 1960s, most scientists still classified all organisms as either plant or animal. By then, though, it was becoming obvious to many biologists that some organisms do not fit either category. For example, bread mold is not a plant, because it does not have chloroplasts, plantlike cell walls, or plant organs such as roots, stems, and leaves, and it doesn't resemble animals. Today, most biologists classify all living things into five kingdoms. As we learn more, this scheme may change. The kingdoms are as follows:

- Monera: bacteria (prokaryotes)
- Protista: single-celled and other simple eukaryotes
- Fungi: molds, yeasts, and mushrooms
- Plantae: plants
- Animalia: animals

Table 6.1 summarizes the characteristics of the five kingdoms. One way to classify an organism is to answer questions, narrowing your search down to one item in a table. To classify an organism into a kingdom, you answer these questions: Does it have one cell or more? Do the cells have nuclei or are they prokaryotes? Do they have cell walls? Is it a heterotroph or an autotroph, perhaps with chloroplasts? If you're looking at a single-celled eukaryote, it's probably in Protista. If it doesn't have a nucleus, it's a prokaryote and therefore in Monera. Cell walls would exclude an organism from Animalia, while chloroplasts make it a plant or an autotrophic protist. (Remember from Chapter 2 that autotrophic bacteria do not have chloroplasts.)

Another important question for kingdom-level classification is how the organism develops. Monerans, fungi, and protists are mature as soon as they form by cell division, though fungi and a few bacteria and protists form special reproductive structures. Most plants and

	Monera	Protista	Fungi	Plantae	Animalia
Cell type	Prokaryote with cell wall	Eukatyote, some with cell walls	Eukaryote with cell wall	Eukaryote with cell wall	Eukaryote
Number of cells	Most are single-celled	Most are single-celled	Most are multicellular	Multicellular	Multicellular
Nutrition	Heterotroph or autotroph	Heterotroph or autotroph	Heterotroph	Autotroph	Heterotroph
Development	Complete after cell division	Complete after cell division	Form special reproductive structures	Develop from embryos	Develop from embryos
Examples	Clostridium botulinum and Spirillum volutans	Paramecium and Euglenoid	Amethyst Deceiver and Blood Red Cort Cortinarius Mushrooms	Woodland plants	Blue Morpho butterfly and human

Table 6.1: Characteristics of the Five Kingdoms

animals, however, develop from **embryos.** An embryo is a small mass of unspecialized cells, descended from a single cell, that grows and develops into an adult organism (Figure 6.16).

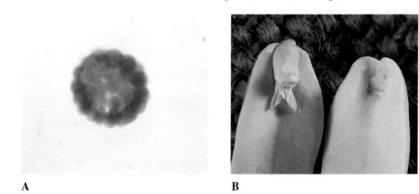

A   B   C

Figure 6.16: *Plant and animal embryos. (A) Early stage sea star embryo. (B) The embryo in a peanut seed and its stored food reserves. The future stem, root, and first two leaves are visible. (C) Frog embryos developing in water.*

FROM ARISTOTLE TO WHITTAKER

CHANGING IDEAS

The ancient Greek philosopher Aristotle classified all organisms as either plant or animal. Aristotle did not think of living things as members of separate species. He wrote, for example, that sheep could be born from melons and worms from dirt. Such beliefs remained common through the Middle Ages.

Carolus Linnaeus, a Swedish biologist of the 1700s, proposed a new system of classification. He divided all organisms into two kingdoms: plant and animal. The kingdoms were subdivided into phyla (plural of phylum). Each phylum was subdivided into classes. Further subdivisions produced a total of seven classification levels, with species as the most basic. Linnaeus grouped species into categories called genera (plural of genus). A genus is a group of similar species. Organisms take their scientific names from their genus and species. For example, the species name for dogs is *Canis familiaris;* their genus is *Canis.* Genus names are always capitalized.

As Darwin's ideas about the origin of today's species became more accepted, classification no longer simply described the "order of nature." Instead, closely related species were thought to have developed from shared ancestors, and classification became a way to describe hypotheses about the origin of groups of species. For example, the black bear (*Ursus americanus*) and the grizzly bear (*Ursus horribilis*) are in the same genus (Figure 6.13 on page 211). The raccoon (*Procyon lotor*) belongs to a different genus in the same class (Figure 6.13). By classifying these animals in this way, biologists express their idea that the ancestors of raccoons separated from the ancestors of bears before the two species of bears separated.

New tools continue to change our ideas about classification. The microscope showed biologists that single-celled organisms are as different as plants and animals. A separate kingdom was set aside for these organisms. By the late 1960s a five-kingdom system, proposed by the American biologist Robert H. Whittaker, was widely accepted. Since then, however, comparisons of the DNA sequences of various species have convinced many scientists that the bacteria called Archaea are so different from other Monerans that they should be a sixth kingdom. The Archaea live in very salty or boiling hot and acidic environments that have no oxygen. Some experts on the Archaea think that these organisms are so unusual that the classification system should include a superkingdom level above the kingdom level. The Archaea would be in one superkingdom, Monera in another, and the four eukaryotic kingdoms in a third.

WORD BANK

archae- = ancient

The Archaea are an ancient group of prokaryotes.

CLASSIFICATION WITHIN THE FIVE KINGDOMS

The classification system is nested like a truckload of eggs. The eggs are grouped by carton, then by box, then by delivery site, and then by truckload: A series of smaller groups are lumped together in larger ones. Scientific classification groups species within a genus, genera within families, and so on. The kingdom is the largest group to which an organism belongs, unless you accept the superkingdom proposal.

Figure 6.17 shows the six major categories below the kingdom level. All of the organisms in the figure are members of the animal kingdom. This kingdom has 18 important phyla. All organisms in the figure except for the butterfly are members of the phylum Chordata. Chordata is divided into nine classes. All the members of Chordata in the figure, except for the bird, are members of the

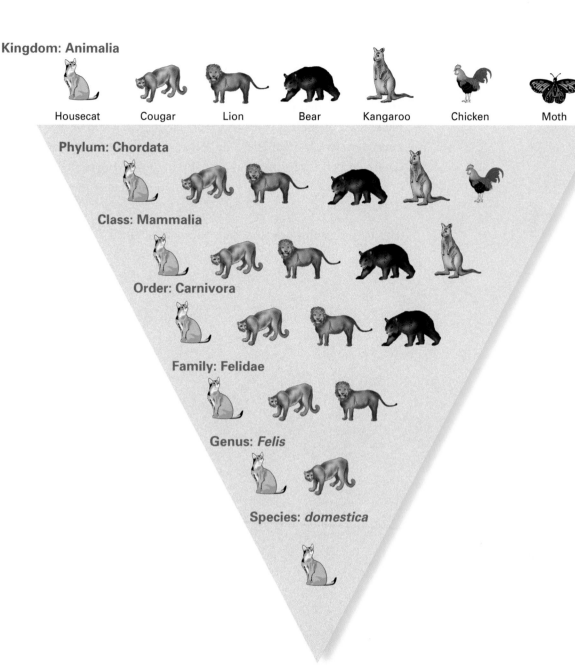

Figure 6.17: *Classification of the house cat,* Felis domestica

class Mammalia. Classes are divided into orders, and orders are divided into families. A family usually has several genera, and each genus includes at least one species and sometimes dozens of species. Remember that a species is defined as a group of organisms that can mate together to produce fertile offspring.

All members of Felidae, the cat family, evolved from a single ancient species, as far as we know. Likewise, each kingdom had a single, extremely early ancestral species. The kingdoms, in turn, are all descended from a single original cell. What evidence is this idea based on? All species share the same genetic code, and many genes, enzymes, organelles, and other chemical and structural details of cells are similar in all five kingdoms. Figure 6.18 shows one version of relations among the kingdoms and some of their important phyla.

The major groups are summarized in Appendix 1.

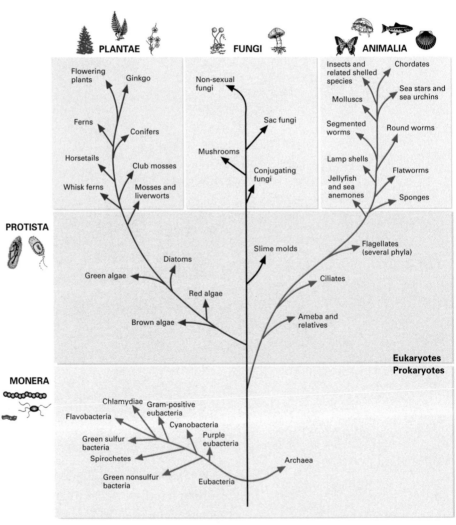

Figure 6.18: *One version of the evolution of the five kingdoms. The names are phyla. Note that the Protista are thought to be the ancestors of all other eukaryotes.*

PROBLEMS IN CLASSIFYING SPECIES

Debates arise over the classification of groups whose origin is unclear. For example, some biologists have suggested that insects, now one class, include three groups of families, each with a different ancestor, and should be three classes.

Some species do not fit all the classification rules. Most protists are unicellular, for example, but kelp and other seaweeds are multicellular. The kingdom Fungi also has mostly multicellular organisms but includes some unicellular species, such as yeasts. The five-kingdom system will change as more evolutionary, genetic, and biochemical information becomes available.

Just defining a species can be as difficult as classifying it. In Chapter 2, you saw yeast reproduce asexually by budding. Many other organisms also reproduce without sex: Monera, most Protista, and some members of the other kingdoms. So the usual definition of species (a population whose members can breed to produce fertile offspring) makes no sense for these organisms. In these cases, biologists rely mostly on similarities in structure and metabolism to identify species. Some populations of sexual organisms are also difficult to divide into species. For example, four types of deer mice live in the Rocky Mountains. All four are now classified as a single species. Two types, however, can't breed with each other, although both can breed with the other two populations. These fuzzy distinctions can make decisions very difficult for people whose jobs involve protection of endangered species or control of agricultural pests.

Banana plants reproduce without sex. How many species of bananas are there? How would you know?

LESSON 6.5 ASSESSMENT

1. No, animals are multicellular.

2. Answers may vary. Classification is often used to identify disease-causing organisms.

3. Aristotle thought that members of existing species could arise from other species or from nonliving materials.

4. No. Since all members of a genus have the same ancestors, a species' descendants can't move to a different genus.

1 Could there ever be a one-celled animal? Give a reason for your answer.

2 How might health care workers use classification of organisms?

3 What is a weakness of Aristotle's classification system?

4 If the Rocky Mountain deer mice become separate species, is it possible that the species will not be in the same genus? Give a reason for your answer.

CAREER APPLICATIONS

The applications that follow are like the ones you will encounter in many workplaces. Use the biology you learned in this chapter to complete the activities. Share your work with the class.

AGRICULTURE & AGRIBUSINESS

Classification on the Job

Using a variety of occupational resources, identify three jobs within agriculture, forestry, or environmental sciences that involve the classification of organisms. Find someone who holds one of these jobs and talk to that person. How much time do people with that occupation spend identifying organisms? To what level do they identify the organism? Do they use an identification key? What are distinguishing characteristics for one or more of the organisms they work with? What harm is done if they identify the organism incorrectly?

Preserving Cheetah Populations

Cheetahs are an endangered species; in other words, the species is threatened with extinction. Individuals in the current cheetah population are descendants of only a few individuals, and the gene pool for the population is limited. Contact a local zoo or conservation group. Find out how the cheetah's limited gene pool has affected its chances for survival and what efforts are being made to save the species. Present your findings to the class.

Doubling Up—Instant Species

In some plant species, chromosomes occasionally fail to separate during cell division, producing plants with twice the usual number of chromosomes. This offspring and plants with the usual chromosomes can't fertilize each other's flowers, so the plant becomes the first member of a new species. Important crops that probably began in this way include wheat, oats, cotton, strawberries, and potatoes. Research chromosome doubling and find out how this happens. Then interview a plant breeder (by phone, e-mail, or letter if necessary) to find out how breeding these species is different from breeding diploid species.

The Business of Protecting Species

Since the 1980s many companies have formed to help fund efforts to protect endangered species. Typically, these companies provide real goods and services but donate some of their profits to environmental groups. Are these companies just cashing in on the public's interest in endangered species? Or do they actually fund activities that help to protect the animals? Check out some of these companies on the Internet. Try the following search words: endangered species, conservation, and wildlife preservation. Research companies that regularly help to protect endangered species. Make a list of the companies, the work they support, and the percentage of their profits that they donate. Give your opinion of the effectiveness of these companies' environmental efforts.

Genetic Variation in the Marketplace

Specific varieties of many crops are grown for different uses and markets. Peppers, for example, include edible spicy and sweet types, varieties that are ground to be sold as paprika, and types that are grown for ornamental plants. There are also sour cherries for baking and sweet ones for eating, and a wide variety of apples, oranges, tomatoes, corn, wheat, and potatoes. Choose a crop and investigate the differences among its varieties. Find out how many varieties are available. Design a marketing brochure for a broker or wholesaler that describes the varieties available, the recommended uses for each, and any other information that you think the customer should have.

Related Produce

Visit the produce section of your local store. Pick a fruit or vegetable. How many varieties of that fruit or vegetable are available? Write their names in your logbook. Look around for other fruits or vegetables that you think are related to your original choice. Record their names also. Using reference books at your library or the Internet, find the scientific name for each of the fruits or vegetables you recorded. How closely related are your choices?

Indoor Plantscaping

Choose a room in your home or school, or use a room plan from a magazine. Determine the light, temperature, and humidity conditions for this room. Develop a plan for incorporating different types of plants into the room. Be sure to avoid selecting poisonous plants and any species to which people may be allergic. In what area(s) of the world do you think plants would have evolved that will thrive in your growing conditions? As you select plants for the room, try to find out where they originated. Were your ideas correct? Outline a plan for the care of the plants. Consider choosing plants that will naturally thrive in the room and modifying the environment with growing lights or by frequent watering to allow plants that evolved in other environments to survive.

HEALTH CAREERS

Antibiotic Resistance

Antibiotics are drugs that are used to fight bacteria that cause diseases in humans and other animals. Some types of bacteria now resist the antibiotics that once killed them. Find out about antibiotic resistance from your state or local health department, the library, or sites on the World Wide Web for the World Health Organization, the American Society of Microbiology, or the Centers for Disease Control and Prevention. Find out how antibiotics work, why antibiotic resistance occurs, and which antibiotics are being resisted. What new strategies are being used to treat diseases caused by antibiotic-resistant bacteria? A good place to start is by looking at the strategies that are used to treat leprosy or nosocomial infections—those that people acquire in hospitals.

INDUSTRIAL TECHNOLOGY

Dating a Fossil

Scientists are using two new techniques—thermoluminescence and electron spin resonance—to determine the age of fossils. Find out how these two techniques work and for which fossils they're most useful.

Fishing Machines

Investigate the mechanization of fishing. Find out how the ability to catch very large numbers of fish at a time has depleted fish populations and affected their ability to reproduce. How might natural selection in response to this technology affect the fish species involved?

CHAPTER 6 SUMMARY

- Populations are genetically diverse. Sources of genetic variation include mutation, sexual recombination, and crossing over.

- Allele frequencies can change as a result of migration, mutation, nonrandom mating, different effects of the environment on different genotypes, or effects of genotype on survival and reproduction.

- Competition among genetically diverse members of a population leads to natural selection. Alleles that increase fitness—the ability to produce fertile, surviving offspring—tend to increase in frequency, while those that reduce it become rare or disappear.

- Evolution can occur not only through natural selection, but also through genetic drift, founder effects, and nonrandom mating.

- When two populations of a species are separated by a reproductive barrier, differences in their gene pools or environments may cause them to become separate species.

- Adaptive radiation often leads to the development of many species from one ancestral species when competition for some resource is intense.

- The occurrence of evolution is supported by evidence from fossils, DNA sequences, comparisons of homologous structures in related species, and observations of populations in the wild and in controlled experiments.

- The origin of life is not well understood. The action of lightning, sunlight, and volcanic heat on simple substances in the atmosphere and oceans probably led to the formation of the more complex chemicals that make up living cells. Photosynthesis produced oxygen gas, which makes aerobic respiration possible, and the ozone layer that protects organisms on land from the Sun's ultraviolet radiation.

■ Species are divided into five kingdoms: Monera, Protista, Fungi, Plantae, and Animalia. Some biologists now divide bacteria into two kingdoms: Monera and Archaea.

■ Each kingdom is divided into six levels of classification, from phylum down to species.

■ The use of the classification system has changed since Linnaeus' time. It now represents the relationships among species in terms of their shared ancestry.

CHAPTER 6 ASSESSMENT

Concept Review

See TRB, Chapter 6 for answers.

1 Explain why some people who are vaccinated against a disease can still get it.

2 Why do plant breeders sometimes treat seeds with chemicals or radiation that cause mutation?

3 Why is it easier to produce a new variety of a sexually reproducing vegetable, such as a melon, than an asexual species, such as bananas?

4 Can fishing prevent the gene pool of a fish population from following the Hardy-Weinberg law? Give reasons for your answer.

5 Use natural selection to explain why some species of tree-dwelling monkeys can use their tails to grasp and hang onto things.

6 If the climate in an area becomes drier over a short period of time, will the next generation of trees there be able to tolerate dry conditions better than the previous one could?

7 Would a recessive lethal allele disappear from a gene pool faster than a dominant one? Give reasons for your answer.

8 Do new species arise mostly by separation of a population from the rest of an existing species or by one entire species gradually changing into another?

9 If you wanted to breed high-jumping dogs, how would Lamarck and Darwin advise you differently? What problem is there with Lamarck's approach?

10 Flower petals may have evolved from leaves that became specialized in shape and color. Would it be correct to say that leaves and petals are homologous?

11 Would it be correct to say that, just as complex chemicals formed from simple chemicals on the young Earth, more complex organisms evolved from simpler organisms?

12 Would an increase in the food supply make a species of animals more or less likely to undergo adaptive radiation?

13 All birds are grouped in one class. Does this mean that all bird species are thought to be descendants of one ancestral species?

14 Which are more closely related: members of the same phylum or members of the same genus?

Think and Discuss

15 In what ways do you think humans will be different in another 100,000 years? Use what you have learned in this chapter to justify and explain your ideas.

16 Which lived more recently: the first species of tree or the first species of pine tree?

17 Many species of rodents (rats, mice, gophers, and related animals) live in the southwestern United States, each with its own species of flea. Why is this considered an example of adaptive radiation among fleas?

18 Why is it unlikely that photosynthetic autotrophs arose before heterotrophs?

In the Workplace

19 Suppose you sprayed your garden with an insecticide to control caterpillars that were eating your tomatoes. If a few caterpillars survive, does this necessarily mean that you missed them? Give a reason for your answer.

20 Breeders want to increase the quantity of fruit produced in their orchards. They could artificially select the trees that had the highest rates of photosynthesis or those that converted the most energy captured through photosynthesis into fruit production rather than into growth of woody parts. Which strategy would lead to more fruit? Explain your answer in terms of natural and artificial selection.

21 Does using antibiotics increase the chance that bacteria will mutate to become antibiotic resistant? Give a reason for your answer.

Investigations

The following questions relate to Investigation 6A:

22 Over time, many dinosaur species increased in size. Is it possible that natural selection could have caused some species to become smaller at the same time? Explain.

23 Why might seed companies need to continue artificial selection when growing plants of an old variety?

The following questions relate to Investigation 6B:

24 Describe how coacervates and microspheres resemble living cells. What are some important ways that they differ from living cells?

25 Are microspheres most likely to form in the deep sea, on dry land, or on the shore of a volcanic island?

CHAPTER 7

WHY SHOULD I LEARN THIS?

Bacteria and viruses—Are they heroes or villains? Outside of a comic book, it's not always easy to separate the good guys from the bad. Yes, there are viruses that make us sick. But what about the gene therapy that viruses make possible? Sure, some bacteria spoil food. But what about the even larger heaps of garbage we would have without bacteria? You'll soon see that sometimes the most feared traits of bacteria and viruses are the same traits that make them useful to humans.

BACTERIA AND VIRUSES

WHAT WILL I LEARN?

1. the general characteristics of bacteria and viruses
2. the importance of the natural roles of bacteria and viruses
3. how we use bacteria and viruses to treat diseases, improve agricultural production, and produce chemicals
4. how infectious diseases are transmitted, prevented, and controlled
5. ways to guard against infections caused by bacteria and viruses

Did you know that microorganisms are used to treat raw sewage? If you went to work at a sewage plant, would you know how to keep microorganisms alive, control their growth, relocate them, and keep them growing free of other organisms? How would

you know whether you were culturing the right microorganism? People who have this knowledge study microbiology, paying special attention to bacteria and viruses. In this chapter you will learn some basics to prepare you to work with these organisms or at least to appreciate what they do for us—and against us.

LESSON 7.1 BACTERIA

Consider the sandwich you had for lunch yesterday. Do you owe its great taste to the chef who prepared it and to the farmers and ranchers who grew its ingredients? Well, first consider the bacteria that gave flavor to that piece of cheese or enriched the soil where the tomatoes and lettuce grew. This lesson tells you how tremendously useful bacteria can be and how they are becoming more useful as genetics labs redesign them to eat oil spills, dissolve grease, and help to pump oil from wells.

FERMENTATION IN FOOD PROCESSING

Do you enjoy sourdough pizzas with extra cheese? Hot dogs with pickle relish or sauerkraut? Chinese food with soy sauce? Without the help of bacteria these foods would be pretty tasteless. They're created by fermentation: biochemical reactions expertly carried out by bacteria. Without fermentation we wouldn't have sourdough bread, cheese, yogurt, pickles, sauerkraut, soy sauce, or vinegar, among other things.

Some unfermented foods also owe their taste, texture, or appearance to bacteria. Bacteria produce aspartame, a sugar substitute, and monosodium glutamate, a flavoring agent. "Friendly" microorganisms are sprayed onto produce on their way to market; their growth crowds out other microorganisms that spoil food.

In the very near future, bacteria will process food waste. Bioreactors stocked with bacteria will accept whey, waste sugars and starch, bottling and candy wastes, molasses, and cellulose wastes. The bioreactors will produce useful substances, such as ethanol, which the U.S. Department of Energy thinks could completely replace imported oil as an energy source. And our landfills would fill up less quickly without about 60,000,000 mL of food waste each year.

Bacteria may be relatively simple organisms, but many industries value them highly for their important metabolic processes.

Without using bacteria, foods such as bread and cheese would not taste very good.

ACTIVITY 7-1   SAMPLING BACTERIA IN FOOD

Bacteria occur naturally in some foods, and food processors may add them to others. The bacteria carry out biochemical reactions that either preserve the food or add to its flavor or texture. You will sample bacteria in foods, describe the growing conditions, and relate these conditions to the metabolism of the bacteria.

Using a pipette, transfer some liquid from the food to a clean microscope slide. Add a coverslip. Mount the slide on the stage of your microscope and observe first at low power, then at the highest power (either 400× or an oil-immersion lens). In your logbook, describe the bacterial cells that you see. Note their shape and any external and internal features.

Using a clean pipette, transfer a small drop of liquid from the food to a strip of pH paper. Record the pH of the liquid in your logbook. Read the container for the food and list the main ingredients (carbohydrates, protein, etc.). Bacteria may be using some of these ingredients as sources of energy or nutrients. Also list any preservatives, including salt. Determine the storage temperature of the food.

Swap slides with other groups, examine the bacteria, and record the pH of other foods. Answer the following in your logbook:

- Describe the environment for the bacteria that you observed. Were these bacteria alive? How do you know?

- What features of the food result from bacterial metabolism?

STRUCTURE

Like all living things, bacteria are composed of cells (Figure 7.1). As a member of the kingdom Monera, a bacterium is just a single cell, though such cells often associate in colonies. A colony of bacteria

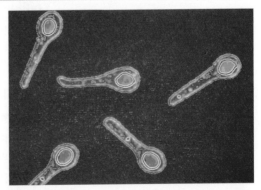

Figure 7.1: *Electron micrographs of bacterial cells*

(Figure 7.2) has billions of cells, each living independently and carrying out its own set of biochemical reactions. These reactions are guided by chemical instructions in a DNA blueprint that is present in a single chromosome and a few smaller circles of genetic material called plasmids.

Figure 7.2: *Tetanus and polluted H₂O*

INVESTIGATION 7A   MAKING YOGURT

See Investigation 7A in the TRB.

Students inoculate milk using store-bought yogurt as a starter culture. They make plain yogurt, add mashed fresh fruit to it, and freeze it. Mashing can be done in a self-sealing plastic bag while students are waiting for yogurt to thicken.

Humans have been using the biochemical abilities of bacteria and other microorganisms for centuries, sometimes without knowing it. Various strains of bacteria have been used to turn milk into buttermilk, cheese, and yogurt. You will investigate how a group of rod-shaped bacteria called lactobacilli help to make yogurt. The chemical reactions of these bacteria in milk are an example of fermentation.

CELL ORGANIZATION

From what you have learned so far, bacteria look pretty primitive. But they have managed, as a group, to survive for more than 3 billion years. If you had a powerful microscope that could show you great detail, you would see that prokaryotes are quite complex. In fact, some of this complexity helps a microbiologist to distinguish different species of bacteria.

A microbiologist can separate bacteria into broad groups by knowing just a few things about their cell structure. This level of information is often enough to tell a doctor what organism is causing an infection or a water-quality technician whether certain disease-causing bacteria are in a pond.

Most bacterial cells are surrounded by a cell wall. A cell wall supports the bacterial cell and protects it from losing its contents. The cell wall of most bacteria is one of the two types shown in Figure 7.3. Both types contain a rigid layer of **peptidoglycan,** made of amino acid chains and other polymers based on sugar subunits.

You can distinguish these two cell wall types with a technique called the Gram stain. Cell walls that hold a violet stain after washing in alcohol are Gram positive. Those that lose the stain and take up a pink counterstain are Gram negative. (Gram-positive bacteria retain the violet stain because their thicker peptidoglycan layer is not affected by alcohol. Gram-negative bacteria have a thinner layer that allows the stain to leak out.)

The Gram stain is one of the first tests that microbiologists run to identify a bacterium. The treatment of bacterial infections with antibiotics depends on whether the target bacteria are Gram positive or Gram negative.

Bacterial cell walls also have surface substances that help bacteria adhere to one another and to foreign objects. Washing with soap and water often helps to disrupt this attachment. A bacterial cell may also have a **capsule,** a protective layer outside the cell wall. A capsule keeps the cell from losing water and protects it from a host's defenses. Bacteria that cause serious infections and resist certain antibiotics often have capsules.

A

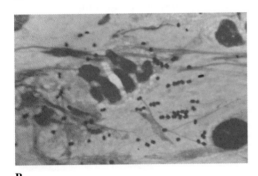

B

Figure 7.3: *Gram stain is one of many techniques used to classify bacteria. (A) Gram-positive stain of bacteria. (B) Gram-negative stain of bacteria.*

Some species of bacteria protect themselves against unfavorable conditions by transforming their cell into a spore when food is scarce or when they begin to dry out (Figure 7.4). The spore has thick walls that protect the cell. The spore is highly resistant to heat, a feature that makes it difficult to sterilize contaminated objects. Bacteria can survive as spores for up to an hour in boiling water. The only sure way to destroy bacterial spores is

Cell   Spore

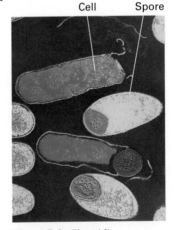

Figure 7.4: *Clostridium Botulinum, the bacteria that causes botulism, survives the acidic conditions of the stomach by transforming into a spore.*

the use of an autoclave (Figure 7.5). An autoclave makes pressurized steam, at temperatures above the boiling point of water, to kill microorganisms. People who use a pressure cooker in canning vegetables also use pressurized steam to kill microorganisms.

Figure 7.5: *Autoclaves use extremely high temperature to kill microorganisms that contaminate laboratory equipment.*

SHAPES AND CELL ARRANGEMENTS

The shapes of bacterial cells are used to classify bacteria. A bacterium may be rod-shaped, spherical, or spiral-shaped (Figure 7.6). A rod-shaped bacterium is referred to as a **bacillus.** A spherical bacterium is called a **coccus.** A spiral-shaped bacterium is a **spirillum,** if it is rigid, or a **spirochete,** if it is slender and more flexible.

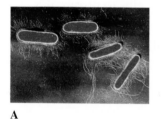
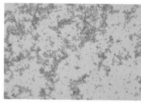
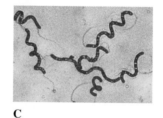

A   B   C   D

Figure 7.6: *There are only four possible shapes for bacterial cells: (A) bacillus, (B) coccus, (C) spirillum, and (D) spirochete.*

Bacterial cells are also classified by the groups that their cells form. Cells that form pairs (Figure 7.7A) have the prefix *diplo-* before their shape. Examples are diplobacilli and diplococci. Cells that form long chains (Figure 7.7B) have the prefix *strepto-* before their shape. Examples are streptobacilli and streptococci. Cells that exist in three-dimensional clusters (Figure 7.7C) have the prefix *staphylo-* before their shape. Examples are staphylobacilli and staphylococci.

If you know how a bacterium stains, what shape it is, and how the cells group, you have narrowed the identification of the bacterium to one of a few families.

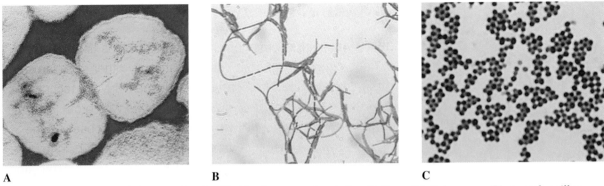

A   **B**   **C**

Figure 7.7: *Each bacterial shape can be classified by the groups that their cells form. (A) diplococcus, (B) streptobaccillus (2000×), and (C) staphlococcus (800×).*

MICROBIOLOGY TECHNICIAN

CAREER PROFILE

Peter J. is a microbiology technician for a company that grows microorganisms in containers the size of trash cans. A culture of microorganisms this size can produce a considerable amount of proteins or other cell products. Sewage treatment plants, concrete-cleaning companies, and septic tank owners buy the products of these cultures.

Peter's company works closely with customers to determine where the product will be used—for instance, in a pond or in the soil. Peter makes follow-up visits to check the product's performance. He explains, "I work mostly in a laboratory, but the environment is really the laboratory—it's filled with microorganisms."

When we asked whether he cultures a single species of bacterium in each vat, Peter answered, "No, we grow mixtures of bacteria as would be found in nature. It takes more than one kind of bacterium to clean up an oil spill or decompose sewage." What Peter feeds the bacteria and how their reproduction is controlled are trade secrets that were developed by a bacteriologist and an oceanographer.

Peter has an associate's degree in applied science from a community college. During his two-year program he studied life sciences, chemistry, physics, mathematics, and computer science.

ACTIVITY 7-2   PRACTICING ASEPTIC TECHNIQUE

Obtain a culture of nonpathogenic eubacteria in screw-cap tubes. Do not send students out to capture wild bacteria around the school. Demonstrate how to transfer bacteria from the broth in these tubes onto petri dishes.

Procedures for this activity can be found in Chapter 7 of the TRB.

WORD BANK

aero- = air

-sol = fluid

An aerosol is a suspension of liquid droplets in air.

In their work with microorganisms, lab technicians use procedures that keep cultures free from contamination by other organisms and protect the technician from infection. Microbiology lab technicians often culture bacteria in petri dishes or in tubes with broth. After allowing a short time for growth, the lab technician observes the petri dishes and tubes for certain growth patterns that help to identify bacteria. Petri dishes and tubes may become contaminated, whenever lids or caps are removed, by microorganisms that are carried on droplets of moisture and dust in the air. These droplets in the air, called aerosols, should be avoided. In this activity you will use aseptic technique to transfer harmless bacteria from a culture containing a liquid medium onto petri dishes containing a solid medium called agar. The agar contains nutrients the bacteria need to grow and reproduce.

INVESTIGATION 7B   GRAM STAINING BACTERIA

See Investigation 7B in the TRB.

Take a look at the plates you incubated in Activity 7-2. Do you see any evidence of microorganism growth on the plates? The agar that is used on these plates contains nutrients that favor growth of a wide variety of bacteria. If bacteria are growing, you can begin to identify them with a Gram stain.

PROKARYOTIC CELLS REVISITED

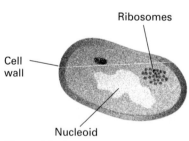

Figure 7.8: *Bacteria lack nuclei and organelles surrounded by membranes.*

Bacteria have much less detail inside their cells than do eukaryotes, yet their cytoplasm is a flurry of biochemical activity (Figure 7.8). Transcription of mRNA takes place from the DNA on the single, large bacterial chromosome near the center of the cytoplasm. Proteins are manufactured on hundreds of ribosomes, smaller than those of eukaryotes. Although bacteria have no organelles surrounded by membranes, such as chloroplasts and mitochondria, they carry out reactions such as photosynthesis and cellular respiration on free membranes in the cytoplasm or on folded regions of the cell membrane (see Figure 7.9).

Bacteria have a number of structures that move them around in their environment. Many bacteria use flagella, threadlike

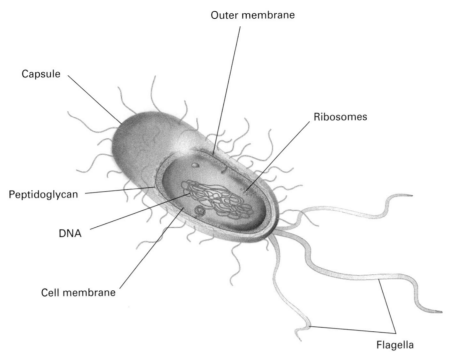

Capsule

Outer membrane

Ribosomes

Peptidoglycan

DNA

Cell membrane

Flagella

Figure 7.9: *Within the cytoplasm of prokaryotic cells, ribosomes and other free membranes are used to carry out reactions such as protein synthesis.*

appendages that extend from the cell wall and move like a boat propeller. Other bacteria move by means of gliding or spinning movements that usually involve special filaments attached to the cell membrane. Bacteria respond to chemicals in their environment, moving toward chemicals such as sugars and amino acids and away from toxic substances.

ENVIRONMENTAL TOLERANCE

Bacteria grow in a wide range of environments. As a group, bacteria can grow at a pH between 0.5 and 11.5, at salt concentrations ranging from zero to almost 350 parts per thousand, and at temperatures ranging from −10° to 110°C (14°F–230°F). Individual species, of course, tolerate narrower ranges. A primitive group of prokaryotes known as the Archaea inhabit the most extreme conditions on Earth. Some live in hot springs, with water near the boiling point. Others live at pH values as low as 1. Still others live in waters that are up to 10 times saltier than seawater.

In recently proposed classification schemes, Archaea are placed in a kingdom separate from other prokaryotes.

As you know, organisms carry out cellular respiration to gain energy from food. Aerobic organisms rely on oxygen in cellular

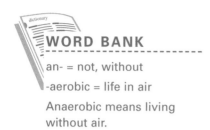
respiration to harvest lots of ATP. Some bacteria, however, either don't need oxygen (although they may grow better with it) or can't tolerate oxygen at all. Both types are examples of anaerobes. Anaerobic bacteria live in oxygen-deprived environments such as the ocean bottom, sulfur-rich marshes, sewage treatment tanks, the digestive tracts of animals, or nodules on the roots of certain plants.

Reproduction for most bacteria is a simple process. Because bacteria have just one chromosome, they don't need complex processes that eukaryotes such as mitosis and meiosis use to transmit DNA to their offspring. Most bacteria simply duplicate their chromosome and split in half (Figure 7.10).

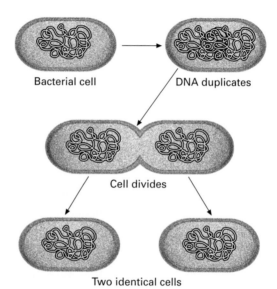

Figure 7.10: *Method of cell division within a bacterium*

ECOLOGICAL ROLES OF BACTERIA

Because bacteria as a group are adapted to so many different environments, they fill a wide variety of ecological roles. For instance some bacteria are important autotrophs. Some species photosynthesize with energy from the sun; these photoautotrophs are important producers of organic material in oceans, lakes, and streams. For several decades, scientists have been trying to carry out photosynthesis using inanimate materials such as clays. Bacterial photoautotrophs have served as models for this research because of their simplicity. One possible use of such artificial photosynthesis, if it works, is to produce energy for power plants.

Figure 7.11: *Archaea bacteria are capable of surviving extreme environments, such as the hydrothermal vent shown here.*

Other bacteria, called chemoautotrophs, get energy directly from inorganic materials. In the process, they change inorganic compounds from one form to another, releasing minerals that are essential to other species. Abundant chemoautotrophs are the Archaea, some of which extract energy from sulfur compounds that pour out of the Earth's crust during volcanic activity on the ocean floor (Figure 7.11). The Earth's crust harbors an immeasurable number of chemoautotrophs that live on the inorganic material in rocks. Some scientists have estimated that the amount of biological activity within the earth's crust is many times the amount of activity of all above-ground species.

LEARNING LINK EXTREME ENVIRONMENTS

Thermal vent ecosystems have warm temperatures and are rich in dissolved minerals needed by chemoautotrophic bacteria.

Research the location of hydrothermal vent ecosystems and the environmental conditions found there that promote the growth of bacteria. Internet sites that can be used are Ocean Planet and Life in Extreme Environments.

Bacteria are also extremely abundant in soil. A single gram of fertile soil may contain more than a million bacterial cells. Soil bacteria decompose organic material (plant debris, animal wastes, carcasses of animals) which then enters the soil. During this process, these decomposers release and recycle minerals needed for plant growth.

A smaller number of bacteria carry out important chemical reactions that convert nitrogen in the atmosphere into inorganic compounds such as ammonia. Farmers make use of some of these bacteria when they plant legumes such as alfalfa and rye. Legumes have nodules on their roots that harbor bacteria that are called nitrogen fixers because they bind nitrogen atoms into compounds (such as ammonia) that can be converted to other forms that plants use (Figure 7.12). Farmers also inject a type of nitrogen-fixing bacteria called *Rhizobium* directly into the soil. Photosynthetic prokaryotes called cyanobacteria are part of a group of nitrogen fixers that live freely in the soil or

Figure 7.12: *Root nodules of a rosary*

water rather than inside plant roots. They're important to soil fertility in areas where legumes don't grow well.

Many types of bacteria live in close association with a eukaryotic species. Such a relationship between two dissimilar species is called symbiosis. Some types live in the stomachs of animals that eat only grasses and other plants. There, the bacteria digest cellulose, which makes up the cell wall of grasses, into energy-rich sugars. As a result of this symbiosis, cattle and other grazers can use cellulose as a source of energy, even though they lack the enzymes to digest cellulose. Some animal species, including humans, have symbiotic bacteria in their digestive tracts that supplement the host's diet with nutrients such as fatty acids, amino acids, and vitamins.

Figure 7.13: *Termites are able to use the cellulose found in wood as an energy source because they harbor symbiotic microorganisms in their guts.*

The relationship mentioned above between nitrogen-fixing bacteria and legumes is another example of symbiosis. The bacteria provide nitrogen compounds for the plant; the plant provides nutrients for the bacteria. Symbiosis in which both species get some benefit is a relationship known as **mutualism** (Figure 7.13).

A small number of bacteria that participate in symbiosis damage their host. This type of symbiosis is known as **parasitism.** Most parasites don't kill their hosts but coexist as a population that the host tolerates and passes on to other hosts. You'll learn more about parasites in Lesson 7.4.

LESSON 7.1 ASSESSMENT

1. The cell wall is a rigid layer of peptidoglycan, made up of amino acid chains and other polymers based on sugar subunits.

2. The bacterial culture might have died, or it could have changed to produce different and undesirable by-products.

3. Heat, such as autoclaving, or a chemical disinfectant.

4. The type of cell wall, the shape of the cells, and their arrangement (single, pairs, chains, clusters).

1 Describe the bacterial cell wall structure that a researcher has to be aware of when developing an effective antibiotic.

2 What might poor flavor and texture indicate about the bacteria in a fermented food product?

3 Describe two ways in which a technician can sterilize a lab instrument to ensure that it is bacteria-free.

4 You are examining Gram stained bacteria under a microscope. What can you tell about the bacteria?

LESSON 7.2 VIRUSES

Because certain species of bacteria cause nasty infections such as Lyme disease and tuberculosis, bacteria in general have a bad reputation that they don't deserve. That same statement can't be made for viruses. The great majority of viruses are trouble for whatever organisms they infect. The reason for this involves the lifestyle of viruses, especially the way they reproduce.

MONONUCLEOSIS

It's time for final exams, and all over the country, high school and college students are staying up late talking on the phone, studying, and eating junk food. Many of these students will be visiting Health Services or their doctor before they even finish their exams. The major complaints will be headaches, fatigue, muscle aches, rashes, and fever. Most will arrive thinking that they've caught a cold or are just stressed out from exams, only to find out that they have been infected with Epstein-Barr virus (EBV).

EBV doesn't often make the news, but you've probably heard about the disease it causes: infectious mononucleosis, or "mono." Mono is called "the kissing disease" because people most often catch it by kissing someone who has it.

You're more likely to get mono when you lower your body's defenses by staying up late; being overstressed from balancing home, school, and social activities; and not practicing good hygiene. Good hygiene includes frequent hand washing and not sharing drinks, eating utensils, toothbrushes, and the like. Once you're infected, you can't take a pill to kill the virus. You'll recover in two to six weeks, however, by getting plenty of bed rest and drinking lots of fluids. This will give your body a chance to build up its defenses and fight off the virus.

ACTIVITY 7-3   VIRUS ATTACK

Viruses can't live and reproduce on their own. They require a host cell. In this activity you will grow a culture of host cells and then add a virus to determine the effect on the host.

Supply each lab group with a sterile tube of nutrient broth. Make available to all groups two or three cultures of nonpathogenic *Escherichia coli* and two or three phage cultures. The phages that you obtain should be specific to the *E. coli* strain. Keep all cultures in the refrigerator before the start of the lab.

Both experimental (student) and control cultures should appear cloudy because of the growth of *E. coli*.

The experimental culture turns clear because the phages have attacked and lysed (destroyed) the *E. coli* cells. The control culture remains cloudy, indicating viable *E. coli* cells.

The activity of the virus accounts for the clearing in the experimental culture.

Obtain a sterile tube of nutrient broth. Using the aseptic technique that you practiced in Activity 7-2, do the following: Using an inoculating loop, remove a small bit of the *Escherichia coli* culture from the agar slant. Put it into a sterile tube of nutrient broth. Incubate the tube for 24 hours at a temperature of 37°C (98.6°F). After 24 hours, observe the culture and describe it in your logbook. Also describe the control culture.

Using an inoculating loop, remove one drop from near the bottom of the virus culture and put it into your growing *E. coli* culture.

Observe the culture after 1–2 hours and compare it to the control culture. Write descriptions of both cultures in your logbook.

- What are some factors that may account for the differences you observed in the two cultures?

STRUCTURE

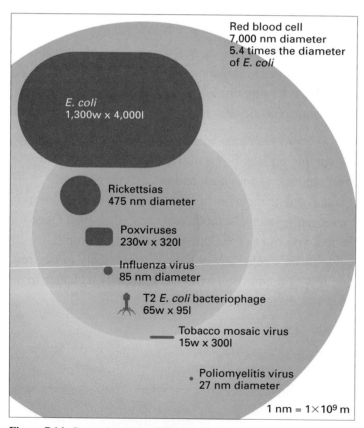

Red blood cell
7,000 nm diameter
5.4 times the diameter of *E. coli*

E. coli
1,300w x 4,000l

Rickettsias
475 nm diameter

Poxviruses
230w x 320l

Influenza virus
85 nm diameter

T2 *E. coli* bacteriophage
65w x 95l

Tobacco mosaic virus
15w x 300l

Poliomyelitis virus
27 nm diameter

1 nm = 1×10⁹ m

Figure 7.14: *Some viruses are 200 times smaller than the cells they infect. Notice the various sizes of the viruses.*

One reason that viruses are hard to detect is their extremely small size. Many viruses are no larger than the ribosomes of eukaryotic cells, and even the largest ones are only as big as the smallest bacteria (Figure 7.14). Viruses aren't made of cells; they are particles of DNA or RNA surrounded by a coat of protein. Their DNA or RNA blueprint is much smaller than that of a bacterium, and they do not have enough DNA to carry out all the life processes that cells perform. In fact, because they don't do a lot of things that living organisms do, most biologists don't consider viruses to be living organisms. For instance, viruses don't carry out their own biochemical processes, don't exchange substances with their environment, don't extract energy from food, and don't grow or divide. They do, however, reproduce (though not on their own) and adapt to changing conditions, two characteristics that make them quite troublesome.

But if viruses aren't alive, what does their DNA or RNA do? Do viral DNA and RNA produce proteins? If so, what do these proteins do? Scientists were able to answer these questions only when they studied viruses inside the cells and tissues of other organisms. It seems that just about every living creature—plant, animal, protist, mushroom, and even bacterium—is infected with viruses. In fact, scientists first learned how viruses infect cells by studying viruses that they called **bacteriophages,** which infect and destroy bacteria (Figure 7.15).

Many bacteriophages infect their host bacteria by first attaching to a site on its cell wall. The bacteriophage then pushes its DNA core through the cell wall and the cell membrane. The protein coat is left outside the host cell. Right after the viral DNA enters, the host cell begins to transcribe it into mRNA, just as if it were the host's own DNA (Figure 7.16). This is a big mistake! The viral mRNA codes for proteins that only a virus can use. And while the host cell is turning out foreign mRNA and translating it into viral protein, most of its own processes are shut down. The virus takes command of the host's nutrients, using its amino acids to make coat proteins and enzymes that the virus will use against the host.

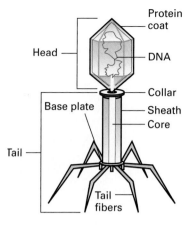

Figure 7.15: *A bacteriophage. This virus will only infect bacteria.*

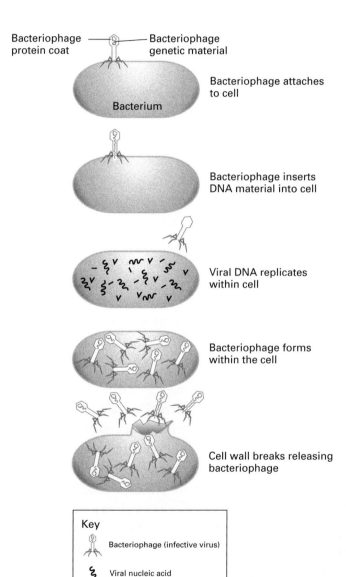

A blackline master of Figure 7.16 can be found in the TRB.

Bacteriophage protein coat — Bacteriophage genetic material

Bacteriophage attaches to cell

Bacterium

Bacteriophage inserts DNA material into cell

Viral DNA replicates within cell

Bacteriophage forms within the cell

Cell wall breaks releasing bacteriophage

Key

Bacteriophage (infective virus)

Viral nucleic acid

Viral protein

Figure 7.16: *The life cycle of a virus. The bacteriophage inserts its DNA into a bacterium, reproduces, and is scattered to invade other bacteria when the cell wall of the bacterium breaks.*

In 1952 Hershey and Chase completed their bacteriophage experiments showing that during viral infection of *E. coli* DNA rather than protein enters the bacterial cell.

One type of bacteriophage was important in the discovery that DNA is the hereditary material of organisms. Find references on the research of Martha Chase and Alfred Hershey, and demonstrate to the class how viral infection played a part in their discovery.

VIRAL REPRODUCTION

Once the host cell has been tricked into making lots of viral DNA and protein, these materials assemble into new viruses. The whole process, called a lytic cycle, takes about 15 minutes (Figure 7.17), culminating when viral enzymes destroy the host cell membrane and cell wall, releasing hundreds of new viral particles. Each particle can infect a new host cell. Viruses may not be alive, but they sure are rapid and efficient reproducers!

In some cases the infecting virus doesn't destroy the host cells. Sometimes, the injected viral DNA integrates into the host chromosome and remains dormant—the host cell doesn't transcribe it. However, when the host cell divides, it replicates the viral DNA along with its own DNA (Figure 7.17). This slower method of reproduction, called **lysogeny,** is sometimes the best course to take if conditions aren't right for the release of new virus particles into the environment. (Viruses don't survive long in the environment; they are better off inside a host.) Environmental factors such as ultraviolet light and mutagenic chemicals can trigger a host cell to enter the lytic cycle.

WORD BANK

retro- = in reverse

A retrovirus transcribes in a way that is the reverse of other organisms' method.

RNA viruses use a variation on the theme above when infecting a host. One type of RNA virus, called a **retrovirus,** makes an enzyme that transcribes its RNA into DNA once it is inside the host. Not surprisingly, the name of this enzyme is **reverse transcriptase;** it carries out transcription in reverse. HIV, responsible for AIDS, is an example of a retrovirus. Early therapy for AIDS patients was a drug that blocked the activity of reverse transcriptase so that viral DNA was kept out of the host cell. Unfortunately, HIV mutated to forms that resisted the effects of this drug, and so it is no longer used.

Viruses are very specific about what host they attack. A virus has coat proteins that recognize and attach to specific molecules on its

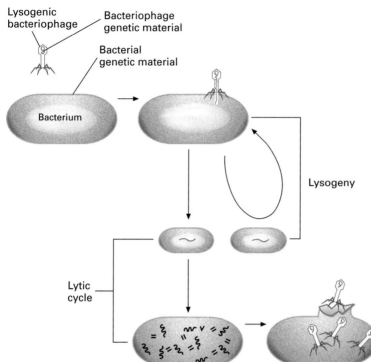

Lysogenic bacteriophage

Bacteriophage genetic material

Bacterial genetic material

Bacterium

Lysogeny

Lytic cycle

Figure 7.17: *In the lytic cycle the virus reproduces; however, in lysogeny the virus remains dormant until the environment factors trigger a host cell to enter the lytic cycle.*

host's cell membrane. For this reason a bacteriophage of *E. coli* can't infect the bacterium *Streptococcus,* and a plant virus can't infect an animal cell. Influenza, or flu, viruses are exceptions to this rule. For instance, swine flu attacks pigs and humans, and avian flu can be transmitted from chickens to humans.

Viruses are classified into families on the basis of what hosts they infect as well as whether they are DNA or RNA viruses, what is in their protein coats, and what diseases they cause.

LESSON 7.2 ASSESSMENT

1. The doctor could take a sample of body fluid and look for bacterial cells. An absence of bacterial cells could mean that you have a viral infection. Blood tests might also verify a viral infection.

2. This baby has a lysogenic infection. EBV will remain dormant until conditions activate it.

3. A virus has certain proteins in its protein coat that recognize and attach to molecules on a host's cell membrane.

4. Viruses don't carry out their own biochemical processes, they don't exchange substances with their environment, they don't grow or divide, and they don't move or respond to stimuli.

1 If you visited your doctor with a sore throat, a runny nose, and a high fever, how could the doctor tell whether you have a viral infection or a bacterial infection?

2 You are a doctor and you discovered that a newborn baby was infected with EBV but appeared perfectly healthy. Three years later, the child still appears to be healthy. How is this possible?

3 Viruses are very selective about what type of host they will infect. How do viruses recognize the correct host cell?

4 You are a microbiologist. The biology department at a local university has asked you to lead a discussion on the topic "viruses are nonliving things." What evidence would you use to support this topic?

LESSON 7.3 BACTERIA AND VIRUSES IN MODERN BIOTECHNOLOGY

"Biotechnology" is a term that you hear almost every day on the news. It means the manipulation of living organisms to form useful products, something that's been done for centuries. Microorganisms, especially bacteria and yeasts, have always played a key role in biotechnology. Now biotechnology may involve using microorganisms and viruses to transfer genetic information from one organism to another. An outgrowth of this work is the ability to replace defective genes in an organism with healthy ones, a technique called gene therapy. Once thought to be mere science fiction, gene therapy is now entering the real world.

VIRUSES AS LEAN GENE DELIVERY MACHINES

BIOLOGY IN CONTEXT

Gene therapy promises to deliver helpful genes to specific cells, supplementing defective genes with good ones. The challenge of the technique is how to send the right genes to the right spot.

Nori Kasahara, assistant professor of pathology and biochemistry at the Institute for Genetic Medicine, develops miniature packages that ferry genes to a particular destination. His strategy involves nesting the genes inside harmless viruses, allowing the viruses to infect target cells. "Viruses are 'machines' that have evolved over millions of years specifically to put their DNA into a host cell," said Kasahara. "So it's beneficial to take advantage of their natural properties."

Viruses that are designed to enter only certain types of cells make for safer therapies. Kasahara recently developed a way to target specific cells using retroviruses. His method uses a lock-and-key system, like that of an enzyme and its substrate. When certain molecules are attached to the outside of a retrovirus, it can infect only cells that have sites that match those molecules.

Retroviruses tend to enter cells only when they are dividing but gene therapy often aims at nondividing cells. What Kasahara eventually hopes to find is a virus that will infect only particular cells, even if they aren't actively dividing.

GENE-SIZED ENGINEERS

An important area of modern biotechnology is genetic engineering, more precisely called recombinant DNA (rDNA) technology, since it involves recombining the genes of two organisms. Table 7.1 lists some early benefits of rDNA technology. Its potential seems almost

Application	Benefit	In Use Today	Projected for the Future
Medicine	Useful protein	Insulin; vaccines; antibodies	New vaccines
Industry	Making raw materials for the chemical industry, such as alcohol, acetone, glycerol	Currently possible with biotechnology techniques that engineer enzymes for use in producing industrial chemicals	
Energy	Inexpensive alcohol fuel	Currently being made from sugarcane juice, a renewable raw material. Amylase is inserted into yeast to enable the production of alcohol fuel from starch, a raw material found in virtually all plants.	
Environment	Cleaning up oil spills and disposing of waste; mixtures of natural microbes and enzymes used for breaking down paper mill waste and oil engineering a microbe that will replace mixtures of several natural microbes	Engineered oil-eating bacteria	Breaking down waste plastics
Agriculture	Help crops to resist drought, frost, and pests; enable plants to make their own fertilizer Improve animals and their products; produce new medical products in cow's milk	Pest resistance Currently limited to selective breeding, using such techniques as in vitro fertilization and artificial insemination and altering an animal embryo to produce disease resistance, bigger animals in a shorter period of time, less fat in hogs, and farm animals whose milk contains medicines such as blood clotting Factor IX for use by hemophiliacs; diagnosing disease and improving vaccines	Applying a frost-resistant strain of bacteria to plant surfaces; adding a natural pesticide to plant roots (bacterial mutant); genetically altering plants to be drought-resistant; genetically engineering corn cells with bacterial cells to produce corn plants that self-fertilize

Table 7.1: Sampling of Present and Future Benefits from Genetic Engineering

limitless, and it is sure to affect medicine, agriculture, the food industry, and the environment, among other areas.

Why are simple organisms such as bacteria major players in rDNA technology?

One reason is the accessibility of their DNA. Unlike eukaryotic DNA, bacterial DNA isn't surrounded by a nucleus and contains little protein. Though most of a bacterium's DNA is in a single chromosome, some is in small rings of DNA in the cytoplasm called **plasmids.** Plasmids can pass through the bacterial cell membrane, a big advantage. On their plasmids, bacteria may have genes for enzymes

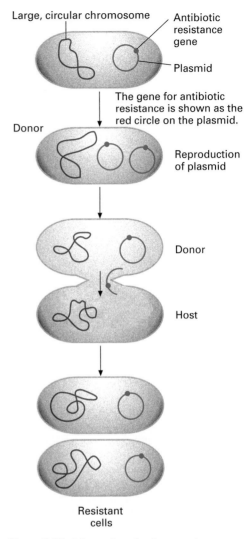

Large, circular chromosome

Antibiotic resistance gene

Plasmid

Donor

The gene for antibiotic resistance is shown as the red circle on the plasmid.

Reproduction of plasmid

Donor

Host

Resistant cells

Figure 7.18: *A bacterium that is not resistant to an antibiotic (acting as a host) can absorb a plasmid from a bacterium (acting as a donor) that has a resistance gene. The host bacterium may inherit the gene for antibiotic resistance.*

that break down antibiotics such as penicillin. One reason that many antibiotics are no longer effective is that bacteria share plasmids. They do this in a process called conjugation. Figure 7.18 shows how this works. Because plasmids are replicated along with the chromosome, genes on plasmids can spread quickly through a population of bacteria.

ACTIVITY 7-4   GENE SPLICING

Obtain a gene-splicing kit. Check it to be sure that each lab station will be able to simulate the entire sequence of events in bacterial production of insulin. The student handout in the TRB gives instructions for how to use the kit materials to simulate gene-splicing techniques.

The ability of plasmids to transport DNA between organisms has important consequences for humans. In this activity you will simulate the techniques that researchers use to make plasmids transfer genetic information from one species to another.

Get a gene-splicing kit and the handout for bacterial production of insulin. Follow the instructions in the handout. After you have completed all the steps, answer the following questions in your logbook:

To provide large quantities, bacteria are placed in large vats under optimal conditions for reproduction.

Different proteins would have to be separated from the same culture. The two proteins produced might be difficult to separate, or one of the proteins might not be produced.

- How would you scale up the production of insulin by genetically altered bacteria to provide the largest supply possible?

- What problems might arise in trying to get one bacterial culture to produce two different proteins by genetic engineering?

THE RIGHT STUFF FOR GENETIC ENGINEERING

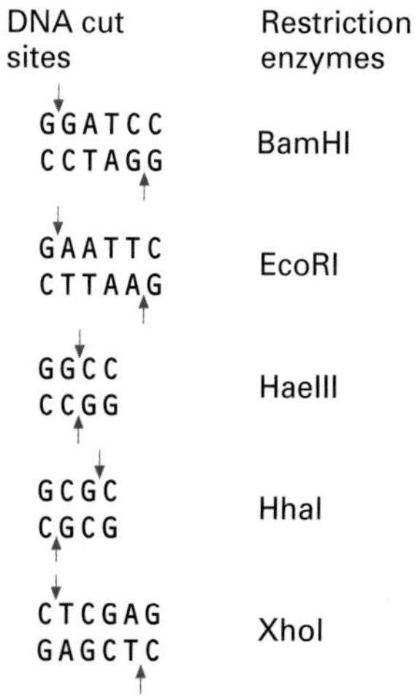

DNA cut sites	Restriction enzymes
G↓GATCC CCTAG↑G	BamHI
G↓AATTC CTTAA↑G	EcoRI
GG↓CC CC↑GG	HaeIII
GCG↓C C↑GCG	HhaI
C↓TCGAG GAGCTC↑	XhoI

Figure 7.19: *Example of cleavage sites on DNA cut by specific restriction enzymes*

Another reason that bacteria are ideal organisms in rDNA technology is a certain group of enzymes that they produce. These **restriction enzymes** can split DNA molecules at specific base sequences (Figure 7.19). Bacteria use restriction enzymes to remove the DNA of bacteriophages that have infected their plasmids or chromosome. Researchers use this natural ability to open plasmid rings at specific DNA sequences and insert a foreign gene. This is the technique known as gene splicing. The researchers might insert the gene that produces human growth hormone (Figure 7.20). As the bacteria transcribe and translate this gene, they make a human protein, which can be purified and collected for use.

A third reason that bacteria are so useful in rDNA technology is that they reproduce very rapidly. As a transformed bacterium reproduces, it quickly creates thousands of identical bacteria, each transformed to produce the protein dictated by the foreign gene.

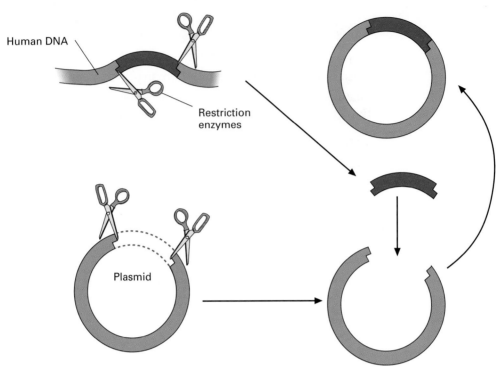

Figure 7.20: *Gene splicing is a technique used by researchers to insert a foreign gene into plasmid rings. Restriction enzymes are used to cut open plasmid rings and to cut out DNA fragments for gene insertion into a plasmid ring.*

A culture of bacteria that has an inserted gene for human growth hormone can produce a supply of the hormone for people who can't produce enough of it themselves. The hormone that the bacteria produce acts exactly like the hormone that is produced in a human cell. Table 7.2 lists products of genetically engineered bacteria and their uses.

Product Made by Bacteria	Application of the Product
Somatostatin (human growth hormone)	Treating dwarfism
Insulin	Treating diabetes
Interferon	Treating cancer
Tissue plasminogen activator	Breaking up or reducing the risk of forming blood clots

Table 7.2: Bacterial Products and Their Uses

Plasmids can transport genes into species that are economically important. One such bacterium, *Agrobacterium tumefaciens,* infects some crops. During an infection, *A. tumefaciens* releases plasmids into plant cells. Some of the genes on these plasmids direct the formation of tumors in the plant tissue (Figure 7.21). Researchers have inserted different genes into plasmids of *A. tumefaciens,* depending on what characteristic of the crop they want to change. To create herbicide-resistant cotton plants, you start with *A. tumefaciens* plasmids that have incorporated herbicide resistance genes. You transfer them to cotton plants by culturing cells of cotton plants and infecting them with the plasmids. Now you have cotton that resists the herbicides that are applied to kill weeds in the cotton field.

Figure 7.21: A. tumefaciens *normally causes plants to develop gall tumors, as shown here.*

Alternative Investigation 7C, Bacterial Transformation, can be found in the Lab Manual.

Scientists working in the field of gene therapy hope that they can use viruses to transport genes into human cells, replacing genes that don't work right. If this were successful, it might lead to a cure for disorders such as cystic fibrosis, muscular dystrophy, and many other genetic disorders.

LESSON 7.3 ASSESSMENT

1. Researchers use restriction enzymes to open plasmid rings. Once these rings are open, a foreign gene can be spliced into the plasmid. The bacteria that take up the plasmid will be genetically transformed and may make human proteins during transcription and translation.

2. A vector that will infect only specific cells can be targeted to certain cells for a specific task. These vectors can be designed to produce proteins that may be otherwise deficient.

3. New copies of the genes are made and passed on to each daughter cell.

4. The bacterium is transformed; the gene that causes tumor formation is removed and replaced by the gene that causes herbicide resistance.

1 Why are restriction enzymes important to researchers using rDNA technology?

2 You're a research scientist working at a genetic engineering laboratory. Your supervisor asks you to design a vector that infects specific cells even when they're not actively dividing. What is the importance of this project?

3 When microorganisms reproduce after acquiring a foreign gene, what happens to the foreign genes that they carry?

4 *Agrobacterium tumefaciens* normally produces tumors in plants that it infects. How are researchers able to use this bacterium to develop herbicide resistance in cotton without also developing tumors in the cotton plant?

LESSON 7.4 AGENTS OF DISEASE

Pathogens, also known as germs, are the bacteria, viruses, and other microorganisms that cause disease in plants, animals, and other eukaryotes. People who track down pathogens include public health workers, veterinarians, horticulturists, and medical researchers. They try to answer questions such as: What pathogen is causing the problem? How does it spread? How can we control it or stamp it out? They inform us about the disease and how to protect against it. Not all bacteria and viruses are pathogens, but in this lesson you'll learn about some that cause important animal and plant diseases.

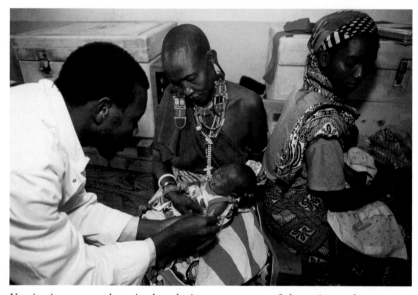

Vaccinations are used to stimulate the immune system to fight against pathogens.

DISEASE DETECTIVES

BIOLOGY IN CONTEXT

"In national news, state health officials reported that people in at least 13 states have reported getting sick after eating ice cream produced in Minnesota . . . FDA investigators confirmed that they have found salmonella bacteria in a second sample of the ice cream, which matches the state health department's result on earlier samples. The US Centers for Disease Control and Prevention has confirmed reports of salmonella poisoning in three states, but said it was too early to know the outbreak's extent . . . The company in the meantime has issued a recall for all of its ice cream and frozen desserts and moved its ice cream production to another plant while investigations continue to identify the source of the contamination."

The incident referred to in the news reports occurred in October 1994. Some two years later, researchers who worked on the case said that at least 224,000 people had become sick from eating bacteria-tainted ice cream, making this the biggest documented case of food poisoning traced to a single source. One investigator noted that a big outbreak of food poisoning used to involve a few dozen people getting sick from eating bad potato salad at a potluck supper. He called the 1994 incident a warning of things to come because of the potential for spreading a single tainted ingredient to millions of people.

Hundreds of doctors, lab technicians, and local, state, and federal health workers began to identify the food-poisoning victims and reported their findings to a team of investigators. The team established a national surveillance program and surveyed customers of the ice cream manufacturer. The steps that were used to make the ice cream were compared to the steps used to make similar, uncontaminated products at the same time. Samples of bacteria were obtained from ice cream, the ice cream plant, and tanker trailers that transported the ice cream base (premix) to the plant. The contamination was traced to tanker trailers that had carried raw eggs just before hauling the pasteurized premix to the plant. The team recommended that food products that are not destined for repasteurization should be transported in dedicated containers.

ACTIVITY 7-5   PREVENTING THE SPREAD OF DISEASE

The procedure for this activity can be found in the TRB.

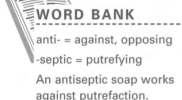

WORD BANK

anti- = against, opposing

-septic = putrefying

An antiseptic soap works against putrefaction.

Restrooms are stocked with various types of hand soaps that are supposed to kill bacteria and other pathogens. Which of these soaps works best? Does ordinary hand soap do the job? In this activity you'll test the ability of different soaps to act as antiseptics, chemical agents that prevent infection by killing pathogens.

FOREWARNED IS FOREARMED

Diseases that are largely under control in the United States, in part because of immunization (more on this in Chapter 15), were omitted from this list. However, these diseases may still be problems in developing countries that lack immunization programs and sanitary procedures.

Washing hands is one way to reduce the chance of catching a cold or flu. It's pretty simple, at least when you've got clean running water and soap. Unfortunately, many other pathogens with more serious consequences can't be controlled so easily. Some of these pathogens cause diseases that you may have heard about in the news, such as AIDS, tuberculosis, herpes, hepatitis, meningitis, and Lyme disease. This is a scary list, but learning about pathogens can help you to avoid becoming their host. You should understand which pathogen is responsible for a disease, how the disease is transmitted, the symptoms, and what you can do to avoid getting it or spreading it. Table 7.3 lists common infectious diseases for humans, the pathogens that cause them, and how these diseases are transmitted.

Disease	Pathogen	Primary Ways Disease Is Transmitted
Meningitis	Various bacteria	Sneezing
Tuberculosis	A mycobacterium	Sneeze droplets
Strep throat	A streptococcus	Contaminated objects or respiratory droplets
Lyme disease	Spirochete bacterium	Tick bite
Cholera	*Vibrio cholerae* bacterium	Contaminated water or food
Typhoid fever	Salmonella bacterium	Contaminated water or food
Traveler's diarrhea	Often *E. coli*	Contaminated water or food
Syphilis	Spirochete bacterium	Sexual contact
Common cold	Rhinoviruses	Hand contact or via respiratory droplets
Genital herpes	Herpes simplex type 2	Sexual contact
HIV infection	HIV	Contact with infected blood, semen, or vaginal fluid
Mononucleosis	Epstein-Barr virus (EBV)	Kissing an infected person, sharing drinks
Rabies	Rabies virus	Bites from infected animals
Hepatitis A	Hepatitis A virus (HAV)	Oral-fecal contact
Viral gastroenteritis	Numerous viruses	Oral-fecal contact
Hepatitis B	Hepatitis B virus (HBV)	Contact with blood of an infected person

Table 7.3: Routes by Which Some Common Human Infectious Diseases Are Transmitted

Sharing food and coming in contact with sneeze droplets are common ways of transmitting pathogens.

DISCOVERY OF DISEASE AGENTS

CHANGING IDEAS

In the late 1600s, Anton van Leeuwenhoek first saw microorganisms with his homemade microscope. But a century before that, the Italian physician Girolamo Fracastoro argued that many diseases are caused by organisms that are too small to see. When microscopes finally showed bacteria and protozoa on boiled extracts of hay and meat, scientists began to explain how they got there. One theory, called the theory of spontaneous generation, proposed that the organisms arose mysteriously out of the organic matter in the extracts.

Louis Pasteur later proved that microorganisms entered the extracts from the surrounding air and reproduced there. He used the apparatus shown in Figure 7.22, putting nutrient solutions in the flasks. Airborne microorganisms were trapped along the glass necks, and nothing grew in the nutrient solutions. When Pasteur broke off the long necks close to the flasks, the nutrient solutions soon turned cloudy with growing microorganisms.

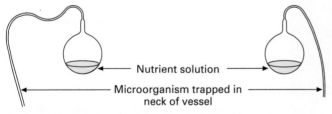

Nutrient solution

Microorganism trapped in neck of vessel

Figure 7.22: *Louis Pasteur's swan neck flasks that were used in his spontaneous generation experiments*

In the late nineteenth century, German physician Robert Koch (Figure 7.23) established an undeniable connection between microorganisms and disease. Koch injected healthy laboratory mice with fluid from cattle with a disease called anthrax. The mice got sick. Koch also grew disease-

causing bacteria in a culture from a sick cow's spleen. When Koch injected new mice with bacteria from this culture, they got sick. These methods are still used to establish a connection between a disease and a pathogen. Koch's postulates, simply stated, are as follows: The pathogen must be present in every case of the disease but is never present in healthy organisms; the pathogen must be isolated and grown in culture; when the isolated

Figure 7.23: *Robert Koch observing a specimen in his laboratory*

pathogen is injected into a healthy organism, it gets the disease; and the same pathogen can be isolated from the infected organism.

VECTORS

Most pathogens live inside a host, although many can survive for a time on skin and inanimate objects. Remember those spores that some bacteria produce? Spores can remain alive because they're protected from adverse conditions. In a **communicable disease,** one host transmits a pathogen through direct contact or via an object like a thermometer or a spoon to another host. Livestock infections, such as mad cow disease, are often transmitted through contaminated feed (Figure 7.24). Plant infections can spread among shrubs and trees during pruning. Tree pruners should wash chain saws and shears in alcohol to sterilize them. We humans pass along respiratory infections, caused by bacteria and viruses, when we sneeze or cough. Bacteria or viruses adhere to these droplets, which land on objects including our hands, from which they can be picked up by others.

Organisms called **vectors** help to transmit infectious disease. For instance, ticks and mosquitoes, which feed on the blood of mammals, can transmit pathogens in their saliva. Malaria is a tropical disease that is

Figure 7.24: *Communicable diseases are spread in cattle through contaminated feed.*

transmitted by a mosquito (Figure 7.25). Rocky Mountain spotted fever is transmitted by a tick. Plant pathogens also travel on the legs and mouthparts of leaf-eating or stem-boring insects.

Sometimes a pathogen travels from one species to another. Rabies, caused by a virus, and Lyme disease, caused by a bacterium, can pass from a variety of mammals to humans (Figure 7.26). A disease that is transmitted from animals to humans is a **zoonosis.** Some pathogens carry out different stages of their life cycle in their different host species, as you will see in Chapter 8.

Natural selection helps pathogens to coexist with their host, rather than killing it. A pathogen that kills its host may not be able to find another host and will be selected against. Occasionally, pathogens get out of control and cause an epidemic. An **epidemic** is a rapid and extensive increase in the occurrence of a disease above an expected level. AIDS is an epidemic, caused by the spread of the HIV virus. Public health officials consider that tuberculosis has become an epidemic in many large U.S. cities. It is spreading rapidly in crowded areas among people who don't get adequate health care.

Figure 7.25: *Mosquitoes are vectors for malaria.*

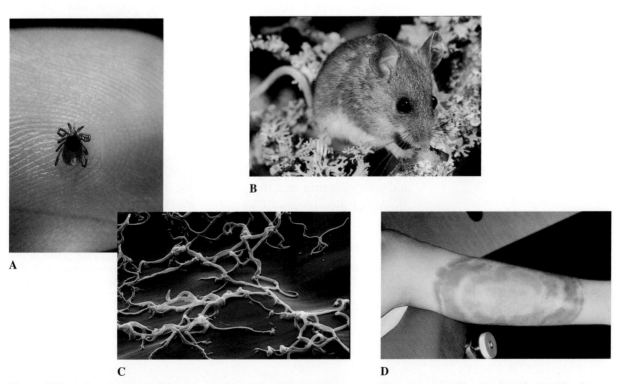

B

A

C

D

Figure 7.26: *A deer tick (A) that feeds on the blood of an infected deer mouse (B) can transmit Borrelia burgdorferi (C), the bacteria that cause Lyme disease. (D) Bacterial arthritis, characterized by a rash that develops during Lyme disease.*

NEW PATHOGENS

Additional Activity, Starting an Epidemic, can be found in the TRB.

As a result of mutations, some pathogens become more **virulent,** that is, better able to break down the host's protective mechanisms. The host species often can't defend itself against a newly altered strain of pathogen. Mutations are why each winter a new strain of influenza, caused by a virus, becomes an epidemic somewhere in North America.

Knowing the pathogen and how it is transmitted helps public health workers to decide how to prevent or control the spread of a disease. Diseases from contaminated food can be controlled by improved sanitation programs. Health departments establish such programs, and health inspectors enforce them by routinely inspecting restaurants and cafeterias. To control diseases that are carried by vectors, we try to reduce the population of the vector. Spraying against mosquitoes has helped to control malaria in some countries. However, this method is often just a stopgap, because mosquitoes mutate and develop resistance to insecticides.

Sewage treatment and chlorination of water supplies have done much to control gastrointestinal diseases such as cholera and dysentery in developed countries. Workers at sewage and water treatment plants closely monitor the treatment to be sure it remains effective. Effective disease control also involves developing and administering vaccines, drugs that give immunity against a disease. In Chapter 15 you will learn more about how people acquire immunity to disease.

To prevent or control the spread of pathogens, health inspectors inspect commercial kitchens to make sure food is handled properly.

LESSON 7.4 ASSESSMENT

1. The pathogen and the method of transmittance must be identified.

2. No, more time is needed to determine the severity of the disease.

3. Because the farmer knows which crops the pathogens will attack, he or she can control the pathogen by alternating the crops planted each year. Or the farmer may choose not to plant the susceptible crops.

4. Try to determine what conditions are encouraging the infection and correct them.

1 You are a public health official. An unknown disease appears to be spreading within your city. Within a week, over 50 people have been diagnosed with it. What do you need to know about this disease to prevent it from spreading further?

2 Can the disease in Question 1 be classified as an epidemic?

3 A pathogen affects potatoes and tomatoes but does not attack squash. How would this information help a farmer to control the pathogen?

4 You work for a grower of ornamental plants. When you're transplanting plants from a tray, you remember that they haven't been looking too healthy lately. You have discovered that the plants have become infected with bacteria. What action should you take?

CAREER APPLICATIONS

The applications that follow are like the ones you will encounter in many workplaces. Use the biology you learned in this chapter to complete the activities. Share your work with the class.

AGRICULTURE & AGRIBUSINESS

Silage

Contact a feed company or use the Internet to find out how to produce silage and how it differs from some other livestock feeds. What are the advantages and disadvantages of making silage from feed crops? Is it feasible for farmers to produce silage on the farm? If so, how widespread is this practice?

Bacteria and Viruses on the Farm

Investigate the use of bacteria and viruses as commercial preparations that can be applied directly to soil or crops. Find out how to apply the microorganisms and why they are used. Use the following resources to help you with this activity: a seed and fertilizer salesperson, a county agricultural extension agent, a local farmer or grower, an agricultural supply catalog, the Internet.

BUSINESS & MARKETING

Microorganisms and Your Community's Economy

Use your local yellow pages to identify categories of businesses that involve microorganisms (particularly bacteria and yeast) in some way. If you work in teams, each group should work with categories that begin with a different letter of the alphabet. List possible categories and the ways in which microorganisms might be important to them. Contact one or more businesses under each category to find out whether your assumptions are correct. After talking to a cross section of businesses in different categories, create a table. For each category, give the businesses' relationship to microorganisms. (For example, do they try to get rid of them? Do they use them in their day-to-day operations?) Give the most specific level possible of the particular microorganisms. (For example, do they try to control all bacterial growth? Do they use lactobacillus to make cheese?) Give a brief description of how the businesses accomplish the tasks. Make a class notebook from the tables of all the groups.

Indicator Microorganisms

Contact your local health department and find out how the local streams, lakes, and other waterways are monitored for pollution that may affect human health. (The health department might not do this monitoring, but they can tell you who does.) What indicator organisms in your area are warning flags to public health officials who are concerned about certain types of pollution? What are acceptable numbers for these organisms? What procedures are followed when a waterway is dangerously polluted?

Famous Pathogens

Working in a group, research an epidemic caused by bacteria or viruses. How was the responsible organism identified? Was the outbreak brought under control? How? With your group, tell this story to the class or dramatize it in a skit. Some possible subjects are influenza (pick a particular strain), cholera, Venezuelan equine encephalitis, Ebola, Legionnaires' disease, Lassa fever, Marburg disease, and AIDS. After each group has presented a disease, make a class chart that traces similarities and differences in the type of pathogen, symptoms of the disease, how it is transmitted, control method, prevention strategies, and statistics on how widespread the disease was or is.

Daycare for Microorganisms

List what child-care workers do in daycare centers to kill potentially harmful microorganisms and prevent the spread of diseases among young children. Include techniques such as hands washing, boiling, and using antibacterial cleaning products. What rules and regulations must child-care workers follow? Contact a microbiologist, a public health expert, or another knowledgeable person to find out how effective these control methods are. Are any of these methods being questioned because of harmful side effects, such as killing off beneficial microorganisms or encouraging resistant strains of pathogens?

Skin Care

Examine products from your bathroom or kitchen that you routinely apply to your skin. List the active ingredients of each product. List questions that you have about microorganisms that live on your skin, such as which ones might cause skin disorders like acne. Also list questions about how the components of the various products might affect the skin and the microorganisms living there. Compile a class list of questions and sort them into categories. Let one or two students contact a dermatologist and someone at the cosmetology department at your local community college. Find out whether they will answer your questions or point you to references where you can find the answers. After getting answers from the professionals and reviewing the references that they suggest, put together a pamphlet describing the normal role of microorganisms on the skin, those that can cause skin problems, and the effects of various products that contact your skin on a regular basis.

INDUSTRIAL TECHNOLOGY

Microorganisms for Manufacturing

Form a group of three or four students. Imagine that the members of your group are managers of a manufacturing plant and will choose the product or products that you make. Discuss how genetic engineering involving microorganisms might help your plant to streamline production, produce a better or less costly product, and/or reduce waste.

Engineering for Safety

Groups such as the Centers for Disease Control and Prevention, medical research institutions, and military research organizations that are looking for ways to protect us from biological warfare often work with potentially deadly microorganisms. They use equipment and protocols to protect against the release of dangerous microorganisms into the environment. Research the four laboratory levels of biological containment. How many Level 4 labs are in the United States? List the equipment that is used at each level to protect the researchers and describe how it works. How are harmful organisms killed and disposed of when the researchers have completed their studies?

Chapter 7   Summary

- A bacterium is a single cell. Bacteria normally live in colonies and may be grouped in pairs, chains, or clusters.

- The DNA of bacteria is found on a single circular chromosome and on small rings called plasmids.

- A cell wall of peptidoglycan protects bacteria. Some cell walls are surrounded by a capsule, which provides protection from a host's defenses and from loss of water. Some bacteria withstand unfavorable environments by forming a spore.

- Species of bacteria can be distinguished by the shape of their cells, the way the cells are arranged, differences in the cell wall (as indicated by a test called the Gram stain), and various other tests.

- Bacteria grow in a wide range of environments. Some species, such as those in the Archaea, grow in extremely acidic, salt, or hot environments.

- Bacteria have important ecological roles such as photosynthesis, recycling of minerals, breakdown of organic matter, and nitrogen fixation. Some bacteria also live in symbiosis with a eukaryotic species; this relationship may benefit the host, or it may damage or sometimes kill the host.

- Viruses are particles, rather than cells, made up of DNA or RNA and a coat of protein. They can't carry out biochemical processes or reproduce without first infecting a host cell. Because viruses meet few of the criteria for life, most biologists consider them to be nonliving.

- When a virus infects a host cell, it injects its nucleic acid into the cell. Viruses recognize certain molecules on the host cell membrane and usually infect only one type of host.

- A viral infection may occur as a lytic cycle in which the host replicates viral DNA or RNA and transcribes viral RNA to produce viral proteins. New viruses self-assemble, lyse the host cell membrane, and attack new host cells. In lysogeny the virus enters the host cell and replicates only when the host cell replicates.

■ Biotechnology manipulates living organisms to form useful products. This technology includes changing the genetic makeup of organisms; in recombinant (rDNA) technology, bacteria and viruses transfer genetic information between species.

■ Because plasmids of bacteria can pass through cell membranes, they are used as vectors in rDNA technology.

■ In rDNA technology, restriction enzymes split DNA molecules, especially plasmids, at specific base sequences.

■ Foreign genes can be spliced into plasmids and inserted into bacteria that then produce a foreign protein. Through these techniques of rDNA technology, bacteria are producing useful products such as insulin and growth hormones.

■ Viruses are used as vectors in rDNA technology. In gene therapy, virus vectors carry healthy genes into genetically defective cells so that these cells can resume normal function.

■ A relatively small number of bacteria and most viruses cause disease in animals (including humans), plants, and other organisms. These bacteria and viruses are pathogens.

■ To prevent the spread of infectious disease, a person should know the pathogen that is responsible, how the pathogen is transmitted, the symptoms of the disease, and how the disease can be prevented or controlled.

■ Disease vectors such as mosquitoes, flies, ticks, and mites transmit disease from one host to another. For some pathogens this transmission occurs between hosts of different species.

■ Most pathogens infect their host populations at low levels. Occasionally, this stable arrangement gets out of control and results in a sudden increase in the occurrence of disease, known as an epidemic.

CHAPTER 7 ASSESSMENT

Concept Review

See TRB Chapter 7 for answers.

1 You are helping a lab technician to observe a sample of microorganisms. You stained the cells to see their DNA. Immediately, you notice that the stained areas of DNA are not surrounded by membranes inside the cells. How would you classify these cells?

2 If you looked very closely at the samples in Question 1 with a high-powered microscope, what would their DNA look like?

3 How can you tell whether a sample is Gram negative or Gram positive?

4 How can you ensure the complete destruction of bacterial spores?

5 List the basic shapes for all bacteria.

6 A laboratory technician was viewing slides of various samples of bacteria. She discovered a slide that was not labeled properly, so she described the bacteria so that she could identify them later. She wrote in her lab notebook that she saw clusters of long, blue, rod-shaped cells. How would you classify this sample?

7 Eukaryotic cells use organelles such as mitochondria and chloroplasts to carry out reactions that are needed for the cells' survival. Describe how prokaryotic cells carry out these same reactions without organelles.

8 A lab technician must include an inorganic energy source in a growth medium. What type of organism is the technician culturing?

9 Describe how bacteria reproduce.

10 What type of bacteria can live in an oxygen-deprived environment?

11 What role do plasmids play in recombinant DNA technology?

12 Give two reasons why bacteria are used in recombinant DNA technology.

13 Before you can insert a foreign gene into a piece of DNA, you have to split the DNA. What type of enzyme would you use to split the DNA?

14 As a result of mutations a few bacterial cells in a population may become resistant to an antibiotic. When the population is later exposed to that antibiotic, what happens?

15 A rancher is going to introduce five new cows to a herd of cattle. The rancher is worried that the cows might have a communicable disease that could be passed to the healthy cows. What should the rancher do?

16 What biochemical process uses bacteria in making breads, dairy products, and alcoholic beverages?

17 How does viral infection spread from one cell to another?

18 You have developed a sinus infection, and your doctor gives you a prescription for antibiotics that should last for 10 days. Because you begin to feel better, you stop taking your medication after five days. But your sinus infection returns, and your doctor tells you that your first prescription might not get rid of your infection this time. Why?

Think and Discuss

19 What physical characteristic(s) of bacterial cells keeps the Gram stain in the cell after it is washed with alcohol?

20 A species that can grow on medium X cannot grow on medium Y. There is no apparent difference between media X and Y. Why can't the species grow on medium Y?

21 Research at a local agricultural experiment station involves removing genes for drought resistance from one variety of crop and transplanting them to a second variety. What is the purpose of this research?

22 A new disease is breaking out in areas that are heavily infested by mosquitoes. Most new cases are reported several weeks after mosquito breeding takes place. If you were the public health official investigating the cause of the disease, what would you conclude?

In the Workplace

Questions 22–26 refer to one case.

23 You are an epidemiologist in the state department of health. *Salmonella enteritidis* cases have been on the rise in your state. The cases were reported in six towns and cities around the state. What steps would you take to find out whether there was some connection between the cases?

24 After collecting case histories, you find that five patients had just returned from summer camp, two were counselors at the same camp, and three were parents who had gone to the camp to pick up their children. What would you do next?

25 All of the patients were at the grand finale celebration, which included a picnic of hot dogs, hamburgers, fried chicken, potato salad, chips, baked beans, cookies, cake, and homemade ice cream. The drinks that were served were canned soft drinks, punch, and iced tea. After analyzing what each patient ate, you found that the only items that every patient consumed were the homemade ice cream, chips, and canned soft drinks. Which items would you suspect and why? What could you do to confirm your suspicions?

26 Your investigations show that the residents of each cabin made ice cream, following their own recipe. One cabin used raw eggs to make a custard base. The base was not cooked. The other cabins either used a cooked custard base or recipes that did not call for eggs. Only the people who ate the ice cream made from raw eggs got sick. What do you need to do to finish your investigation? Make a final recommendation to prevent future problems.

Investigations

The following questions refer to Investigation 7A:

27 Why do you think the milk was heated to 82°C and then cooled?

28 What is the ideal temperature range for growing a yogurt culture? Give your answer in Celsius.

The following questions refer to Investigation 7B:

29 You are a clinical microbiologist hired to identify two strains of pathogenic bacteria. One is a large cluster of rod-shaped bacteria that stained purple and the other a chain of circular bacteria that stained pink. How would you classify these bacteria?

30 After Gram staining a slide of a microorganism and mounting it under the 40× objective, you see the following picture:

(stained pink)

*Gram stain slide
of microorganism*

Describe the type and characteristics of the microorganism that you see.

CHAPTER 8

WHY SHOULD I LEARN THIS?

You won't see most protists without using a microscope, but these tiny eukaryotes are important to all kinds of people. Health care workers combat the protists that cause malaria, sexually transmitted diseases, and nervous disorders. The fisheries industry charts the wanderings of the large masses of marine protists that fish eat. A water quality specialist looks for protists as indicators of clean or polluted water. As we learn more about protists, they will become important in even more occupations. Perhaps one day you'll work with them yourself.

KINGDOM PROTISTA

WHAT WILL I LEARN?

1. the importance of Protista as food for many species, as consumers and decomposers of organic matter, and as symbionts

2. how to observe protists under the microscope and relate their structures to their life cycles

3. how to analyze the life cycles of selected protists and understand how asexual and sexual reproduction helps them to adapt to changing environments

4. how to guard against protists that contaminate the environment or that cause infections of humans and the animals and crops they cultivate

5. the relationships of protists to members of the other four kingdoms

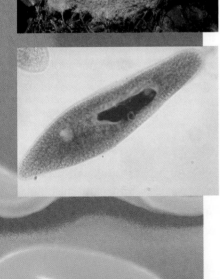

If you've been keeping up with the news, you've read or heard a lot about protists. The stories may have had titles such as "Red Tide Contaminates Shellfish," "Can Farming Algae Meet World Protein Needs?," and "New Epidemic of Sleeping Sickness Hits West Africa." Every environment on Earth is home to at least a few of the more than 50,000 species that make up this unusual assortment of organisms—seaweeds, amebas, slime molds, paramecia, and many others. In this chapter you'll meet some common protists and learn how they have adapted to many environments.

LESSON 8.1 SURVEY OF THE KINGDOM

Our bodies are complex. We have many types of specialized cells. But you don't have to be big to be complex. No group of organisms on Earth has developed more highly specialized organelles than the protists have. If you were small enough to dive into a protist's structure, you might find yourself helplessly pushed along in a stream of cytoplasm, dodging whirling oars, or trapped in a membrane. But thanks to the microscope between you and the protists, what you'll find instead are cilia, flagella, and vacuoles.

ACTIVITY 8-1   PROTISTS UP CLOSE

Make available a culture of mixed freshwater algae and of mixed protozoans for students to sample. Other equipment needed: compound light microscope, slides, coverslips, transfer pipettes, methylcellulose slowing reagent (Protoslo). Students will need drawings of indicator species that are provided in manuals such as *Standard Methods for the Examination of Water and Wastewater.*

Organisms with green pigmentation (an indication of chlorophyll) are probably autotrophs, although you may see chlorophyll inside a heterotroph that has consumed an autotroph. Organisms that lack this pigmentation are probably heterotrophs, but some algae use pigments other than chlorophyll. An organism that is engulfing another is most likely a heterotroph.

Most of the organisms will be unicellular, but there may be a few colonial algae.

Answers will depend on samples examined.

Specialized organelles make many protists important players in aquatic environments. In this activity you will examine protists' diversity and observe their complex structure.

Pipette a drop of algae culture onto the center of a slide. Add a drop of slowing reagent. Gently lay a coverslip over the drop. Look at the slide first at low power and then at high power, finding as many types of organisms as you can. Draw or describe each type in your logbook. Make a similar slide and observations of the protozoan culture.

If you have pictures or lists of protists that indicate clean or polluted water, try to match them with the protists on your slides. Discuss the following questions in your logbook:

- Which of the organisms that you observed are autotrophs? Which are heterotrophs? Which organisms are you unable to decide about? Why did you classify the organisms this way?

- Which of the organisms that you observed are unicellular? Which organisms have more than one cell?

- How would you classify the water quality on the basis of the organisms you observed?

FOCUS ON WATER QUALITY

BIOLOGY IN CONTEXT

If your water supply is a lake or reservoir, lots of things are living in it. Jerry M., the laboratory supervisor at a water treatment plant in a midsize Texas city, is more aware of what is floating in lakes and reservoirs than most people are. Part of his job is to identify and count the protists and other microorganisms that live in the water that enters the treatment plant.

A lab technician takes water samples from the city's reservoir at four points where rivers and creeks enter it. At the plant, the samples are poured through a filter. Any microorganisms that are trapped by the filter are rinsed off into a small volume of water.

This lab technician is collecting water samples from the reservoir to determine water quality.

Jerry puts the filtered sample onto a microscope slide with a very shallow rectangular well. He looks at 10 areas of the well and counts and records individuals of different species. He pays special attention to problem species, which can multiply uncontrollably and ruin the quality of the drinking water coming out of the treatment plant.

Jerry recalls one summer when a species of green alga took over the reservoir. The presence of the species indicates clean water, but residents complained that their tap water tasted and smelled bad. Jerry said that the algae or microscopic algae eaters produced compounds that fouled the water.

By tracking the microorganisms in the reservoir, Jerry can predict when the city's drinking water will start to taste and smell bad. At that point, he tries to correct the problem. He might add copper sulfate to the lake to reduce algal growth or increase carbon flow to the plant's filters to handle the increased load of algal by-products.

Copper sulfate is used with caution. It may be outlawed in some areas.

OUTSTANDING FEATURES

Alternative Investigation 8C, What's Living in the Water, may be performed here. See the Lab Manual for this investigation.

Protists have some features in common with three kingdoms that you will study later: Fungi, Plantae, and Animalia. What makes protists stand out? Their cell organization, how they are nourished, their locomotion, and their reproductive methods are a few characteristics that separate the kingdoms.

Protists have diverse forms: some, like *Euglena* (Figure 8.1A), are a single cell, while others, like the alga *Volvox* (Figure 8.1B), are a **colony** of cells. (A colony is a group of cells or organisms of the same species that live in close association.) Still other protists, like the seaweeds and kelps (Figure 8.1C), are multicellular.

A

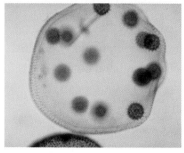

B

C

Figure 8.1: *Some common types of protists: (A)* Euglena, *a unicellular protist; (B)* Volvox, *a colony of photosynthetic cells; (C) Seaweed and kelp, multicellular marine algae.*

Protists have a number of different types of boundary layers around their cells. Some protists, such as algae, have cell walls. Many protists that lack cell walls have a rigid protein layer, called a **pellicle,** just inside their cell membrane. Some have just a cell membrane.

Protists have many ways to get energy and nutrients. Algae are phototrophs, and protozoans are heterotrophs. A few protists can be either phototrophs or heterotrophs. Some protists form symbiotic associations with fungi, plants, animals, or other protists.

Many methods of locomotion are seen among the protists. Some small aquatic protists remain suspended near the surface of a water body and float with the currents. Some of these protists contain oil droplets that make them less dense than water, which means that they will be buoyant (will float on water). Many protists have special organelles that allow them to make rapid and coordinated movements. *Paramecium* (Figure 8.2A) explores its environment using cilia that project from its cell membrane. *Ameba* (Figure 8.2B) moves slowly, oozing cytoplasm into fingerlike regions called **pseudopodia.** *Trypanosoma* (Figure 8.2C), the parasite that causes African sleeping sickness, propels itself through blood with flagella, which operate like whips.

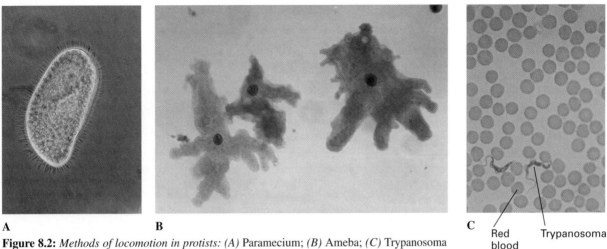

Figure 8.2: *Methods of locomotion in protists: (A)* Paramecium; *(B)* Ameba; *(C)* Trypanosoma

C | Red blood cell | Trypanosoma

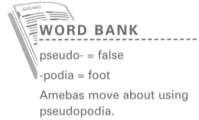

In Lessons 8.2–8.4, students will study the protists as three large groups. These groups are not taxonomic categories. They treat the protists as plantlike, animal-like, or funguslike. This assumes that a particular group of protists shares more similarities with one of the three kingdoms mentioned than with the other two. Some organisms, such as *Euglena,* don't fit well in any of the groups.

Like bacteria, all protists reproduce by asexual means. Asexual reproduction in the protists often involves one cell forming two cells through mitosis. Many protists can also reproduce sexually. For instance, the life cycle of the green alga *Chlamydomonas* (Figure 8.3) alternates between a sexually reproducing stage and an asexual stage. This alternation is an adaptation to changes in the environment. Remember that sexual reproduction increases the genetic variation in the next generation. New phenotypes may arise that are better adapted to the new environment than were previous phenotypes.

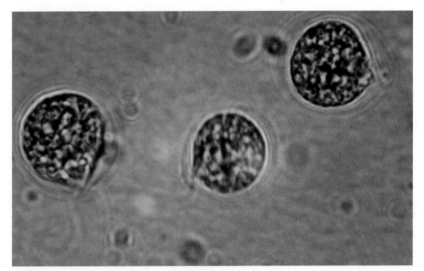

Figure 8.3: Chlamydomonas, *a unicellular alga*

MISFITS AND OUTCASTS OF THE LIVING WORLD

Before the 1960s, protists were grouped with either the plant kingdom or the animal kingdom. If a protist was photosynthetic, it was lumped with the plants; if a protist was heterotrophic, it was lumped with the animals. But taxonomists recognized that animals and plants are multicellular organisms, and it didn't seem right to include unicellular organisms in Plantae or Animalia. What's more, a few protists can be phototrophs *or* heterotrophs, depending on environmental conditions. Species such as *Euglena* have characteristics of both plants and animals. To solve some of these discrepancies, when the five-kingdom system was adopted, the unicellular eukaryotes got their own kingdom, the Protista, even though these organisms don't all have a lot in common.

Botanists recognized many colonial and multicellular species that don't fit well in Plantae. Their reproduction is different, and their photosynthetic pigments don't match those of true plants. But where to put them? It seemed like a good idea to lump them with the protists. Another of the new kingdoms, Fungi, was also in turmoil. Its caretakers, the mycologists, wanted a category to dump some of the misfits that had strange life cycles, such as the slime molds, which have stages that move like amebas.

WORD BANK

myco- = fungus

-logist = one who studies

Mycologists study fungi.

The kingdom Protista quickly became a taxonomic dumping ground for organisms that biologists had long considered outcasts. By the 1980s the kingdom had more than 30 phyla. Now some biologists think that Protista has large groups of organisms that should themselves be kingdoms. One plan is to break the protists into three separate kingdoms. This reshuffling is part of a scheme that would divide life into eight kingdoms. This is a good example of how scientists' ideas about organizing what they observe change over time.

LESSON 8.1 ASSESSMENT

1. Unicellular, colonial, and multicellular.

2. Locomotor organelles such as pseudopodia, flagella, and cilia; photosynthetic pigments; the shape and size of cells.

3. All protists are eukaryotes and are capable of asexual reproduction. Organisms in the fungus, animal, and plant kingdoms are also eukaryotes. Some organisms in all other kingdoms can reproduce asexually.

4. Protozoans are heterotrophs, so any answer that would imply that an organic energy source is required in the growth medium is acceptable.

1 What are the three forms of cellular organization that are found in protists?

2 You are a technician in a microbiology lab. Your supervisor has prepared several slides of protists for you to identify. List some features that you would look for in making your identifications.

3 Although protists are a diverse group of organisms, what features do all protists share? Do organisms in other kingdoms share these features?

4 You are working in a lab that cultures protozoans. What type of energy source should be in the growth medium?

LESSON 8.2 PLANTLIKE PROTISTS

Many protists are plantlike. They share one or more of the following characteristics with members of Plantae: They are phototrophs, they have cell walls, and they trap sunlight using chlorophyll. Here, we'll look at these similarities as well as the differences that place protists in a kingdom separate from plants.

ACTIVITY 8-2   *EUGLENA*: NEITHER PLANT NOR ANIMAL

The procedure for this Activity is provided as a blackline master in the TRB.

Euglena can be phototrophic or heterotrophic. When there isn't enough light for photosynthesis, they lose their chloroplasts and search for food. In this activity you will use two methods to try to persuade a phototrophic culture of *Euglena* to become heterotrophic.

FISHING WITH A SATELLITE

BIOLOGY IN CONTEXT

Imagine that you are on a research boat off the coast of the beautiful island of Tasmania. The sun is shining above, and a cool breeze sails you across the blue ocean, as you study the coastal fish population. The captain uses an odd map that looks like the weather map that you see during a forecast. It's a satellite image of the coast of Tasmania. The captain uses the map to find areas that have high concentrations of fish food.

The amount of fish food in the ocean is very important to the fisheries industry, marine biologists, and environmentalists. The National Aeronautics and Space Administration (NASA) and the National Oceanic and Atmospheric Administration (NOAA) have developed the coastal zone color scanner (CZCS), used on the *Nimbus-7* satellite, to measure the amount of energy given off by the ocean and the fish food in it.

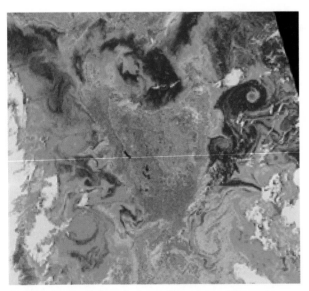

Satellite image of phytoplankton off the coast of Tasmania (center). The colors in this satellite image represent the various amounts of phytoplankton on the ocean surface. Yellows and reds indicate a large amount of phytoplankton, while blues and greens indicate small amounts of phytoplankton.

The fish food is plantlike cells called **phytoplankton** that live near the ocean surface. Phytoplankton contain chlorophyll, which

Sunlight, nutrients, and changes in water temperature are the four factors that will enhance the growth of phytoplankton. Pollution containing any form of organic matter or a change in current will provide nutrients for phytoplankton growth.

absorbs energy from the sun; CZCS can detect that energy. The fisheries industry uses satellite images of phytoplankton as the first indicator of change in the ocean environment. Scientists can use this information to monitor pollution, to determine the effects of El Niño and monsoons, to measure the ocean current, and to study fish.

PROTISTS—WHERE TO FIND THEM

Table 8.1 lists the major phyla of protists and some representative organisms in each phylum. Plantlike protists may be unicellular, colonial, or multicellular. They comprise the first five phyla in Table 8.1.

Common Name	Phylum	Practical Importance	Examples
Plantlike Protists			
Diatoms	Bacillariophyta	Form deposits of diatomaceous earth	Plankton species
Dinoflagellates	Dinoflagellata	Release toxins during red tides	Plankton species
Green algae	Chlorophyta	Aquatic food production	*Chlamydomonas, Volvox*
Red algae	Rhodophyta	Aquatic food production	Nori and other seaweeds
Brown algae	Phaeophyta	Aquatic food production	Seaweeds and kelps
Animal-like Protists			
Zooflagellates	Zoomastigophora	Animal parasites	Trypanosomes
Amebas	Rhizopoda	Zooplankton and intestinal	*Ameba* parasites
Apicomplexans	Apicomplexa	Animal parasites	*Plasmodium, Toxoplasma*
Ciliates	Ciliophora	Zooplankton	*Paramecium*
Funguslike Protists			
True slime molds	Myxomycota	Decomposers, turf pests	*Physarum*
Cellular slime molds	Acrasiomycota	Decomposers, turf pests	*Dictyostelium*
Water molds	Oomycota	Fish parasites, crop pests	Downy mildews

Table 8.1: Common Phyla of Protists

Unicellular plantlike protists are the main photosynthetic organisms in oceans, lakes, ponds, and streams. Their activity supports the entire aquatic food chain, so they are called primary producers. They are part of a diverse group of organisms that make up the plankton. The plankton are suspended at or near the surface of water bodies. For the most part they float or drift with the currents, although some organisms in the plankton move independently. They are a rich food source for many aquatic species, including small shrimp, young fish, and some whales.

WORD BANK

phyto- = plant

zoo- = animal

-plankton = drifting

Phytoplankton and zooplankton drift with the water current.

Unicellular algae, along with some photosynthetic bacteria, make up the part of the phytoplankton. Phytoplankton are consumed by other small floating organisms called **zooplankton.** Zooplankton include animal-like protists and a number of small animal species.

The algae in the phytoplankton include diatoms with their ornate shells (Figure 8.4A), dinoflagellates with their unusual

WORD BANK

dino- = force or energy

-flagella = whip

Dinoflagellates use whiplike "oars" to turn themselves.

flagella (Figure 8.4B), and unicellular and colonial green algae (Figure 8.4C), which have a lot in common with plant cells. Notice how green they are!

Although some species live in fresh water, diatoms are the most

A

B

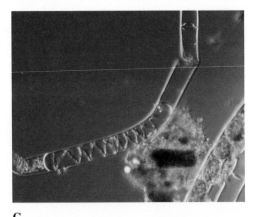

C

Figure 8.4: *Some common algae in the phytoplankton: (A) diatoms; (B) dinoflagellates (200×); (C) green algae*

abundant phytoplankton in the oceans. One liter of water from the surface of an ocean might contain as many as a million diatoms. You can identify the many species of diatoms by their shells. The shells, which have a shape characteristic of the species, are cell walls laced with calcium or silica taken up from seawater.

ACTIVITY 8-3   TRACKING DOWN DIATOMS

Obtain 1–2 oz of diatomaceous earth powder, which is sold commercially as aquarium filter powder. Allow students to work with very small quantities of the powder. Students should keep the powder away from their eyes and noses, for it contains silica crystals, which irritate eyes and can cause silicosis of the lung. A mere speck of diatomaceous earth powder contains thousands of diatom shells. Provide each lab station with a very small amount of powder on the tip of a lab spatula. Other equipment needed: a compound light microscope, slides, coverslips, distilled water. Have dust masks available for students who may want them.

When diatoms die, their shells sink to the bottom of the ocean. In some areas, diatom shells accumulate to several hundred feet in thickness (Figure 8.5). These accumulations, called diatomaceous earth, are mined from the ocean bottom and used in many different products, such as polishes, filters, insulation, insecticides, and fertilizers.

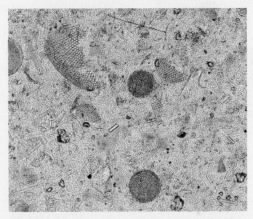

Figure 8.5: *A deposit of diatomaceous earth (50×)*

Place a tiny amount of diatomaceous earth powder on a slide. Add a drop of water and gently lay a coverslip on it. Examine the slide first at low power and then at the highest power. Focus until you clearly see a few whole diatoms (refer to Figure 8.4A). Describe and draw in your logbook the diatoms you see. Then answer the following questions in your logbook:

Differences in shells represent different species of diatoms.

The tiny shells make good filters and insulation. The sharp angles of the hard shells give them insecticidal properties (they scar the exoskeletons of insects).

- How many different sizes, shapes, and patterns of shells did you see? What do these differences represent?

- What properties of the diatoms that you observed suit them for the different uses described above?

DINOFLAGELLATES

Dinoflagellates are active members of a group of organisms propelled by two flagella that are oriented roughly at right angles to one another (refer to Figure 8.4B). They are the second most numerous protists making up marine phytoplankton, after diatoms.

Figure 8.6: *A red tide off the coast of Baja California.*

Dinoflagellates have red pigments that are most noticeable during algal blooms called red tides. An **algal bloom** is an overgrowth of algae, usually due to warm temperatures and/or an oversupply of nutrients, often from runoff containing nitrogen from fertilizers. During red tides, dinoflagellates release toxins that kill fish and contaminate shellfish (Figure 8.6). The toxins don't harm the shellfish, but the shellfish store up large concentrations of the toxins in their tissue. If you eat the shellfish, you could get severe food poisoning and even die.

Some dinoflagellates are strict heterotrophs, consuming organic particles or parasitizing animals.

GREEN ALGAE

Green algae include a wide variety of species, most of which live in fresh water. Some are unicellular. Some are colonial: The cells group together in loose associations. Cells of a colonial organism can survive on their own because each performs all of the functions needed to survive. In a true multicellular organism, cells are specialized, and each type of cell performs some unique function, such as sensing light or digesting food.

Chlamydomonas (Figure 8.3) is a unicellular green alga. Its two flagella move it toward light at the surface of the water. Its chloroplast has chlorophyll pigments that trap the sun's energy for photosynthesis. Like aquatic plants, protists of the phytoplankton get carbon from carbon dioxide, dissolved in water as carbonate ions (CO_3^{-2}). Ocean-living phytoplankton capture more than 70% of all the carbon taken up by Earth's phototrophs.

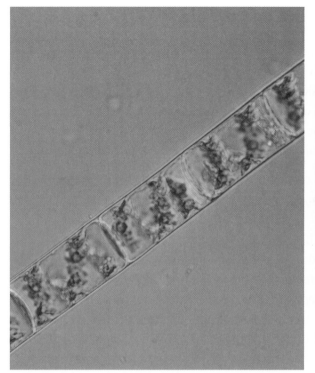

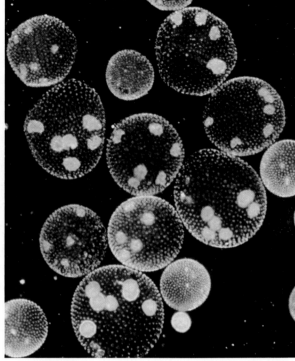

Figure 8.7: *Spirogyra is a filamentous colonial green alga.*   **Figure 8.8:** Volvox *colonies*

Many species of green algae have cells that resemble those of *Chlamydomonas* living in colonies. The colonies may be filaments, sheets, or spheres of cells (Figure 8.7). Strands of cytoplasm often connect adjacent cells of a green algal colony. In organisms such as *Volvox* (Figure 8.1B), the flagella of individual cells beat at the same time, moving the colony as one entity through the water. Some cells in a *Volvox* colony (Figure 8.8) function as gametes for sexual reproduction, an indication that *Volvox* may be a multicellular organism.

ACTIVITY 8-4 | COLONY OR MULTICELLULAR ORGANISM?

The procedure for this activity is provided in the TRB as a blackline master.

As a group, green algae put cells together in a lot of ways. Sometimes the cells are similar, and the alga is considered colonial. Other algae have vegetative or reproductive cells and specialized areas within the cell mass. There is no distinct line between what is a colony and what is a true multicellular organism. In this activity you will examine several forms of algae and try to determine whether they are colonial or multicellular.

Not all plantlike protists are free-living autotrophs. Some species live in symbiosis with small animals called corals that form reefs. A green alga that lives inside the cells of corals makes some reefs green. This coexistence is an example of mutualism, a symbiosis that benefits both partners. The corals protect the algae, while the algae provide oxygen and help to build the coral's skeleton with calcium deposits.

Reproduction in unicellular algae is largely asexual, like cell division in an animal or a plant. *Chlamydomonas,* for instance, replicates its DNA, then undergoes mitosis and cell division to produce spores that mature into offspring. When stressed by unfavorable conditions, *Chlamydomonas* will reproduce sexually by producing gametes, as shown in Figure 8.9. Gametes of different parent cells fuse to produce a zygote that produces a protective coat and lies dormant until conditions improve. The zygote will produce, by meiosis, four individuals that are released at germination. The resulting cells represent new phenotypes.

Sexual reproduction also takes place in colonial algae such as *Volvox*. Some cells of the colony act like haploid gametes that

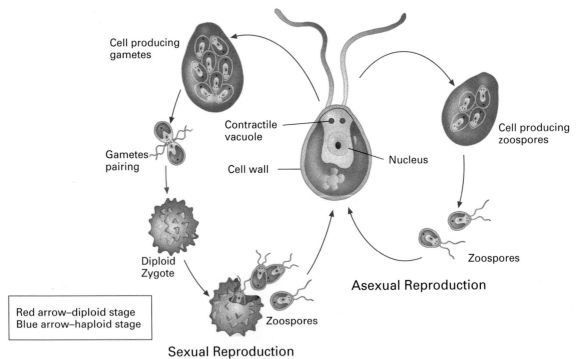

Figure 8.9: *Reproduction in* Chlamydomonas. *Chlamydomonas exhibit both sexual (left) and asexual (right) reproduction. In sexual reproduction two gametes of different parents fuse to produce a zygote. The zygote will produce four gametes by meiosis. In asexual reproduction Chlamydomonas undergoes mitosis to produce zoospores.*

fuse to produce diploid zygotes. The zygotes can divide by meiosis to produce daughter colonies of haploid cells. The parent colony breaks apart to release the daughter colonies to live on their own.

MULTICELLULAR PLANTLIKE PROTISTS

A number of algae are multicellular, including some green algae, brown algae, and red algae. These algae crudely resemble land

plants. They have a flat or a branching plant body called a **thallus,** made of filaments of cells (Figure 8.10). These filaments resemble some colonial species of algae such as *Spirogyra* (Figure 8.7). A stalk supports the thallus, and its other end grows a

A   **B**

Figure 8.10: *Multicellular algae: (A) red alga and (B) brown alga*

holdfast that attaches the algae to the bottom or some submerged object. Some large algae don't form a holdfast but instead float freely in the water.

WORD BANK

aqua- = water

-culture = to grow or cultivate

Aquaculture creates farms for fish and other aquatic creatures in fresh or salt water.

Red algae are the chief primary producers in coastal zones, especially in tropical waters. They are important components of many food chains for both marine animals and humans. A red alga called nori is served in soups and salads or used to wrap sushi. It is now grown by **aquaculture** on seaweed farms off the coast of

Maine. Red algae are also economically important because of the polysaccharides in their cell walls. Agar and carrageenan are polysaccharides that are extracted from red algae and used to thicken foods such as soups, ice cream, ketchup, and pudding. Agar is also used as a medium for bacterial growth.

Figure 8.11: *A stand of brown algae*

Brown algae grow along rocky coasts and are the chief primary producers in deep, cold seawater, where they often form large underwater "forests" (Figure 8.11).

Raising seaweed in offshore beds is not new. The Japanese and Chinese have done it for years, supplying the world with a tasty seaweed called nori. Thanks to Dr. Ike L., a marine biologist and seaweed farmer, nori is the newest aquaculture crop in Maine. Ike's company, Coastal Plantations International (CPI), has nori cultivation and processing facilities. Ike also has helped to develop markets for the nori.

To raise nori, Ike takes advantage of the two phases in its life cycle. Nori, a seaweed that normally inhabits cold coastal waters, alternates each year between a seeding phase and a blade phase. During the seeding phase, nori is a microscopic filament that produces "seeds." CPI grows "seeds" in oyster and clam shells in special tanks, controlling temperature and light.

The nori "seeds" are not actually seeds like those in true plants; they are spores.

The processing plant presses nori into sheets for wrapping sushi or for making soups, salads, teas, and candy. Pharmaceutical companies buy nori and extract pigments from it, to be used in medical procedures.

Some of the nori is used to begin the next crop. In the lab, blades of mature seaweed are induced to release male and female gametes. The gametes fuse to produce spores, which are collected to grow into the seeding phase in flasks or are allowed to settle directly onto clam, scallop, or oyster shells.

Ike lifts the edge of a nori net.

If you're not enthusiastic about growing seaweed, there are many other careers in aquaculture. For example, you could raise finfish and shellfish in tanks, cages, or other enclosures in natural or artificial bodies of water.

Several Internet sites have detailed information on nori aquaculture.

ALTERNATING LIFE CYCLES

The life cycle of multicellular algae may alternate between a haploid stage and a diploid stage. Some cells of the haploid form become gametes that fuse to form a diploid zygote (Figure 8.12). The diploid zygote germinates to become a multicellular stage. Some cells in the diploid stage undergo meiosis to produce haploid spores. These spores germinate to form the haploid stage. This alternation of stages, or alternation of generations, is characteristic of protists and simple plants.

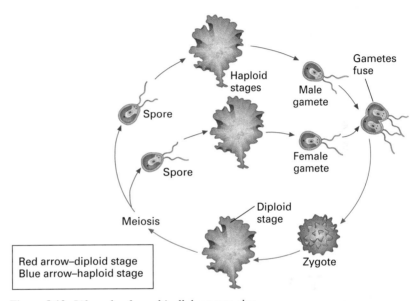

Figure 8.12: *Life cycle of a multicellular green alga*

EVOLUTIONARY RELATIONSHIPS

Green algae form an important branch on the evolutionary tree. They share a number of characteristics with plants: chlorophyll, cellulose cell walls, and starch as storage for energy. Protists like modern multicellular algae may have been the ancestors of the first true plants. You'll learn more about them when you begin studying plants.

LESSON 8.2 ASSESSMENT

1. Unicellular plantlike organisms make up the phytoplankton, which are the primary producers in the aquatic food chain.

2. Placing *Euglena* in the light and adding streptomycin will cause this effect in about a week. Placing *Euglena* in the dark will take longer to achieve a heterotrophic culture.

3. Look for cells that appear different from the rest, specialized cells, and connections between cells. These would be characteristic of a multicellular organism. Look for single cells of the same type elsewhere in the culture. This would indicate a colonial organism in which cells can survive independently.

4. The Irish moss is used to thicken the pudding.

1 Why would a decrease in the population of unicellular plantlike protists be important to a marine biologist studying the aquatic food chain?

2 Explain how to make a phototrophic *Euglena* change to a heterotrophic *Euglena*.

3 You are viewing a group of cells that appear to be connected. How would you determine whether the organism is colonial or multicellular?

4 You are making seaweed pudding for your class. Your recipe calls for Irish moss (seaweed), milk, salt, and vanilla. What is the purpose of the Irish moss in this recipe?

LESSON 8.3 ANIMAL-LIKE PROTISTS

A number of protists, the protozoans, are animal-like. Protozoans share the following characteristics with members of Animalia: They are heterotrophs; they lack a cell wall; they have specific structures for locomotion, either flagella, cilia, or pseudopodia; and their organelles carry out specific metabolic roles, as organs in an animal do. Table 8.1 on page 285 lists some phyla of animal-like protists.

AFRICAN SLEEPING SICKNESS

BIOLOGY IN CONTEXT

Dr. Michaleen Richer rested her hand on the head of Juliana Jima, a 15-year-old patient who lay curled beneath a wool blanket in the tropical heat. "Juliana!" said the doctor. "Juliana! Wake up!"

Like the other patients in this ward of metal-framed beds and whitewashed walls, Jima's central nervous system is being consumed by **trypanosomiasis,** the disease more commonly known as sleeping sickness. If she had not been treated, Jima would have lapsed into a coma and died.

Unchecked, sleeping sickness is extremely deadly. Transmitted by the bites of blood-sucking tsetse flies, the parasite multiplies and kills its host in about two years. Meanwhile, each infected person becomes a reservoir of parasites that spread throughout the human population. When a tsetse fly bites an infected person, it takes in parasites with the victim's blood. When it bites its next victim, it can inject the parasite with its saliva.

Workers for CARE International and the International Medical Corps have found that more than 20% of people tested in the remote corner of southern Sudan have sleeping sickness—an estimated 6,000 victims. In Ezo village, near the center of the 1997 outbreak, nearly half the population tested positive.

Health officials say it is impossible to control the tsetse fly, a honey-bee-sized insect that can travel great distances. Instead, they attack the disease by treating its victims—which means finding and treating everybody in the bush who is infected.

African sleeping sickness can be transmitted by the bite of the blood-sucking testse fly.

Doctors inject patients who are in the first stage of the illness with pentamidine. The drug is about 99% effective. Other patients in more advanced stages are less fortunate.

As the parasite multiplies and invades the central nervous system—causing lethargy and finally coma—the disease is treated with nine intravenous injections of the antiparasite drug melarsoprol. Melarsoprol is so toxic that 5–10% of patients treated with the drug die from the cure rather than the disease.

Modified and reprinted with permission from *The Philadelphia Inquirer*, July 18, 1997.

Pentamidine inhibits phospholipids, proteins, DNA and RNA synthesis by interfering with nuclear metabolism. Exactly how pentamidine inhibits nuclear metabolism is not known. Melarsoprol is believed to cause a mutation in the parasite by inserting itself into the parasite's DNA.

LEARNING LINK THE LIFE CYCLE OF A PROTOZOAN PARASITE

Have students work in pairs for this activity. The FDA Bad Bug Book is on the Internet. The Centers for Disease Control (CDC) and National Institute of Health (NIH) have large Web sites. The Merck Manual is on the Web with information on protozoan diseases. Text-based versions of much of this information is available, as well as home medical encyclopedias. Unless your class is extremely large, no more than two groups should have to take the same protozoan.

Using library and Internet resources recommended by your teacher, research the life cycle of a protozoan that is a human or animal parasite. Make a poster of the parasite's life cycle, showing sample hosts. Indicate which stage(s) of the life cycle causes infections and which stages are targeted by various prevention or treatment strategies. Present your poster to the class.

ZOOFLAGELLATES

WORD BANK

zoo- = animal

-flagella = whip

Zooflagellates are animal-like protists with flagella.

Protozoans are classified by their method of locomotion. Members of one of the most-studied groups, the **zooflagellates,** use flagella to move. Zooflagellates have a thick layer of protein called a pellicle inside their cell membrane (Figure 8.13). Protruding from the pellicle are one or more flagella used in locomotion.

Zooflagellates are known worldwide as parasites of humans and other animals. Depending on the species, they can infect the intestine, genital organs, blood, and other organs of their host. Outside of their host, you can find them in contaminated water and soil.

An intestinal zooflagellate that you may know too well is *Giardia lamblia*. It thrives in water that has lots of organic matter, often from sewage and animal waste. It is a problem for bathers who use natural

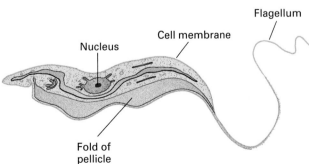

Figure 8.13: *Structure of a typical zooflagellate*

swimming holes and campers who drink untreated stream water. If you ingest *Giardia,* you'll experience symptoms that include cramps, nausea, and diarrhea. You might have to take a powerful drug to get rid of the organisms.

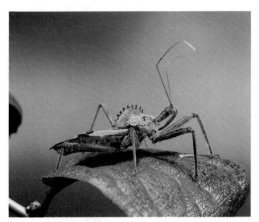

Figure 8.14: *Called a kissing bug, this insect carries a zooflagellate that can infect humans with a serious disease.*

Zooflagellates that enter the blood and move to tissues are a more serious problem. A group of zooflagellates called **trypanosomes** cause diseases such as African sleeping sickness and Chagas' disease, both of which can eventually kill the human host. In Central and South America, as well as the southern United States, people who have been bitten by a "kissing bug" (Figure 8.14) may become infected with a species of trypanosome. The kissing bug is a blood-sucking insect that tends to bite its victims in the tender skin around their mouth. This species, which causes Chagas' disease, can invade a host's heart muscle and eventually cause heart failure.

AMEBAS AND THEIR RELATIVES

An ameba is a protozoan that moves by changing the shape of its cell. The cytoplasm extends into fingerlike projections called pseudopodia (Figure 8.2B) that surround food particles. An ameba uses endocytosis to absorb the food.

Amebas live primarily in fresh water and in the intestines of various animal hosts. When an intestinal ameba enters its victim, it is in a protected **cyst,** a resistant cover (Figure 8.15). The cysts pass undamaged through the acid of the stomach. Once in the intestine, the amebas emerge from their cysts and attach to the intestinal wall. They cause nausea and abdominal discomfort. If amebas penetrate the intestinal wall, they can cause **dysentery,** an intestinal disease with symptoms such as bloody diarrhea, nausea, vomiting, and sometimes liver damage.

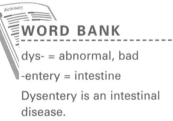

WORD BANK

dys- = abnormal, bad

-entery = intestine

Dysentery is an intestinal disease.

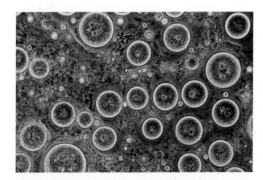

Figure 8.15: *Cysts of an intestinal ameba*

Intestinal amebas are a problem in areas with poor sanitation. Ameba cysts are found in water or soil that has been contaminated by feces, and they can survive in chlorinated drinking water.

Some relatives of the ameba have shells made of calcium or silica and live in the ocean (Figure 8.16). Long, spinelike pseudopodia protrude through the shells and trap food particles in a sticky secretion. The secretions contain enzymes that begin digesting food outside the cell. The food is eventually taken into the cell by endocytosis and completely digested.

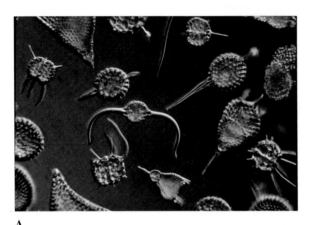

A

B

Figure 8.16: *Some ameboid protists have intricate shells. (A) radiolarians and (B) foraminiforans (50×)*

INVESTIGATION 8A   MICROFOSSILS IN LIMESTONE

See Investigation 8A in the TRB.

When shelled protists die, their shells sink to the bottom of the ocean and accumulate as deposits of limestone and chert, a rock containing quartz and silica. The white cliffs of Dover, England, are formed of such deposits. Biologists who study fossils have been able to recover these shells and use them to date deposits from around the world. In this investigation you will recover shells and compare them to the diatoms that you observed in Activity 8-3.

APICOMPLEXANS

Apicomplexans used to be called sporozoans.

Apicomplexans have no organelles for locomotion (Figure 8.17). They are parasites that invade the blood and other tissues of their animal hosts and cause some nasty problems.

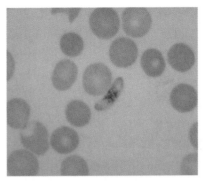
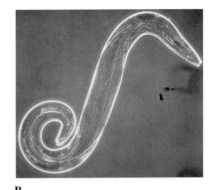

A **B**

Figure 8.17: *Cyst (A) and motile (B) forms of* Plasmodium

dictionary

WORD BANK

api- = apex, tip

-complexan = referring to a system of components

The term "Apicomplexan" refers to a system of organelles at the tip of the organism, used to penetrate host cells.

An apicomplexan alternates between two or more host species, usually reproducing asexually in one species and sexually in another. Cysts of some apicomplexans pass from host to host in feces.

The apicomplexan known as *Toxoplasma* can be a problem for cat owners. An infected cat releases cysts in its feces. The person who cleans the litter box can pick up these cysts either through inhaling them or by ingesting them (Figure 8.18). *Toxoplasma* cysts are most dangerous for a pregnant woman, who could pass the cysts to her unborn child. The baby could be born mentally retarded, visually impaired—or even dead.

The most widespread threat to human health among the apicomplexans are the *Plasmodium* parasites that cause malaria. These parasites reproduce sexually in certain species of host mosquitoes. When an infected mosquito bites a person, apicomplexans enter the person's blood and travel to liver cells, where they multiply but do little harm.

Figure 8.18: *Pregnant women should avoid contact with cat litter boxes.*

After a week or two, the liver suddenly releases the apicomplexans into the blood. There, they attack and rupture the victim's red blood cells. The symptoms—high fever, shaking chills, headache, backache, nausea, and muscle soreness—come and go about every

WORD BANK

pneumo- = lung

-cystic = like a bladder or sac

Pneumocystic organisms attach to the lining of the air spaces in the lung.

48 hours for two weeks. Attacks can recur for two to three years. Malaria infection impedes blood circulation; moving blood clots form, and blood vessels weaken or rupture. One species of *Plasmodium* can kill or put its host in a coma in a few hours.

Another widespread apicomplexan, *Pneumocystis carinii,* causes a rare form of **pneumonia,** a lung infection, in people whose immune systems are weak. AIDS patients often suffer from pneumocystic pneumonia because their immune systems have been damaged.

CILIATES

Ciliates are protozoans that move by beating cilia, arranged in rows along their surface (Figure 8.19). Most species are zooplankton in salt or fresh water. Some are mutualistic symbionts inside the bodies of animals, where they assist in digestion, and some are parasites. One example of a mutualistic relationship is the ciliate that lives in the termite. Termites are well known for eating wood structures. Although termites ingest wood, ciliates in the termites' gut actually digest the wood components. Without the ciliates, a termite would starve to death, because it doesn't

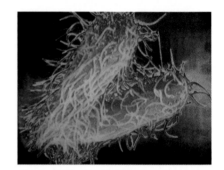

Figure 8.19: *The hairlike structures on the surface of this ciliate are called cilia.*

produce enzymes that break down the cellulose in wood. Many pesticides that are designed to kill termites actually destroy their gut ciliate populations.

Ciliates are large for protozoans and are structurally very complex. Like the trypanosomes, they have a pellicle (Figure 8.20), which gives the cell a definite shape and flexibility of movement. *Paramecium,* for example, can move forward or backward, turn, and even revolve about itself.

Ciliates are unique because of their specialized organelles. One organelle that you may have seen in Activity 8-3 is the **contractile vacuole.** Since most paramecia live in fresh water, a hypotonic environment, they take up water rapidly through the cell membrane. The contractile vacuole pumps out excess water.

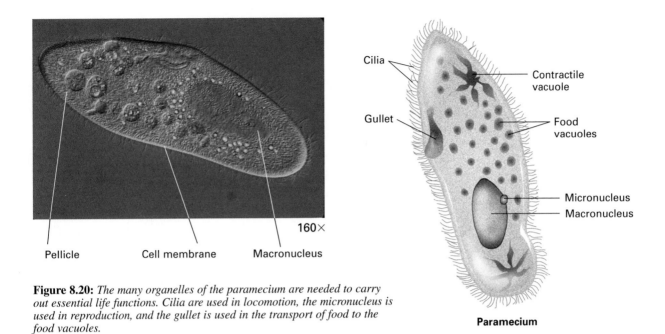

Pellicle   Cell membrane   Macronucleus

160×

Cilia
Gullet

Contractile vacuole
Food vacuoles
Micronucleus
Macronucleus

Paramecium

Figure 8.20: *The many organelles of the paramecium are needed to carry out essential life functions. Cilia are used in locomotion, the micronucleus is used in reproduction, and the gullet is used in the transport of food to the food vacuoles.*

Another organelle is the trichocyst (Figure 8.21), which contains needle-like fibers that can shoot out from the cell membrane. Paramecia use trichocysts for defense.

Some ciliates form stalks to hang onto rocks and soil particles. Their cilia sweep a constant stream of food particles toward the oral cavity. This technique, filter feeding, is a tactic that many aquatic animals use. Food particles reaching the oral cavity enter the cell through endocytosis. Food vacuoles engulf the particles and digest them.

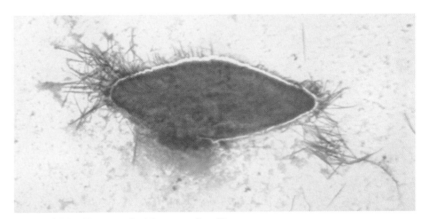

Figure 8.21: *Close-up of trichocysts of a ciliate*

LESSON 8.3 ASSESSMENT

1. Because of the spread of this parasite and other sexually transmitted diseases (STDs), partners should have protected intercourse or abstain altogether. Anyone who is diagnosed with the STD should seek treatment immediately.

2. A zooflagellate has a pellicle and flagella.

3. Ameba.

4. Locomotion, digestion, and elimination of wastes are three answers that should be obvious even before students study animals.

1 A zooflagellate called *Trichomonas* is transmitted by sexual intercourse. The vagina is infected, leading to a discharge, inflammation, and discomfort. An infected male usually does not show symptoms. If you were a public health official, what steps would you recommend to control the spread of *Trichomonas*?

2 Your instructor gave your class a tray of water samples taken from a nearby lake. The class was asked to check the water samples for zooflagellates. How would you identify the zooflagellates that are present in the sample?

3 A soil sample was taken from a farmer's pigpen. Microscopic analysis of the sample revealed a protozoan that changes its shape when it moves. It was also noted that the protozoan forms a cyst under acidic conditions. What would you call this protozoan?

4 What activities carried out by protozoans are also carried out by an animal?

LESSON 8.4 FUNGUSLIKE PROTISTS

Now it's time to look at some of the funguslike protists. Funguslike protists share these characteristics with species in Fungi. They are heterotrophic, they grow by producing a threadlike mass, and they are decomposers or parasites. Table 8.1 on page 285 lists some phyla of funguslike protists.

SLIME MOLD PROBLEMS

BIOLOGY IN CONTEXT

A sudden cloudburst on a cool spring day sends you running for shelter. The same weather really wakes up a slime mold. In the rain, slime mold cells begin to move along the ground like protozoans, consuming decaying organic matter and bacteria as they go. Before long, the first pair of cells collide and fuse to form a zygote.

The zygote grows rapidly in rich decaying matter, such as a fallen log. It may become a colorful, slimy mass called a **plasmodium** (Figure 8.22).

Figure 8.22: *A slime mold is a nuisance to people who grow sod and other plants.*

While a slime mold may add mystery and color to the surroundings, it doesn't thrill a turf grass manager. People who maintain lawns, golf courses, and sod farms know that slime molds can cover their grasses, making them vulnerable to pathogens. Turf managers spring into action when they see a plasmodium sending up stalks, each capped with new spores. The cap will burst and release spores, which disperse and repeat the cycle. Unless the mold is raked up or treated chemically, bright yellow slime can cover an entire lawn.

ACTIVITY 8-5   SLIME MOLD PLASMODIUM

The procedure for this activity is provided in the TRB as a blackline master.

Slime molds grow best in moist places with lots of nutrients. A slime mold will thrive on an agar plate if you add some nutrients to the agar. In this activity you will examine a slime mold under a microscope.

SLIME MOLDS

A slime mold forms a slimy mass on dead leaves, rotting logs, and wet ground. This plasmodium can form in one of two ways, depending on the type of slime mold that produces it.

A true slime mold produces a plasmodium from haploid spores that differentiate to form either ameboid cells or cells with flagella (Figure 8.23). These cells also serve as haploid gametes, capable of sexual reproduction. When gametes of two mating types meet, they fuse to form a diploid zygote. This zygote begins to consume organic matter around it by phagocytosis and grows outward along the ground. It grows through repeated mitosis, resulting in a single-celled plasmodium with many nuclei to control its activities.

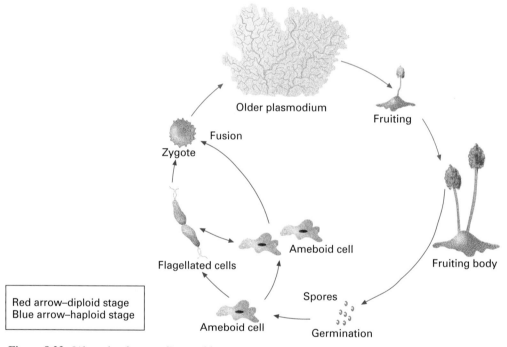

Figure 8.23: *Life cycle of a true slime mold*

In the second type of slime mold, haploid spores develop into ameboid cells that divide rapidly as they move (Figure 8.24). The cells produce cyclic AMP, an important chemical messenger in many organisms. The message in this case tells the cells to clump into a slug-shaped plasmodium. As the slug moves slowly along, it consumes organic matter by phagocytosis and grows. Because its plasmodium is cellular (made of individual cells), this protist is

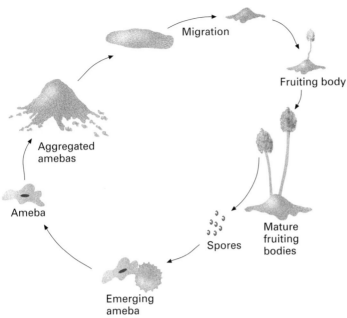

Figure 8.24: *Asexual phase of a cellular slime mold*

called a cellular slime mold. Cellular slime molds can reproduce sexually if ameboid cells of different mating types meet up and fuse.

Often a plasmodium runs—or oozes—into an environment that does not have enough moisture or organic matter to support further growth. As a defense mechanism, the plasmodium stops growing and develops **fruiting bodies,** reproductive structures in which spores form. When conditions are right, the fruiting bodies rupture and release their spores. The spores develop into flagellated or ameboid cells, which continue the life cycle.

INVESTIGATION 8B GROWTH OF A TRUE SLIME MOLD

See Investigation 8B in the TRB.

The growth of a slime mold depends on its environment. Under the right conditions a slime mold can cover many square feet. In this investigation you will manipulate the environment in which a slime mold is growing. You will test one or more factors that you think might promote growth (if you like slime molds) or inhibit growth (if you plan on a career in turf management).

WATER MOLDS

A water mold is a network of branching filaments (Figure 8.25) that penetrates the tissues of other organisms, either living or dead.

Water molds are important parasites of plants and animals and decompose algae, fish, and other organisms in aquatic environments.

Water molds are easily mistaken for true molds and mildews and were once classified as Fungi. However, they have a unique form of reproduction in which the tips of their filaments produce asexual flagellated spores. These spores disperse and form cysts that may lie dormant.

Figure 8.25: *Growth form of a water mold*

When a cyst lands on the body of a host, it germinates and begins to form the network of filaments that damage or, in the case of decomposers, break down the host tissue.

Some of the best-known water molds attack plants. Such infections, often referred to as mildew, decrease agricultural production of potatoes, grapes, and other crops (Figure 8.26). Water molds may also grow on fabrics and may attack camping gear that is not dried thoroughly before it is stored.

Figure 8.26: *Downy mildew, a type of water mold, is a threat to vineyards.*

EVOLUTIONARY RELATIONSHIPS

Although the threadlike growth habit of some funguslike protists seems identical to that of fungi, these protists are not considered close relatives of fungi. For one thing, funguslike protists sometimes have cells with flagella, which fungi never have. Many make cell walls of cellulose. Other evidence suggests that ancient red algae may have been the ancestors of some types of fungus.

LESSON 8.4 ASSESSMENT

1. It's a water mold.

2. Fruiting bodies develop from a plasmodium when the food supply has been exhausted.

3. The true slime mold has a threadlike plasmodium that contains only one cell. The cellular slime mold has a sluglike plasmodium that is made up of individual cells.

4. A parasitic water mold infects its host by penetrating it with a network of branching filaments. A water mold that is a decomposer does the same thing to dead organic matter.

1 Phytophthora blight is a common fruit infection that produces a white funguslike growth on the surface of plant lesions. Fruits eventually shrivel up but will remain on the plant for several days. How would you classify this protist?

2 At what point in the life cycle of a slime mold do fruiting bodies develop?

3 During class, your instructor asked you to examine several slime mold specimens. Explain how you would distinguish between true slime molds and cellular slime molds.

4 What do water molds that are parasites have in common with water molds that are decomposers?

CAREER APPLICATIONS

The applications that follow are like the ones you will encounter in many workplaces. Use the biology you learned in this chapter to complete the activities. Share your work with the class.

AGRICULTURE & AGRIBUSINESS

Diatomaceous Earth Can Control Parasites

Diatomaceous earth can be used to control certain plant pests and internal and external parasites on pets and livestock. Using a variety of sources such as animal care or gardening books and magazines, veterinarians, nursery workers or landscape contractors, and the Internet, find out what types of pests and parasites diatomaceous earth can control, how it is used, what precautions should be taken, and how it works. Make a brochure describing its uses. Include a chart that lists specific pests or parasites, the plant or animal being protected, and various uses of diatomaceous earth.

Preventing Crop Losses

Talk to an agricultural extension agent or a farmer about crop damage caused by protists. Alternatively use books, periodicals, or the Internet to find out about slime molds and water molds. Find out how much of a problem there is with slime and water molds in your area, how the problem is diagnosed, how it is treated, and steps that can be taken to prevent the growth of slime molds and water molds.

Termite Control

Talk to an exterminator about various methods of treating and preventing termites. Which methods involve killing the gut ciliates that digest the wood? Are these methods still used? What method does the exterminator you talk to prefer to treat existing termites? to prevent termites? Prepare a chart of methods and tell how each works, how often must the treatment be repeated, and how much it costs.

BUSINESS & MARKETING

Selling Algae

Using the Internet, books on algae, and other sources, research the range of products that are made from algae. Select one type of algae (remember that blue-green algae are not Protista) and create a marketing strategy for someone who is harvesting that type of algae. Come up with a name for the company you are representing and a slogan. What categories of companies or industries might you target? What processing will you have to do before delivery?

Red Tide

Research the impact of red tide on the shellfish and finfish industries. Write an article for the business section of the newspaper describing the short-term and long-term effects on the economy.

FAMILY & CONSUMER SCIENCE

Seaweed in Your Food

Three seaweed derivatives are commonly used as stabilizers, thickeners, and colorants in our foods. They are alginates, carrageenan, and beta carotene. Research these products on the Internet to find out which of the three main groups of seaweed—brown algae, green algae, and red algae—they come from and what they are used for. Examine the food packages you have at home and list any that contain one or more of these food additives. Collect the labels of several products (after the products have been used) and make a class collage of the labels.

Tonic Water

Research the history of tonic water and why quinine is included as an ingredient. From what plant does quinine come? What disease caused by a protist is quinine used for? Is tonic water still prescribed today? Write up a brief report. Include a sketch of the plant quinine comes from.

Toxoplasmosis

Prepare a brochure for pregnant women with pet cats about the dangers of toxoplasmosis and how it can be prevented. Describe the life cycle of the organism, how it can be transmitted, and the problems it can cause in an unborn child. Describe why toxoplasmosis is generally not a problem for healthy adults.

Sexually Transmitted Diseases

Investigate sexually transmitted diseases caused by protists. Prepare a brief report on their symptoms, treatment, prevention, and short- and long-term effects.

Wilderness Water

On an extended wilderness trip, you might not be able to carry enough drinking water to last the entire trip. You might have to drink water from the local streams or lakes. Discover and compare the various ways in which you can make safe drinking water by removing or killing bacteria and protozoans. Brainstorm criteria, such as effectiveness and taste, that you will use in comparing the methods. Prepare a brochure for hikers and campers describing why water should be treated and outlining various methods. Include information on diseases that you can get from drinking untreated water, their symptoms, and first-aid tips.

INDUSTRIAL
TECHNOLOGY

Comparing Harvesters

Compare the equipment used to harvest a crop like nori from the ocean to the equipment used to harvest a crop grown on land. What are some of the considerations that must be taken into account when building equipment that is used in the ocean environment, which is quite corrosive to many materials? Create a poster showing the differences. Present your poster to class.

Industrial-Strength Diatoms

Diatomaceous earth is used by a wide variety of industries as a filter, a filler, insulation, and a mild abrasive. Use the Internet to find out about one or more of these uses and the industries involved. Locate a company either locally or via the Internet that uses diatoms in one of the ways you researched. Is the diatomaceous earth part of a process or part of a product? If it is part of a process, how often does it have to be replaced? What happens to used diatomaceous earth? Inhaling diatomaceous earth is a health hazard. How are workers protected from this hazard? Write your findings in your logbook and report them to your class.

CHAPTER 8 SUMMARY

■ Protista is a kingdom of unicellular eukaryotes and some colonial and multicellular species that don't fit into other kingdoms.

■ Protists include phototrophs, heterotrophs, and a few species that may be either phototrophs or heterotrophs. Some protists form symbiotic associations with fungi, animals, plants, or other protists.

■ All protists reproduce asexually, and many can also reproduce sexually. The latter have a life cycle that includes an asexual stage alternating with a sexual stage.

■ Some protists are plantlike. They are phototrophs with light-absorbing pigments like those of plants, and many have cell walls. The protists that are known as green algae may be related to the ancestors of plants.

■ Protists make up much of the phytoplankton of oceans and bodies of fresh water. As primary producers they supply food for aquatic food chains.

■ Some types of algae, especially green algae, form colonies of cells. Each cell resembles a unicellular alga, is unspecialized, and could survive on its own.

■ Multicellular algae have a branching plant body composed of filaments of cells. Multicellular green algae are important primary producers in fresh water; red and brown algae are important primary producers in salt water.

■ Some protists form mutualistic and parasitic relationships with another species.

■ Some protists are animal-like. They are heterotrophs, lack a cell wall, and have specific structures for locomotion. Their organelles function like the tissues and organs of an animal. Animal-like protists are called protozoans.

■ Some zooflagellates are parasites of humans and other animals.

■ Amebas move by means of pseudopodia. They are important zooplankton in fresh water and intestinal parasites of animals.

■ Apicomplexans are protozoans that have no organelles for locomotion. They are important parasites in the blood and tissues of animals.

- Ciliates are important zooplankton in salt and fresh water and can be symbionts in animals. They have a number of highly specialized organelles, such as a contractile vacuole.

- Some protists are funguslike. They are heterotrophs and grow as a threadlike mass. They are important as decomposers (slime molds) or as parasites (water molds).

- A slime mold produces a slimy mass called a plasmodium, which decomposes organic materials as it moves.

- A water mold is a network of branching filaments that is either a decomposer or a parasite of aquatic animals or plants.

- The evolutionary relationships within the kingdom Protista are not clear.

CHAPTER 8 ASSESSMENT

Concept Review

See TRB, Chapter 8 for answers.

1 How are apicomplexans passed on to a host?

2 What characteristic do you use to classify protozoans?

3 What is the function of pseudopodia in amebas?

4 There is an algal bloom in the reservoir of your city's water supply. What problems might occur at the water treatment plant?

5 Why are unicellular algae ecologically important?

6 By what process do phytoplankton convert sunlight into chemical energy?

7 Several of you have been on an extended camping trip. You used all of the drinking water you brought with you and refilled your water bottles from the beautiful clear lake near the campsite. You didn't treat the water before drinking it. Now everyone in the camping party is having severe diarrhea and is becoming dehydrated. Choosing from the organisms studied in this chapter, what protist might be causing the problem?

8 Untreated sewage that enters lakes and ponds has several unpleasant consequences, one of which is the formation of algal blooms. Why does sewage promote algal growth?

9 An epidemiologist goes to a remote village in Africa to try to control an unknown disease that has killed half the village. She has determined that the cause of the disease is an intestinal protozoan that has contaminated the village drinking water. After the drinking water is treated with chlorine, other villagers became infected with the protozoan. Why didn't the chlorine treatment kill the protozoan?

10 List ways in which protozoan parasites might enter their host.

11 Discuss the differences between zooflagellates and ciliates.

12 *Euglena* have enzymes that can't replicate chloroplast DNA at high temperatures, but the enzymes that replicate nuclear DNA do function at high temperatures. *Euglena* grows at a temperature of 25°C. If you attempted to grow *Euglena* at 37°C, what would happen?

13 Why is the kingdom Protista considered a taxonomic dumping ground for species?

14 What is the function of fruiting bodies of slime molds?

15 What is the ecological importance of water molds?

16 A microbiologist describes the appearance of a group of cells as a slug-shaped mass that moves slowly along the ground. What kind of protist is the microbiologist describing?

17 Mosquitoes can pass malaria to humans. How are the mosquitoes infected?

18 How do phytoplankton indirectly feed humans?

Think and Discuss

19 *Trichonympha* is a cellulose-digesting protist that breaks down wood in the intestines of termites and wood roaches. Testing the effects of high levels of oxygen on the mutualistic relationship between *Trichonympha* and wood roaches, a researcher placed a wood roach in a bell jar full of oxygen. She put wood in the jar as a source of food. The wood roach died of starvation after several days. Why did the wood roach starve? (*Hint:* The high level of oxygen increased the level of oxygen in the organism.)

20 Describe the advantages and disadvantages of sexual and asexual reproduction of a protist.

21 The great blooms of the dinoflagellate *Gonyaulax polyhedron* (red tide) contain a toxin that can cause paralysis and even death if ingested in large amounts. A person swimming in a red tide could not possibly swallow enough *Gonyaulax* to be harmed. Why does eating shellfish from areas infected with red tide cause serious illness?

22 In the late 1980s a very salty environment developed in the Florida Bay, because of low rainfall and the diversion of fresh water from the Everglades. Turtle grass grew throughout the bay, forming dense, lush beds. The increased salinity also increased slime mold growth that killed the turtle grass. Suggest how the Florida Department of Environmental Protection could control slime mold growth in the Florida Bay.

In the Workplace

23 You have a lawn business. One of your customers is complaining that snails or slugs are attacking his lilies and other plants. He has small children who play in the yard, and he doesn't want you to use chemicals on the plants. What might you use?

24 Workers at a water treatment plant take samples around the reservoir at four locations close to where creeks and rivers enter. Would you expect the algae populations to be similar at all locations? Give a reason for your answer.

25 You are a travel agent. One of your clients is concerned about dysentery. You have a brochure that describes precautions such as not drinking water unless it has been boiled, not drinking bottled drinks poured over ice, and not eating raw fruits and vegetables. Explain why you would take each of these precautions.

Investigations

The following questions are related to Investigation 8A:

26 Limestone is made up primarily of calcium carbonate. What protists contribute to the calcium in limestone?

27 Describe how formaminiferans, radiolarians, and diatoms are different in shell structure.

The following questions are related to Investigation 8B:

28 On the basis of your experiments with the slime mold *Physarum,* in what conditions would you expect slime molds to be the greatest nuisance?

29 How can you supply a slime mold's requirements for moisture and nutrients if you want to grow it on an agar surface?

CHAPTER 9

WHY SHOULD I LEARN THIS?

How do you keep mold out of the bread and jam in your kitchen and mildew out of your clothes? If you learn about the weak points of these fungi, you can fight them, as the workers do who control fungi that cause diseases in people, livestock, and crops. People have also learned to put fungi to work in baking, brewing, food processing, antibiotics, and other medicines. Fungi can be friends or foes, but they'll always be part of your life.

FUNGI

WHAT WILL I LEARN?

1. how fungi grow, reproduce, and obtain nutrients
2. the differences among the four types of fungi
3. fungi's roles in natural communities
4. how various industries and occupations use and control fungi

---

What do you think of fungi? Does the word *fungus* make you think of mushrooms? Of the mold on some stale bread? Both are fungi, and there are plenty of other types. Fungi grow just about anywhere, so they have to be versatile and diverse. Next time you take a walk in the park, try to spot the fungus. It's in the soil and on the roots of tree seedlings; without funghi, survival would be a lot harder for these seedlings.

Visible or invisible, fungi are vital inhabitants of this planet—and you should know something about them.

LESSON 9.1 WHAT ARE FUNGI?

The kingdom Fungi includes mushrooms, molds, yeast, and the microscopic organisms that cause athlete's foot. It's hard to see what characteristics these organisms have in common just by looking at them. In this lesson you will learn what traits all these organisms share and why they are all classified as Fungi.

ACTIVITY 9-1   FUNGI UP CLOSE

Make certain that you and your students do not have any mold or mushroom allergies before performing this activity. Provide dust masks, disposable gloves, and goggles for students. Each pair of students will need a whole common edible mushroom and a sample of moldy bread or other food. You may prepare the latter by sprinkling slices of bread with water and keeping them in a plastic bag in a warm, dark place for several days before conducting this activity, or you might prefer to have students do this themselves. Use homemade or bakery bread, which do not contain antifungal preservatives. **Caution students not to eat anything in the lab and to wash their hands before leaving.**

If your refrigerator contains a package of mushrooms, a very old lemon encrusted with green mold, and a wrapped-up bit of blue cheese, then you're the proud possessor of at least three species of fungi! In this activity you'll find similarities between some very different-looking fungi. You will also learn to recognize a fungus when you see it in your refrigerator.

Working with a partner, use a hand lens or stereo microscope to observe the mold on a piece of food. Describe and draw the mold in your logbook. Then use forceps to transfer a bit of the mold to a microscope slide. Add a drop of water, and use the forceps or a dissecting needle to pick the mold apart in the water drop. Add a coverslip and observe the mold at low power and then at high power. Draw and describe the mold in your logbook.

Examine a mushroom and record your observations in your logbook as you work. Separate the stalk and the cap. Split the stalk lengthwise with your hands, and observe how it breaks. Crumble a piece of stalk and observe the shape and direction of its fibers. Does the stalk break more easily lengthwise or crosswise?

When you have finished your observations, clean up and return your materials, and wash your hands with soap and water. Answer the following questions in your logbook:

- What structures make up a mass of mold?

- How could a solid structure such as a mushroom be formed by a mold?

- What features do mushrooms and molds share?

- What other features do fungi share?

Molds consist of thin, threadlike fibers.

Mushrooms are formed by tightly packed fibers.

The fibers are a common structural unit that fungi share.
All fungi are heterotrophs, and most are decomposers.

Conservators care for and preserve the collections in museums, libraries, and historical societies. Because molds can severely damage paper and other materials, conservators take these fungi very seriously.

Saundra B. is a conservator who works in a library that houses many rare and valuable documents. The library holdings range from fragments of second-century papyrus to manuscripts written by modern authors. The collection includes artworks, textiles, photographs, puppets, and much more, but most of the materials are books and papers.

"Molds need a freely soluble carbon source to grow," says Saundra, "and that's found in many papers—in the paper itself, in the sizing, and in impurities in the paper."

Saundra shows us a document she's been working on: a manuscript by a well-known twentieth century author. She says, "The importance and the beauty of the literary works here make my job what it is. We can preserve such works for many years, but only if we control conditions inside the library such as temperature, relative humidity, and pH. Relative humidity is a factor because nutrients can diffuse through the cell membranes of molds only if there is moisture to carry them through. Good air circulation helps to keep down mold populations, too.

"Sometimes, the care of paper entails procedures such as washing it to enhance its chemical stability or removing some types of glue or tape. Ultraviolet light is often used in research labs to kill molds, but we don't use that as a control method because light can damage paper, books, and other artifacts in other ways."

Many colleges offer art and book conservation courses that may last 6 to 24 months. Graduates usually spend a year as an apprentice before they begin to work independently.

STRUCTURE AND GROWTH OF FUNGI

Most fungi look like hair. Their long, thin cells form strands called **hyphae** (plural of **hypha**) (Figure 9.1). As you saw in Activity 9-1, even thick structures such as mushrooms are simply tightly packed

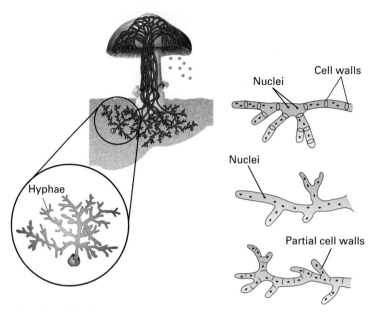

Figure 9.1: *Threadlike hyphae are the most typical fungal cells. Cells in a hypha may be fused together or have more than one nucleus. Large, thick fungal organs such as mushrooms are often compressed bundles of hyphae.*

masses of hyphae. Fungi grow quickly by elongating their hyphae. They spread out and explore their environment, seeking available food, though they cannot move. Grocers, grain shippers, and millers quickly learn that it is nearly impossible to control fungi by simply removing visible mold. Individual hyphae extend far beyond the visibly infested parts.

When these "pioneer" hyphae grow into areas where food is not plentiful, they get nutrients from the rest of the fungus. Two special features of fungal cells make this possible. First, the walls between fungal cells are often incomplete or missing (Figure 9.1). Cytoplasm and organelles can move freely between cells. Second, a form of active transport called **cytoplasmic streaming** causes nutrient-rich fluid in the cytoplasm to flow toward the growing tips of hyphae. Cytoplasmic streaming can often be seen with a microscope.

Cytoplasmic streaming also occurs in some plant cells and protists.

Fungal cells differ in other ways from members of the other kingdoms. Unlike the cell walls of plants, bacteria, and many protists, many fungal cell walls are made of **chitin,** a tough, nitrogen-containing carbohydrate. Many fungi also have more than one nucleus in each long cell. In species with incomplete or no cross walls, a mass of hyphae may contain a single large body of cytoplasm with many nuclei.

FUNGAL NUTRITION

An Additional Activity, Rotten to the Core, can be found in the TRB.

Most fungi need moist conditions to grow well. Evaporation from their large surface area can quickly kill fungi. Fungi also need water to dissolve the nutrients in their food before they can absorb them. Foods with lots of salt or sugar are hypertonic, so water tends to flow out of the cells of fungal invaders. This water loss makes growth difficult for fungi. Lumberyard operators and home builders in warm, moist climates struggle to prevent fungi from damaging wood and damp basements. Health care workers in humid tropical climates must treat injuries quickly, before they become infected with fungi.

Many fungi grow on dead organisms and animal wastes (Figure 9.2). Before absorbing these materials, they release enzymes that break them down into amino acids, sugars, and other simple substances. This is an example of **external digestion.** (Animals that swallow their food digest it internally.)

Figure 9.2: *The hyphae of this oyster mushroom produce enzymes that break the dead log into small carbohydrate molecules that the hyphae can absorb.*

ACTIVITY 9-2　SPORE PRINTS

After confirming that students do not have allergies to fungi, provide them with fresh edible mushrooms in which the undersides of the caps are still closed or are just beginning to open. Provide dust masks, disposable gloves, and goggles for students. If possible, provide common button mushrooms to some students and other edible species to others. Use only cultivated mushrooms from a supermarket or other reputable source. **Caution students against collecting wild mushrooms and against eating in the lab.**

If you want to grow mushrooms or prevent the spread of mold on food, you should know how fungi reproduce. In this activity you will examine some fungal reproductive structures.

Working with a partner, cut the stalk out of a mushroom without damaging the cap. Examine the underside of the cap without touching it, using a hand lens or a stereomicroscope. Describe your observations and draw the cap in your logbook.

On a piece of white paper, record your names, the date, and the type of mushroom you are studying. Set the paper where it will not be disturbed, with the mushroom cap right-side up on the paper. When you have finished, clean up your work area, and wash your hands with soap and hot water.

After the mushroom cap has dried up (at least 2–3 days), examine the paper and the underside of the cap with a hand lens or a stereomicroscope. Draw and describe your observations in your logbook and compare your findings with those of other groups. When you have finished your observations, clean up and return your materials, and wash your hands with soap and hot water. Answer the following questions in your logbook:

No, they are round, not fibrous.

The small, round structures are spores that can produce a new fungal organism.

Fungi reproduce by producing spores.

Covering food helps to prevent spores from falling on it.

- Did the fungal structures that fell onto the paper look as though they were made of hyphae?
- What part do these structures play in the life of the fungus?
- What can you conclude about how fungi reproduce?
- Can you explain how covering food helps to protect it from fungi?

HOW FUNGI REPRODUCE

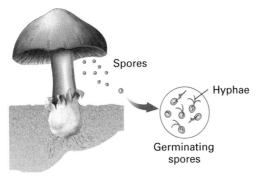

Spores

Hyphae

Germinating spores

Figure 9.3: *Germination of a fungal spore*

The small, round objects that you observed in Activity 9-2 are spores. A **spore** is a tough, dry structure containing a single fungal cell. Spores absorb water and open when they land in warm, moist environments with plenty of nutrients. The cell begins to grow and divide, producing hyphae (Figure 9.3). This process is called **germination.** Fungi can produce spores either sexually or asexually. Mushrooms produce sexual spores, such as the ones you observed in Activity 9-2.

What makes a fungus form a mushroom? If growers understood this process better, they could control it and force fungi to produce mushrooms sooner or in larger numbers. It all depends on the environment. When environmental conditions are good, many fungi produce asexual spores (Figure 9.4). The spores germinate when wind, water, or passing animals carry them to moist food. The new fungi are genetically identical to their parent.

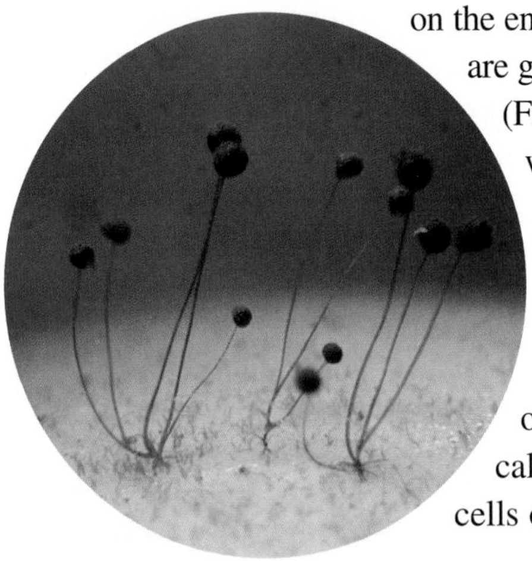

Sexual reproduction occurs most often when fungi are starved for nutrients. Instead of male and female, fungi have mating types, called + and −. Most hyphae are haploid. If cells of the two mating types of a species come into contact, they may fuse, and their nuclei then join to form a diploid nucleus. This cell is a zygote, described in Chapter 4 on page 130. Zygotes undergo meiosis, producing haploid spores (Figure 9.5).

Figure 9.4: *This common black bread mold produces stalks that bear round asexual spores (magnification 32×).*

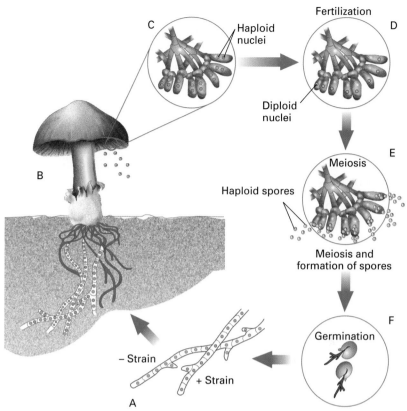

Figure 9.5: *Sexual reproduction of a fungus. Haploid cells of + and – mating types merge (A) to form cells with two nuclei that make up a mushroom (B and C). In some cells, the two nuclei fuse (D) to produce diploid cells that undergo meiosis to produce haploid spores (E). Each spore can germinate (F) to produce new hyphae of + or – mating type.*

INVESTIGATION 9A   REPRODUCTION AND LIFE CYCLE

See Investigation 9A in the TRB.

Everyone has seen moldy bread, but have you ever paid much attention to it? White bread mold can give you a window into the life cycle of a common fungus that causes large economic losses in kitchens everywhere, as people are forced to throw out spoiled bakery products in stores, restaurants, and homes. In this investigation you will see hyphae spread and produce spores that germinate and begin to produce new hyphae.

Bakers and their customers struggle constantly against mold.

LESSON 9.1 ASSESSMENT

1. Fungi that live in water float through a food supply and do not need the large surface area of hyphae to absorb it.

2. Sexual reproduction produces new combinations of alleles that add to variation, increasing the chance that a new, better-adapted phenotype will appear.

3. No, mushrooms are reproductive structures that can be produced only by hyphae that obtain nutrients by growing through food.

4. Fungi, plants, bacteria, and some protists all have cell walls. Fungal and animal cells are heterotrophic.

1 Some single-celled fungi live in water. These cells are rounder than hyphae. How is the unusual shape of these fungi connected to their life in water?

2 Why do most fungi reproduce asexually when conditions favor their growth and sexually when conditions are unfavorable?

3 Could a fungus consist entirely of mushrooms? Give a reason for your answer.

4 What do fungal cells have in common with the cells of plants, bacteria, and some protists? What do they have in common with animal cells?

Lesson 9.2 Kinds of Fungi

On the spur of the moment, a few friends decided to get together for lunch. Everyone brought something from home. Ari had a fresh loaf of bread, Stu brought leftover stir-fry (onions, tempeh, peapods, and tiny corn ears), Pat packed blueberries and some pickled portobello mushrooms, and Bree Ann brought Brie cheese (of course) and a little green salad. While they ate, they realized that their lunch included all four divisions of the Fungi kingdom: and they weren't even trying! Do you recognize the fungi? You will as you continue through the chapter.

Activity 9-3   No Hair . . . No Fungus?

Dissolve a package or cake of bread yeast in 200 mL of warm water. Divide the suspension equally between two beakers and add 5-10 g of sucrose to one beaker. Allow the mixtures to stand at room temperature for 1 hour before class. Explain to the class the difference between the two beakers.

Are you beginning to see why molds and mushrooms are considered similar enough to be in the same kingdom? Do some fungi still seem too different to be placed in the same kingdom? Maybe you need to look more closely at them. In this activity you'll see why yeast cells are classified as fungi.

Using a dropper pipette, place a drop of the dissolved yeast suspension without sugar on a microscope slide. Add a drop of Neutral Red stain, if available, and a coverslip. Examine the yeast cells under the microscope, first at low power and then at high. Draw and describe your observations in your logbook. Rinse off your slide and examine a sample of the yeast suspension with sugar, adding a drop of Neutral Red and a coverslip, as before. Also examine the two yeast solutions in the beakers and describe each in your logbook.

Clean up and return your materials, and wash your hands with soap and hot water. Then compare observations with a partner and answer the following questions in your logbook:

Yeasts are not composed of hyphae; they are rounded.

Sugar promoted cell respiration and growth.

Evidence that they are heterotrophs includes the increased rate of respiration (gas production) and greater growth in the presence of sugar.

Like other fungi, yeasts are heterotrophic, walled eukaryotes.

- Are yeasts composed of hyphae?

- How did sugar affect the yeast?

- Are yeasts heterotrophs or autotrophs?

- Why classify yeasts as fungi?

Paul D. works at a bakery, making breads, cakes, pies, tarts, croissants, cookies, and—Paul's specialty—bagels. He describes how he makes bagels and the help he gets from yeasts, a kind of fungus.

"It's funny that you stopped by today to ask about yeast because yesterday we had some 'super yeast' that I still can't explain. It was rising so fast that I barely got it mixed with the flour, and the dough was jumping out of the bowl and crawling off the table."

"Yeast, even out of the same case, doesn't always act the same. It depends on the weather to some extent, and the strain of yeast.

"The yeast we get comes in a case of 1-pound packages, which are vacuum-packed. The case sits in the refrigerated cooler, and we open one bag at a time as we need it. If we take out a bag that has a hole in it for some reason, we usually toss it. Exposure to air will kill the yeast, because baker's yeast is anaerobic. But in the airtight bags, it can sit in the walk-in cooler for months."

Paul begins by mixing the yeast with lukewarm water. "I leave the yeast in the water while I mix the dry ingredients—unbleached flour, sugar, and salt. The sugar is probably the most important for getting the yeast going. I mix the dry ingredients with the yeast and water and let the dough sit on the table, covered with a kitchen towel, for the first rising."

Paul then cuts up the dough and lets it rise a second time. Then he shapes each bagel by hand and boils them. "It's kind of like cooking doughnuts, but there's no grease. They stay in the water for less than 5 minutes. Then I take them out and put them on a tray to bake."

FOUR KINDS OF FUNGI

Correct identification is especially important when people collect wild fungi to eat. Some deadly species look very much like edible ones, so an expert must check wild mushrooms before they are sold. Never eat wild mushrooms without having them checked first by an expert on fungi. In this lesson you will learn to tell apart members of the four divisions of the fungi. (In Fungi and Plantae,

Common Name	Division	Practical Importance[1]	Example	
Sac Fungi	Ascomycota	Baking, brewing, plant parasites	Yeasts, cup fungi, mildew	
Club fungi	Basidiomycota	Food, plant parasites	Mushrooms, puffballs, shelf fungi	
Molds	Zygomycota	Plant parasites, food spoilage	Bread mold	
Imperfect (asexual) fungi	Deuteromycota	Plant parasites, food spoilage	Penicillium	

[1]All four divisions of Fungi include species that are used to produce medicines, such as antibiotics, or other valuable chemicals.

Table 9.1: Divisions of the Kingdom Fungi

the term "division" is used instead of "phylum.") Table 9.1 gives the scientific names of these groups. This chapter will use their common names: sac fungi, club fungi, molds, and imperfect fungi.

THE SAC FUNGI

Yeasts are a type of **sac fungi.** Cup fungi, found most often on rotting logs in forests, are another group of sac fungi. So is mildew. Perhaps you've eaten truffles or morels (Figure 9.6), which are also species of sac fungi. Why are these fungi grouped together?

Sac fungi produce both sexual and asexual spores (Figure 9.7). After meiosis the haploid sexual spores stay within the cell wall of their parent zygote until they are released. This cell wall is the sac that gives these fungi their name. The cups of some sac fungi, shown in Table 9.1, consist of thousands of these spore sacs packed closely together.

Figure 9.6: *The morel, common in the forests of the eastern U.S., is the sexual structure of a sac fungus. The spore sacs are in the netted brown cap.*

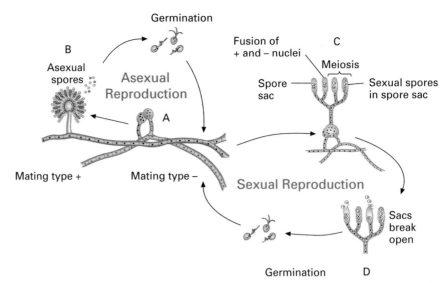

Figure 9.7: *Reproduction in sac fungi. Haploid hyphae of either mating type (A) can produce haploid asexual spores (B) that germinate to produce new hyphae. Fusion of + and – cells produces a large cell with many nuclei. Divisions of this cell produce smaller cells with two nuclei, one + and one – (C). Fusion of these nuclei is followed by meiosis, which produces haploid spores (D) that germinate to produce new hyphae.*

Mildews, yeasts, and cup fungi are classified together because they all reproduce in this way. You might know mildew as the mold that grows on damp fabrics and shower walls. Other mildews are parasites of many important crops. Yeasts are normal inhabitants of our skin and internal surfaces, where competition with bacteria keeps their populations under control. Long periods without bathing or a new brand of shampoo or soap can disrupt this balance, resulting in a fungal skin disease, or an internal fungal infection such as thrush and yeast infections.

ACTIVITY 9-4   WHAT GOOD IS A MUSHROOM?

Provide students with several types of fresh edible mushrooms that have caps that are just beginning to open or are about to open. **Stress the importance of not eating lab materials, and make sure that no student will be exposed to any materials to which he or she is allergic.**

You've already seen that mushrooms are spore-bearing reproductive structures. What makes them so different from the cups of the sac fungi? In this activity you will examine the spore-bearing parts of mushrooms, and compare them to sac fungi.

Working with a partner, remove the stalk from a mushroom. Carefully cut a piece about 1 mm from the edge of one or more of the **gills**—the thin, flat sheets on the lower surface of the cap—and place it on a microscope slide. You may put more than one sample close together on the same slide. Add a drop of water and a coverslip, and observe under the microscope at low and then high magnification. Draw and describe your observations in your logbook. Examine samples of gills from other types of mushrooms with the microscope and record your observations.

Clean up and return your materials, and wash your hands with soap and hot water. Answer the following in your logbook:

- Compare the mushroom spores you observed to the spores of sac fungi.

- Do mushrooms seem to have spore sacs like those of cup fungi?

- How are mushrooms different from sac fungi?

Mushroom spores are separate and not enclosed in a sac.

No, mushroom spores are not enclosed in a sac.

Sac fungi produce spores in sacs on the upper surface of their reproductive structures, while mushrooms bear separate spores on the lower surface of reproductive structures.

THE CLUB FUNGI

Mushrooms are the sexual reproductive structures of **club fungi.** This group is named for its club-shaped spore-forming cells. When sufficient water is available, rapid cell division and growth can form a mushroom in hours. Meiosis in the gills produces a huge number of haploid spores—about 10 billion in a mushroom and trillions in a puffball (Figure 9.8). Unlike other fungi, club fungi do not produce asexual spores.

Figure 9.8: *Puffballs are club fungi that often produce trillions of spores.*

Many rural people supplement their income by collecting and selling edible mushrooms. An interesting growth habit makes mushrooms easy to find. The center of a fungal mass dies as it uses up nutrients in the soil, while its outer edge produces a ring of mushrooms called a fairy ring (Figure 9.9). The dead zone in the center expands and the fairy ring becomes wider each year. Old, wide rings lead collectors to more mushrooms.

Figure 9.9: *A fairy ring. Note the lush growth of grass inside the ring, where decay of the dead fungus releases nutrients to the soil.*

ACTIVITY 9-5   MOLD UP CLOSE

Provide students with moldy bread and other foods, prepared as described in Activity 9-1 but producing spore-bearing structures above the hyphae on the surface. Provide a variety of samples; not all food molds will produce visible reproductive structures. As always when working with fungi, do not expose susceptible students to allergenic molds, and caution students against getting any of the fungus in their mouths. Have dust masks, disposable gloves, and goggles on hand for students. **Students must wash hands before leaving the lab.**

Zygomycote fungi produce small spore-bearing structures. Some are shaped like mushrooms but produce spores on their upper surfaces.

The mushroom-like structures produce spores.

Why don't mushrooms grow on moldy food? How do fungi spread their spores without mushrooms? In this activity you will examine a food mold and look for spore-bearing structures.

Working with a partner, examine with a hand lens the mold on a food sample. Draw and describe in your logbook any structures that might be specialized for spore production. Use forceps to transfer a small piece of the fungus to a microscope slide. Add a drop of water and a coverslip, and observe under low and then high power, noting any specialized structures. Note your observations in your logbook, including the magnification at which you saw things.

Clean up and return your materials, and wash your hands with soap and hot water. Discuss the following questions with your partner and record your answers in your logbook:

- Did any structures resemble the reproductive structures of club or sac fungi?

- Did any part of the fungus look as though it might be a spore-producing structure?

SIMPLE MOLDS

Mold spores are carried nearly everywhere by air and water. They have even been found in air samples 100 miles above Earth! Bakers, produce wholesalers, seed companies, and grain shippers are constantly fighting fungi that spoil their products. Antifungal preservatives added to bakery products and seeds, gases that disinfect stored grain, and frequent cleaning and drying of equipment all help to control spoilage molds. The molds of division Zygomycota have the smallest, simplest sexual spore-bearing structures of all fungi. This simplicity implies that they may be the living species closest to the first fungi.

Asexual reproduction in Zygomycota involves tiny, often microscopic structures with haploid asexual spores on their upper surfaces (Figure 9.10). Most species depend on wind to carry these spores to moist, dead plant or animal material or animal wastes, where they germinate.

Figure 9.10: *Spore-bearing structure of a zygomycote fungus. (Color added. Magnification × 800)*

Sexual reproduction in these molds usually occurs in response to a stress such as freezing, drying, or starvation (Figure 9.11). A tough outer wall protects the spore. When conditions improve, the + and – nuclei in the spore fuse, and meiosis occurs. The germinating spore produces a haploid spore-bearing structure similar to the ones that are produced asexually.

Zygomycote fungi are simple but well adapted. *Pilobolus,* a genus that lives on manure, demonstrates this dramatically. Each spore of this fungus is on a separate stalk that bends to follow the sun, keeping he spore aimed upward. Turgor pressure within the stalk shoots the spore up and away from the manure pile and onto fresh grass. Grazers swallow the spores as they eat, eventually planting them in a fresh pile of nutritious manure.

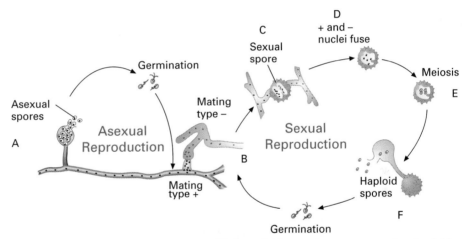

Figure 9.11: *Reproduction in zygomycota. Hyphae of either mating type can produce haploid spores (A). Cells of + and – hyphae that come into contact (B) each produce many haploid nuclei, and are enclosed together in a spore (C). Fusion of + and – nuclei (D) and meiosis (E) occur in the spore. When the spore germinates, it releases the haploid spores (F), which germinate to produce new hyphae.*

THE IMPERFECT FUNGI: THE GRAB BAG

Some species of fungi are called **imperfect** because they have lost the ability to reproduce sexually. These species are in the division Deuteromycota, although they probably include descendants of the other three divisions. Remember that classification above the level of species is artificial. Whenever a species is discovered to produce sexual spores, it is reassigned to another division.

A few species of imperfect fungi are predators. Some of their hyphae form rings. When a microscopic worm or other tiny animal swims through the ring, the fungal cells suddenly contract, trapping the prey (Figure 9.12). The fungus then digests the prey and absorbs its nutrients.

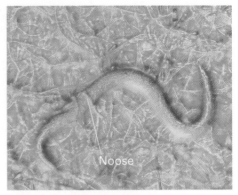

Noose

Figure 9.12: *Some imperfect fungi feed on small worms that they capture in rings of hyphae (noose) (magnification 200×).*

LICHENS: ENVIRONMENTAL WARNING SIGNS

Environmental quality managers and others who are concerned about the environment often examine what look like spots of green paint peeling off of rocks, trees, and buildings. These spots are living **lichens** (Figure 9.13), symbiotic mixtures of fungi (usually sac fungi) and green algae or cyanobacteria. The fungus kills and digests some of the algal cells. It may provide the algae with minerals extracted from the underlying stone or tree bark and help to protect them from drying out; but in many cases the fungus is simply a parasite. Because they obtain most of their minerals from airborne dust and rainwater, lichens are easily damaged by air pollution. A decline in lichen populations may indicate an air quality problem.

If lichens are available, have students tease a piece apart in a drop of water on a microscope slide and observe the algae and hyphae.

Figure 9.13: *The lichens on these rocks are symbiotic mixtures of fungi and algae.*

Lichens can survive in environments that are too cold or dry for plants. These "pioneer species" are among the first to enter a new environment. Lichens produce acids that help to break down exposed rock into new soil. In the Arctic, they provide food for animals. The brightly colored spore sacs of some Arctic lichens are the source of an acid-sensitive dye called litmus that you may have used in science classes to test for the presence of acids and bases.

LESSON 9.2 ASSESSMENT

1. Fungi are soft and moist and do not leave hard parts behind as fossils.

2. No, mushrooms are club fungi, which do not produce asexual spores.

3. Fungal mating types are equivalent; one does not fertilize the cells of the other, as in plants and animals.

4. Lichens are mixtures of two unrelated species.

1 Why is there so little evidence of the origin and evolution of fungi compared to that for animals with hard shells or bones?

2 If you wanted to start a business that sells starter samples of fungi to mushroom growers, could you offer a choice of hyphae, sexual spores, or asexual spores? Give a reason for your answer.

3 How are fungal mating types different from the two sexes that are seen in plant and animal species?

4 What problem is there in classifying lichens as other living things are classified?

LESSON 9.3 FUNGI IN THE ENVIRONMENT

We usually think of fungi as the species that affect our lives and work—edible mushrooms, mold on food or crops, mildew on clothing, and so on. But most fungi live in nature as decomposers. Fungi contribute to nearly every community of organisms, breaking down dead materials and animal wastes to release their nutrients and living symbiotically with trees, insects, and other plants and animals. The way fungi live in nature determines the role they play in human activities.

ACTIVITY 9-6 TESTING FOR SEED CONTAMINATION

The day before performing the activity, soak each batch of seeds prepared earlier (see TRB for preparation instructions) in 1 M (40 g/L solution) NaOH to soften the seed coats. Use caution in preparing and using the NaOH solution. **Wearing rubber gloves and safety goggles,** drain and rinse each batch of seeds with several changes of tap water. Provide them to the class in containers labeled "cool, dry" and "warm, humid," according to how they were stored. Provide students with small dropper bottles of Aniline Blue stain.

Fungi can be a benefit or a problem for people, but no matter how we feel about them, they do the same things to us, our food, and our crops and livestock that they do in nature. In this activity you will study the effect of fungi on stored grain.

The seeds that you will study have been stored under cool, dry conditions or warm, humid conditions. They were softened in sodium hydroxide (NaOH). Working with a partner, use clean forceps to remove the seed coat from a seed. Record in your logbook which batch of seeds your sample was from.

Place one drop of Aniline Blue stain on a microscope slide. Don't get it on your fingers or clothing. Use the forceps to put the seed in the drop of stain. Use the flat side of your forceps or a scalpel to gently mash the seed. Add a coverslip and gently press on it so that it rests flat on the slide. Observe the sample at 400–600×. Any hyphae that are present will absorb the blue stain. Draw and describe your observations in your logbook.

Stain and examine a seed that was stored differently, and record your observations. Compare your results with your partner's and with those of other groups. Answer the following questions in your logbook:

Fungi grow best in warm, humid environments.

They attack seeds just as they would dead organisms.

- What kind of environment favors the growth of fungi?

- How might this fungus survive in nature, away from stored seeds?

As decomposers

Waste management and in breaking down garbage, sewage, and plant debris from lawns and farms.

- How would you expect fungi to contribute to the communities of organisms in which they live: as producers, as consumers, or as decomposers?

- What uses of fungi might depend on their role in natural communities?

WINE TESTING

BIOLOGY IN CONTEXT

Keith R. works for a wine-testing lab. Wineries send samples of their wine to be tested for microorganisms that spoil wine. Although Keith can usually detect spoiled wine by its odor when he uncorks it, the main test is microbiological.

Food and beverage analysis can involve chemical testing, shown here, or checking for fungi and other spoilage organisms.

Keith explains, "Microorganisms are an important element in wine making—primarily yeasts but also bacteria. They carry out fermentation, and they contribute to the wine's flavor and aroma. However, some yeasts and bacteria can spoil wine. Some yeasts can cause cloudiness in the wine and excess carbon dioxide. Two types of bacteria can affect wine negatively: acetic acid bacteria, which can turn the wine to vinegar; and halo-lactic bacteria, which can give it other off-flavors and undesirable smells."

How does Keith test for wine spoilage microorganisms? He filters the wine and puts the filtrate on a medium that has specific nutrients. "We add nutrients to the medium that are suited to the wine spoilage organism for which we are looking. The medium is selective for these organisms. We plate the filtrate of the wine on the medium, incubate it, and observe several days later to see whether wine spoilage organisms are present. If they are, we count them and report the counts to the wine company."

What does the wine company do if wine spoilage organisms are present? "They have to make a decision," he explains. "They can uncork the bottles and refilter the wine, then rebottle it. Or they may take off their label and sell it to a company that makes wine vinegar. It's their decision."

FUNGI AS DECOMPOSERS

By now, you know that most fungi are decomposers. Without them the world's nutrients would soon be tied up in a deep layer of dead organisms, and life on Earth would probably end. We become aware of decomposer fungi when they attack our food, clothing, paper, and other materials (Figure 9.14). Builders willingly pay more for lumber that is treated with **fungicides,** chemicals that kill fungi. Seed producers also use fungicides. They add a pink stain to seeds before selling them to show that the seeds have been treated and shouldn't be eaten. Because most fungi need a moist environment, dry storage conditions help to protect seeds, food, and clothing (Figure 9.14).

Figure 9.14: *Grain inspectors examine samples of all grain sold for food use in the United States. They also use microscopes and chemical tests to check for fungi, insects, and other contamination.*

ACTIVITY 9-7   A GOOD PLACE TO MAKE BREAD

Perform this activity in a commercial foods classroom or in the cafeteria kitchen. **Remind students of the dangers of eating food that has been on the surfaces in the lab.**

Explain what you are doing as you dissolve 1.5 tablespoons (or 2 packets) of dry yeast and 4 tablespoons of sugar with a half-cup of warm (35–40°C) water. Divide the mixture into two equal portions, and add 2 tablespoons of white vinegar to one portion. Explain that vinegar is an acid. Label the containers "with acid" and "without acid." Designate half the lab groups to add a tablespoon of salt to their dough. Designate separate baking sheets on which students should put each of the four kinds of dough (with and without salt and vinegar). Let the doughs rise for 30 minutes, and then bake them for 20 minutes at 350°F.

Making bread doesn't seem like decomposition, but the decomposing action of yeast makes it useful in baking. In this activity you'll see that in a hospitable environment, yeast helps to produce bread and other baked goods.

Working with a partner, make two small batches of dough. In separate bowls, combine a half-cup of flour, 3 tablespoons of sugar, and 2 tablespoons of water. If your teacher tells you to add salt, add 1 tablespoon of salt to your bowl. Agree on which partner will add a half-teaspoon of the yeast mixture without acid to the dough. The other person should add a half-teaspoon of the yeast mixture with acid. Observe and compare the two batches of yeast. Smell them both. Mix the contents of your bowl with a spoon to make a dough. Observe your own dough and your partner's as you mix, noting any differences.

After you set the dough to rise, record your observations in your logbook. Observe and smell the four kinds of dough again after 30 minutes and after baking. Record your observations, and discuss the results with the class. Answer the following questions in your logbook:

The yeast acts as a decomposer, breaking down the sugar.

It requires specific conditions of pH and salinity to function well.

- How did the yeast act as a decomposer?

- What environmental factors affect the growth of yeast?

USING FUNGI AS DECOMPOSERS

Baking bread might not look like decomposition, but the sugar that you add to dough gives the yeast cells food to break down. Bubbles of the CO_2 gas that they produce make the dough rise. Without yeast your bread would be a heavy, unappetizing block. Bakers adjust the pH, salt concentration, and temperature of their dough to encourage yeast growth.

People in many occupations use fungi as decomposers (Figure 9.15). Solid-waste managers let fungi and bacteria break down waste materials, such as stems and leaves left after crops are harvested, manure from large farms, grass clippings, autumn leaves, and solid sewage residue. This forms **compost,** decomposed organic matter used as fertilizer. It reduces the amount of garbage buried in landfills or burned, and the compost can be spread on gardens and farmers' fields to feed the next season's crops.

Figure 9.15: *A garden compost bin. Fungi and other decomposers break down dead leaves, stems, and other plant debris to compost. Growers plow this nutrient-rich material back into the soil.*

Yeasts and other fungi are used to decompose foods into forms that are more valuable or easier to store (Figure 9.16). Fungi can tolerate the low oxygen levels deep in the tanks of partly digested food products. There, they ferment milk into cheese, grain and fruit into alcoholic beverages, and soy beans into soy sauce. Ethyl alcohol, lactic acid, and other products of fermentation add flavor, aroma, and color to the products. Baking kills the yeast in bread and evaporates any alcohol that they produce. In nature, fungi survive in wet, low-oxygen environments. They thrive in moist soil and decaying materials.

Figure 9.16: *Using fungi in food processing. Food industry workers use yeasts and other fungi in brewing (shown here) and to produce fermented foods such as cheeses and soy sauce.*

Fungal fermentation is also useful to pharmaceutical companies. Some fungi produce medically useful compounds that are hard to synthesize artificially. Other species have adapted to competition with decomposer bacteria for nutrients by producing antibiotics such as penicillin, tetracycline, and streptomycin that kill bacteria. Environmental conditions such as temperature and pH also affect the balance between fungi and bacteria. Fungi tend to dominate at higher temperatures; people who drill oil wells in humid tropical areas soon learn to keep their cuts and scrapes clean, covered, and treated with antifungal cream (Figure 9.17). Because antibiotics upset the balance between bacteria and fungi, they are often given together with antifungal medications to prevent internal yeast infections.

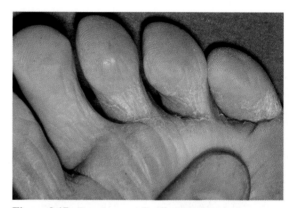

Figure 9.17: *Fungi are medically important both as sources of antibiotics and as infectious pathogens. Athlete's foot (shown here) is a fungal skin infection.*

LEARNING LINK PRODUCTS OF THE FUNGUS AMONG US

Select one of the following products and investigate the role of fungi in its production. Use the library or the Internet or write to a company that makes the food.

- Cheeses (especially blue, Brie, Camembert, or Roquefort)

- Antibiotics

- Homemade and brewed soft drinks (especially root beer, ginger beer, and birch beer)

- Vinegar (find out the differences between the various types)

- Soy sauce, tempeh, or other East Asian fermented foods

Record your findings in your logbook. Present them to the class with a flowchart showing the process that is used to make these products and any by-products.

INVESTIGATION 9B — COMPETITION BETWEEN FUNGI AND BACTERIA

See Investigation 9B in the TRB.

Competition between bacteria and fungi is important wherever they depend on the same food source. This is true on our skin, inside our digestive and reproductive organs, and in compost piles and sewage treatment plants. Any change in the environment that disrupts the balance between bacteria and fungi can lead to all sorts of problems, including fungal skin diseases, putrid compost, and waste treatment plants that don't work. In this investigation you will explore the effects of the environment on competition between these decomposers.

Antibiotics that kill disease-causing bacteria can also tip the balance in the competition between healthy bacteria and fungi, resulting in skin irritations, digestive upsets, or other medical problems.

FUNGI IN SYMBIOSIS

Lichens and many other symbiotic relationships are flexible. They may shift from mutualistic to parasitic when conditions favor one member more than the other. The ability of fungi to form symbiotic relationships, like their activities as decomposers, is one reason they are important, both in nature and in human activities.

Many fungi form close relationships with plants. Farmers spend a great deal of time and money on control of fungal parasites of plants, especially during warm, humid weather (Figure 9.18). Fungicides, removal of infected and dead plants, and adequate soil drainage are all part of agricultural fungus control. Fungi that grow on plants are not all harmful. Corn smut can be cooked and eaten (Figure 9.18B). Soil fungi and the roots of most plants form mutualistic relationships called **mycorrhizae,** though this mutualism seems to be most important for conifers (needle leaf trees such as pines). The plants provide the fungi with carbohydrate, and the fungi add surface area to the roots, increasing their ability to absorb water and mineral nutrients from the soil. Fungicides can interfere with mycorrhizae, hurting the very crops that they were meant to help.

WORD BANK

myco- = fungus

rhizo- = root

Mycorrhizae are the relationships of plant roots with fungi growing on and into their surfaces.

A

B

C

D

Figure 9.18: *Fungi that attack crops include members of all four divisions. (A) Mold on a fallen apple in an orchard. (B) Corn smut, a club fungus, on sweet corn. (C) Powdery mildew on a leaf, caused by a sac fungus. (D) Riesling wine grapes infected by an imperfect fungus.*

Some fungi grow inside the bodies of insects. Some insects even have special organs that house the fungi, a fact that suggests that the insects benefit from the relationship. Other fungi are insect parasites. Their spores are sold to control insects that attack farm and garden plants.

FUNGI AS FOOD

See the TRB, Chapter 9, for an Additional Activity, Cultivated Mushrooms.

Pigs were used to find truffles because the odor of truffles resembles a pig pheromone. They have been largely replaced by dogs because pigs tend to eat the truffles they find.

Most edible fungi are mushrooms. Morels (Figure 9.6) are formed by a sac fungus. Truffles are the sexual structures of mycorrhizal sac fungi. Because they have a symbiotic relationship with trees, truffles are almost impossible to grow commercially. They are very expensive because they must be collected with the aid of animals that are trained to sniff them out. If you discover a way to cultivate truffles, you could make a lot of money!

Some fungi become part of a food. The blue streaks in Roquefort and Danish blue cheeses are hyphae of the imperfect fungi that converted the milk to cheese. The edible rinds of Camembert and Brie cheeses are also mostly fungus. The fungus that converts soybeans to tempeh, a popular East Asian food, makes up a large part of that product.

LESSON 9.3 ASSESSMENT

1. Warm conditions favor the growth of most fungi, including yeast.

2. Though both of these groups work in hot climates, fungi are favored by a humid climate and inhibited by dry conditions.

3. Fungi produce tempeh by partly decomposing soy protein.

4. No, the fungus in blue cheese is imperfect; it does not produce sexual structures such as mushrooms.

1 Why must bakers keep yeast dough warm while it is rising?

2 Why are fungal infections more common in rainforest lumberjacks than among the cotton farmers of hot, dry west Texas?

3 How is the production of tempeh related to the role of fungi in nature?

4 Could you produce blue cheese that grows edible mushrooms? Give a reason for your answer.

CAREER APPLICATIONS

The applications that follow are like the ones you will encounter in many workplaces. Use the biology you learned in this chapter to complete the activities. Share your work with the class.

AGRICULTURE & AGRIBUSINESS

Grain Spoilage

Using a variety of sources such as the Internet, books, periodicals, and interviews, find out about fungi that destroy grains and other crops such as peanuts. What are the primary spoilage organisms? What are the symptoms? What prevention and treatment are used against the fungi? What impact do fungi have on farmers? Report your findings to the class.

Mushroom Farming

Use the library, the Internet, and interviews of local food marketers and producers to investigate how various types of exotic and gourmet mushrooms are produced and stored for shipment. Make a table comparing the growth of various mushrooms. Include a picture of each variety. Explain what is needed to produce various kinds of mushrooms, and why the various kinds vary so widely in price. Report your findings to the class.

BUSINESS & MARKETING

Marketing Exotic Edible Fungi

A gourmet food shop is developing brochures to market its products. You have been asked to create an informational brochure about exotic edible fungi. The brochure will encourage customers to try new foods. To achieve this purpose, include in your brochure information about various types of mushrooms, morels, tempeh, corn smut, and truffles. Describe taste, texture, recipes, cultivation, seasonal availability, and other relevant features. You can find the information by contacting gourmet shops or produce managers at food stores and by using the Internet, cookbooks, and encyclopedias. Study the advertising for these products in stores, in food magazines, in cookbooks, and on the Internet. Find out how marketers describe and segment this market. Prepare a report to the class on your findings.

Regulation of Food Product Safety

Fungal contamination of food products is regulated by the U.S. Food and Drug Administration (FDA). Use resources such as your local public library, the FDA's Web site, or the home economics specialist at your county's cooperative extension office to find out about the safety standards for fungal toxins and how they are enforced. Peanut butter is a good product to start with. You might investigate whether there is any difference between how often the FDA seizes unsafe "natural" peanut butter and other types.

FAMILY & CONSUMER SCIENCE

Keeping Mold and Mildew at Bay

Mold and mildew can become a problem around the house, particularly in bathrooms and basements that have poor air circulation and high humidity. In some parts of the country the outside of a house may become discolored and slowly deteriorate from fungal attacks. Get together a group of students to ask people what the major fungal problems are in your area and how they can be treated and prevented. Possible sources include professional cleaning services, building contractors who specialize in remodeling older homes, allergists, home extension agents, and retailers of cleaning supplies. Your group should report its research to the class. If the equipment is available, you might tape a documentary on preventing fungus growth in or on the house.

Food Additives

Look at the ingredients lists of a variety of bakery products. List any ingredients that you think might be food additives. Talk to a commercial baker or search the Internet for food additives and find out the purpose of each one. Which additives inhibit bread mold? How do they work? For products that are baked without additives, find out the recommended shelf life, handling, and storage procedures that inhibit the growth of bread mold.

HEALTH CAREERS

Artificial Nails

Some people who wear artificial nails develop a nail fungus. Talk to a cosmetologist to find out how common the problem is, what can be done to prevent it, and how it is treated. Make a brochure for people who are getting artificial nails. Describe the problem, list symptoms for early detection of nail fungus, suggest preventive measures, and recommend treatment. Be sure to answer the kinds of questions that people are likely to have about these infections, such as whether they can be spread by borrowing a friend's nail clipper or nail polish and whether professional manicurists are at greater risk for getting and spreading nail fungus.

Fungal Diseases

Use library or Internet resources to learn about fungal diseases of humans and animals. Find out what kinds of conditions can lead to fungal infections and how people can avoid getting this kind of disease. Then interview a doctor, nurse, veterinarian, or other health professional about one or more of these diseases. Report to the class what causes and spreads the disease, what the symptoms are, and how to prevent and treat it.

INDUSTRIAL TECHNOLOGY

Construction Materials

Lumber that will be used to construct buildings is often treated with fungicide to keep it from rotting. What other construction materials are treated with fungicides? How are they applied? Why is lumber sometimes treated in high-pressure tanks? In which parts of the country and which climates is fungal rotting of building materials an important problem? You might start by interviewing a local building contractor or building supply manager. Information about construction methods and materials may also be available at your public library.

Manufacturing Mildew Resistance

Contact a company that produces mildew-resistant products such as shower curtains or nonslip bathtub mats. How is mildew resistance built into their product? How do boat builders and sailors protect boats and the things that they carry from getting moldy? Are there mold-resistant paints for ships, bathrooms, locker rooms, and other damp places? If so, how do they work? How does washing or cleaning affect the mildew resistance? How long does the mildew resistance last? What safety precautions do people working with the chemicals need to take?

CHAPTER 9   SUMMARY

- Fungi are heterotrophs. Most fungi release enzymes that digest dead plant and animal materials, and the hyphae absorb the digested substances. Fungi compete with bacteria as decomposers of these foods.

- Dry conditions easily kill fungi. Most fungi grow best under warm, moist, dark conditions.

- Most fungi consist of strands of elongated cells with walls made of chitin.

- Walls between cells in hyphae may have gaps or be completely absent. Some fungal cells contain more than one nucleus.

- Fungi reproduce by spores that spread mostly by air circulation. Spores may be produced asexually by a single organism or sexually when cells of two mating types merge.

- Sexual reproduction in fungi occurs mostly in response to environmental stress such as cold, dryness, or starvation. This response is thought to aid survival by increasing genetic variation during times of stress.

- The kingdom Fungi has four divisions (equivalent to phyla) that differ mainly in the details of reproduction. Club fungi reproduce only sexually, and imperfect fungi reproduce only asexually. Sac and zygomycote fungi produce spores both sexually and asexually.

- Many fungal species form symbioses with animals, plants, or algae. Some of these symbiotic fungi are parasitic, some are mutualistic, and others may benefit or harm their partners, depending on environmental conditions.

- Fungi that are important to people include symbionts on human skin and internal surfaces; parasites of humans, crops, and livestock; decomposers in waste management and food processing; and producers of antibiotics and other medically useful compounds.

CHAPTER 9 ASSESSMENT

Concept Review

See TRB, Chapter 9 for answers.

1 Comment on the following statement. If you agree, explain the statement and give examples. If you disagree, explain why and revise the statement to make it correct.

Most fungi are parasites that cause disease.

2 How do most fungi find food?

3 What aspect of the environment is most important for fungal growth?

4 How do fungi transport nutrients internally?

5 Compare fungi to animals in the way they break down their food to obtain nutrients.

6 What do all fungal reproductive structures have in common?

7 Why would a fungus shift from asexual to sexual reproduction?

8 What makes the imperfect fungi different from all the other fungi?

9 How are yeast species different from most other fungi?

10 Describe how to use a microscope to tell the difference between a sac fungus and a club fungus.

11 Which division includes the mushrooms?

12 Which kingdoms include species that are symbiotic with fungi?

13 Explain how fungi can both help and hurt the growth of crops.

14 Describe at least three practical uses of fungi.

Think and Discuss

15 Compare the roles of haploid and diploid stages in the life cycles of fungi with their occurrence in the life cycles of plants and animals.

16 Do all four divisions of fungi represent evolutionary family trees? Give a reason for your answer.

17 Describe at least two ways in which fungi have affected your life in the last week.

18 Why do pickles stay mold-free longer than fresh cucumbers do?

In the Workplace

19 Would you be more likely to force a fungus to start producing edible mushrooms by giving it extra nutrients or by starving it? Give a reason for your answer.

20 In walk-in refrigerators for storing seeds, why would you put the bags of seeds on wooden platforms?

21 How does the action of fungi as decomposers make them useful in food processing? In waste management?

22 What organisms would you test when looking for natural antifungal medicines? Give reasons for your choices.

Investigations

The following items relate to Investigation 9A:

23 How do fungi benefit from having their spores germinate only when both water and nutrients are present?

24 How can fungal spores appear in food that contained none when it was packaged?

The following items relate to Investigation 9B:

25 Describe how temperature affects the balance between fungi and bacteria.

26 Should a pharmaceutical company grow its bacteria and fungi mixed together, to save on the cost of containers, or separately? State any assumptions that you make, and give reasons for your answer.

CHAPTER 10

WHY SHOULD I LEARN THIS?

That stray dog you took to the animal shelter is a mobile zoo! When you called to check on it, the shelter's vet listed all the animals you turned in: a malnourished dog, several plump ticks, two kinds of intestinal worms, ear mites. . . . The kingdom Animalia includes both hairy dogs and some very different organisms. What do dogs have in common with ticks, worms, and mites? You'll find out in this chapter.

Chapter Contents

INTRODUCTION TO ANIMALS

WHAT WILL I LEARN?

1. what distinguishes animals from other living things
2. how body shapes and mobility help animals to survive
3. functions of the major organs and systems in animal bodies
4. similarities and differences among major phyla of animals
5. the likely relationships among major animal phyla

- -

People in many occupations work with animals or their products. Much of the food industry catches or raises animals or makes animal products. The fashion and clothing industries use feathers,

leather, fur, and animal-derived glues. Breeders, trainers, and pet shop owners work with animals that are guides for the blind or companions. Just about every human animal has ties to other animals. In this chapter you will learn about many kinds of animals and how they live.

LESSON 10.1 WHAT IS AN ANIMAL?

Would you recognize an animal if you saw one? Are you sure? People haven't always agreed on which species are animals or even on how to define the term "animal." In this lesson you will learn about the characteristics shared by all animals, and how animals survive and adapt to nearly every environment on Earth.

What's the difference between animals and other organisms? Which of these are animals?

ACTIVITY 10-1   WHAT MAKES AN ANIMAL AN ANIMAL?

Provide a variety of specimens for students to examine, including some that are clearly animals and nonanimals, and some that are harder to classify, such as protozoa, rotifers, tardigrades, sea fans, hydra, nematodes, sponges, and photos of people. Other examples could be video segments or photos of sea anemones, coral organisms, marine flatworms, and motile cellular slime mold "slugs." Distribute copies of blackline master 10-1, which can be found in the TRB.

Make clear to students that this is a speculative brainstorming activity. They are not expected to come up with detailed, authoritative answers to the discussion questions.

Which species are animals, and why? In this activity you will examine some organisms. As you study each one with your group, share your knowledge of each organism and how it lives. Also share your reasons for thinking that it is or is not an animal. Then discuss the questions on the handout that your teacher distributes, and record your answers in your logbook.

When you have finished working with the handout, discuss the following questions with your group and record your answers in your logbook. Be prepared to defend your ideas to the class, giving examples, evidence, or other reasons for your answers.

- What made it hard to decide which organisms are animals?
- What do all animals have in common?
- What are the important differences between animals and other organisms?

PICKY EATERS GO FOR MASHED SQUID

You've met picky eaters, but you haven't seen anything until you've met the diners at the Monterey Bay Aquarium in California. This aquarium holds more than 300,000 animals from 571 species. Each animal needs a specific environment and diet.

The aquarium's three sea otters—Hailey, Goldie, and Roscoe—are among the fussiest of the bunch. They love squid but not livery-tasting Pacific squid. Every morning, their chef (the animal food technician) prepares just what they like: Atlantic squid, sliced, with the liver removed.

The shorebirds of the sandy shore exhibit demand live mealworms, crickets, and fly larvae. These birds won't eat the local insects. All their food must be shipped live from Ohio, Arizona, and other states.

But that's not all. Every morning, Herb, the aquarium's 350-pound green sea turtle, eats a taco made of squid wrapped in lettuce. Keeping him company are the meat eaters: soup-fin sharks, yellow-fin tuna, California barracuda, and bonita. These meat eaters snap up salmon steaks and squid-smelt gelatin. The aquarium's veterinarians put these meat eaters on a low-fat diet after several tuna died of heart attacks.

In the wharf exhibit are the animals that aren't so picky about what they eat: sea anemones, sea stars, corals, sponges, and an assortment of fish. The technician pours squid puree into the tank, which already contains plankton, creating a floating feast. The fish grab pieces of squid, while the others pick plankton and other tidbits out of the water.

The Monterey Bay Aquarium represents many of the feeding habits in the animal kingdom. Humans aren't the only animals that are particular about what and how they eat. Just ask Herb.

RECOGNIZING ANIMALS

Many animals are important to humans. Sheep wool varies greatly in length and strength, so tailors, clothing manufacturers, and fashion designers carefully examine the quality of wool before

Muscle cells

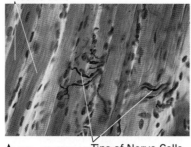

A   Tips of Nerve Cells

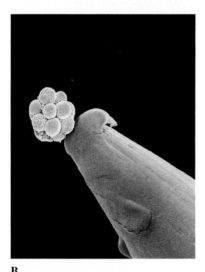

B

Figure 10.01: *(A) Animals are made up of many cell types, such as these nerve and muscle cells. (magnification ×200) (B) This embryo of 16 apparently identical unspecialized cells developed from a single fertilized human egg cell. Eventually, through growth, cell division, and cell specialization, it will develop into an adult human. (magnification ×130)*

buying it. Bakers know that their choice of animal fat, such as butter or lard, will affect the quality and texture of their pastries. Butchers, food distributors, and chefs judge meat quality by looking at muscle and fat tissues. And people who work in fisheries or wildlife management must learn how a great variety of animal species live and interact with each other. What defines all these valuable species as animals?

All animals share certain characteristics that set them apart from members of the other four kingdoms (Figure 10.1):

- Animals are multicellular eukaryotes made up of many specialized cells without cell walls.

- Animals are heterotrophic.

- Adult animals develop from **embryos:** small masses of unspecialized cells.

Although nonanimals may have one or two of these characteristics, only animals have all of them.

Most animals reproduce sexually and therefore develop from single-celled zygotes. Pet and livestock breeders carefully select parents in these species. Professional breeders can sell some varieties of dogs for hundreds of dollars if they can guarantee the animals' ancestry.

Some worms and other simple animals reproduce asexually, and some use both sexual and asexual reproduction. A few species of more complex animals, including certain insects and lizards, have lost the ability to reproduce sexually.

Oyster fishermen learned the hard way about animal reproduction. Their nets often caught sea stars (Figure 10.2), which feed on oysters and their relatives. To protect the oyster population, they cut up the sea stars they caught and threw the

Figure 10.2: *Sea stars reduce harvests by preying on oysters and clams. This bat star is eating a crab.*

pieces back in the water. But sea stars, like many simple animals, can **regenerate,** or grow back missing parts. A piece of a sea star can regenerate the missing arms. The oyster fishermen were actually increasing the number of sea stars that preyed on oysters!

Some worms reproduce asexually by splitting in two and regenerating the missing parts in each piece. Although more complex animals have more limited regeneration, it's the process that heals your cuts, scrapes, and broken bones.

ACTIVITY 10-2   SYMMETRY AND STRATEGY

You might need to explain Figure 10.3C. Students should understand that buoys are designed to remain anchored upright in one spot but may "face" any horizontal direction.

As you know, animals have many shapes and ways of surviving. Is there any connection between the shape of an animal and how it lives and obtains food? In this activity you'll compare the shapes and functions of some animals to mechanical devices.

Working with two other students, examine Figure 10.3. Discuss with your group the questions that your teacher asks about Figure 10.3A. Record your answers in your logbook.

A

B

Ask the class the following questions, or post them on a chalkboard or overhead projector: Imagine making an imaginary cut through the truck in Figure 10.3A.

- How many different cuts could you make that would divide the truck into two parts that are mirror images of each other?

- Does your answer relate to how a truck works and moves? For example, must the front and back ends of a truck be different? Why or why not?

- What about the sides? Are they usually very similar or different? Are the top and bottom similar or different?

- Why are these similarities and differences necessary for the truck to move and work in a useful way?

C

D

Figure 10.3: *Forms and functions. How is a truck (A) like a shark (B)? How is a buoy (C) like a jellyfish (D)?*

Discuss and evaluate the device in Figure 10.3C as you did the truck. Record your conclusions in your logbook.

Discuss the following questions with your group and record your answers in your logbook:

- Compare Figures 10.3B and 10.3A. Does a shark need different front and back ends? Why? Do the animal's back and belly have to be different? What about its sides? How are the shape and movement of a shark like those of a truck?

- Compare Figures 10.3C and 10.3D. Does a buoy need a front and back that are different? A top and bottom? Does a jellyfish? What connections can you see between the shapes of the buoy and the jellyfish and the way they work?

Animals and vehicles that move forward need a front and rear that are different and a top that is different from the bottom; balanced, symmetrical sides are an advantage.

Floating and stationary animals and structures need different tops and bottoms so that they can remain upright or attached, but all horizontal directions are equivalent to them.

ANIMAL BODY PLANS AND WAYS OF FEEDING

The general structure of an organism is its **body plan.** Most animal body plans are more like a truck than like a buoy. They have an **anterior** (front) end and a **posterior** (rear) end (Figure 10.4A). Except for the simplest animals, the anterior end usually consists of a head with sense organs such as eyes and smell or taste sensors, a mouth for taking in food, and a brain that receives sensory information and controls responses such as movement. A similar plan is used in vehicles such as cars and aircraft, in which the driver and controls are in the front, with a clear view as the vehicle moves forward. Successful engineering often imitates nature.

The left and right sides of most animals are nearly mirror images. Their body plan is bilaterally symmetrical (Figure 10.4A). Their anterior and posterior ends are not the same. Neither are their **dorsal** (upper or back) and **ventral** (belly) surfaces. **Bilateral symmetry** works well for animals that move forward much of the time as they seek food, shelter, and mates. If a body part is damaged, the animal can rely on an identical part on its other side. Bilateral symmetry provides balance that aids movement. Anterior and dorsal defenses such as bones, shells, and horns protect delicate internal organs.

Some simple animals, such as jellyfish and sea anemones, are **radially symmetrical,** like a wheel (Figure 10.4B). They spend most of their time floating like a buoy or attached to rocks.

WORD BANK

bi- = two

-lateral = side

radius = the distance from the center to the edge of a circle

sym- = same

metri- = measurement or size

Symmetry refers to a pattern of similar body parts. Bilateral symmetry means that left and right sides have similar size and shape. Radial symmetry means that the areas on both sides of a radius have similar size and shape.

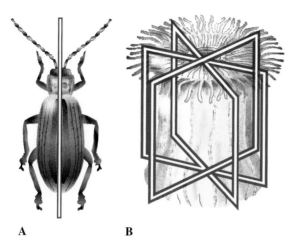

Figure 10.4: *Types of symmetry in animal body plans. (A) Most animals have bilateral symmetry. (B) Some simple animals have radial symmetry.*

A   B

Because the small animals they consume may swim or float by in any horizontal direction, radial symmetry is a successful feeding adaptation. Differences between the dorsal and ventral surfaces allow jellyfish to float upright; sea anemones grip rocks with their ventral surfaces and collect food with their specialized dorsal surfaces. Architects and engineers use radially symmetrical designs for structures such as fire hydrants and lighthouses so that the structures will be accessible or visible from any horizontal direction.

Organisms such as sea anemones that live rooted to one spot are **sessile;** those that move around are **motile.** Even most sessile animals can move at least part of their bodies. Their nervous system senses the environment and uses the information to control muscle movement. Movement is important for feeding and for self-defense.

Figure 10.5: *Tiny parasitic wasps attack a crop-destroying caterpillar.*

Animals vary greatly in how they obtain food, though they are all heterotrophs. Businesses have used studies of animal feeding styles to develop products and services. For example, studies have revealed that some tiny wasps lay their eggs on large caterpillars. When the eggs hatch, the wasp larvae feed on the caterpillar. This information enabled businesses to raise and sell the wasps to farmers and plant nurseries for control of insects that prey on valuable crops (Figure 10.5).

WORD BANK

herb = plant

carni- = meat

omni- = all

-vore = consumer

Herbivores eat plants, carnivores eat meat, and omnivores eat both plants and meat.

Animals that eat plants are called **herbivores.** Meat eaters are **carnivores;** those that kill their own food are **predators.** Rats and raccoons eat both plants and animals; they are **omnivores.** Many species of worms and insects are decomposers or scavengers that consume dead plant or animal material. Ticks and tapeworms are examples of animal parasites. Some animals feed by filtering small organisms and bits of dead material out of flowing water. This method is ideal for animals that are sessile or simply don't move much. Clams, for example, bury themselves in the muddy ocean floor and pump water in one body opening and out another. Some species of whales are motile filter feeders; they take huge mouthfuls of water and then spit it back out through a dense layer of hairlike baleen that lines their mouths.

Great adaptability of body plans and a variety of feeding styles allow members of Animalia to survive in virtually every environment on Earth.

LESSON 10.1 ASSESSMENT

1. No, all animals are multicellular.

2. Sessile animals depend on water to bring prey into range. This adaptation would not be successful out of water.

3. Humans are omnivores, though some voluntarily limit themselves to a vegetarian diet. A dietitian would need to include both meat and plant foods such as fruits, vegetables, and baked goods in the menu.

4. Humans have bilateral symmetry, so we have two arms. A person with a sprained wrist can rely on the other hand to some extent, but we have no backups for our stomach and heart.

1 Are there any one-celled animals? Give a reason for your answer.

2 Scuba divers learn to avoid brushing up against sea anemones and other sessile animals that capture prey with poisonous stings. Why don't surveyors and loggers, whose work involves walking through fields and forests, need to watch out for sessile animals with poisonous stings?

3 Are humans carnivores, herbivores, or omnivores? How would this affect the work of a dietitian who is planning a cafeteria menu?

4 A person with a sprained wrist can usually return to work more quickly than a person with a stomach or heart problem can. What does this have to do with bilateral symmetry?

LESSON 10.2 ORGAN SYSTEMS

What makes some cats better pets than others? Why do farmers prefer particular varieties of cattle or pigs? Most of the differences among breeds lie in the animals' systems of specialized cells. Hair-producing cells make the difference between an easily groomed short-haired cat and a long-haired cat that requires frequent brushing. Slower-growing fat cells in livestock lead to the leaner meats that today's consumers desire. In this lesson you will learn how animals' cells differ and how cells work in harmony as organs and larger systems.

ACTIVITY 10-3   BUILDING-BLOCK CELLS

How do soft, often shapeless cells combine to produce a bird with feathers and bones? In this activity you'll learn about important kinds of animal cells and their contribution to the structure and function of animal bodies.

The ability of muscle cells (Figure 10.6A) to contract makes them ideal for moving body parts, such as legs, and for moving food, blood, and other substances through the body. Nerve cells (Figure 10.6B) transmit sensory information from skin, eyes, ears, and other sense organs to the brain and transmit instructions to move from the brain to the muscles. Bone-forming cells (Figure 10.6C) build a strong, durable framework that supports and protects internal organs and enables body parts to move.

Working with two other students, examine Figure 10.6 and read the description of each cell type. Think about the tasks performed by the parts of an animal. What contribution could each cell type in Figure 10.6 make? Could it help the animal move? Could it help to hold the animal together or support its weight? Discuss your ideas with your group. In your logbook, describe the importance of each cell type, including the characteristics that suit it to a job.

A

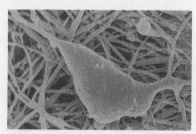

B

C

Figure 10.6: *Some types of cells found in nearly all animals. (A) This elongated cell is packed with many long parallel protein fibers. The fibers can slide past one another and latch together temporarily, causing the cell to contract. (B) This kind of cell can have many extensions and branches that reach a meter or longer. When it is stimulated, electric pulses travel from one end of the cell to the other. The pattern of pulses depends on the intensity of the stimulation. (C) This cell surrounds itself with protein fibers that become embedded with minerals, forming a strong, hard structure.*

Imagine an artificial heart pumping blood to artificial lungs that remove CO_2 from blood and replace it with oxygen. The oxygenated blood returns to the artificial heart, which pumps it to other artificial organs. Replacing real organs with artificial ones once seemed impossible. But within the last 40 years the development of artificial organs has made great advances.

Artificial hearts are used temporarily until a heart donor can be found. By 1989 an artificial heart had kept a patient alive for 620 days. But blood clots forming in the artificial heart are still a problem, so engineers are lining artificial hearts with a smooth layer that reduces the risk of clotting. Many scientists believe that a permanent artificial heart could be developed by 2015.

Artificial lungs have developed more slowly than artificial hearts. Problems include inflammation, blood clotting, and inadequate oxygen supply during exercise. In 1996, artificial lungs were able to keep a pig alive for 24 hours.

Artificial organs still need a lot of work but progress continues. Diabetics may soon be able to wear an artificial pancreas that automatically measures and controls blood sugar levels. And thousands of people may survive liver disease when doctors can implant an artificial liver containing healthy liver cells. Artificial organs will no longer be science fiction.

TISSUES, ORGANS, AND SYSTEMS

Have you ever looked inside a computer? Thousands of electronic components, most of them too small to see, form a central processing unit, a power supply, and other systems. When all these systems work together, the computer runs smoothly.

Animals are also composed of systems made up of microscopic units. These units are cells. As an embryo develops, repeated cell divisions produce millions of cells. These cells **differentiate,** becoming specialized in various ways. Layers or masses of differentiated cells form **tissues** suited for particular tasks. For example, your ability to move depends on the muscle cells (Figure 10.6A) that make up each of your muscles.

Some tissues consist of more than one kind of cell. Blood vessels, the tubes that carry blood throughout your body, have three tissue layers. Each layer has a different mix of cell types. Muscle cells can make vessels narrower; smooth, flat cells line the vessel's inner surface; and other cells build a network of tough, elastic protein fibers. Health care workers, nutritionists, public health officials, and medical equipment manufacturers are among the many people with an interest in blood vessels. Fat deposits can clog and stiffen vessels, raising blood pressure and possibly leading to a blood clot or a stroke.

A body part such as a blood vessel that has several types of tissues is an **organ.** Each of your bones and muscles and your heart, brain, and skin are organs. Your heart, blood, and blood vessels make up an organ system that makes blood circulate through your body, delivering food and oxygen and removing wastes. Animal bodies have several organ systems that cooperate to maintain the animal's survival. Table 10.1 lists the major organ systems that are found in nearly all animals. The organ systems of worms, insects, and people accomplish the same tasks in very different ways, but their major types of cells and tissues are quite similar.

System	Major Functions	Typical Major Organs
Skeletal	Support, protection	Bones, shells, cartilage
Muscular	Movement	Muscles
Digestive	Digestion of food and absorption of nutrients	Mouth, stomach, intestine
Circulatory	Distribution of nutrients and oxygen; removal of wastes	Heart, blood vessels, blood
Respiratory	Absorption of oxygen; removal of CO_2	Lungs, gills
Excretory	Removal of wastes	Kidneys
Nervous	Perception, control of movement, control and coordination of organ system activities	Brain, spinal cord, nerves
Endocrine	Control and coordination of organ system activities	Glands
Immune	Defense against disease-causing organisms	Blood cells, glands, skin
Reproductive	Production of new organisms	Ovaries, testes

Table 10.1: Major Animal Organ Systems

SKELETAL AND MUSCULAR SYSTEMS

Skeletal systems provide mechanical support, help to maintain a shape, and protect soft internal organs. The rigid cast that an orthopedist puts on a broken arm or leg is a kind of temporary external skeleton that maintains the shape of the limb while the bone heals. Most animals have either an internal skeleton of bones or an external skeleton, or shell.

Living cells are only a small part of most skeletons (Figure 10.7). Skeleton-forming cells produce a network of protein or carbohydrates outside the cells. Our bones and the external shells of clams, snails, and lobsters and their relatives are strengthened by compounds of calcium and other minerals. Skeleton-forming cells collect and deposit these minerals on the network.

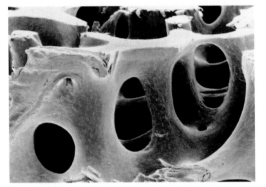

Figure 10.7: *A thin slice of a human leg bone, seen through a microscope. Bone is made of mineral-coated protein fibers. The small dark spaces contain bone-forming cells. (Magnification ×7.3. Color added.)*

Muscles connect parts of the skeleton, making movement possible. These muscles can contract, moving the parts of the skeleton. Physical therapists help people to recover muscle strength after an injury to the muscles or bones. The exercise becomes more challenging at a rate that is slow enough to be safe but fast enough to stimulate recovery and strengthening of the injured organ. Some muscles that aren't attached to the skeleton help to move food through the digestive system. Others pull hairs upright when you are cold or fluff out a cat's fur to look big and threatening. Muscle layers in the walls of blood vessels adjust their diameter and help to move blood through the vessels. Most of your heart is a mass of muscle that contracts and relaxes to pump blood through your circulatory system. Athletic trainers and health care workers encourage people to exercise, which builds up and strengthens the heart's muscle just as it does skeletal muscle.

DIGESTIVE SYSTEMS

Most animals are many cells thick, and only the cells on the surface can absorb water and nutrients directly from the environment. To get nourishment to all their cells, animals have organ systems that **digest** (break down) and absorb food. Differences among animals' digestive systems have important practical consequences. Dairy farmers and kennel owners know that health, fertility, and milk production require a healthy diet that matches the abilities of the animals' digestive systems. Cattle can digest and absorb nutrients in plants that dogs can't.

Small, simple animals, such as jellyfish and flatworms, digest food in an internal cavity with one opening. After the animal absorbs the nutrients, the remaining waste is pushed back out through the same opening. Nearly all the animal's cells are in contact with the digestive cavity, which may be highly branched (Figure 10.8). Larger animals, however, can't depend on the slow process of diffusion to carry nutrients from their digestive systems to the rest of their cells. They need a more efficient system. Most have a hollow internal tube with an opening at each end. A mouth takes in food at one end, and an **anus** at the other end releases the waste that remains after the animal absorbs some of the nutrients.

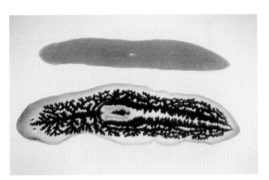

Figure 10.8: *Planaria, small flatworms a few millimeters long, have extensively branched digestive cavities (colored black here) that extend throughout the body.*

CIRCULATORY SYSTEMS

The thin bodies of the smallest animals absorb nutrients directly from the digestive cavity and exchange oxygen and CO_2 gases directly with the surrounding water. Larger animals need circulatory systems that pump fluid throughout the body. The fluid carries nutrients from the digestive system to the other organs and removes their wastes. It also distributes oxygen to the tissues and removes CO_2.

Your occupation can affect the health of your circulatory system. Drivers, office workers, and sales clerks don't get enough exercise at work to maintain a strong, healthy heart. Military and acrobatic pilots have to limit their acceleration to avoid producing forces that interfere with circulation. Interrupting the brain's oxygen supply can cause a pilot to lose consciousness—not recommended!

In smaller animals, such as insects, snails, and their relatives, a large blood vessel runs the length of the body and is open at both ends. It may also have small openings along its length. This vessel pumps fluid throughout the body. Because most of the fluid in such an open circulatory system is outside of this vessel, it circulates very slowly (Figure 10.9).

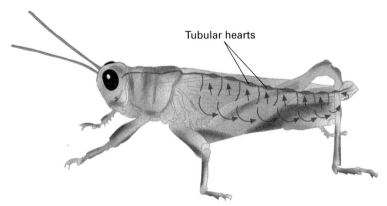

Tubular hearts

Figure 10.9: *The open circulatory systems of insects carry food and wastes, but not much oxygen.*

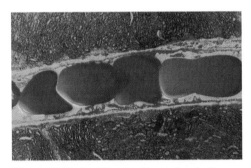

Figure 10.10: *Red blood cells can barely fit through the narrowest vessels of a closed circulatory system, shown here. (Magnification ×1,650. Color added.)*

Humans, like most large animals, have a closed circulatory system. Blood is the circulating fluid. Unlike the fluid in open systems, blood travels only within the blood vessels. A heart continually pumps the blood around the system. Repeated branching of the vessels produces a fine mesh of narrow passageways that approaches nearly every cell in the body (Figure 10.10). The thin walls of the finest branches are only one cell thick, so oxygen and nutrients diffuse quickly from the blood into the tissues, and CO_2 and other wastes diffuse from the tissues to the blood.

RESPIRATORY SYSTEMS

Large animals need large surfaces for oxygen to enter and CO_2 to leave. Respiratory systems perform this function. Many work environments can endanger this system. Dust and fumes are common respiratory health hazards. Exterminators, coal miners, welders, textile mill workers, painters, firefighters, and auto body repair technicians often need to wear dust masks, gas masks, or other safety equipment. Occupational health and safety laws regulate the exposure of workers' lungs to dangerous chemicals.

All respiratory systems have two parts: a large surface that is exposed to the environment and a way to make air or water flow past that surface. Small water dwellers and air breathers such as earthworms that have moist skin exchange gases through their body surfaces. Larger water-dwelling species absorb dissolved oxygen from the water through internal or external **gills** (Figure 10.11). As fish and other large water animals swim, water sweeps over their gill surfaces. This helps them to absorb oxygen and release CO_2.

A   **B**

Figure 10.11: *Water-dwelling animals exchange gases with their environment through gills. (A) External gills of a Christmas tree worm. (B) Internal gills of a rainbow trout.*

Most air breathers have internal respiratory systems. The system connects to the outside through a small opening that can be closed temporarily for defense or to conserve water.

Insects have a branched system of air-filled internal tubes. Orchard owners often spray special oils on their trees to clog the breathing passages of pest insects.

Larger land animals have **lungs,** which are internal organs composed mostly of thin, moist tissues containing very fine blood vessels. Muscles rhythmically push air in and out through a tube connecting the lungs to the mouth or nose. Heavy breathing during exercise can make you thirsty partly because of evaporation from your lungs. It's important to drink plenty of water to replace what evaporates.

EXCRETORY SYSTEMS

Like all organisms, animals produce wastes and need to get rid of them. The smallest water-dwelling animals rely on diffusion to help wash these often harmful substances away. Larger animals have excretory organs such as **kidneys** that remove wastes from the blood or other body fluids. The kidneys concentrate the wastes in a small amount of water and release them into the environment, usually as urine. This process ensures that the concentrations of water, minerals, and other substances in the blood stay at normal levels.

High blood pressure, which can lead to ballooning or bursting of blood vessels, often results when kidneys fail to regulate salt and water concentrations in blood. Until problems become obvious, many people have no idea that their blood pressure is too high.

When high blood pressure is discovered, a doctor might advise cutting down on salt or might prescribe medicine that helps to regulate the kidneys' activities.

Biotechnology may soon create more jobs that involve excretory systems. Genes that encode valuable medicines can be inserted into plants and animals. These organisms produce leaves or milk containing the medicines, which can be extracted with some difficulty. Drug companies are now experimenting with the possibility of producing animals that would release medically useful substances in their urine. It's much easier to extract the substances from this simple combination of water and minerals than from milk, and both male and female animals can be used.

NERVOUS SYSTEMS

Nervous systems collect sensory information from skin and sense organs, such as eyes and ears, and return instructions to contract the muscles or change the activities of glands. In all but the simplest animals a central cluster of nerve cells or a brain receives the sensory information and sends out the signals to muscles and glands. Some automatic responses, or reflexes, such as blinking to protect your eyes, don't involve the brain. More complex behavior and learning, however, require a brain.

Figure 10.12: *Bees are efficient builders and communicators, but their behavior is genetically determined; it isn't a sign of intelligence.*

The small brains of insects are nearly incapable of learning. Beekeepers take advantage of bees' genetically determined behavior (Figure 10.12). Once they install a swarm of bees in a hive, the animals build a wax comb and fill it with honey, find food, and care for their larvae. Larger animals with more complex brains are more intelligent, or able to learn. Animal trainers prize intelligence in the animals that they work with. Horses can learn to accept a rider or a load, dogs are trained as companions for the blind, and dolphins can learn to perform at aquariums.

Nervous systems are composed almost entirely of one type of cell (see Figure 10.6B, page 365). Nerve cells have long extensions that may stretch from one end of an animal to the other. The brain is a network of such cells. Transport of ions through the cell membrane creates an electrical pulse that travels the length of a nerve cell. When the pulse reaches a junction with another nerve cell, it releases a chemical that travels to the other cell and stimulates it to produce a similar pulse. The pattern of pulses and their destination determine their meaning.

ENDOCRINE SYSTEMS

The endocrine system consists of several types of glands in various parts of the body. A **gland** is an organ that produces a particular substance. In animals with closed circulatory systems, endocrine glands deliver their products directly to the bloodstream. These substances, called **hormones,** bind to specific protein receptors, usually on the surfaces of cells, and stimulate or inhibit cell activities. Many glands connect to nerves that stimulate or inhibit the production of hormones. Together, the nervous and endocrine systems affect virtually every organ, controlling and coordinating the activities of the other organ systems.

Hormones have many practical uses. Fish hatchery workers and dairy farmers use them to provoke egg production in fish and cattle. Sprays containing insect "juvenile" hormones are sometimes used in homes and businesses and on farms to prevent pest insects from maturing and reproducing. Growth hormones are used to stimulate milk production in cattle and sometimes growth in unusually short children. Some animals release into the environment hormones called **pheromones** that attract mates or alert other members of their species to danger or food. Artificial pheromones attract insects to traps.

Hormones regulate and coordinate growth and activity. They affect how active an animal is and whether it will attempt to eat, drink, or mate.

IMMUNE SYSTEMS

Some animal cells can tell the difference between an animal's own cells and foreign cells. These special cells are part of the **immune system,** which protects animals from disease-causing microorganisms and other parasites.

Little is known about the defenses of simpler animals. In animals with backbones, skin, scales, and shells offer some protection from parasites, and a great variety of defensive cells and proteins guard the body, especially at its entry points. Some cells of the immune system circulate in the blood, consuming any bacteria or other foreign cells that they find.

One defensive protein is an enzyme on your skin that breaks down the RNA of viruses and foreign cells. Most bacteria in your food are killed by enzymes in your saliva or by the strong acid in your stomach. Other proteins, called **antibodies,** bind to foreign cells and viruses, marking them for destruction by the cells of the immune system. Antibodies bind only to very specific substances. Medical lab technicians use them to identify quickly disease-causing microorganisms, hormones in urine that indicate pregnancy, and the blood and tissue types of people who need or donate blood or organs for transplants. A chemical technician first attaches a radioactive or fluorescent (glowing) substance to the purified antibody molecules. This allows the medical technician to see easily whether the antibody binds to the test material (Figure 10.13). Toxic or strongly radioactive substances can be attached to antibodies that bind to cancer cells.

Figure 10.13: *Adding fluorescent antibodies to test samples. Samples that glow contain the antibody's target substance.*

A doctor can then inject these modified antibodies into a cancer patient. The toxic or radioactive material concentrates in the cancer cells, while the rest of the patient's body is spared.

REPRODUCTIVE SYSTEMS

Farmers and breeders who raise animals for pets or for research labs aren't the only ones who manage animal reproduction. About one in five married couples seeks medical help to have children. Assisting these people are health care workers, technicians, and engineers who design the tools that are used to treat infertility.

Many simple animals reproduce asexually by budding, or breaking off pieces of themselves. However, most animals reproduce sexually, and some species use both mechanisms. Sexual reproduction has the advantage of producing new—and potentially better-adapted—genetic combinations. Asexual reproduction often occurs in species that have difficulty finding mates because they are sessile or rare.

In sexual reproduction, animals produce two kinds of gametes: small, motile male sperm cells and larger, female egg cells containing stored nutrients that will nourish the embryo. In some simpler species, such as earthworms, each individual produces both kinds of gametes; pairs of worms fertilize each other's eggs. More complex species generally have two separate sexes.

Sexual reproductive systems include organs that produce gametes and organs that encourage fertilization and survival of the offspring.

A

B

Figure 10.14: *(A) These salmon release millions of gametes that will receive no parental care. (B) Mammals, such as these sheep, have few offspring, but provide each with extensive care.*

Species differ in how they balance producing as many gametes (and offspring) as possible with devoting resources to the care and survival of each offspring (Figure 10.14). At one extreme are animals such as oysters that release millions of unprotected sperm and egg cells into the water around them. Few of these cells find each other and produce embryos that survive to maturity. At the other extreme are mammals: animals with hair and milk-producing glands, such as bats, dogs, and humans. Most mammals produce five or fewer offspring at a time and devote much energy to their care. Males deposit sperm inside the female's reproductive organs, increasing the chances of fertilization. Development of the embryo inside the female increases the offspring's chances of survival, as do nursing and other parental care. Among the more intelligent mammals, parents teach the young to find food or to hide.

HOW ORGAN SYSTEMS COOPERATE

Every organ in an animal is an adaptation that enables the species to survive. These organs all work together to help the animal maintain homeostasis. Let's examine a set of organ systems in action.

What happens when a police dog named Scamp sniffs out illegal drugs? When Scamp's handler shows her a sample of the drug that is being sought, the dog's brain interprets the pattern of electrical pulses in the nerves from her eyes to form an image and directs her legs to move close to the sample. Muscle contractions draw air through the dog's nose and into her lungs. Nerves carry the scent signal to the brain. Parts of the brain that are responsible for hunting become active.

Muscles contract in Scamp's legs as she uses her leg bones to climb through an airline cart filled with luggage. As she picks up the scent, her brain matches it with the memory of the sample. Nerves stimulate glands to release hormones that regulate the dog's level of activity; she gets excited, barks, and moves more quickly. Her heart and lungs become more active, delivering more oxygen to hard-working muscles. Hormones stimulate cells in the muscles and liver to rapidly release stored energy from carbohydrate reserves. Finally, Scamp barks once more and sits in front of one bag, as she was taught, to indicate the source of the scent.

The officer removes the bag and praises the dog, giving her a treat. Scamp's digestive system breaks down the food and absorbs its nutrients. They pass into her bloodstream, which distributes them to the other organs. Cells, enzymes, and antibodies of the immune system protect the dog from bacteria and viruses in the snack. Drinking provides water to carry any wastes absorbed with the nutrients through the excretory system. After eating, Scamp relaxes, as hormones signal the circulatory system to direct more blood to the digestive organs and less to the muscles.

A well-fed dog is more likely to become pregnant after mating and to give birth. All her organ systems cooperate so that she remains healthy, well-nourished, and able to produce and care for pups.

INVESTIGATION 10A ORGAN SYSTEMS UP CLOSE

See Investigation 10A in the TRB.

You've read about the major organ systems and how they vary in many kinds of animals. Are you ready to identify the major organs in an animal? Health care workers, veterinarians, meat inspectors, and child-care providers need to know enough about organ systems to recognize problems in them. In this investigation you will use a variety of resources to compare the organ systems of several animals.

Some animals don't have every organ system that you learned about in this lesson. For example, sponges are simple animals that don't have clearly different types of tissues and organs, but they do have ways of carrying out all the functions of organ systems, such as digestion, internal transport, and reproduction. And certain kinds of worms that live in tubes around undersea volcanic vents have no digestive systems. They never eat! Instead, they rely on symbiotic bacteria within their bodies to produce their food. Can an animal be considered a whole organism if it doesn't eat? Use the Internet or the library to research these animals, call pogonophorid worms.

LESSON 10.2 ASSESSMENT

1. An arm that can't move may mean an injury to the muscular, skeletal, or nervous system. If neither movement nor sensation is present, the nervous system has probably been injured.

2. The patient's immune system would attack the animal organs.

3. Because blood is the main pathway for substances moving through the body, it indicates what is in the body. Drugs and their breakdown products are eliminated from the blood by the excretory system and appear in the urine.

4. The immune system is responsible for allergic reactions.

1 Suppose someone you work with is injured on the job. If the person can't move an arm, which three organ systems are most likely to be injured? If the person can't move the arm and can't feel it either, which organ system has probably been injured?

2 What problems might occur when a surgeon transplants an animal's organ into a human patient to replace a diseased organ? Which organ system would be involved?

3 Many workers must have blood or urine tests to prove that they don't use illegal drugs that could endanger them or others on the job. Use your knowledge of the circulatory and excretory systems to explain why these body fluids can be used for such a test.

4 Office workers sometimes suffer from sick building syndrome. They may have terrible allergy attacks at work because of poor ventilation that allows dust, mold, and other irritating substances to accumulate in the air. Which organ system is responsible for these allergic reactions?

LESSON 10.3 ANIMAL DIVERSITY

More than one million species of animals exist. There might even be several million! Knowledge of animal diversity can come in handy. Maybe you'll supply suitable animals to pet shops. What you learn about a species can help you decide whether it will make a good pet. In this lesson you will learn about the largest animal phyla and their importance to people.

ACTIVITY 10-4   ANIMALS IN THE WORKPLACE

List the following occupations on the chalkboard or an overhead transparency: architect, cafeteria manager, city planner, customs inspector, fashion designer, groundskeeper, shipyard worker, public health inspector.

List the following animals on the chalkboard or an overhead transparency: alligators, beetles, coyotes, dogs, dolphins, monkeys, raccoons, wasps.

See Activity 10-4 in the TRB for more information.

Most people work with some animal products. In this activity you'll discover the wide variety of animals that are involved in various occupations.

With a partner, review the list of occupations that your teacher provides. Come up with as many ways as you can in which each occupation involves work with animals or animal products or uses knowledge about them. Record your conclusions in your logbook.

Who would need to know about or work with the animals that your teacher lists or with their products? Discuss each of them with your partner, and write how each occupation involves one or more of these animals. Compare your findings with those of other groups. Develop a class list of occupations involving as many kinds of animals as possible.

CAREER PROFILE   WORM FARMER

Jay M. of Rabbit Hill Farm is an expert in vermiculture (earthworm raising) and vermicomposting. Jay explains, "Vermicomposting is using worms to eat anything that has ever been alive (for example, leaves, grass clippings, food scraps, manure, paper, cardboard) and turning it into humus, the organic component of soil."

Worms do not have teeth, so they swallow food whole. Then the worm's digestive system softens the food with saliva. Small stones that the worm has swallowed grind the food up. The worm absorbs the nutrients through its intestine. What remains is humus. The humus passes out of the worm's intestine and becomes part of the soil.

According to Jay, "The quality of the humus depends on the nutrients in the starting products." Jay's worms produce a rich humus because they eat droppings from his rabbits, along with leaves, hay, yard waste, food waste, cornmeal, and rough compost. Jay sells vermicompost and earthworms to organic plant nurseries. Many vermiculturists also sell worms as fishing bait.

Becoming a vermicomposting farmer doesn't require a degree. But it helps to have courses in biology, business, and marketing and practical training in running a vermicomposting farm.

THE FAMILY TREE OF THE ANIMAL KINGDOM

Animalia is divided into about 35 phyla. (Not everyone agrees on how to divide some groups.) The nine largest phyla are listed in Table 10.2. Scientists have had a hard time determining the ancestry and relationships of these phyla, partly because most animals don't have hard skeletons to leave as fossils. However, they have found important clues in animals' development from fertilized eggs to embryos to adults.

Most animal embryos have three layers of cells (Figure 10.15). The inner layer becomes the lining of the digestive system. The outer layer becomes the skin, or other outer covering, and the nervous system. The middle layer develops into the organs, such as muscles and blood vessels, that lie between the skin and the digestive system. In more complex animals the middle layer splits

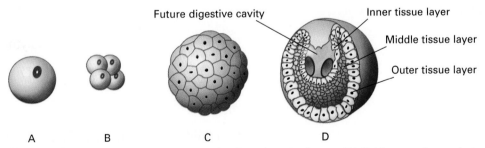

A   B   C   D

Figure 10.15: *Early development in an animal embryo. An animal zygote (A) divides to produce a cluster of cells (B) that becomes a hollow ball (C). Infolding of embryo on one end and formation of a middle cell layer (D) produce the three tissue layers that will develop into all of the animal's organs.*

Scientific Name	Common Name	Examples
Porifera	Sponges	Vase sponge, finger sponge
Cnidaria		Jellyfish, sea anemones, corals
Platyhelminthes	Flatworms	Planaria, liver flukes, tapeworms
Nematoda	Roundworms (nematodes)	Pinworm, hookworm, trichina
Mollusca	Mollusks	Clam, snail, slug, octopus, squid
Annelida	Segmented worms	Earthworms, leeches, sea worms
Arthropoda	Arthropods	Insects, lobsters, spiders, barnacles, centipedes
Echinodermata	Echinoderms	Sea stars, sea urchins, sea cucumbers, sand dollars
Chordata	Chordates	Fish, frogs, snakes, birds, mice, humans

Table 10.2: Largest Animal Phyla

and forms a fluid-filled cavity. The fluid cushions the organs that develop in the cavity and gives them room to move and grow. Comparing DNA sequences has provided evidence of the relationships among the phyla. Still, some of the ideas that are expressed in Figure 10.16 are uncertain.

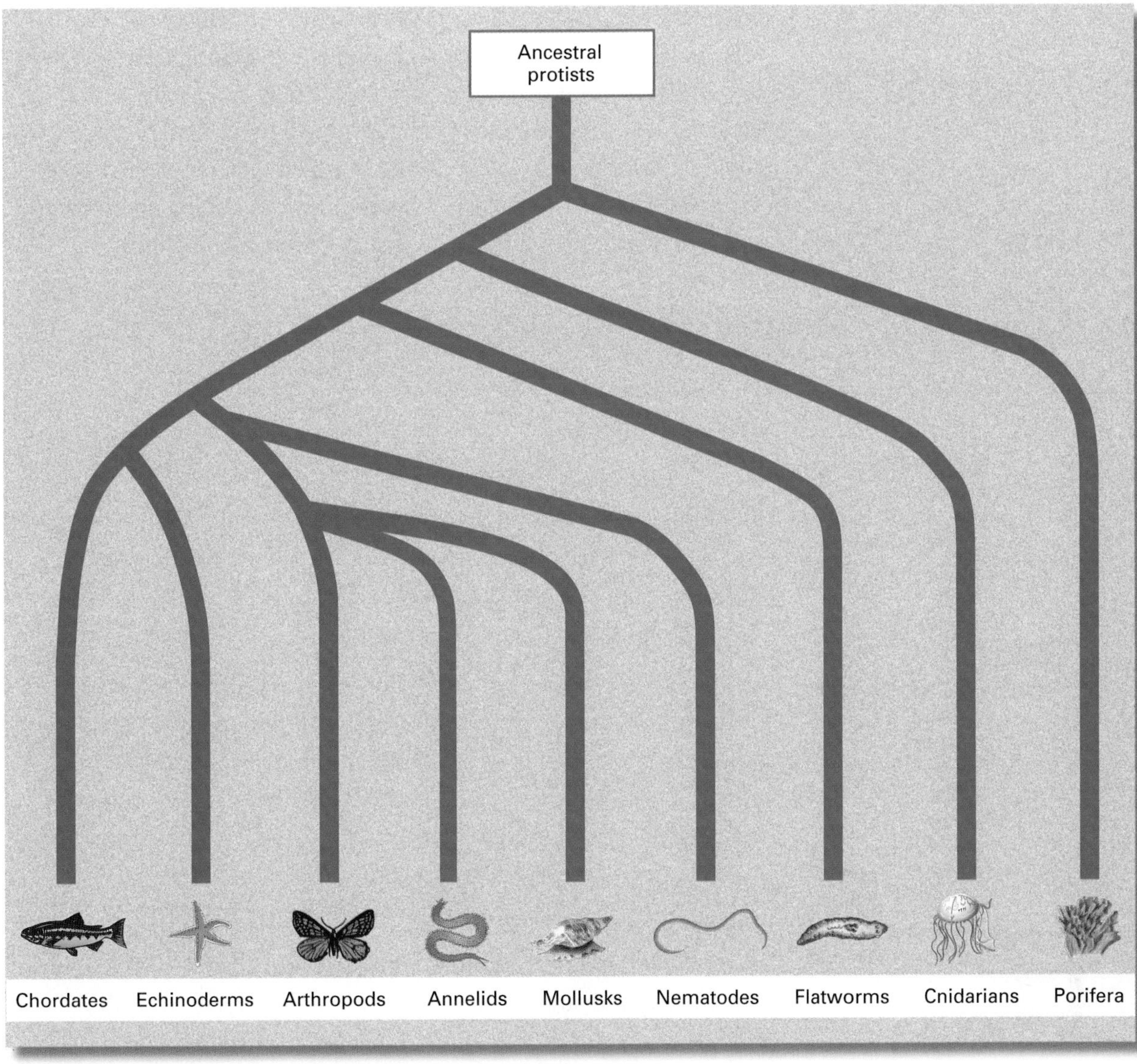
Ancestral protists
Chordates
Echinoderms
Arthropods
Annelids
Mollusks
Nematodes
Flatworms
Cnidarians
Porifera

Figure 10.16: *The origin of the major animal phyla, according to a popular theory.*

Sponges and Cnidaria

Sponges are so primitive and different from other animal phyla that some biologists consider them barely animals. They don't have true organs or nerve or muscle cells. These hollow animals live attached to rocks and other objects under water. They get food from water that is pumped through their hollow bodies by cells with flagella. Divers collect and sell sponges; the tough, elastic internal skeletons of some species are great for household cleaning, though most of the "sponges" you use at home are made of plastic.

Sponge divers learn to avoid jellyfish and other cnidarians (nee-DARE-ee-unz). Species in this phylum use stinging cells for defense and for capturing prey. These cells are usually on tentacles that surround the opening of the digestive cavity. Some adult cnidarians, such as corals and sea anemones, are sessile. Others, such as jellyfish, are floaters. A simple nervous system coordinates swimming and other movement, but cnidarians can't learn.

ACTIVITY 10-5   FINDING TINY WORMS

Students may need help in identifying planaria (flatworms) and nematodes. Planaria can be found under flat rocks in streams or can be collected on a small piece of raw meat suspended on a string in a stream for 30 minutes. Keep them in a container of stream water. Alternatively, both planaria and nematodes can be ordered from a biological supply house. A more elaborate but effective way to collect nematodes involves a funnel with a clamped-off piece of rubber tubing attached to its neck. Place a screen or piece of cheesecloth in the funnel and fill it with fresh, moist garden soil. Shine a bright, hot lamp on top for 24–48 hours. Then open the clamp and collect the water, which will contain nematodes driven down by the lamp.

Make certain that students wash their hands well with soap and hot water after working with the materials in this activity.

The worms seek favorable conditions by avoiding light and acid.

This implies that their nervous systems are more complex than those of sea anemones.

Needlelike mouthparts occur in many parasitic roundworms.

The most common types of animals live all around us, and we hardly ever notice them. In this activity you will examine two common animals: flatworms and roundworms.

Examine the flatworms in a shallow dish of water with a hand lens or a dissecting microscope. Sketch the worms in your logbook. How do they respond to light coming from one side? Record your observations in your logbook.

Put a drop of water from an aquarium filter or a small pinch of garden soil in a drop of water on a microscope slide and add a coverslip. Examine the slide with a microscope at low power. Look for small, round worms. Sketch and describe your observations in your logbook, including any differences among the worms (especially in the details of their mouthparts).

Place a small drop of vinegar (an acid) on one edge of your coverslip so that it will diffuse into the sample from one side. Record whether the worms move toward or away from the vinegar.

After cleaning up and returning all your materials, wash your hands thoroughly with soap and hot water. Answer the following questions in your logbook:

- What evidence did you see that flatworms or roundworms can respond to the environment?

- Would you predict that the nervous systems of these worms are more or less complex than those of sea anemones?

- Roundworms include both parasites and decomposers. How could you tell these types apart? (*Hint:* Think about their mouthparts.)

UNSEGMENTED WORMS

Roundworms and flatworms are among the simplest bilaterally symmetrical animals. Flatworms don't need respiratory or circulatory systems; their thin, flat bodies absorb oxygen and release CO_2 and wastes directly into the surrounding water. Food diffuses from the digestive cavity to the whole body. Some parasitic flatworms, such as tapeworms, are little more than reproductive systems. They live in the digestive systems of other animals, absorbing their hosts' food through their skins. Tapeworms often infest people where poor sanitation allows the worms' eggs in human waste to enter drinking water.

Most nematodes, or roundworms, are less than a millimeter long. There are loads of nematodes in soil and water, where some are decomposers and others are parasites of animals or plants. Important parasitic nematodes include species that infest crops, livestock, and pets. Pinworms and hookworms in soil burrow into the skin of people who go barefoot outdoors. Trichina worms infest people who eat undercooked pork or wild game. Trichina burrows through the walls of the digestive system and moves into the muscles, where its movements cause great pain. If they invade heart muscle, they can cause death.

SEGMENTED WORMS

Earthworms, leeches, and other segmented worms live in water or damp soil. Leeches were once used to suck out people's "excess" blood and reduce harmful high blood pressure. Today, they are used to produce anti-blood-clotting medicines, to suck blood from bruises, and to stimulate blood circulation in severed limbs that have been surgically reattached.

You can easily see the many segments that make up an earthworm's body. Each segment is separated from its neighbors by a membrane and has its own excretory system and branches of the main nerves and blood vessels that run the length of the animal.

Both segmented and unsegmented worms have definite anterior and posterior ends. Food travels through the digestive system in one direction: from anterior to posterior. A cluster of nerve cells at the anterior end serves as a simple brain. Reproduction occurs by splitting or by mutual fertilization.

MOLLUSKS

The phylum Mollusca includes snails, clams, slugs, squid, and their relatives (Figure 10.17). Mollusks have soft bodies with three parts: a mass that contains most of the organs, a muscular "foot" that is used in movement, and a thick flap called a mantle, which covers the body and in most species produces a heavy shell of calcium compounds.

Most mollusks pump water through an internal gill. Two-shelled species such as clams and oysters filter small food particles from this water. Squid and octopuses use the pump for jet propulsion through the water in search of prey. Their activities are coordinated by a large brain. Octopuses are about as intelligent as mice.

A   B

Figure 10.17: *(A) Sea slugs are related to snails, but have no shells. (B) This tropical cone snail has an attractive shell, but packs a painful poisonous stinger that it uses to capture small fish and other prey.*

Growing demand for seafood is shrinking natural populations of edible mollusk species. Farming of edible clams, scallops, and other species is increasing. Oysters are raised for food and pearls. When a grain of sand, a parasite, or another small object falls inside an oyster's shell and irritates its mantle, this organ surrounds the object with a layer of shell. This is the pearl. Oyster farmers put grains of sand inside their oysters so that they will produce plenty of pearls.

Other mollusk species create environmental problems. Throughout the Midwest, freshwater zebra mussels are a nuisance. These mollusks, which were accidentally introduced into American rivers, crowd out native species, encrust the hulls of boats, and clog pipes that carry drinking water or sewage.

ARTHROPODS

Members of Arthropoda, the largest animal phylum, have jointed external skeletons. Over one million species, such as crabs,

WORD BANK

arthro- = joint

-pod = leg

The external skeletons of arthropod legs have joints that allow movement.

shrimp, spiders, and scorpions (Figure 10.18), live in virtually every environment on Earth. Insects, the most diverse and successful animals, are arthropods. As they grow, arthropods **molt,** shedding their rigid shells. They swell with water or air to expand the new shell before it hardens. Crabbers prize newly molted soft-shell crabs, which bring a high price.

Arthropods have heads with many sensory organs. These include simple eyes that detect only light intensity, complex eyes that form images, and antennae that "smell" chemical substances in the environment. Many biting insects such as mosquitoes respond to the water vapor, CO_2, and warm air that rises from our skins. Chemists aided by technicians develop insect repellents that confuse insects' chemical senses.

A B C

Figure 10.18: *Arthropods are extremely diverse. (A) Spiny lobster. (B) Leaf insect. (C) Giant African millipede.*

Most arthropods reproduce sexually. Sperm are released inside the female's body, not in water. This adaptation lets arthropods live in dry environments. Most species lay huge numbers of eggs, few of which produce offspring that survive to adulthood. Larvae of many species develop into very different adults, a process called **metamorphosis.** For example, moths and butterflies seem completely unlike the caterpillars from which they developed. These differences often include diet, which lessens competition between generations and allows insects to use more resources.

Why are arthropods, particularly insects, so successful? The development of resistance to insecticides demonstrates how quickly they adapt to a changing environment. Short generations and many offspring increase the chance that random mutations will produce a few resistant individuals. When a new insecticide is used, these insects and their offspring survive and multiply quickly.

New insecticides are often sold for only a few years before they lose effectiveness. The development of safe, effective, inexpensive insecticides is a continuing challenge to chemists. In a similar way, natural selection has modified insects' mouthparts and their many **appendages** (limbs) to produce species that have adapted to survive in different environments by walking, hopping, or holding tight and biting, chewing, piercing, or sucking.

Some arthropods eat our crops or bite or sting us. Others are important as food, disease carriers, and pest predators. Lobsters, crabs, and shrimp are raised or caught for food. Bees produce honey and pollinate crops. Growers buy praying mantises and ladybird beetles that consume crop-eating insects. Declining numbers of insect larvae in a stream are often the first sign of water pollution. Microorganisms in the saliva of certain mosquitoes, flies, and other biting species cause malaria, sleeping sickness, and other serious diseases.

ECHINODERMS

Echinoderms (ee-KINE-o-dermz) are ocean creatures such as sea stars and sea urchins. In parts of Europe and East Asia, sea urchins and sea cucumbers are popular food. Sea cucumbers that are collected from waters off Washington state bring high prices in East Asia.

WORD BANK

echino- = spiny or prickly

-derm = skin

Echinoderms are protected by the spines or prickles on their skins.

Echinoderms reproduce sexually. Sperm and eggs are released into the water, where they meet and join. Echinoderms move by pumping seawater into and out of a system of internal tubes. The ventral surfaces of sea stars are covered with tiny "tube feet" that swell or collapse like balloons as they fill with water or empty out. When tube feet collapse while touching a rock or other surface, they grasp it like suction cups, protecting the soft ventral surface. A sea star uses this grip to pry open oysters and other mollusks. It then turns its stomach inside out and releases enzymes that digest the mollusk in its own shell (Figure 10.19).

Figure 10.19: *Sea star attacking an oyster*

Most members of the phylum Chordata are vertebrates—fish, amphibians, reptiles, birds, and mammals. The vertebrates and the arthropods together make up the majority of all animals. Humans use other vertebrates for many purposes: food, wool, oils, medicines, leather, transportation, and companionship. Many people raise vertebrate animals. Vertebrates are important in medical research, to national park rangers and others who work in wildlife and fisheries management and oceanography, and to all the tourists and other consumers who benefit from these people's work.

All chordate embryos share four characteristics (Figure 10.20):

1. A stiff dorsal rod helps to organize the embryo's development. In most species, only traces of this **notochord** persist in the adults. The jellylike disks that pad the bones of our spinal columns are composed of notochordal tissue.
2. The central nervous system (brain and spinal cord) is tubular.
3. Their sides have slits just behind the head. These **pharyngeal slits** (pharynx means "throat") become the gill slits of adult fish. In air-breathing chordates, they develop into various organs such as internal parts of the ears.
4. They have a tail; in humans it's the tailbone, or coccyx, which curls internally.

Table 10.3 shows the seven living classes of vertebrates. An internal skeleton of bone, cartilage, or both includes the vertebrae that give these animals their name. The **vertebrae** (singular: **vertebra**) are

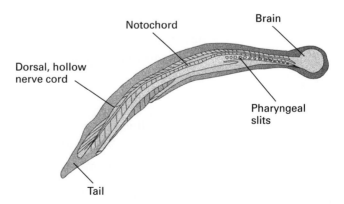

Figure 10.20: *A typical vertebrate embryo. Embryos of different vertebrate species look alike in their early stages.*

Scientific Name	Common Name	Examples	
Agnatha	Jawless fish	Lampreys, hagfish	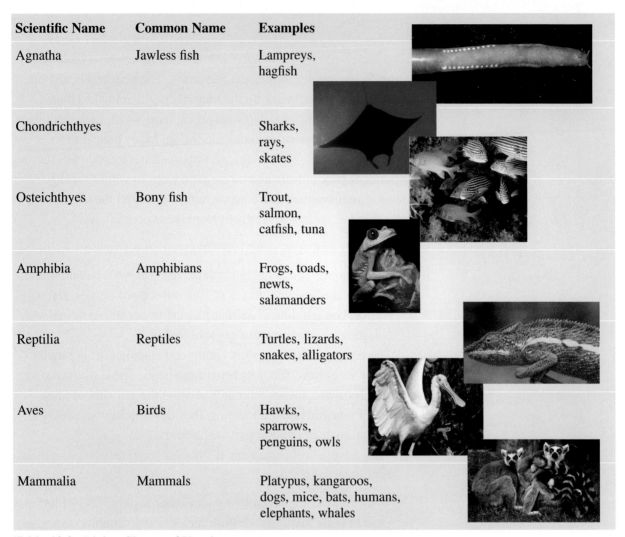
Chondrichthyes		Sharks, rays, skates	
Osteichthyes	Bony fish	Trout, salmon, catfish, tuna	
Amphibia	Amphibians	Frogs, toads, newts, salamanders	
Reptilia	Reptiles	Turtles, lizards, snakes, alligators	
Aves	Birds	Hawks, sparrows, penguins, owls	
Mammalia	Mammals	Platypus, kangaroos, dogs, mice, bats, humans, elephants, whales	

Table 10.3: Living Classes of Vertebrates

the bones of the vertebral column (backbone) that encloses and protects the spinal cord. Adaptations of reptiles, birds, and mammals to life on land include dry skins covered with scales or a thin layer of oil, internal lungs, and the ability to reproduce out of water.

Reproduction in vertebrates is nearly always sexual. Many land vertebrates have few offspring and expend lots of energy feeding and caring for them. Most lay large eggs with rich nutrient reserves. Others nourish and protect the developing embryos in their bodies.

Reptiles and birds lay eggs surrounded by a shell that blocks evaporation of water but allows passage of oxygen and CO_2.

Birds use up lots of energy collecting food for their young. In most species of mammals the embryo develops within the female's body. After birth the mother feeds her young with milk.

Birds and mammals regulate their body temperatures. This is not important for aquatic animals; water temperatures change little and slowly. On land, however, where temperatures often change drastically from day to night, it's more important. Mammals and birds thrive from the polar regions to the tropics; reptiles live mostly in warm climates.

INVESTIGATION 10B   WHAT IS IT?

See Investigation 10B in the TRB.

Biologists and other explorers discover a few new animal species every year. Some of these species turn out to be useful as predators of pest species, as sources of food or medicines, or even as tourist attractions. Biologists always try to classify a new species. In this investigation you will classify some unfamiliar animals and predict which of their characteristics would be important to anyone who is looking for ways to use these species.

IS IT AN ANIMAL?

CHANGING IDEAS

Picture an animal.

Did you think of something with hair? Most people do. In fact, many people use the word "animal" to describe mammals (animals with hair) or vertebrates. But the word has a much wider meaning.

The first scientific classification of living organisms had only two kingdoms: plants and animals. An organism was an animal if it moved on its own. But reality is more complicated: In the past, many people thought that sponges were plants. The two-kingdom system persisted until the mid-1900s. Then animals were defined as "motile eukaryotic organisms." Under the five-kingdom system proposed by Robert H. Whittaker in 1969, animals are defined as multicellular eukaryotic heterotrophs. Their cells form complex tissues. Biologist Lynn Margulis added the fact that animals develop from embryos. The five-kingdom system is currently the most popular, but things are changing.

Some biologists have questioned whether sponges belong in the same kingdom as other animals. They have suggested that the other animal phyla descended from one ancient protist species and the sponges came from another. Others point to evidence that the insects may consist of three groups of orders with different ancestors. A similar argument has been made about the reptiles. Comparisons of DNA sequences are helping to resolve some of these questions. As we develop new techniques for analyzing relationships between species, classification will continue to change. Perhaps people will one day consider the concept "animal" an old-fashioned idea.

LESSON 10.3 ASSESSMENT

1. Clams do not move around much. Squid need more space for swimming.

2. The internal vertebrate skeleton grows with an animal. The arthropod exoskeleton must be replaced as the animal grows. Vertebrates have an advantage because molting wastes nutrients and energy used in growing the discarded shell, and recently molted animals are vulnerable to predators.

3. Arthropods such as crayfish and crabs are active and motile and would make more interesting pets than sea anemones and jellyfish would.

4. Mammalian pests such as mice and rats are chordates that wouldn't survive as eggs outside the female adult's body.

1 Why would raising clams be easier and less expensive than raising squid?

2 Explain why arthropods molt but vertebrates do not. Is this difference more of an advantage for vertebrates or for arthropods? Give a reason for your answer.

3 As the manager of an aquarium store, would you expect to sell more live cnidarians or live arthropods? Give a reason for your answer.

4 Some agricultural and household pests survive as eggs when chemical sprays kill the larvae and adults. What household pests can an exterminator be sure will not lay eggs? To which phylum do these animals belong?

CAREER APPLICATIONS

The applications that follow are like the ones you will encounter in many workplaces. Use the biology you learned in this chapter to complete the activities. Share your work with the class.

AGRICULTURE & AGRIBUSINESS

Economically Important Animals

Using various resources, make a list of animals that people raise. Make a list of the animals' primary uses, such as food, leather, or transportation. Using Appendix E, assign the animals to the proper phyla. Include at least one member of each phylum, and several members of the large phyla. Describe animals in different phyla that serve similar purposes in different cultures.

Native Animal Zoo

Your community wants to build a special zoo that will house animals that are native to your state or region. Using Appendix E and a list of native animals, see how many phyla you could represent. A different group of students should research animals from each phylum and make placards for each exhibit, describing the animal. What environment will the animals need to survive? What will they eat? Make a model of the zoo display and combine all the displays to show the entire zoo.

BUSINESS & MARKETING

Furs and Fashion

For a long time, humans have used animal skins, furs, and feathers as clothing and accessories. The fashion industry has contributed to the extinction or endangerment of some species. Today, some people want to ban furs for fashion. They say that there is no need to use furs when so many synthetic materials are available and some fur-bearing species are endangered. In many instances, furriers use pelts from animals that are raised on farms much like other domestic species; they argue that they help to protect the species. Divide the class into groups to research both sides of the issue and hold a debate. Find out how the size of the fur market has changed. Prepare a report on your findings. Survey consumers and include your findings in your report.

From Feed to Food

Use the library and interviews with your county cooperative extension agent or a sheep or cattle producer to find out how the unique digestive systems of ruminants (animals with four stomachs) work, and how they are different from the digestive systems of other mammals. Why are these species so popular as sources of meat, milk, and other products? Are they easier to raise than chickens, pigs, and other livestock? Are they more efficient producers; that is, do they produce the same quantity of meat or other products as other animals from less food? From lower-quality, cheaper food?

FAMILY & CONSUMER SCIENCE

Symmetry in Interior Design

Look for pictures of interior designs in sources such as architecture magazines. Classify the objects that are used to decorate the rooms according to the type of symmetry each has. Include examples of objects with bilateral, radial, and other kinds of symmetries. How does the symmetry or lack of symmetry contribute to the beauty and function of the object?

Cheese, Please!

Research how cheese is made. Use the library, the Internet, or interviews with cheese makers to learn about the process. Find out how different kinds of cheese are made and what kinds of substances are used to curdle the milk. How do they work? Is all cheese made from cow's milk, or are other species used, too? What is rennet and where does it come from? Get some rennet and make some cheese to show the class. Be sure to check with your teacher before eating the cheese or giving it away.

HEALTH CAREERS

Medical Specialties

Using O*NET or the Dictionary of Occupational Titles, make a list of medical specialties. Include the organ system that is the focus of each specialty. (For example, cardiologists specialize in the heart and circulatory system, and neurologists treat diseases of the nervous system.) Select a specialty area, and use the Internet or other resources to write a summary of a person's work in that specialty. Possible sources of information are professional organizations for the specialty, interviews with people who work in it, and health care books. What are common conditions, diseases, or injuries that the specialty deals with? Where do people in the specialty work?

Stream Watch

Find out about Project Streamwatch or other opportunities for volunteers to help keep track of water quality. You may need to contact your state's department of environmental management or natural resources, or a teacher or organization involved in environmental education. These programs rely on people like you to check occasionally the numbers of insect larvae and other small animals in rivers and streams. Changes in their populations often serve as early warnings of environmental problems. What kinds of animals are used? Explain the program to your class and organize a stream watch in your community.

INDUSTRIAL TECHNOLOGY

Bionics

Research bionics and the various types of bionic systems. You might investigate heart pacemakers, electronic sensory systems for people with hearing or vision loss, electronic brain stimulation for control of seizures or violent behavior, or other kinds of systems. Which organ systems have been duplicated, enhanced, or combined in a bionic system? Choose one organ system and make a poster showing the bionic components. Did the engineers imitate a natural system or use a different approach? Label and briefly describe the function of the artificial components. Compare the artificial system and the natural one it is meant to assist or replace.

Cultured Pearls

Use the library, the Internet, or interviews with jewelers or people who raise or fish for oysters to learn how pearls are cultured, or produced. How long do pearls take to grow? Do all mollusks have them? What makes some pearls more valuable than others? What is mother-of-pearl? Find out the differences between natural and cultured pearls, and report your findings to the class.

- Animals are multicellular, heterotrophic eukaryotes made up of many specialized types of cells. Adult animals develop from embryos.

- Animals, especially the simpler ones, can heal injuries by regeneration. Some can also reproduce by breaking up and regenerating missing parts.

- Some simple sessile or floating animals are radially symmetrical. More complex animals are bilaterally symmetrical. Body symmetry is an adaptation to a floating, sessile, or motile existence.

- The various types of specialized cells in an animal make up tissues, which form organs. Organs make up organ systems that carry out specific functions such as digesting and absorbing food or transporting substances through the body.

- As cells differentiate, they lose the ability to change function and to regenerate lost parts.

- Most animals have internal or external skeletons that support and protect soft internal organs.

- Muscles attached to skeletons enable animals to move their bodies. Other muscles move substances such as blood and partly digested food through the body.

- Digestive systems break down and absorb nutrients from food.

- Circulatory and respiratory systems are important in larger, more complex animals. Air breathing requires a large internal surface.

- Excretory systems remove waste and help to maintain normal internal concentrations of water and minerals.

- Complex animals have a central nervous system and an anterior head with a mouth and sensory organs. The nervous and endocrine systems coordinate and control body movements and functions.

- A skin or other covering, antibodies, and immune system cells are important animal defenses.

- As animals become more complex, reproduction shifts from asexual toward systems involving two sexes.

- Respiratory, circulatory, or digestive systems may be very simple or absent in small, simple animals such as flatworms and nematodes.

- The arthropods and chordates are the largest, most successful animal phyla. They are the dominant forms of animal life on land because they have the most successful adaptations to life out of water.

CHAPTER 10 ASSESSMENT

Concept Review

See TRB, Chapter 10 for answers.

1 What makes animals different from other organisms?

2 Do most animals reproduce sexually or asexually?

3 Is your belly a dorsal or a ventral surface?

4 Explain why there are almost no sessile land animals.

5 What is the main advantage of a closed circulatory system over an open one?

6 Could bees be trained to build useful objects other than hives? Give the reasons for your answer.

7 Which organ system is probably responsible for a condition that causes people to grow to a "gigantic" size?

8 Would herbivores, carnivores, or omnivores have the best chance of surviving in a new or changed environment?

9 Is your hand an organ, an organ system, or something else? Give a reason for your answer. If it is not an organ, what organs does it contain?

10 Which organ system consists mostly of nonliving material?

11 Which organ is your first line of defense against disease-causing microorganisms?

12 Why can't desert lizards conserve water by closing their mouths and breathing through their skins?

13 Some fish seem to be constantly swallowing water; others never stop swimming or close their mouths. How does this behavior help to keep the fish alive?

14 How do your muscles help your respiratory system to function?

15 Which organ system could relieve a condition in which a person retains too much water, becoming swollen and puffy?

Think and Discuss

16 Is asexual reproduction more common among simpler animals or more complex ones?

17 What is the difference between regeneration and asexual reproduction?

18 How is the body plan of a ship like that of a fish?

19 Think of a radially symmetrical device that is not mentioned in this chapter and explain how its symmetry relates to its function.

20 Why do most sessile animals produce motile gametes?

21 Could a very large, thin, floating sea animal survive without most of the organ systems that you studied in this chapter? Would your answer change if it lived on land? Give a reason for your answer.

22 Why aren't oils that are sprayed on orchards to block insects' breathing passages as harmful to birds and mammals?

23 Explain why most intelligent, large-brained animals are predators.

24 Is the great success on land of arthropods and vertebrates evidence of a close relationship between these two groups? Give a reason for your answer.

25 Earthworms don't have true respiratory systems. Would you expect longer annelids, such as the 2-meter-long Australian earthworm, to have a respiratory system? What about a thicker worm? Give the reasons for your answer.

26 Compare the numbers of gametes and offspring produced, and the care and resources given to each, in arthropods and mammals. What does it mean to say that the reproductive success of one of these groups depends on quantity and that of the other on quality?

In the Workplace

27 Is a scuba diver in more danger of being attacked by a radially symmetrical animal or a bilaterally symmetrical animal?

28 Why are animals without skeletal systems more common in water than on land?

29 A consumer protection investigator suspects a store of mislabeling meat. The investigator plans to examine samples of the suspect meat with a microscope. Will this test be more likely to determine the kind of tissue the meat is made of (muscle, fat, stomach, etc.) or the kind of animal (cow, pig, horse, etc.) that it came from? Give a reason for your answer.

30 Give an example of a task or job that a person with an artificial arm might be better able to do than a person with two intact arms.

Investigations

The following questions relate to Investigation 10A:

31 A large ridge containing blood vessels protrudes into the digestive system of earthworms and runs the length of the intestine. What is its function?

32 Tadpoles undergo metamorphosis to become adult frogs. The digestive system is much shorter in frogs in relation to the size of the animal. What does this tell you about the diets of frogs and tadpoles?

The following question relates to Investigation 10B:

33 Which phyla would you expect to be closely related to a soft-bodied, segmented animal without a skeleton that walks on many short legs?

CHAPTER 11

WHY SHOULD I LEARN THIS?

You use your bones, muscles, and other parts of your musculoskeletal system in every activity, even while you sleep. Most of the time, you don't give them a thought. But lifting a pile of books, playing basketball, or riding a bike poses a risk to your musculoskeletal system. You can avoid most injuries by learning about musculoskeletal systems, their capabilities, and their limitations.

Chapter Contents

MUSCULOSKELETAL SYSTEMS

WHAT WILL I LEARN?

1. to predict how an animal will move, on the basis of its skeleton

2. how a musculoskeletal system helps to maintain homeostasis

3. the cellular processes and organelles that are involved in muscle contraction

4. the capabilities and limits of your musculoskeletal system

5. how exercise affects the development and maintenance of your musculoskeletal system

6. common musculoskeletal injuries and how medical personnel treat them

How do animators for video games and movies create realistic on-screen motion? They instruct a computer to simulate the movements of real animals. They have to know how an animal's muscles move the bones of its limbs when it runs, walks, hunts, eats, and stretches. Animators have to know what muscles and bones can't do, too. If they want a human or an animal to perform an activity, they have to identify what parts of the body coordinate to create movement.

LESSON 11.1 SKELETONS AND MUSCLES: AN OVERVIEW

WORD BANK

musculo- = muscle

skeletal = dried-up

The musculoskeletal system is the framework of muscles and dry, hard bones that supports, protects, and moves the body.

What's so great about bones and muscles? Well, without them you would be tiny: Large multicellular organisms need a skeleton to support their body mass. You would also be less well educated: You couldn't open a book or turn a page, much less carry a book bag. And forget about driving a car, passing a football, or dancing—all would be beyond your reach. On a more basic level, animals with well-developed musculoskeletal systems have the mobility to search for food, escape from predators, look for mates, and relocate to more favorable environments.

ACTIVITY 11-1 ROUTINE MOVEMENTS

Instruct students to divide into groups of two or more. Be sure there is enough space for students to demonstrate simple body movements without risk of injury. Provide objects that can be safely handled and won't break if dropped: a joystick, tennis racket, plastic cup, coin, tennis ball, clean comb or brush (do not reuse combs and brushes).

WORD BANK

kinesis- = motion

-ologist = one who studies

A kinesiologist studies the principles of mechanics and anatomy and how they are related to human movement.

A **kinesiologist** studies the body's movements in sports, dance, and exercise. In this activity you will make some simple movements and analyze how they happen.

Pick a task that is common in a sport, job, or household activity. Use an object to demonstrate the movements that are needed to perform this task. After each student's demonstration, identify the activity and the parts of the body in which muscles are active. Have a backup idea in case another student demonstrates the movement you chose. Answer the following questions in your logbook:

- Which demonstrated activity might lead to injury if it were repeated many times in an hour?

- Is there a way to make the activity more efficient so that repeating it would be less likely to cause fatigue or injury?

THE EQUINE ATHLETE

BIOLOGY IN CONTEXT

A serious leg injury can threaten the life of a racehorse. Accidents or improper exercise can fracture bones or tear and weaken tendons and ligaments. While these injuries heal, which can take a long time, the horse's movement has to be restricted. At the very least, serious leg injuries keep a horse out of competition. At the worst, they can result in humane killing of a horse and a serious economic loss to the owner.

Swimming allows this horse to strengthen its musculoskeletal system without causing any additional damage to its bones or muscles.

New techniques can detect musculoskeletal problems in horses at an early stage, before they disable the animal. A horse with a suspected injury is sent to an equine medical center like the one at Santa Anita Park, California. There, vets use imaging techniques to diagnose the horse's problem. In nuclear scintigraphy, for example, a vet injects the horse with a radioactive substance that concentrates in inflamed areas. These hot spots mark damage to bone and surrounding soft tissue, which may be repaired with surgery or rest. With proper treatment and care, a racehorse can return to the track healthy and ready to face the competition.

SKELETAL SYSTEMS

A skeleton plays three major roles in any animal. It supports the animal's weight, it protects the internal organs, and it translates the forces of muscles into movements. **Invertebrates** (animals that have no backbones), although they lack true bones, have various types of skeletons. For example, a caterpillar uses the watery fluid inside its body cavity as a **hydrostatic** skeleton. "Hydrostatic" means that the body water exerts pressure on the animal's internal compartments. By using its muscles to change the shape of the fluid, the caterpillar can move its mouthparts, legs, and body as a whole. Other invertebrates, such as adult mollusks and arthropods, secrete organic substances and minerals that harden on the outer body surface, forming a hard layer called an **exoskeleton** (Figure 11.1). Although an exoskeleton may have joints that allow an animal to move its limbs,

Figure 11.1: *The scorpion has a jointed exoskeleton.*

its rigidity doesn't permit growth. Arthropods deal with this problem by periodically shedding their exoskeleton. Growth takes place before the next exoskeleton hardens (Figure 11.2).

An exoskeleton is an improvement over a hydrostatic skeleton, but it works best for animals that are no more than a few inches long. Aquatic animals, such as the lobster, can be exceptions to this size limit, because the water in which they live helps to support their weight. Without the support of water, a large animal must have an internal skeleton made of bone, called an **endoskeleton.** Bone is a product of cells that secrete protein fibers in which minerals such as calcium and phosphate become embedded. The vertebrate endoskeleton grows with the organism. You are a vertebrate, with an endoskeleton of bones and joints. A joint is a junction between bones that allows movement of body parts.

Figure 11.2: *Cicadas shed their old exoskeleton before the new one hardens.*

FRAME SIZE

Frame size is important in figuring one's ideal weight for a given height or in picking out clothes that come in bulky sizes.

Have students develop a chart showing the frame size of all of the students in the class.

Bone size varies from individual to individual even in people of the same age and height. Industries that make clothing and footwear must recognize that people with similar body dimensions often wear markedly different sizes of clothes and shoes. You can observe this variance in bone size by taking a simple measurement.

Put the thumb and little finger of one hand (Figure 11.3) around the narrow part of your other wrist. If your thumb and little

Figure 11.3: *Estimating frame size by grasping wrist*

finger overlap, you are small-boned, or have a small frame, for your height. If they don't touch, you are large-boned, or have a large frame. If they just meet, you are medium-boned, or have an average frame.

HOW MUSCLES WORK

When advising athletes about how to build muscle, professionals in sports medicine are concerned with muscle fibers. A muscle fiber is a highly specialized cell filled with filaments of protein. The filaments slide along each other to shorten or lengthen the fiber. A fiber that shortens is contracting; a fiber that lengthens is relaxing.

The activity of muscles often supports movement of the organism. But in many animals, muscle activity is also vital to the following functions:

- transporting internal fluids such as blood
- moving food along the digestive tract
- eliminating wastes such as urine and feces
- secreting substances such as sweat, oils, and venom
- laying eggs or bearing live young
- generating body heat
- inhaling and exhaling air

EXAMPLES OF MUSCULOSKELETAL SYSTEMS IN DIFFERENT ANIMAL PHYLA

Simple animals such as jellyfish do not have a musculoskeletal system. A jellyfish, however, can close off its digestive cavity and use the seawater inside it as a hydrostatic skeleton. Although they lack muscles, jellyfish have cells in their body linings containing fibers that contract and relax. Contraction and relaxation of these fibers change the shape of the gastrovascular cavity, much as you would change the shape of a water-filled balloon by gently pushing and pulling on it. Cnidarians called hydrozoans use such a system to turn somersaults (Figure 11.4).

Some animals with hydrostatic skeletons, such as planarians and annelids, employ bands of muscles in locomotion. These muscles line the worm's body wall in lengthwise and circular bands. When lengthwise muscles contract and circular muscles relax, they force the body fluid against the body wall, and the worm's body gets short and thick. When circular muscles contract and lengthwise muscles relax, the worm's body gets long and thin.

Each body segment of an earthworm has both lengthwise and circular muscles that work independently. An earthworm can lengthen some segments while shortening others to produce waves of muscle action, called **peristalsis.** With help from bristles that hold onto the soil, an earthworm uses peristalsis to burrow through the ground. In more complex animals, peristalsis helps to move food through the digestive tract.

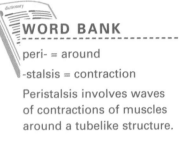

Figure 11.4: *Microscopic view of a small* Hydra *bending and somersaulting*

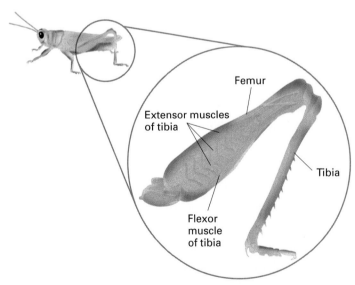

Femur

Extensor muscles of tibia

Tibia

Flexor muscle of tibia

Figure 11.5: *Hind leg of a grasshopper. Notice the size of the extensor muscles that provide the force to cause the grasshopper to jump.*

An arthropod is an animal that has an exoskeleton with movable joints. Muscles are attached inside the exoskeleton, rather than to the outside of bones, as in vertebrates. When muscles on one side of a joint contract, a body segment bends at the joint. For instance, when muscles of a grasshopper's lower leg extend and muscles of the upper hind leg contract (Figure 11.5), the insect body catapults into the air, sometimes as high as 1 foot or more.

VERTEBRATES

Additional Activity, Invertebrates in Motion, can be found in the TRB.

A vertebrate has hundreds of muscles that are responsible for movement. Even the slightest movement requires at least two muscles working against one another. For example, when you bend your arm to drink from a cup, the muscles on the inside of your arm contract while the muscles on the outside relax. When you straighten your arm to put the cup down, the outside muscles contract and the inside muscles relax. In either case the arm is bending around a joint: the elbow.

LESSON 11.1 ASSESSMENT

1. The answer to this question should compare the skeleton's ability to support an organism's weight, protect internal organs, and translate the forces of its muscles to movement.

2. The answer should include items that will represent the skeletal muscular system in form and function.

3. The waves of muscle action are used to push food through the digestive tract.

4. Both are made up of fibers.

1 Compare the functions of a musculoskeletal system to the function of one or more mechanical devices.

2 You are designing a robot. What common household items could you use to make a model of the musculoskeletal system of the robot?

3 Peristalsis moves food through your gastrointestinal tract. Is the food that is passing through the gastrointestinal tract being pushed or pulled?

4 How is skeletal muscle similar in structure to a piece of cloth?

LESSON 11.2 MUSCLES

Why do some people excel at sprints and some at long distances (in swimming or running, for example)? There are different types of muscle cells. Some work best in bursts of intense activity, and others work well on the long haul. If you win foot races only if they're under 400 meters, you probably have more sprint muscle. If you're more likely to win a marathon, you have more endurance muscle. In this lesson you'll learn about several differences in muscle cells.

BUILDING BETTER MUSCLES

So you're going to shape up, get toned, and build your muscles. Where do you start? Calisthenics, aerobics, weight training, or isometrics? For building strength and stamina, weight training is probably most efficient. Will you use free weights or machines? The choices seem endless, but a knowledgeable trainer can help you to design the right workout.

Actin and myosin are the proteins that make muscles contract

What happens when you lift weights to exercise a muscle? You put more and more load on the muscles, stretching muscle fibers. If they stretch too much, they tear. After exercise, the body repairs the broken fibers, a process that you know is taking place because your muscles feel sore. The proteins that make muscles contract are produced in greater quantities, which increases the muscles' capacity for work. The muscles grow and take on more definite shape.

If lifting weights builds muscles, then lifting weights every day should build them really fast, right? Wrong. Muscles need 48–72 hours to rebuild fibers. If you exercise more often, the muscles don't have time to recover, and you can seriously damage them. If you want to work out every day, exercise different muscle groups on consecutive days. You could exercise your upper body one day and your lower body the next. But plan your workouts carefully, because some exercises affect more than the targeted muscle group, and you could unintentionally overtrain some muscles.

Weight training shapes muscles and builds strength.

INVESTIGATION 11A MUSCLE CONTRACTION

See Investigation 11A in the TRB.

What would happen if you put muscle fibers in a test tube and added some ATP? Would the fibers contract, relax, spasm, or just vegetate? In this Investigation you'll work with a strip of muscle from a rabbit and learn more about the things that affect muscle activity.

SKELETAL MUSCLE

If you look at a round steak in the meat case at your supermarket, you'll see that it contains muscle tissue and a cross section of bone. This arrangement is like the muscles and bone in your own body. Beneath your skin is a layer of fat and connective tissue that forms a webbing that reinforces the muscles. In the steak you can see several small circles of muscles, each surrounded by another layer of connective tissue and some fat. If you pulled apart a smaller muscle bundle, you would see that muscle fibers (cells) make up the bundle. Each fiber, an elongated cell packed with protein filaments, is enveloped by connective tissue. Embedded in the bundles of fibers are networks of nerve cells. The nerve cells, called **neurons,** carry electrical impulses that tell the muscles served by each neuron to contract (Figure 11.6). A band of connective tissue extends beyond the muscle to form a **tendon,** which connects to a bone.

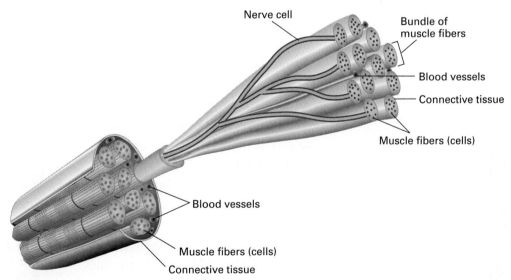

Figure 11.6: *A muscle bundle has nerves and blood vessels.*

How Skeletal Muscles Contract

Inside skeletal muscle fibers are two types of protein filaments that are responsible for contraction. **Myosin** is a thick filament that has many short hooks extending from it. **Actin** filaments are thin filaments that are twisted together in pairs (Figure 11.7). The overlapping positions of myosin and actin filaments inside a muscle cell create a pattern of light and dark bands that are clearly visible in stained muscle tissue under a microscope (Figure 11.8).

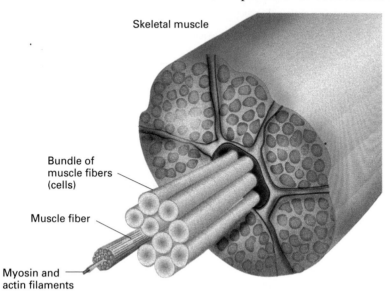

Skeletal muscle

Bundle of muscle fibers (cells)

Muscle fiber

Myosin and actin filaments

Figure 11.7: *Skeletal muscle showing myosin and actin filaments*

Please note that unlike the rope in this analogy, the muscle fibers will not develop slack.

Picture the contraction of muscle fibers by thinking of how you might use a rope to dock a boat. You pull the rope hand over hand, the boat moves closer to you, and the taut part of the rope gets shorter. If you drop the rope, the boat floats away, increasing the length of rope between you and the boat.

When a chemical impulse reaches a muscle, a stimulating impulse is released into the cytoplasm. The stimulating impulse will cause calcium ions to diffuse into the cytoplasm toward the thin filaments and bind to them. This binding will expose binding sites on the thin filaments that attract regions of the thick filaments, forming bridges. These bridges allow the thick filaments to pull the thin filaments along the length of a muscle fiber. This pulling action makes the fiber shorten, or contract. When the chemical impulse to a muscle ceases, calcium ions move out of the cytoplasm by active transport. After ATP binds to the thick filaments, the bridges break between

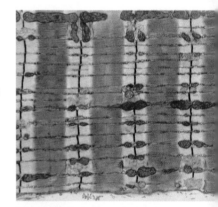

Figure 11.8: *Actin and myosin filaments of a skeletal muscle appear as bands under a microscope.*

the thick and thin filaments. The thin filaments return to their normal position, and the muscle relaxes. Because the thin filaments appear to slide along the thick filaments, this explanation of muscle contraction and relaxation is called the sliding filament theory.

USE IT OR LOSE IT

The nerves that serve your muscles signal only the muscle fibers that are needed for a task. Fewer fibers are needed when you lift a pencil, more fibers when you lift a book. The longer you hold the pencil or the book, the more fibers join in the contraction. But at some point, your muscles become overloaded. Say you can bench-press 200 pounds but want to lift more weight. As you exercise to build muscle strength, the size of the muscle increases. The fibers make more protein filaments. As the muscle fiber enlarges, it incorporates nuclei from nearby cells that seem to be waiting around just in case nuclei are needed. If all these "spare" nuclei are used during muscle-building exercise, the body produces additional muscle fibers.

LEARNING LINK WORKOUTS IN SPACE

In space, astronauts work in weightless conditions. Although this means that they don't use as much energy as they would doing the same work on earth, working in zero gravity poses problems for the human musculoskeletal system. Using the Internet and library resources, research what these problems are and how astronauts combat them.

CHANGING IDEAS

TAKING THE MYSTERY OUT OF MUSCLE

In early cultures, hunters and butchers learned about the body when they slaughtered animals. The ancient Greeks dissected various animals and even human cadavers. This led Galen, who lived in Greece about 1800 years ago, to explore muscles and to find out that they work in opposing pairs. But he wrongly believed that muscle activity was caused by animal spirits moving along the nerves that enter the muscles. When the Black Death, a form of bubonic plague, swept through Europe in the fourteenth century,

dissection became an important tool for trying to understand the cause of infectious disease. Observations that contradicted traditional knowledge were disregarded.

In the sixteenth century, Andreas Vesalius of Belgium challenged the traditions that contradicted his observations. His work focused on detailed descriptions and exact drawings. His book was the first guide to the musculoskeletal system and was supported by Leonardo da Vinci and other artists, who used dissection to study and draw muscles, bones, and other systems (Figure 11.9). Leonardo applied principles of motion and mechanics to the muscles that he drew, often building models to demonstrate movement.

See Additional Activity, Muscle Fatigue, in Chapter 11 of the TRB.

See Additional Learning Link, Carbohydrate Loading, in Chapter 11 of the TRB.

Figure 11.9: *da Vinci's drawing depicting a shoulder, an arm, and a foot*

CAREER PROFILE

ATHLETIC TRAINER

Unlike many athletic trainers who work in team sports, Lynn G. works at a sports medicine and rehabilitation center. In the mornings, he has a free clinic for high school students who have sports injuries. He has just evaluated a 16-year-old boy who hurt his knee while playing football yesterday. Although the boy used ice and compression immediately, his knee is swollen, and he is having trouble walking.

Lynn tests the boy's knee to evaluate the extent of injury. He measures the injured knee and compares it to the good knee. He measures its width, range of flexion, and extension; looks for discoloration; and tests nerves and circulation in the injured leg. Lynn's impression is that the student probably has torn and sprained ligaments, which are common athletic knee injuries. He refers the student to a physician for follow-up

Athletic trainers are present during sporting events to prevent an athlete from getting seriously injured.

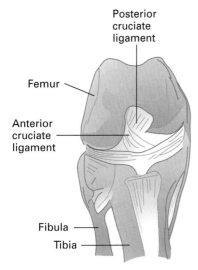

Posterior cruciate ligament

Femur

Anterior cruciate ligament

Fibula

Tibia

Posterior view of a knee joint

See Additional Activity, Three Types of Muscle, in the TRB.

evaluation and treatment. Later, Lynn may help the boy through a rehabilitation program for the knee. When asked how injuries can be prevented, Lynn says that the best way is to include flexibility and strength training in any conditioning program.

Athletic trainers have a four-year degree in athletic training or in a related field with an internship. They must have 800 hours of supervised experience. They are licensed in the state where they practice, and they may be certified by the National Athletic Trainers' Association (NATA). To maintain his certification, Lynn completes 80 hours of continuing education every three years. He recommends that anyone who is interested in becoming an athletic trainer contact NATA for more information. He also recommends that students volunteer to work with a certified trainer, either at their school or in a clinic.

CARDIAC MUSCLE

Cardiac muscle makes up the walls of the heart. Cardiac muscle is striated like skeletal muscle because of an orderly arrangement of thick and thin filaments. Cardiac muscle cells branch and interconnect. The branching pattern provides strength to the heart muscle, allowing it to beat steadily for years. The ends of the cells interlock for added strength.

Lock the knuckles of your hands together so that the fingers of one hand are sandwiched between the fingers of the other (Figure 11.10A). By squeezing slightly, the fingers lock together. That's how the cell ends work. These specialized ends allow electrical impulses to cross from one cardiac muscle cell to another, so all cells contract and relax as a unit (Figure 11.10B). Cardiac muscle depends on aerobic respiration. When heart muscle has no oxygen, as happens in a heart attack, the cells quickly die, and the damage to the heart is permanent.

Figure 11.10A: *Clasped hands resemble interlocking heart muscle cells.*

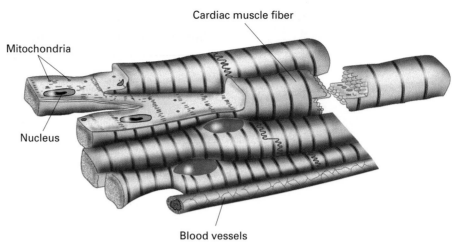

Figure 11.10B: *Close-up of cardiac muscle cells*

SMOOTH MUSCLE

Smooth muscle cells are football-shaped. Like skeletal muscles, smooth muscles operate in pairs; one set of muscles contracts while another relaxes. Actin and myosin filaments are more randomly ordered in smooth muscle than in skeletal muscle, so there is no pattern in the filaments (Figure 11.11).

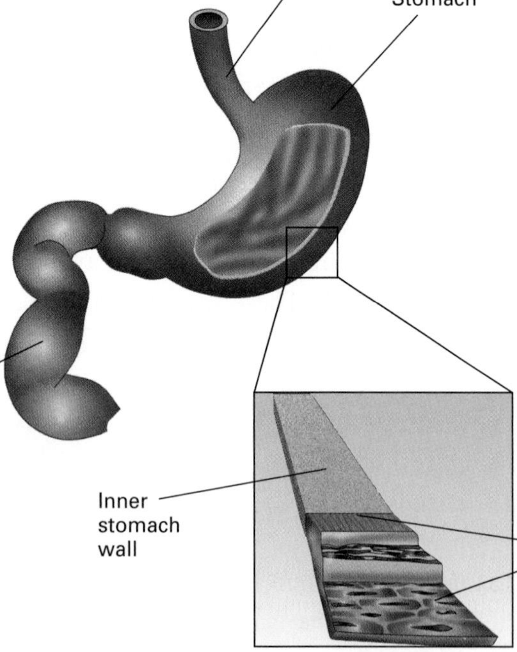

Smooth muscle plays a major role in maintaining homeostasis of the body's internal environment. For instance, the pulse that you can feel in your wrist comes from the contraction and relaxation of smooth muscles in the lining of a blood vessel. You use smooth muscle contraction when you swallow food. (You'll learn more about this in Chapter 13.) Smooth muscle controls the size of passageways in the body. When you breathe, for

Figure 11.11: *Smooth muscles within the digestive system use peristalsis to move food through the digestive tract.*

example, smooth muscles around the air passages in your lungs relax, letting in air. The muscles then contract, pushing air out. When someone has an asthma attack, the muscle contracts but doesn't relax (a condition that is referred to as a muscle spasm). The airway is obstructed, and the person can't breathe properly. Asthmatic people use drugs that open the air passages by relaxing the smooth muscle in the lungs, allowing them to breathe normally.

See Additional Activity, Tubifex Worms, in Chapter 11 of the TRB.

LESSON 11.2 ASSESSMENT

1. The muscles are alternately contracting and relaxing.

2. The iris dilates to gather as much moonlight as possible.

3. The muscle atrophies, losing actin and myosin filaments because it isn't used. Exercise will refill the muscle cells with proteins, increasing the size of the muscle.

4. The pulse is a peristaltic motion that keeps pushing the oxygenated blood through the body.

1 When a small muscle twitches, what are the skeletal muscles doing?

2 You walk from your bright living room to the porch on a moonlit night. How do your eyes adjust?

3 Why does a person's leg seem to shrink after wearing a cast for six weeks? How can that person rebuild the leg to its size before casting?

4 Why can you feel the pulse in your wrist when the heart that is contracting is so far away?

LESSON 11.3 BONES

Pick up a pair of tweezers and press on them. You are holding a lever system, powered by your fingers. If you imagine a muscle attached to the inside of the tweezers' legs and contracting, in place of your fingers, you have a model of the way your muscles and bones work together. Most bones are levers that use muscles to supply the force. The result is a simple machine.

ACTIVITY 11-2   X-RAY VISION

Collect student X rays or ask radiology departments to give you several when they dispose of old ones each year. Be sure that the patient's name is taped over or removed for privacy. Tape the X rays to a well-lit window or view them on a light box.

The major bones in your arms and legs are called long bones; they grow quickly during puberty (Figure 11.12). Smaller bones in your hands and feet can be thin and flat, sesamoid (shaped like a sesame seed), short, or irregular. In this activity you will use X rays to identify various types of bones and their features.

Long bones include the femur, tibia, and fibula in the leg and the humerus, radius, and ulna in the arm. Short bones include the carpals in the wrist and tarsals in the feet. Sesamoid bones include the patella (kneecap—develops in the tendon that holds it in place). Flat bones include the bones of the of skull and ribs. Irregular bones include the bones of the face, such as the sphenoid.

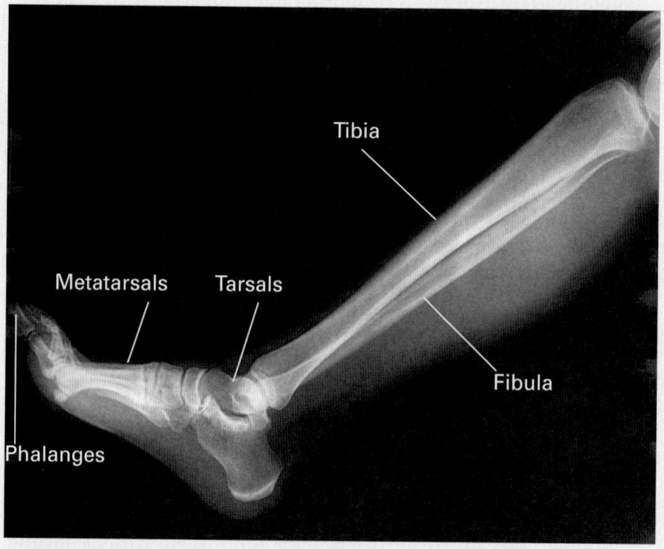

Figure 11.12: *This X ray of the lower leg and foot shows the tibia, fibula, tarsals, metatarsals, and phalanges.*

Students may identify the following: compact bone that forms the outer shaft, the lighter center that is the marrow, where pits and grooves that are for the muscle tendon to attach or blood vessels to enter, or the growth plate at the ends if the person is still growing. This open area is the cartilage that doesn't show up in X rays. Breaks and healed breaks may be visible.

As you view the X rays, study the shape and position of the bones. Consider where muscles attach to these bones and how the bones will move. Examine the bones more closely to observe detailed features. Why is the center of the long bone lighter? You may notice pits and grooves on your X ray of your long bone. How are the pits and grooves important? Record your observations of each bone in your logbook.

THE LEGACY OF BONES

BIOLOGY IN CONTEXT

WORD BANK

anthropo- = human being

-ologist = one who studies

Anthropologists study the origins of humans and the development of their society.

What happens when someone reports finding bones in suspicious circumstances? A forensic **anthropologist** helps to decide whether the bones are animal or human. In a human skeleton a heavy skull, a bony ridge above the eyes, and a narrow pelvic structure may indicate a male. A short face with squared cheekbones may indicate a Native American or Asian. Africans usually have a wider nose bridge and a distinctive jaw. The length of the bones and presence or absence of a growth plate (found in the bone ends of children) help to determine the person's age.

Forensic anthropologists can also recreate faces with clay to help identify a murder victim. Studies have determined the approximate thickness of skin, underlying muscle, and fat on various areas of the face. The person doing the recreation puts measured pegs in these areas of a skull and molds clay into a face, covering the pegs (Figure 11.13).

In 1974, anthropologists discovered "Lucy," a 3-million-year-old humanlike skeleton, in Africa. Using techniques such as those of the forensic anthropologist, researchers have concluded that Lucy walked upright on two feet. Her stance is reflected in the pelvis, leg, and foot bones. Her height was about 3 feet, 6 inches, and she weighed about 65 pounds. Lucy had a small brain but a large face. Her worn teeth and jaw indicate strong chewing muscles, needed for a mainly vegetarian diet.

Figure 11.13: *Reconstructing a face from a skull*

Anthropologists use their knowledge of bones to unravel ancient and recent mysteries. They're helped by forensic dentists, pathologists (who study disease), and entomologists (who study the insects that infest decaying bodies).

MORE BONES

Alternative Investigation 11C, Chicken Bones, may be performed here. See Chapter 11 of the laboratory manual for this investigation. See also Chapter 11 of the TRB.

A long bone such as the femur of a chicken has a shaft made up of layers of dense matrix. The shaft is a bundle of **osteons,** which are cylinders of layered bone matrix and cells. Osteons give strength to bones.

An osteon has layer on layer of concentric cylinders of fibrous matrix. The matrix traps cells called **osteocytes,** the bone forming cells, in pockets. The matrix is made up of minerals such as calcium and the protein collagen. The center of each osteon is a canal with blood vessels and nerves running through it (Figure 11.14). As in cartilage, a network of even smaller canals allows nutrients and waste to enter and leave the osteocytes.

At the ends of long bones the matrix and cells are arranged like a honeycomb, referred to as spongy bone. Spongy bone contains many spaces that are filled with connective tissue and blood vessels. Spongy bone is constantly re-forming to withstand pressure. For example, if you began jogging regularly, the spongy parts of your leg bones would reshape to withstand the increased stress and pressure. The open structure of spongy bone is a lightweight yet strong support system. Spongy bone is also found in short, flat, and irregular bones.

You can demonstrate the way in which osteons make bones strong. Glue several punched circles of paper (representing the bone cells, or osteocytes) to a sheet of notebook paper (representing the matrix surrounding the cells). When flat, the paper bone is rather weak. When rolled up, it is very strong. Glue several rolls together, side by side, and you'll see why bones are so tough.

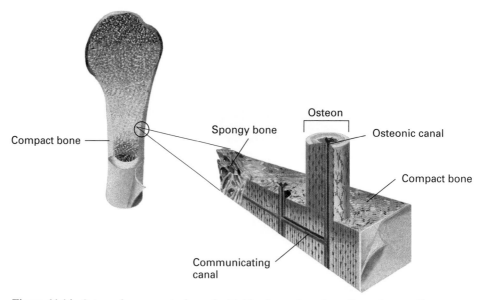

Figure 11.14: *Osteons have a central canal with blood vessels and small canals extending from osteocytes.*

BLOOD PRODUCTION

At birth, the centers of your bones were filled with red marrow. Marrow holds the cells that divide to produce red blood cells and many white blood cells. When the blood cells mature, they enter the blood vessels and circulate throughout your body. Every day the liver filters and destroys approximately 120 million worn-out red blood cells. The red marrow replaces these cells.

As you get older, the blood-forming marrow decreases and is restricted to the skull, the sternum (breast bone), the vertebrae that form the spine, the hip bones, and the ends of the long bones. The space that once held red marrow is now filled with yellow marrow, which is fat.

Bones, except near the joints, are covered by a tough membrane made mostly of collagen fibers. Connective tissues called tendons are woven into this membrane. Tendons anchor muscles to strong ridges on the bones.

CARTILAGE

Have students feel the cartilage rings in the trachea just under the skin.

The musculoskeletal system also has a connective tissue called cartilage. Cartilage forms the ridge of your nose and your outer ear as well as the rings of tissue that you can feel in the front of your neck.

Cartilage is like bone but more flexible because the matrix doesn't contain minerals. Under a microscope, cartilage looks glassy with cells in pockets (Figure 11.15). The glassy matrix is a firm gel embedded with collagen fibers,

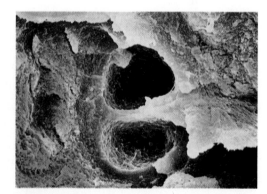

Figure 11.15: *A microscopic view of hyaline cartilage*

rather like tire rubber embedded with polyester fibers. One important function of cartilage is to pad bones where they meet in a joint. Growth plates of cartilage in the ends of bones enable bones to grow longer. By age 18–20, most cartilage has been replaced by osteocytes that secrete minerals, forming hard bone. At this time, bones have reached their maximum length; you stop growing.

BONE IS DYNAMIC

At birth, much of your skeleton was cartilage. This cartilage made you very flexible, but as you grew, cartilage was replaced by bone. Cells that were trapped in the bone matrix became osteocytes. By puberty the main growth sites are the cartilage plates in the ends of the growth plates. When you go through a growth spurt, maybe growing several inches in a year, your body releases hormones that will change these growth plates into bone. At some point, you won't grow any taller. This happens by age 18–21 for most boys and by age 16–18 for most girls.

Even though your bones eventually stop growing longer, they don't stop changing. When you gain or lose weight or change your exercise habits, your bones change. This reshaping is equivalent to forming a new skeleton every 7 years. As some cells lay down new bone and matrix, other cells reabsorb existing matrix. This process is what heals a broken bone. Bone-forming cells infiltrate the blood clot at the break. A callus, or thickened area, forms. The callus is gradually reshaped, and new compact bone forms.

THE HUMAN SKELETON

When a figure skater spins in a circle, the skater's body turns around on an axis. The parts of the skater's skeleton that are close to this imaginary axis compose the **axial skeleton** (Figure 11.16). The limbs and the bones that connect them to the axial skeleton (shoulder and pelvic girdles) make up the **appendicular skeleton.**

THE AXIAL SKELETON

The axial skeleton consists of skull bones, **vertebrae,** ribs, and sternum. The human skull looks like a 3-D puzzle made of flat bones that fuse together by age two, forming immovable joints called **sutures.** Before they fuse, you have to be careful of soft

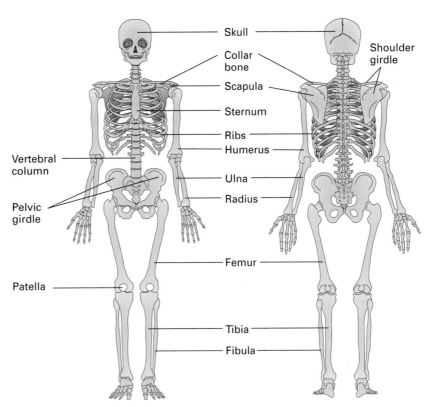

Figure 11.16: *Front and back view of a human skeleton*

spots in a baby's skull. Eight bones form a protective case for the brain. The only bone in the skull that moves is the mandible, or jaw, allowing you to talk and chew. It forms a hinge joint with a bone at the side of the skull near the ear (Figure 11.17).

As you learned in Chapter 10, a vertebrate has a central nerve cord, the spinal cord, which is protected by bones called vertebrae. The vertebrae are stacked to form the vertebral column. They provide a canal for the nerve cord and spaces for the nerves that branch from the spinal cord. Because the vertebrae contain delicate tissue, they must be held very securely in place and cushioned. A cartilage disk is sandwiched between each pair of vertebrae. When you turn or flex, the disk compresses or stretches slightly to give you limited motion. The disk also absorbs shock when you run. The topmost vertebra pivots on an extension from the second vertebra. This special joint gives extra range of motion, so you can turn your head.

Ligaments stabilize the spinal column. Ligaments are bands of collagen (connective tissue) that reinforce bone-to-bone connections. A pair of ribs attach to each of the vertebrae in the chest region. Most ribs have a cartilage connection to the sternum (breastbone).

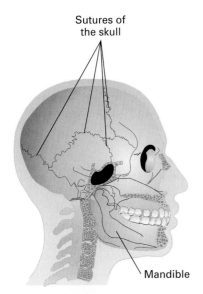

Figure 11.17: *Human skull, showing sutures and mandible*

THE APPENDICULAR SKELETON

The appendicular skeleton consists of the limbs and the shoulder and pelvic girdles, which attach them to the axial skeleton. The skeleton of the upper arm consists of the humerus, which connects to two bones in the lower arm: the radius on the thumb side and the ulna on the other. The lower arm also includes the small bones of the wrist, hand, and fingers.

Your upper leg has a femur (the largest human bone) connecting to two bones in the lower leg. The larger of the two is the tibia, and the smaller is the fibula. The patella (kneecap) protects the knee and prevents it from bending forward. The ankle and foot have bones that are similar to those in your wrist and hand.

OTHER VERTEBRATE MUSCULOSKELETAL SYSTEMS

The musculoskeletal systems of vertebrates show an amazing number of adaptations. Snakes have a unique hinged jaw (Figure 11.18). A special bone between the lower jaw and the skull creates a very flexible joint. That's how snakes can swallow their prey whole. The cheetah's spinal column is straight and very flexible. It curves and extends with each stride to maximize speed. The turkey's wishbone consists of fused collarbones that help to stabilize the wing joint during flight. Both bats and birds have a sternum that is adapted to act as a keel, or ridge, to which the flight muscles attach (Figure 11.19).

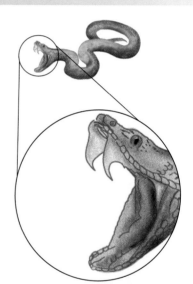

Figure 11.18: *Snakes have a unique hinged jaw.*

Commercial poultry production breeds for large keels so there is more white breast meat on the birds.

Women have a slightly different angle on their coccyx, which accounts for why women are more likely to break or bruise their tailbone.

The human pelvic girdle is unusual because it must accommodate walking upright. It protects and supports the abdominal organs. Three fused bones form the hips. A woman's pelvis is shallower and wider than a man's.

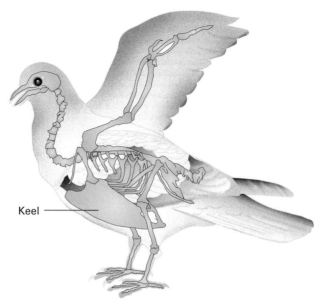

Figure 11.19: *The sternum of this pigeon shows the characteristic keel-shaped adaptation that allows for the attachment of flight muscles.*

Keel

Figure 11.20: *Many vertebrates have a tail like the black lemur's in this photo.*

Many animals have a tail that includes the lowest vertebrae (Figure 11.20). The kangaroo uses its massive tail for balance as it stands upright on its hind feet. Squirrels use their tails to stabilize their body as they leap through the trees. Monkeys use their tails to grasp and hang from branches.

Like the axial skeleton, parts of the appendicular skeleton have become specialized in different animals. The moles that tunnel through a lawn have a very wide humerus and a large sternum for anchoring the powerful digging muscles (Figure 11.21). Mole palms face backward, so they can push dirt away. Bats have long digits ("finger bones") that support wings that join to the side of their

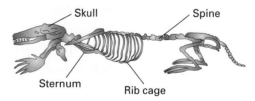

Skull Spine

Sternum Rib cage

Figure 11.21: *A mole's skeleton is adapted for digging and burrowing.*

Figure 11.22: *Forelimb of a bat*

body and legs (Figure 11.22). Cheetahs have well-developed femurs and long, thin legs. Most of the muscle is in the upper leg, close to the body, and the tendons that anchor the muscles are in the lower foot. This construction allows long, powerful strides as cheetahs chase their prey. In frogs, which are adapted for jumping, the tibia and fibula are one bone. The gibbon, a member of the ape family, has an opposable big toe like a human thumb, which is handy for moving from tree to tree.

MUSCULOSKELETAL INJURY AND DISEASE

Like any other living tissue, bones are subject to injury and disease. A fracture can be tiny cracks in the surface of a bone or a complete break. Sprains occur when a ligament tears or a joint moves beyond its normal range. In a car crash, a common injury is whiplash, which is a sprained neck from the jarring impact. Bones can also dislocate, or pop out of their normal position.

Bone disorders include osteoporosis and osteomalacia. In osteoporosis the bone matrix is reabsorbed and loses calcium and mineral salts. Thinner bones can't always support the body's weight, and that means broken bones and falls. Eating calcium-rich foods and exercising regularly throughout your life help to build and maintain bones that resist these changes. Both men and women develop osteoporosis, but women are especially susceptible when they stop producing female hormones, such as estrogen, that help to maintain bones.

Osteomalacia involves a loss of minerals but not the matrix. It occurs mostly in adult women. Young children who lack vitamin D, which is needed to absorb calcium from food, develop a similar condition called rickets. The bones become soft and deform when they can't support the body's weight.

LESSON 11.3 ASSESSMENT

1. Both have cells that are trapped in a collagen gel, both give support, both are part of the skeletal system, and both obtain food and waste, which diffuse across the matrix. They differ because the matrix in cartilage is not filled with minerals to harden it.

2. The thickened area is more likely a healed break and not the site of a blow.

3. The injury will be slow-healing because there isn't a direct blood supply, and food and waste have to diffuse across the matrix.

4. They were well developed.

1 Compare and contrast bone and cartilage.

2 At a crime scene a femur with a thickened area in the shaft is found. The media speculate that the killer struck the victim in this area. Would you agree?

3 While playing basketball, a teenager was hurt, and the orthopedic doctor says that a bone has slipped sideways at the growth plate and will need to be realigned. Will this be a fast-healing injury or a slow-healing one?

4 While walking in the woods, you find an animal bone that has several large, bumpy surfaces. What can you say about the muscles that attached to this bone?

LESSON 11.4 MUSCLES, BONES, AND JOINTS IN MOTION

You run up a flight of stairs. You hop on a bike and speed off to the corner store. You pick up a slide, put it under a microscope, and adjust the focus. Movement is so easy, who needs to give it a thought? The injured athlete, the arthritic person, and the violinist with a sore elbow are a few. Muscles and bones come together at joints (Figure 11.23). Joints are complex structures that are one of the main concerns of **orthopedists,** doctors who are trained to preserve and restore the human musculoskeletal system.

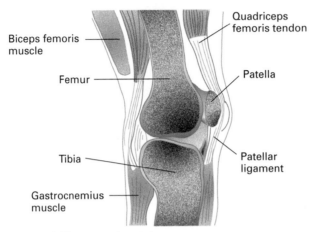

Biceps femoris muscle

Femur

Tibia

Gastrocnemius muscle

Quadriceps femoris tendon

Patella

Patellar ligament

Figure 11.23: *Human knee joint*

HIP REPLACEMENT

BIOLOGY IN CONTEXT

Joints in the body are subject to various disorders such as arthritis and to damage from injury or overuse. When a joint is no longer functioning adequately, an orthopedic surgeon may recommend joint replacement.

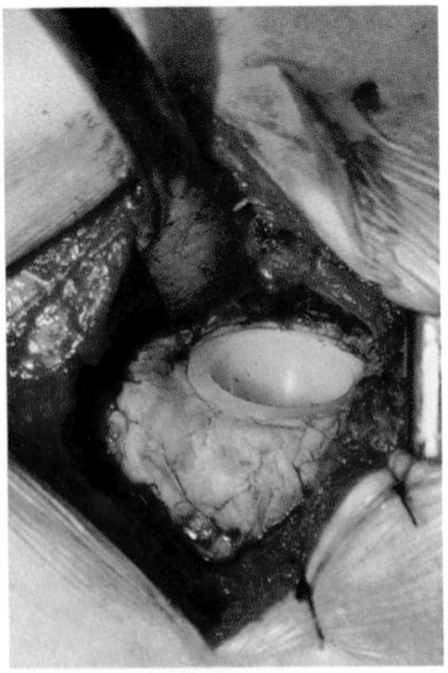

Hip joint replacement became very popular around 1992 when Bo Jackson, the two-sport athlete, underwent hip joint replacement surgery (Figure 11.24). Bo had a condition that is called "avascular necrosis." Avascular necrosis is a condition in which the veins and capillaries feeding blood to a given joint are damaged or destroyed. This condition disrupts the blood flow to the joint. Bo developed avascular necrosis after dislocating his hip during a football game. After his injury, Bo had his hip replaced and retired from football, but he continued to play baseball. He later retired from baseball after playing two seasons with an artificial hip. Bo became the first athlete to play professional baseball with an artificial hip.

Figure 11.24: *Hip replacement surgery showing socket side of joint*

ACTIVITY 11-9 OPPOSITES CONTRACT

How do muscles and bones come together at the moveable joints? In this activity you'll build a model of your upper and lower arm, with the biceps and triceps muscles moving the forearm.

The closer the dots are, the more contracted the muscle; the farther apart they are, the straighter the joint. The wires or paper clips represent ligaments.

A blackline master of this model can be found in the TRB.

Cut the cardboard tube from a pants hanger into three parts to construct an ulna, a radius, and a humerus. Fasten them together with wire or straightened paper clips. Cut notches for the insertion and origin points of the tendons. Hook a rubber band muscle on the notches, and adjust the tension to make a right angle. On one rubber band, mark dots 1 cm apart. Flex the rubber band muscle, and observe the spacing of the dots. What can you conclude from your observation? What do the wires or paper clips represent in this model?

HOW A JOINT IS CONSTRUCTED

Bones of a joint are connected by fibrous tissue, including ligaments. A wear-resistant layer of cartilage covers the bone ends. Between the cartilage layers is a small cavity coated with synovial fluid. **Synovial fluid,** which has the consistency of egg white, lubricates the cartilage layers of the joint. Fibrous and fatty areas around the joint help to cushion it and distribute the body's weight onto it.

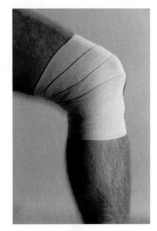

Figure 11.25: *Flexing the knee*

A joint moves when the muscle attached to bones on either side of the joint contracts. For instance, contraction of a muscle in the thigh flexes the knee joint, so the lower leg moves backward toward the back of the thigh (Figure 11.25).

When a muscle contracts and shortens, it acts like a force on a lever, bending the joint and pulling the second bone closer (Figure 11.26). A different muscle moves the joint in the opposite direction. To understand this better, wrap your hand around your upper arm, and move your lower arm up and down. When you raise your arm, you can feel the muscle on the top (your biceps) contract, pulling your arm up. The muscle on the underside (your triceps) is relaxed and extended. When you lower your arm, you can feel the triceps contract. This action is typical of the way skeletal muscles work in pairs.

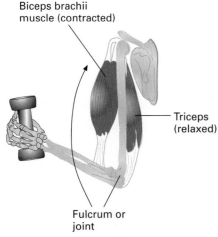

Biceps brachii muscle (contracted)

Triceps (relaxed)

Fulcrum or joint

Figure 11.26: *Action of hinge joint in arm*

The four major joints of the human skeleton are the shoulder, elbow, hip, and knee (see Figure 11.16 on page 409). The shoulder and hip joints are typical ball-and-socket joints. They allow a wide range of movement, especially rotation. The top of the femur (thighbone) is shaped like a ball and fits into a special socket, creating a very strong joint. The complex elbow and knee joints are somewhat alike; they use hingelike and gliding movements. Hinge joints allow you to lift things (see Figure 11.26). Because your lower arm has two bones, you can swivel your elbow, that is, twist it as if you are using a screwdriver. The elbow also allows some rotation.

Figure 11.27: *Trainer applying a knee brace to an athlete*

Special structures reinforce some joints, such as your shoulder and knee. A thick, fibrous sleeve wraps the joints and helps to support them, like the knee braces some athletes wear (Figure 11.27). This sleeve also produces a lubricating fluid that lessens the friction between the bone surfaces. The knee also has a cushion of cartilage called the meniscus that softens the bone-to-bone contact and absorbs the impact of movement. The bone surfaces that meet are covered with slick cartilage that acts like a nonstick coating to keep them from sticking as they move.

Try touching the back of your hand while you wiggle your fingers; you can feel the tendons that attach the muscles in your forearm to the bones in the fingers. Rings of ligaments encircle the tendons, nerves, and blood vessels that enter the hand (Figure 11.28).Some tendons in the knee and shoulder are padded to prevent friction and wear and tear on the joint. **Bursae,** which are shock-absorbing, fluid-filled sacs, are strategically located under these tendons.

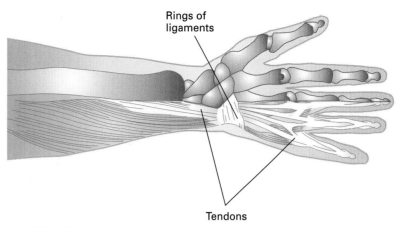

Rings of
ligaments

Tendons

Figure 11.28: *Hand and forearm*

See Investigation 11B in TRB.

INVESTIGATION 11B   MUSCULOSKELETAL PERFORMANCE

Range of motion helps to determine how easily a person can perform an activity. On the basis of the results of such tests, athletic trainers and physical therapists recommend exercise programs for athletes and patients. In this investigation you will measure the range of motion of several joints.

LESSON 11.4 ASSESSMENT

1. Answers should describe the flexion and extension movements of the hinge joint and the rotational movements of the ball-and-socket joints.

2. Shock absorber.

3. Synovial fluid.

4. Get plenty of exercise and be sure you have enough calcium in your diet.

1 Describe the action of the hinge and ball-and-socket joints by analogy to household items.

2 What car part works like the meniscus of the knee?

3 A car uses oil to lubricate engine parts so that they move smoothly. What fluid lubricates the synovial joints?

4 You are an orthopedic surgeon. What general advice would you give to your patients to help strengthen their bones?

CAREER APPLICATIONS

The applications that follow are like the ones you will encounter in many workplaces. Use the biology you learned in this chapter to complete the activities. Share your work with the class.

AGRICULTURE & AGRIBUSINESS

Lifting Safely

Create a poster or video that demonstrates how to use bones, muscles, and joints correctly to lift and move heavy objects. Demonstrate various devices that are available to make lifting safer.

Hip Dysplasia

Contact a veterinarian or dog breeder for information on hip dysplasia. Sketch a normal hip joint and one with dysplasia. What problems does a dog with dysplasia have? What causes the disease? What is done to prevent it? What can be done to correct it?

BUSINESS & MARKETING

Equipped for Exercise

Let's say your class has been asked to create a notebook for the sales staff at a department store that describes the exercise equipment for sale and which muscle groups each piece exercises. Visit the sporting goods section of a store and make a list of all the exercise equipment there. Each student should take one piece of equipment and do a one-page write-up for the notebook. List the item, its cost, the types of exercise that the item can be used for, and what muscles it works. Include a picture if possible. Try to find a review of the item or exercise, and describe the benefits and possible problems if the exerciser doesn't use the equipment correctly. List other benefits of the item, such as ease in setup and portability. After all of the individual items are described, work in small groups to make pages that describe combinations of equipment for a total workout package for a specific client, such as an apartment dweller, a person who travels frequently and needs portability, or someone who has a garage workout area.

**FAMILY &
CONSUMER
SCIENCE**

Musculoskeletal Development

Using resources such as books on early childhood development, interviews with daycare providers or a pediatric nurse, and the Internet, research the development of bone and muscle in children from birth to age 6. Make a guide for parents and baby-sitters that discusses development of the musculoskeletal system, special care needed, and activities and toys that are appropriate for the motor skills of the child at different ages.

Fashions for Every Body

Research body types. Make a poster that shows different body types. For each body type, cut out pictures of clothing that you think would look good and some that you think would look bad. Tell why you would or would not choose those styles. Alternatively, you might stage a fashion show demonstrating body types and clothing styles.

**HEALTH
CAREERS**

Athletic Injuries

Prepare a report or documentary on sports injuries. With a group, contact a source to investigate athletic injuries. One group should contact the National Athletic Trainers' Association to get statistics on sports injuries. What protective equipment is recommended to reduce the chance of injury? One group should talk with an athletic trainer. What is the athletic trainer's role in preventing injuries, responding to an injury, and working to rehabilitate the athlete? Find out what stretching exercises should be done to reduce the chances of injury. Another group, using the same list of injuries, should talk to an emergency medical technician or paramedic to find out what treatment he or she provides on the sports field and during transport. Another group should talk to a doctor who treats sports injuries to find out what the treatment options are for more serious injuries.

Occupational Therapy

Ask an occupational therapist how to train recovering muscles to compensate for lost movement due to injury and disease. In your logbook make a chart of injuries and list compensating muscles and prescribed therapy. Share your findings with the class.

Injuries in the Workplace

Talk with an occupational therapist about the most common work-related injuries in local businesses and industries. What is done to prevent these types of injuries? Make a brochure that describes the injury, the causes, its prevention, and rehabilitation.

Robots in Manufacturing

Visit a manufacturing facility that uses robot arms in an assembly process, or research manufacturing robots on the Internet. Observe how the robot arm functions. Diagram the robot arm, and compare its action to that of a human arm and hand. Compare the joints, muscles, and tendons that are found in the arm and their counterparts, if any, in the robot. Compare range of motion, strength, flexibility, and adaptability to changing requirements.

CHAPTER 11 SUMMARY

■ A musculoskeletal system allows an animal to move its body parts or to move about in its environment.

■ A skeleton helps an organism to support its weight, protects its internal organs, and transmits the forces of its muscles into movements.

■ Animals have a hydrostatic skeleton, in which the body fluid transmits force; an exoskeleton, with attached muscles and sometimes joints that allow coordinated movement; or a bony internal skeleton, or endoskeleton, to which muscles are attached.

■ Cells called muscle fibers make up muscle tissue. Each fiber contains many bundles of twisted protein filaments, which work as a group to contract or relax the fibers. The filaments slide along each other in a way that shortens or lengthens the fiber.

■ Embedded in the muscle fibers are networks of nerves that carry electrical stimuli that provoke muscle contraction. Also embedded are blood vessels that carry nutrients and oxygen to the fibers and carry waste and carbon dioxide away.

- Muscle fibers make up the walls of the heart. These cardiac muscle cells are constructed somewhat like skeletal muscle cells. Both are striated because of the filament arrangement.

- Smooth muscle cells are football-shaped with a single nucleus. Filaments are not aligned, so there is no pattern of striations.

- A bone shaft is made of units called osteons, layer after layer of concentric cylinders of fibrous matrix that have cells trapped in pockets. In the center of each osteon is an opening for the blood vessels and nerves that supply the living tissue.

- Cartilage contains cells in a matrix of collagen fibers. It is more flexible than bone because calcium and other minerals do not harden the matrix. Cartilage is found on the ends of bones that meet at a joint and in the growth plates of bones.

- Bone is constantly remodeled by groups of cells that lay down new matrix and other groups of cells that reabsorb previously formed matrix. This process of reabsorbing and laying down new bone is what heals a broken bone.

- The parts of the human skeleton that are close to the body's central axis compose the axial skeleton. The appendicular skeleton consists of the limbs and the shoulder and pelvic girdles, which attach to the axial skeleton.

- A joint is a junction between two or more bones that allows movement of body parts. Movement at a joint occurs when a muscle that is attached to bones on either side of the joint contracts.

- The four major joints of the human skeleton are the shoulder, elbow, hip, and knee. The shoulder and hip joints are ball-and-socket joints. The elbow and knee are hingelike joints.

- Ligaments connect bones. Tendons attach muscle to the outer membrane of the bone near a joint.

Chapter 11 Assessment

Concept Review

See TRB, Chapter 11 for answers.

1 With your feet flat on the ground, put your hand at the bottom of your gastrocnemius (calf muscle). While keeping the heel of your foot on the ground, tap your foot on the floor several times. When does this muscle contract and relax?

2 Which of the bones in the leg bears the most weight of the leg? (See Figure 11.16 on page 409.)

3 Why would a chicken's thighbone become soft and leathery if you soaked it in acid solution? (*Hint:* The acid removes minerals from the bone.)

4 A classmate is running in the New York marathon. After 1 kilometer, he is out of breath and his muscles begin to feel as if they are burning. Explain to your classmate why he feels this way.

5 Are the biceps muscles in the upper arm voluntary or involuntary?

6 Why is calcium important to the bone?

7 How do food and waste pass through the bone matrix?

8 A lever is a rigid rod that moves around a fixed point called a fulcrum. To produce body movement, certain body parts must function as a lever and others as a fulcrum. What parts of the musculoskeletal system function as fulcrum and lever?

9 Rickets is a disease caused by lack of vitamin D. Young children who have rickets have soft bones that are misshapen because they can't support the body's weight. Why doesn't eating calcium-rich foods cure this disorder?

10 Why are women more susceptible to osteoporosis than men?

11 Cartilage covers the ends of bones and prevents them from grinding against each other. Why doesn't the cartilage wear out as it rubs against other cartilage?

12 As a result of age, the central opening of the osteon becomes blocked. How does this affect the osteocytes?

13 What will happen if ATP is not available after myosin binds with actin in the sliding filament model?

14 How are ligaments different from tendons?

15 Create a chart describing the various types of muscle tissue. Include a drawing of the muscle, its location, its appearance, and its nervous control (voluntary or involuntary).

16 Muscles can be named for their size, number of origins, or shape. Using Figure 11.29, create a chart that names two muscles for each characteristic.

17 Calcium enters and leaves smooth muscle slowly. What effect does this have on smooth muscle?

18 What effect does prolonged weightlessness have on the bones of an astronaut?

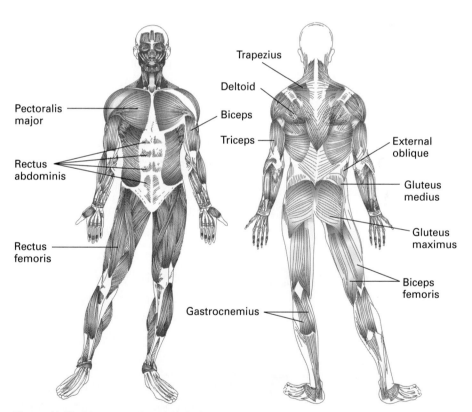

Figure 11.29: *Major muscles of the body*

19 Discuss how you might develop a model of osteons from items you might find at home or in your classroom.

20 Osteoblasts are collagen-secreting cells that are used to build bones. Osteoclasts are cells that destroy the matrix of the bone. The activities of these bone tissues are controlled to maintain proper bone density. Recent studies suggest that alcohol destroys osteoblasts. If that is true, what effect does alcoholism have on bones?

21 Paget's disease increases the activity of osteoblasts and osteoclasts. How does this affect bone?

22 Why is the knee considered the joint that is most vulnerable to damage?

In the Workplace

23 Editorial assistants at a publishing company work eight-hour days. Much of their work is keying in and manipulating text on the computer and printing and proofreading their work. Assistants key one or more documents, stop and proofread the documents, and enter corrections. Besides lunch, they usually get a 15-minute break in the morning and one in the afternoon. Several assistants are starting to notice symptoms of carpal tunnel syndrome and other repetitive stress disorders. What might help to relieve the problems?

24 You are a butcher. A customer asks you why round steaks have the fatty tough fiber dividing the meat into sections. What's your answer?

25 A student athlete is training for a marathon. She stretches and then runs several miles every other day. She has a part-time job at a fast-food restaurant on the days she doesn't work out. If she asks you, her trainer, how to improve this workout, what would you say?

26 You are a biologist studying owl diets. You analyze the pellets of undigested bones, fur, and other material around a nesting site of an owl. How can you tell whether the small bones in the pellet belonged to a mole or a mouse?

Investigations

The following questions relate to Investigation 11A:

27 What solution or combination of solutions caused the muscle to contract?

28 Why is ATP used in this investigation?

The following questions relate to Investigation 11B:

29 Is it true that young men are more flexible than young women? Why?

30 Develop a procedure to measure the angle of extension of the wrist.

CHAPTER 12

WHY SHOULD I LEARN THIS?

Emergency medical personnel act quickly in a cardiac or respiratory emergency. They have to understand how the heart and lungs work together and what causes heart attacks, strokes, and suffocation. What if *you're* the only one on the scene in an emergency? How much do you know about cardiac or respiratory distress? How would you respond?

CARDIOVASCULAR AND RESPIRATORY SYSTEMS

WHAT WILL I LEARN?

1. the role of blood in maintaining homeostasis
2. how blood flows to organs, tissues, and cells
3. how blood obtains and delivers oxygen and gets rid of CO_2
4. how the kidney filters blood and maintains proper levels of substances in the blood

Suppose it's your first visit to a doctor's office. Maybe you've never been ill and your vaccinations were given to you so long ago that

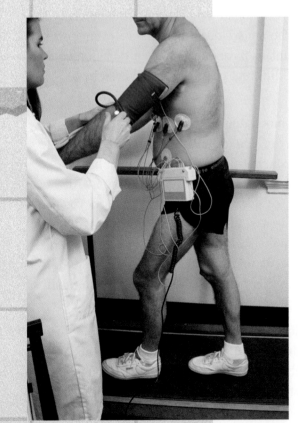

you can't remember them. The doctor's assistant wants to take your blood pressure with a sphygmomanometer, and you start to worry. Will it hurt? Your heart beats faster, you breathe harder, and you start sweating a bit. The assistant puts a cuff on your arm and tightens it until it bothers you, reaches for your elbow, and places the bell of the stethoscope in the crook of your arm. Help! "Gee, your blood pressure's high for someone your age," the assistant says. Then and there, you make up your mind to learn more about blood, blood pressure, and sphygmomanometers.

LESSON 12.1 OVERVIEW OF THE CARDIOVASCULAR SYSTEM

In everyday speech, people express the importance of blood to life: "blood-and-guts competition," "blood-curdling screams," "bloodthirsty enemies." Do they realize why blood is so vital? Maybe not. But you will, by the end of this lesson.

ACTIVITY 12-1   VITAL SIGNS

The procedure for this Activity can be found in Chapter 12 of the TRB.

Some of the first things a medical assistant observes and records in a physical exam are the patient's pulse and breathing rate. In this activity you will record this vital information about your partner. Your teacher will provide you with the procedures to follow.

BLOOD, THE GIFT OF LIFE

BIOLOGY IN CONTEXT

Unlike proteins and DNA, human blood can't be commercially manufactured. People who are seriously injured or ill sometimes need blood that other humans have donated. Your donation could be used by three or four different people—maybe even by a member of your family.

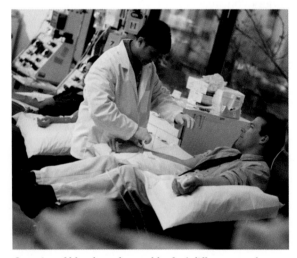

One pint of blood can be used by 3–4 different people.

Because blood donors must be at least 17 years old, many of you haven't yet been asked to donate blood. As a first-time donor, you might have questions about how safe the procedure is. All supplies are sterile and used only once, so you can't catch a disease. Instead, the blood center will check to be sure that *your* blood has no infectious agents. After tests and processing, your blood goes to a transfusion service or waits at a blood center until needed. It cannot be stored indefinitely, however.

Your blood will probably be separated into red cells, plasma, and platelets to supply patients' needs most efficiently. A person who loses blood in an accident or in surgery needs red cells. Burn patients, cardiovascular surgery patients, and organ transplant recipients need plasma, the watery part of the blood, to increase fluid volume and to replace proteins, immunity factors, clotting factors, and hormones. Cancer patients who have had radiation and chemotherapy and cardiovascular surgery patients may need platelets to promote clotting. Groups such as the American Red Cross encourage healthy people to make blood donation a regular part of their lives. When you give blood, you give a precious gift.

LEARNING LINK DR. CHARLES DREW

Dr. Charles Drew was a scholar, surgeon, scientist and humanitarian. Using library and Internet resources discover what contributions Dr. Drew and other scientists have made to the advancement of plasma donation.

Charles Drew

IN CIRCULATION

Diffusion moves substances between the environment and each cell of organisms that live in water, such as jellyfish and flatworms (Figure 12.1A). As you saw in Chapter 10, this exchange takes place throughout these organisms.

A   **B**

Figure 12.1: *In small animals diffusion moves materials between the environment and each cell; larger animals need a circulatory system. (A) Boring sponge and flatworm. (B) Cuban tree frog.*

Diffusion isn't enough for larger animals, which need specialized organs to exchange materials with the environment (Figure 12.1B). Depending on the species, the organs can be skin, lungs, gills, the gut, or organs of excretion. These organs transport nutrients and wastes to and from the cells in the body. Transport requires a circulatory pathway, fluid to convey the materials, and a force to move the fluid.

BLOOD CIRCULATION

Humans and other vertebrates have a **cardiovascular system** for circulation. The system is a network of vessels through which a heart pumps blood. The cardiovascular system of an adult human contains about 4–6 liters of blood. The blood carries nutrients from food, oxygen from the air, CO_2 and nitrogenous waste from cells, and various ions, proteins, and other organic compounds. Some of this material is carried by blood cells, and some is carried by the fluid called plasma.

In vertebrates, red blood cells transport oxygen (Figure 12.2). White blood cells defend against pathogens and foreign substances. Blood cell fragments called platelets help to repair damaged blood vessels and prevent excessive bleeding by clotting.

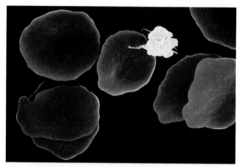

Figure 12.2: *Red cells of most mammals are round, flattened cells with no nucleus. Platelets (yellow object) are fragments of special bone-marrow cells.*

Blood **plasma** is mostly water, mixed with various proteins, salts, blood gases, nutrients, hormones, and other substances. It's not hard to separate plasma from the blood cells. A donor's blood is drawn into a tube that is then placed in a centrifuge. The centrifuge spins the blood at a high speed, settling the dense red cells at the bottom of the tube (Figure 12.3). White blood cells and platelets are layered between the red cells and the less dense plasma above the cells.

Figure 12.3: *During plasma donation, blood is separated into plasma and cells.*

INVESTIGATION 12A   BLOOD SMEAR

See Investigation 12A in TRB.

Blood has several components that you can distinguish under a microscope. In this investigation you will prepare and stain blood and examine some of the different cells in mammal blood. A medical technologist prepares and analyzes a specimen in a similar way.

Blood Flow through the Heart

A heart pumps blood through the cardiovascular system of vertebrates. The vertebrate heart has chambers that are separated by heart valves (Figure 12.4). Strong cardiac muscles make up the chambers and contract to force blood out. Blood leaving a heart chamber passes through a valve that keeps blood flowing in one direction.

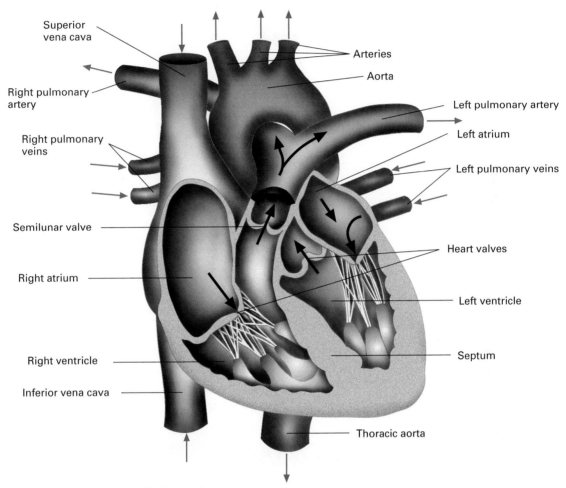

Figure 12.4: *Cross-section of the human heart*

The heart pumps the blood into large vessels called **arteries** that carry the blood away from the heart (Figure 12.5A). Arteries branch into smaller vessels called **arterioles,** and they continue to branch and narrow until they become the smallest vessels, called **capillaries** (Figure 12.5B). Capillaries are so narrow that blood cells go through them single file. In the capillary bed, blood

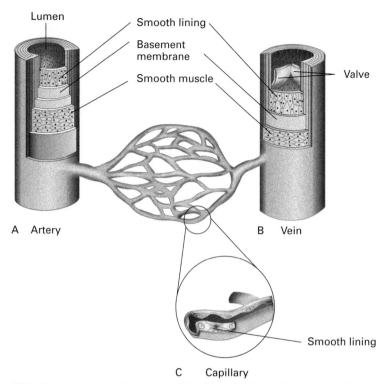

Figure 12.5: *Cross-sections of (A) artery, (B) vein, and (C) capillary. Arteries have thick walls because the blood flowing through them is at higher pressure. Capillaries have thin walls that allow diffusion.*

exchanges substances with the surrounding tissue fluid of body organs. These substances and water diffuse through the selectively permeable walls of the capillaries, moving from regions of high concentration in the blood to an area of lower concentration in the tissue. Blood that leaves from a capillary bed enters a network of small, microscopic blood vessels called **venules** that converge into larger **veins** (Figure 12.5C), vessels that carry blood to the heart.

See TRB for Additional Learning Link, LDL, the Lousy Cholesterol.

DIFFERENT STROKES

Many invertebrates, such as mollusks and insects, have an open circulatory system. Blood vessels empty into a cavity filled with fluid and bathe the internal cells. An open system works well for highly active insects because muscle movement aids circulation. Oxygen is distributed through a separate system of tubes.

The cardiovascular systems of vertebrates and of some invertebrates, such as the earthworm, are closed. All the blood

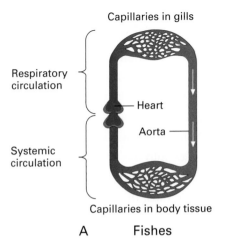

Respiratory circulation

Systemic circulation

Capillaries in gills

Heart

Aorta

Capillaries in body tissue

A   Fishes

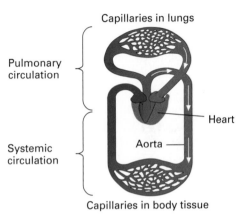

Pulmonary circulation

Systemic circulation

Capillaries in lungs

Heart

Aorta

Capillaries in body tissue

B   Amphibians

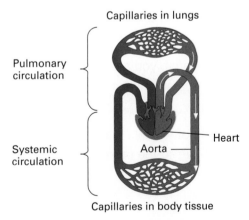

Pulmonary circulation

Systemic circulation

Capillaries in lungs

Heart

Aorta

Capillaries in body tissue

C Birds and mammals

Figure 12.6: *Comparison of the systemic and pulmonary circulation of (A) a fish, (B) a frog, and (C) a bird or mammal*

vessels are interconnected and don't open into the body. The earthworm has a relatively simple arrangement of blood vessels and pumps in a closed circulatory system. In vertebrates, which have special structures (lungs and gills) for taking in oxygen, the cardiovascular system is more complex. In a fish (Figure 12.6A) the heart has two chambers: an **atrium** (plural, atria) that collects the oxygen-poor blood and a **ventricle** that pumps this blood to the gills. The gills absorb oxygen dissolved in the water, and the oxygen diffuses into the blood.

In an amphibian, such as a frog (Figure 12.6B), the atrium has two chambers. Oxygen-rich blood from the lungs enters one chamber, and oxygen-poor blood from the tissues enters the other chamber. Some mixing of the blood occurs in the single ventricle. The ventricle simultaneously pumps a mixture of oxygen-rich and oxygen-poor blood to the body tissues and to the lungs and skin to pick up oxygen.

Birds and mammals have two atria and two ventricles (Figure 12.6C). There is no mixing of oxygen-rich and oxygen-poor blood. Having two upper and two lower heart chambers pumping keeps the blood flowing strongly. This efficient arrangement supports a high metabolic rate, allowing birds and mammals to maintain a constant body temperature and to be more active than many other animals.

OTHER FUNCTIONS OF A CIRCULATORY SYSTEM

The cardiovascular system delivers to cells nutrients dissolved in plasma and oxygen carried by the red blood cells, and it removes wastes from cells. It also has other functions. Blood distributes a lot of heat. In a hot environment an animal can pump more blood to the body surface. The blood conducts heat to dilated capillaries in the skin, where it radiates into the air, helping the animal to maintain a constant body temperature. In cold environments, animals retain body heat by keeping most of the blood in deep tissues.

A cardiovascular system also helps to maintain homeostasis in an animal's body fluids, adjusting the mixture of water and ions. In Lesson 12.4 you will learn about the kidney's role in homeostasis as a blood filter. The white blood cells that circulate play a role in defending the body against pathogens and foreign substances.

Alternative Investigation 12C, Earthworm Dissection, can be found in Chapter 12 of the TRB.

LESSON 12.1 ASSESSMENT

1. The irregularly-shaped red blood cells hinder circulation of blood.

2. Answers will vary: heart disease, stroke, high blood pressure.

3. Any three of the following: transports oxygen, CO_2, and other blood solutes; mimics blood flow; sustains blood cells; can't be toxic to the body; maintains homeostasis of body fluids; defends against pathogens; clots and forms scabs to prevent blood loss.

4. Arteries, arterioles, capillaries, venules, and veins.

1 People who have sickle-cell anemia develop red blood cells that are elongated and crescent-shaped and less flexible than normal blood cells. Speculate how this shape difference affects the circulatory system.

2 Lab tests indicate that someone has a high level of cholesterol in his or her blood. What major health threat might develop from this problem?

3 As a research scientist, you are attempting to design artificial blood. List three characteristics of blood that must be included in your design.

4 List the sequence of vessels that blood travels through while going from the heart to the other organs.

LESSON 12.2 DYNAMICS OF THE CARDIOVASCULAR SYSTEM

Every day, your heart beats more than 100,000 times as it pumps the equivalent of 2000 gallons of blood to body tissues. But what makes the heart beat? How does blood flow through the heart? How does blood lose oxygen and gain CO_2 as it travels from arteries to veins? This lesson will answer these questions.

ACTIVITY 12-2 HEART SOUNDS

The procedure for this Activity can be found in Chapter 12 of the TRB.

When you listen to a heart, you hear two distinct beats, the "lubb" and the "dupp." The lubb is longer and louder. These sounds have a lot to tell about someone's circulatory system. Follow the instructions your teacher gives you to listen to your partner's heartbeat.

GETTING THE BEAT RIGHT

BIOLOGY IN CONTEXT

Does your heart have rhythm? It's a strange question but a serious one. If you have fainting spells, dizziness, shortness of breath, or light-headedness, it could be the fault of an irregular heart rhythm. You might think that you're too young to have heart problems, but Nicholas K. was a senior in high school when his heart stopped beating during a basketball game. Paramedics revived Nicholas with cardiopulmonary resuscitation (CPR) (Figure 12.7). It turns out that Nicholas has an irregular heart rhythm. Within a few weeks, doctors implanted an internal **defibrillator.** Implantable defibrillators are small electronic devices that use an electrical shock to correct irregular heart rhythms. The defibrillator can also record an irregular rhythm, a feature that is useful for diagnosis.

Figure 12.7: *Emergency medical personnel know what to do when someone's heart stops beating.*

Nicholas can live a normal life. He might feel a mild shock from the defibrillator when his heart has an irregular rhythm. Although Nicholas couldn't play basketball for the rest of his senior year, he did receive a full basketball scholarship from Northwestern University.

The blood flow through the heart creates sounds—the lubb and the dupp. The longer, louder lubb sound comes from the valves closing between the atria and the ventricles. This occurs after the atria contract, filling the ventricles with blood. The dupp comes from the valves closing after the contraction of the ventricles, pushing the blood into the large arteries that leave the heart.

Heart sounds can signal heart problems. One problem is **arrhythmia,** an irregularity in the rhythm or force of a heartbeat. Serious cases may require drugs or an external electrical shock to return the heart rate to normal. One type of arrhythmia is tachycardia, in which the heart beats rapidly. If the heartbeat becomes very slow, the arrhythmia is called bradycardia. When the heartbeat suddenly becomes unpredictable, random, and quivering, the arrhythmia is called fibrillation. Fibrillation is life-threatening because, while the heart muscle quivers, no blood is pumped to body tissues.

Another signal of possible problems is a heart murmur. You can hear this abnormal sound at any time during the beat cycle. You can hear some heart murmurs even over the lubb-dupp. An abnormally shaped heart valve can't close completely causing the valves to leak. The leaking valves create the abnormal sound of a murmur. Abnormal valves result from damage or congenital defects (defects that are present at birth). Some murmurs are detected in adolescents who have a rapid growth spurt. Usually, this type of murmur gradually disappears.

WORD BANK

tachy- = rapid

brady- = slow

-cardia = heart

Tachycardia is an abnormally fast heartbeat. Bradycardia is an abnormally slow heartbeat.

THE CONDUCTION SYSTEM

The machine is called a defibrillator.

Have you ever seen a doctor in a TV hospital holding two paddles in the air, shouting "Clear!," and then giving a patient an electrical jolt? This procedure can restart a stopped heart. The electricity stimulates the flow of electrical current in the heart. Special muscle tissue in the heart sends an electrical impulse to all the heart muscles. The electrical impulses make the muscles contract.

The acronym ECG is more correct in English, but the German EKG is the term that is used most often. The SA node is normally the site in which cardiac excitation begins. It is located in right atrial wall.

To produce an electrocardiogram (EKG), a technician puts recording electrodes on a patient's chest and arms and connects them to an instrument called an electrocardiograph. An EKG is a tracing of heartbeats (Figure 12.8). These tracings show the electrical impulses that the heart muscles transmit during and between heartbeats.

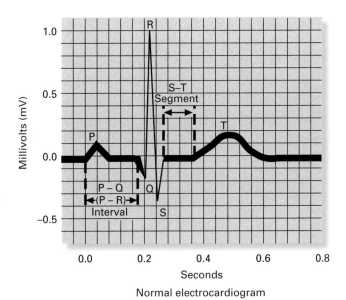

Normal electrocardiogram

Figure 12.8: *An electrocardiogram shows the electrical activity of the heart. The P wave begins at the SA node causing the atria to contract. QRS measures the rapid ventricle contraction. T indicates the ventricles are depolarizing or relaxing.*

The system of specialized muscle tissue that regulates heart contractions is called the conduction system. It's a messenger service that first tells the atria to contract, pushing blood that has returned from the lungs or body tissues into the ventricles (Figure 12.9). The electrical impulse then flows to the ventricles, making the lower chambers contract, pushing blood out of the heart.

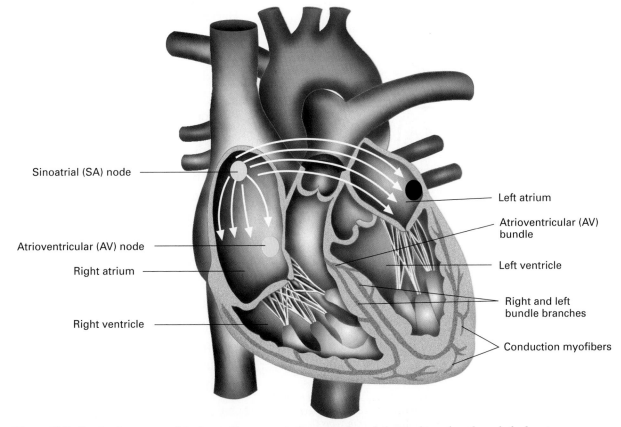

Figure 12.9: *Conduction system of the heart. The arrows indicate the flow of electrical impulses through the heart.*

AN EMOTIONAL SUBJECT YIELDS TO SCIENTIFIC ANALYSIS

People haven't always viewed the circulatory system the way we do today. Some cultures believed that the heart was the center of emotion and intellect; others believed that the heart held the spirit. About 2,000 years ago there were many theories about the circulatory system. The most popular one was developed by the Greek physician Claudius Galen. Galen believed that blood was either nutritive or vital life force. The liver made nutritive blood, which became food for the other organs. The heart made vital blood, which carried vital spirits. According to Galen, the heart also sucked blood from the veins. Although Galen was wrong, his theory seemed plausible and many physicians accepted it.

In 1628, William Harvey, an English physician, published his theory on the function of the heart and blood. Harvey questioned the theories of Galen and tested their accuracy. Through his experiments, Harvey quickly learned that blood flows in one direction and that there is only one kind of blood. Harvey experimentally proved three things: Blood isn't consumed by the organs of the body, the heart is a pump that circulates blood throughout the body, and blood flows in a closed system.

Initially, many scientists rejected Harvey's theories. After years of debate and experimentation his theories were proved correct. Accepting Harvey's theories raised new questions. His work was the beginning of what we now know about the circulatory system.

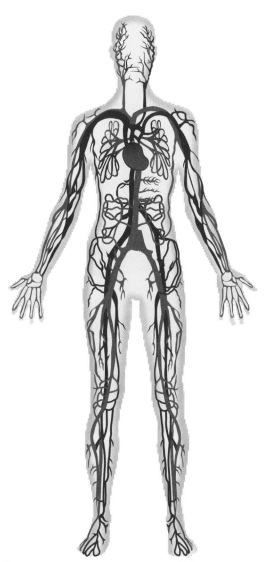

In the human circulatory system, oxygenated blood is carried in the arteries (red vessels) and oxygen poor blood is carried back to the heart in veins (blue vessels).

BLOOD PRESSURE

Additional Activity, The Pumping System, can be found in Chapter 12 of the TRB.

Each heartbeat is a single cardiac cycle. In one cardiac cycle, the atria contract and the ventricles relax; then the ventricles contract and the atria relax. The ventricles' contraction stage of the cycle is **systole,** and their relaxation stage of the cycle is **diastole.**

Place your first two fingers on your wrist just under your thumb. Can you feel a beating against your fingers? The pressure that you feel is blood flowing through your radial artery. This pulse is a fraction of the force that is exerted on the arteries when oxygenated blood leaves the heart. The force that is exerted on the artery walls is blood pressure. A normal blood pressure for a young adult is 120/80 mm Hg. The first number is the systolic pressure (pressure during ventricle contraction), and the second is the diastolic pressure (pressure during ventricle relaxation). Factors that affect blood pressure are how forceful the contraction is, how well the heart functions, flow resistance, blood volume, thickness of blood, and how elastic the arteries are.

The cardiac cycle starts when the atria fill with blood. As pressure increases in the atria, the pressure in the ventricles decreases. When the ventricle pressure is less than the atrial pressure, the left and right atrioventricular valves open, and blood flows from the atria into the ventricles. When the ventricle contracts and the atrium relaxes, both the semilunar valves remain closed until the pressure in the ventricles is greater than the pressure in the arteries (the aorta and pulmonary arteries). Once the semilunar valves open, the blood leaves the heart through the aorta and the pulmonary arteries (Figure 12.10).

If the larger arteries were accidentally cut, blood loss could be life-threatening. The pressure in these arteries is high, and the forceful heart pumping could quickly cause the person to "bleed out." These arteries are deep in the body's tissues, often near bones that protect them. Blood pressure drops as blood moves from arteries to veins (Figure 12.11).

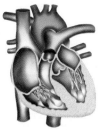
Relaxation

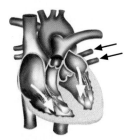
Ventricular filling

Atrial systole

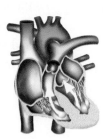
Ventricular contraction

Figure 12.10: *The cardiac cycle*

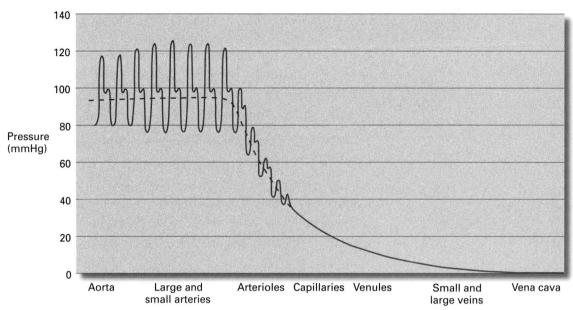

Figure 12.11: *Blood pressure in various systemic blood vessels*

It is lowest in the capillary network, where blood cells move single file. As blood collects in venules and moves into larger veins, pressure is also low. In areas where blood pressure in veins is very low, the contractions of the skeletal muscles help blood to flow back to the heart. As you can see in Figure 12.12, veins lie between skeletal muscles, so the muscle contractions squeeze the veins, pushing blood forward. Valves located in the veins keep the blood from flowing backward when the muscles relax. So walking, flexing your arm, breathing, and stretching all help blood to travel to your heart.

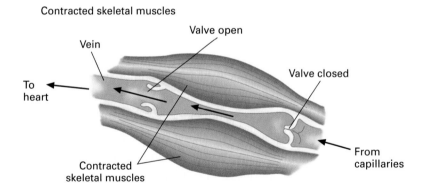

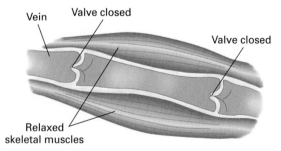

Figure 12.12: *Skeletal muscles and venous valves cooperate to return blood to the heart.*

CAPILLARY EXCHANGE

How can the information provided by a blood sample help to determine why you're sick? Your blood sample goes to a medical lab technician, who analyzes it and gives the doctor information from it. Your doctor looks for positive or negative indicators in the information. An increase or a decrease in any normal substance might indicate that an organ isn't working right. For example, a change in the normal amount of glucose might indicate that your pancreas isn't healthy. A change in the oxygen or CO_2 level is a clue that something might be wrong with your lungs.

These substances leave and enter the bloodstream through the permeable walls of a network of capillaries (Figure 12.13). In the capillaries, blood and the fluids surrounding the body cells exchange nutrients, gases, and other substances. This process is capillary exchange. Oxygen, CO_2, amino acids, glucose, and other small molecules diffuse through the single cell layer that forms the capillary walls. In lung capillaries, CO_2 leaves the bloodstream, and oxygen enters to be carried by the red blood cells. In the muscles, glucose and oxygen pass out of the capillaries to nourish the cells, and waste such as CO_2 enters the blood to be carried back to the heart and lungs.

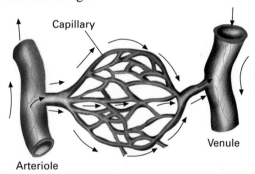

Capillary

Venule

Arteriole

Figure 12.13: *Structure of the capillary network*

A few substances pass through the capillary walls by vesicular transport. In this process, a tiny sac (vesicle) engulfs the substance, carries it across the capillary wall, and releases it on the other side. Vesicular transport takes place only for plasma solutes that can't get through the capillary wall any other way. For example, antibody proteins from a mother's circulatory system are transported to her fetus in this way.

LESSON 12.2 ASSESSMENT

1. Capillaries. The smallness and the extensive branching network of the capillaries drastically decrease the rate of blood flow.

2. Veins. The larger arteries are located deeper in the tissues of the body, often near bones for protection from injury.

3. About 48 times. Bradycardia, because that is below the normal range.

4. Left ventricle.

1 Through which blood vessels does blood flow most slowly? Explain why.

2 Roll up your shirtsleeve and look at your forearm. Open and close your hand rapidly several times until you see the vessels. Are these more likely to be veins or arteries? Why do you think so?

3 Your patient's heart beats every 1.25 seconds while resting. How many times will the patient's heart beat in 1 minute? What type of arrhythmia does this patient have? Give a reason for your answer.

4 Your doctor tells you that, because of damage to your aortic valve, blood is flowing back into one of your heart chambers. Which heart chamber has the backflow of blood?

LESSON 12.3 THE RESPIRATORY SYSTEM

Leave the oxygen-rich atmosphere for more than a few seconds, and you'll stop taking your breath for granted. Divers use scuba gear to get enough oxygen underwater. Astronauts can walk in space only if their equipment feeds them oxygenated air. Most people don't think about their ability to breathe, but individuals with chronic lung disease may find trying to take a deep breath of polluted air life-threatening. Let's find out how your body gets the oxygen that it needs so much.

ACTIVITY 12-3   FULL OF HOT AIR!

The procedure for this Activity can be found in Chapter 12 of the TRB.

Respiratory therapists work with patients who have chronic respiratory illness, measuring the amount of air the lungs hold at different times. Athletic trainers check lung capacity when they measure the volume of air the lungs hold in different circumstances. In this activity you will measure how much air you breathe in and out.

RESPIRATORY THERAPIST

CAREER PROFILE

James C., a respiratory therapist at St. Stephens Hospital, greets an emphysema patient as he is wheeled into the hospital clinic. The therapist will test how well his lungs are working since he began taking new medications and treatments. He explains to the patient that he'll evaluate the bronchial dilator medication that was prescribed the previous week. As he talks, James hands the patient a large, flexible tube with a detachable mouthpiece. The tube is part of a machine called a spirometer (Figure 12.14), which measures lung capacity. The test is difficult for someone with a chronic lung disease.

Figure 12.14: *A spirometer measures the air capacity of the lungs.*

Besides performing diagnostic tests, respiratory therapists administer oxygen and medications and set up and monitor mechanical ventilators according to a doctor's orders. They also teach their clients exercises that will improve their breathing. Respiratory therapists need a minimum of a 2-year degree and are licensed by a professional association.

HOW MARINE ANIMALS BREATHE

These aquatic animals have low energy demands because they are sessile, or slow-moving, and ectothermic.

Although water doesn't hold as much oxygen as air does, aquatic animals such as sponges and some cnidarians get enough oxygen from their watery environment. They're small and have low energy demands. For these animals, enough dissolved gas diffuses in and out through their cells' membranes.

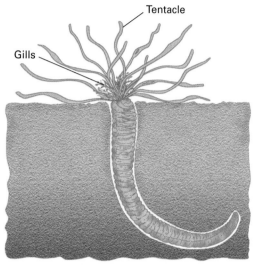
Tentacle

Gills

Figure 12.15: Amphitrite, *a marine worm that exchanges oxygen and CO_2 through its gills*

How do larger aquatic animals get enough oxygen? Marine worms called polychaetes have capillary-rich bristles along their sides that have special surfaces for respiration. Other marine worms, such as *Amphitrite,* have external gills to absorb oxygen and release waste CO_2 (Figure 12.15). In these worms the feathery gills wave in the water. Many aquatic animals, such as crabs, mollusks, fish, and young amphibians, have internal gills. These membranes have a capillary network sandwiched between thin folds. In fish the delicate gills are covered by a movable flap of tissue that opens and closes. Water is pumped past the gill membranes, so oxygen dissolved in the water can diffuse into the membranes and waste CO_2 can diffuse out faster.

BREATHING ON LAND

Animals that live on land have adapted to breathing oxygen from the air. Insects have spiracles, openings on both sides of their body, that lead to tracheae, a system of tubes (Figure 12.16A). These tubes branch and branch again, allowing air to reach all parts of the insect's body.

Other land animals and some marine animals such as whales have evolved lungs to exchange oxygen and CO_2. Lungs are moist membranes surrounded by a capillary network, like gills. On land these air exchange surfaces would dry out if they weren't protected. So they are enclosed in the body. Some animals have other adaptations. Some frogs have vascular skin that exchanges gases with the air, in addition to their lungs (Figure 12.16B). In the respiratory systems of birds, most inhaled air flows into air sacs, which are branches of the lungs. The air remains in the air sacs until the bird exhales. The air then travels from the air sacs to the lungs where gas exchange occurs. In this way a flying bird can

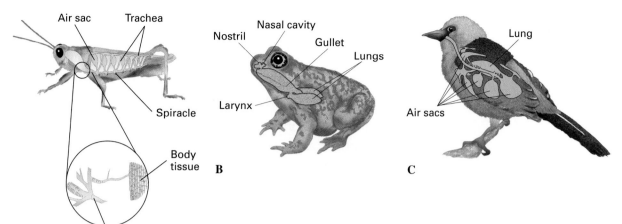

Figure 12.16: *The grasshopper (A) has spiracles to distribute air throughout its body. The frog (B) and bird (C) primarily use lungs.*

maintain a consistently high level of oxygen. Fresh air passes over the lung surface both when the bird breathes in and when it breathes out (Figure 12.16C).

In humans and other mammals, air enters through the mouth and nostrils when the animal inhales (Figure 12.17). The nasal passage is lined with sticky mucous **epithelium,** which traps dust and odor molecules. Nose hairs filter particles from the incoming air. The air flows past the back of the throat—the pharynx—and past the flap called the epiglottis, which closes when you eat to keep food out of your lungs. It passes the larynx (your vocal cords) and enters

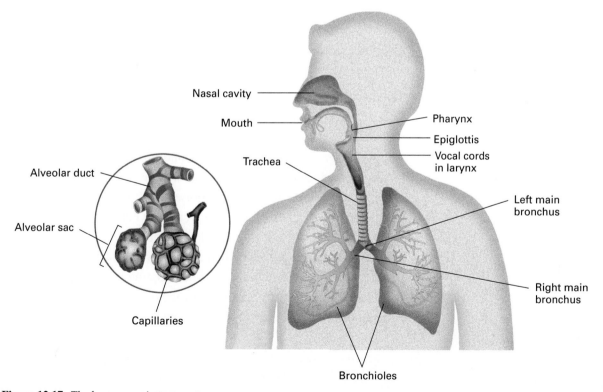

Figure 12.17: *The human respiratory system*

the trachea (your windpipe). When you expel air, you can create sound because your vocal cords vibrate. Feel the effects of air movement on your vocal cords by touching your throat and saying your name out loud. Can you feel the vibrations? Now say your name in a whisper. Can you feel the vibrations now?

The human respiratory system is like an upside-down hollow tree with the trachea as the trunk. The trachea's hard C-shaped rings of cartilage give it shape and strength. When you inhale, oxygen-filled air passes through the trachea, which is lined with special cells. These cells make mucus that trap dust particles and have cilia that extend into the hollow center and wave rhythmically. This beating motion pushes dust out of the lungs and may signal the need to cough to bring up particles.

Air continues into the **bronchi,** two tubes that branch out to the lungs. The tubes continue to branch, becoming smaller and smaller until they reach millions of tiny dead-end spaces that are lined with thin, flat respiratory cells. These clusters of saclike spaces are **alveoli.** As the air travels through these respiratory passageways, it is warmed and moistened.

Lungs look like a sponge with very fine holes. The air spaces in a sponge are similar to the air-filled spaces in the lungs. The solid part of the sponge represents the working cells of the lung—the respiratory cells that exchange gases and the capillary network that surrounds the alveoli.

The alveoli make a protein called **surfactant** that keeps the tiny alveolar sacs from collapsing and staying shut. Without it, water molecules would hold the sides of the sacs together like two pieces of wet plastic wrap, preventing air from entering or leaving. Surfactant breaks the surface tension of the water so that the balloonlike alveoli can expand and fill with oxygen. Fetuses don't produce surfactant until about the seventh month, and premature babies are at risk for lung disease because their alveoli don't expand with the first breath. The damage can be reduced when a commercial foam, called perfluorocarbon, is used in the artificial respirator. The oxygen-rich foam bathes the lung surface. The oxygen diffuses through the respiratory cells, and the CO_2 diffuses out.

INVESTIGATION 12B METABOLIC RATES AND RESPIRATION

See Investigation 12B in the TRB.

Cells that are respiring demand oxygen and produce waste CO_2. When you decrease or increase your level of physical activity, you decrease or increase the rate of cell respiration in the muscles that you're using. In this investigation you'll measure the exhaled CO_2 to determine the rate of metabolism at rest and just after exercise. You'll observe the changes that your respiratory and circulatory systems make to accommodate the change in activity.

MAKE WAY FOR AIR

Additional Activity Take a Deep Breath can be found in Chapter 12 of the TRB.

This principle is known as Boyle's law. The volume of a gas is inversely proportional to the pressure, provided the temperature remains constant.

Imagine that you have 100 molecules of a gas in a balloon. If you make the balloon smaller, decreasing the volume, the molecules become more crowded and pressure increases. When the balloon is expanded, increasing the volume, the pressure decreases because the molecules aren't so crowded and there is room for more.

The diaphragm is a dome-shaped muscle that lies at the base of the rib cage (Figure 12.18). When the diaphragm contracts, it pulls down and flattens, and the chest cavity becomes larger. The increased volume means that pressure decreases in the lungs (which also expand). Because the air pressure in the lungs is lower than the air pressure outside, air is brought deep into the lungs, filling the alveoli.

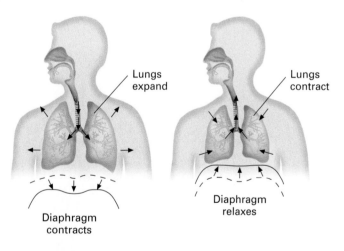

Inhalation   Expiration

Figure 12.18: *Diaphragm during inhalation and exhalation. Both inhalation and exhalation change the air pressure in the lungs.*

AIR CAPACITY

Imagine exhaling as much air as you can, then taking the deepest breath you can. The volume of this big breath is called **vital capacity.** This air would fill about three 2-L soda bottles. A normal breath, the **tidal volume,** is about $\frac{1}{2}$ L of air. People with chronic lung disease such as emphysema have less lung capacity, which makes many ordinary activities fatiguing.

A specialized area of the brain controls your breathing rate. It monitors the CO_2 level in your blood. As the level increases, breathing rates increase and each breath is deeper. The lungs quickly expel the excessive CO_2 and provide increased oxygen.

PULMONARY CIRCULATION

Note that blood, leaving the heart in the pulmonary arteries, is arterial blood even though it is rich in CO_2 and low in oxygen.

Once oxygen diffuses through the respiratory cells and into the capillary network that lines the alveoli, its trip through the body is just beginning (Figure 12.19). The capillaries surrounding the alveoli join to form veins that travel to the heart. Blood that flows to the heart is oxygen-rich when it comes from the lungs. It enters the left atrium, flows to the left ventricle, and, with the next contraction, is forced out of the heart through the aorta, an artery. The blood flows out to the cells in the body through the systemic circulatory system.

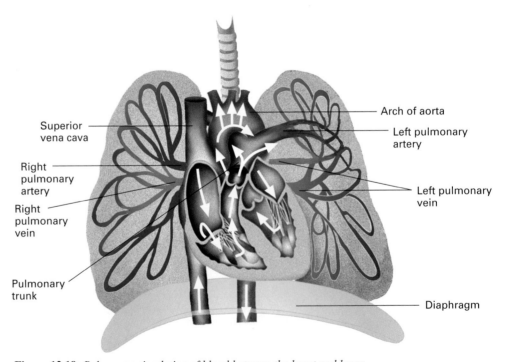

Figure 12.19: *Pulmonary circulation of blood between the heart and lungs*

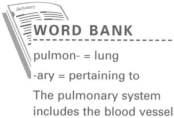

As the nutrients and the oxygen diffuse through the capillary walls to body tissues, waste and CO_2 pass into the blood. Returning to the heart through the venous system, the dark-colored blood enters the right atrium, flows to the ventricle, and is forced out of the heart. The arteries leaving the heart carry the oxygen-poor blood to the lungs to exchange gases. After it travels through the lungs, veins return oxygenated blood to the left side of the heart. This system of arteries and veins that carry blood to and from the lungs is the pulmonary system.

RESPIRATORY PROBLEMS

What causes a cold? It's a virus that infects your nasal passages. The symptoms are depressingly familiar: sneezing, coughing, a runny nose. A more serious respiratory disease is pneumonia, which may be caused by either a bacterial or a viral infection. The alveoli fill with fluid and immune cells that are fighting the infection. The sticky fluids limit the surface area that is available for gas exchange. Without prompt treatment, severe cases of pneumonia can be fatal.

Air pollution also interferes with a person's breathing (see Table 12.1). Many cities post air pollution warnings for those who have health problems. Allergic reactions and asthma may cause breathing

Pollutant	Major Source	Effect
Nitrogen oxides	Automobile exhaust; power plant furnaces	Irritates lungs; can be fatal in high doses; combines with water vapor to make acid rain and ozone
Sulfur oxides	Home, industrial, and power plant furnaces	Aggravates lung disease; reacts with water to form acid rain
Carbon monoxide	Automobile exhaust	Displaces oxygen in the blood; can be fatal
Lead	Leaded gasoline (no longer common)	Affects blood formation; causes reproductive and nervous problems; damages kidney
Particulates	Coal-burning plants; mining; forest fires	Aggravates heart and respiratory problems; may be carcinogenic
Ozone	Combination of nitrogen oxides and hydrocarbons from fuel, cleaning fluids, and solvents	Causes respiratory problems

Table 12.1: Six Major Air Pollutants

problems when they narrow the air passageways. Tobacco smoking damages the respiratory system and other organs. Smoke contains hundreds of substances, but tar, nicotine, and carbon monoxide are the worst. Tar, in smoke, destroys respiratory cells in the alveoli. Over time this limits the amount of oxygen that can be exchanged. Carbon monoxide also prevents the blood from being saturated with oxygen. Nicotine is the addictive agent in the smoke. Tobacco use can lead to cancer and emphysema.

See TRB for an additional Learning Link, Up in Smoke.

LESSON 12.3 ASSESSMENT

1. Breath comes in through the mouth or nose, pharynx, trachea, bronchi, smaller branches, and alveoli.

2. The pulmonary system circulates between the heart and lungs; the systemic system supplies the rest of the body. In both, the arteries leave the heart, and the veins carry blood back to the heart. In the pulmonary system the arterial blood is nonoxygenated; in the systemic system it is oxygenated. The pulmonary venous blood is oxygenated, and the systemic venous blood is not.

3. The pale color and bluish lips are from a significant decrease in oxygen to these tissues. Students may also mention reduced heart and lung function.

4. The volume of air in the lungs increases to allow more air to travel across the vocal cords with more force (volume).

1 Trace the flow of incoming air through the nose and lungs.

2 Compare the pulmonary blood circulation to the systemic blood circulation (the arteries and veins that travel throughout the rest of the body).

3 You happen to be present when a fender bender occurs. The emergency medical technician reports that the victim is in shock and has pale skin color and bluish lips. Hypothesize what causes the paleness and blue tinge in the lips.

4 A choral director urges a singer to breathe from the diaphragm. How might that help the singer?

LESSON 12.4 THE KIDNEY

Your blood carries wastes that must be removed so that they don't poison you. Your kidneys, a pair of blood-washing machines, clean up the wastes. Like a washing machine, a kidney receives "dirty" blood, cleans it, pumps out the waste water, and sends the clean blood on its way. Fortunately, the kidney does this quietly, with no moving parts. And if one kidney stops working, the other can do all the cleaning by itself.

KIDNEY DIALYSIS

BIOLOGY IN CONTEXT

The four options for treatment of end-stage renal disease are hemodialysis, peritoneal dialysis, kidney transplant, and no treatment. If it is not treated, it results in death.

Dale E. knows a lot about how kidneys work—or don't work. When he was young, he developed kidney disease. He was on hemodialysis for 8 months until a kidney became available for transplant.

Dale explains that circulating blood makes a big loop through your body. Kidneys clean metabolic wastes from the blood and maintain homeostasis of electrolytes and fluids. The kidney contains many filters, called nephrons, which remove toxins and excess fluids. As Dale's kidneys slowly stopped working, the nephrons developed holes and could no longer adequately filter his blood. When they deteriorated to less than 10% of their original filtering capacity, he could no longer urinate, and his body built up excess fluid. "I was about to drown in my own fluids," he explains.

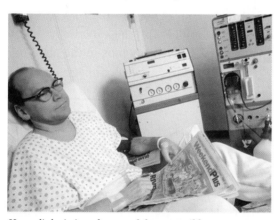

Hemodialysis is only one of three possible treatments for patients with end-stage renal disease.

Electrolytes are dissolved ions in the blood, including calcium, sodium, and potassium.

So Dale went on hemodialysis. A medical procedure expanded a blood vessel in his arm to make room for two large needles that took his blood to and from the hemodialysis machine. The machine passed blood through a permeable membrane tube in a dialyzer, while a glucose and water solution called dialysate flowed around the fibers surrounding the dialyzer. The excess water and toxins from the blood diffused into the dialysate, and

A fistula, the expanded blood vessel, is formed when a patient's vein is cut and attached to an artery. The pressure in the two vessels causes the junction to swell to over 1 cm in diameter.

Dale's clean blood flowed back into his body. "They say that there is never more than a pint of blood outside your body at any one time, but it always looks like more," Dale said.

Dale received hemodialysis for $3\frac{1}{2}$ hours every other day while he waited for a kidney. After 8 months, a cadaver donor was found. The donor died from an accident or other cause and had given permission to donate his or her organs. A living person can also donate a kidney, because it's possible to live normally with only one kidney.

"Organ donation is a wonderful way to get a person back to a seminormal quality of life. You are never cured of kidney failure; even with a transplant I still have what is called ESRD (end-stage renal disease). That's because kidneys don't last forever—even transplanted." Dale volunteers with the American Association of Kidney Patients, which provides information to kidney patients and their families.

KIDNEY FUNCTION

In vertebrates the kidneys are a pair of bean-shaped organs in the abdominal cavity. A large artery connects each kidney to the aorta (Figure 12.20). This artery brings nourishing blood to the cells that make up the kidney, but most of the blood flows into tiny filters, called nephrons. The filter creates urine, a fluid that carries dissolved body waste out of your body. Your kidneys filter about 180 L of blood every day and produce about 1.5 L of urine. Urine is mostly water that you replace via your diet; the rest is waste.

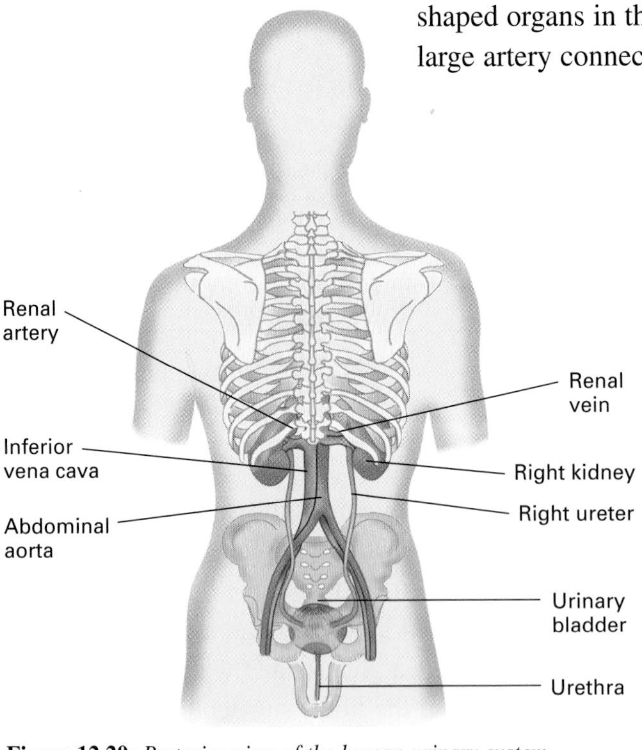

Renal artery

Inferior vena cava

Abdominal aorta

Renal vein

Right kidney

Right ureter

Urinary bladder

Urethra

Figure 12.20: *Posterior view of the human urinary system*

The Internet site Institute for a Drug-Free Workplace has an extensive list of resources about workplace drug testing.

Your consumption of certain substances can be detected in your urine. For instance, if a person has used an illegal substance, it leaves traces in his or her urine. The body breaks down most foreign substances such as drugs into simpler substances, which are excreted in the urine. If you are hired for a job, you might be asked to provide a urine sample for a drug test. Use library and Internet resources to find out what a workplace drug test is likely to look for and how the tests detect drugs.

NEPHRONS

Each of your kidneys has about one million microscopic filtering units called **nephrons** (Figure 12.21). If you sliced through a kidney and microscopically examined a very thin section from the outer layer, you would see nephrons. A nephron looks like a tiny coiled tube with a distinct U-shaped loop. One end of the nephron cups a capillary, collecting waste and water that diffuse out of the

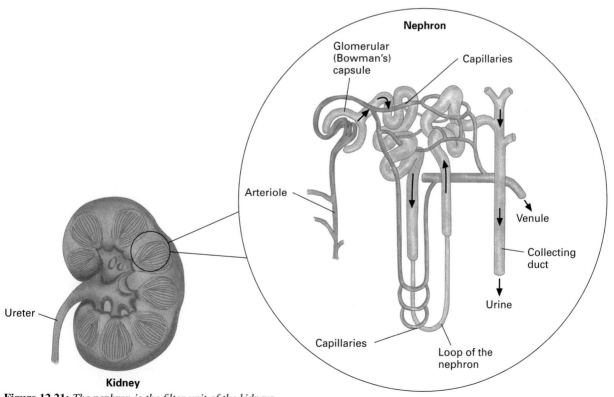

Figure 12.21: *The nephron is the filter unit of the kidneys.*

blood. The loops of all the nephrons extend toward the center of the kidney, where very salty fluid surrounds them. The walls of the nephron loops and of the larger ducts into which the nephrons empty are selectively permeable to water, ions, amino acids, and other substances. Any substance that the body needs is reabsorbed into nearby blood vessels by active transport and diffusion. Remaining substances move through the rest of the nephron and into larger collecting ducts. Now called urine, water and these substances that are dissolved in it travel through a tube called the ureter. It connects the kidney to the bladder. The urine is held in the bladder.

See TRB for an additional activity, What Can Urine Indicate About Your Health?

LESSON 12.4 ASSESSMENT

1. The presence of white blood cells indicates an infection in the urinary system.

2. They are too large to pass through the hemodialysis membrane.

3. Active transport and diffusion.

4. The specific gravity (density) increases.

1 Urinalysis indicates that a patient has white blood cells in his or her urine. What could this mean?

2 During hemodialysis the red blood cells are not filtered out. Why?

3 What processes return many solutes to the bloodstream?

4 How does the specific gravity of your urine change when you don't replace fluids after exercise?

CAREER APPLICATIONS

The applications that follow are like the ones you will encounter in many workplaces. Use the biology you learned in this chapter to complete the activities. Share your work with the class.

AGRICULTURE & AGRIBUSINESS

Fish Breath

Some fish can't get enough oxygen unless water constantly moves over their gills. This constant flow is called ram ventilation. Find the names of several economically important fish that use ram ventilation. Why is ram ventilation a problem for hatcheries that raise fish in recirculating tanks? How do they overcome the problem? Sketch a tank setup that allows ram ventilation.

Waste Water

Ammonia: aquatic species; urea: mammals and amphibians; uric acid (insoluble precipitate): birds, insects, reptiles.

Use encyclopedias, animal books, the Internet, or other sources to find out how animals living in different environments maintain water balance. Look at aquatic species as well as terrestrial species. As a general rule, what kinds of animals excrete ammonia, urea, and uric acid? What other adaptations do animals have for regulating water in the body? Write up your findings in your logbook.

BUSINESS & MARKETING

Blood Donation

Investigate how blood centers that pay for blood donations affect collections at strictly volunteer facilities such as the American Red Cross. If possible, have someone from your class contact a center that pays for blood and a Red Cross center. What are the requirements for donating blood at each facility? What types of donations (whole blood, plasma, etc.) does each facility accept? How is payment calculated? How does each center distribute the blood? Make a chart comparing blood donation at both centers.

Smoke Out

Contact the local branch of the American Cancer Society or investigate the organization through the Internet. Find out about their annual Great American Smoke-Out campaign. How much money is spent locally or nationally on the campaign? How many Americans participate in the campaign each year (locally or nationally)? Of that number, what percentage of those people are successful in permanently quitting? Find out what opportunities exist for volunteers to help get ready for the Great American Smoke-Out, and create a poster describing those opportunities, to be displayed in your school.

Smoke Gets in Your Eyes

Talk to a childbirth class instructor, a pediatrician, a midwife, or someone else who works with parents of small children or parents-to-be; ask whether he or she will distribute the brochures pertaining to second-hand smoke.

Working in a small group, prepare a brochure on an issue related to smoking and the family. Possible issues might be the effects of a mother's smoking on her unborn child, effects of secondhand smoke on young children at home, effects of smoking on the smoker, effects of smoking on the incidence of respiratory problems such as asthma, and ways to minimize exposure to secondhand smoke in the home.

Cardiovascular Fitness, Family Style

Research the benefits of aerobic exercise as a regular part of a healthy lifestyle. What exercises provide these benefits? Create a booklet or video that gives the benefits of aerobic exercise. Illustrate or demonstrate exercises that are suitable for individuals at different stages in their lives such as preschoolers, school age, young adult, middle age, and senior citizen. What are some activities that families could do together that would have aerobic benefits?

Respiratory Therapy

Interview a respiratory therapist (RT). What kind of education and state or national certification must the RT have? What is a typical day like? Ask the RT how blood-gas analyzers, oximeters, and spirometers work. What can each device reveal about a patient's health?

ABCs of First Aid

Talk to an emergency medical technician, emergency room nurse, or first-aid instructor. Find out what the ABCs of first aid are. What vital signs and other tests are performed in the field to get information about the cardiovascular and respiratory systems? What common injuries affect these systems? How are they treated or stabilized in the field? Where can students become certified in first aid and CPR?

Artificial Heart

Research artificial heart valves, left ventricular assist devices (LVAD), and artificial hearts. At what stage of development is each of these technologies? Who is the ideal patient to receive one of these devices? Are the devices considered permanent interventions or stopgap measures to keep a patient alive until a heart transplant becomes possible? What are some of the obstacles that must be overcome in designing a replacement heart or heart parts? Develop a script or storyboard for a documentary on the strides that are being made in developing an artificial heart.

Kidney Stones

Research kidney stones by using medical reference books or the Internet or by contacting a physician or other knowledgeable person. What are the major types of kidney stones? Are they dangerous? What causes kidney stones? Can they be prevented? How are they treated? Write up your findings in a short report.

INDUSTRIAL TECHNOLOGY

Respiratory Protection

Brainstorm a list of occupations in environments that could harm the respiratory system, such as working in air that carries a lot of particulates or among poisonous gases, under water, in space, or in an oxygen-poor atmosphere. Find out about the personal protective equipment that workers use. How is the equipment rated? For how long can people use it?

Air Cleaners and Purifiers

Use consumer reports or talk to an allergist about ratings for various home air filters or purifiers. Analyze the design features of the most efficient models. Write a brochure for consumers that lists desirable features of a home air cleaner. Explain why each feature is important in improving the air you breathe.

Pacemakers

Research cardiac pacemakers. Report to your class when they are used and how they work. Create a visual aid that will illustrate how they work. How are they engineered for reliability? What happens if the battery fails?

- Small aquatic animals, such as sponges and flatworms, rely on diffusion to exchange substances between the environment and each cell of their bodies. Large animals cannot rely on diffusion alone and must exchange materials with their environments via specialized organs such as gills and lungs.

- Humans and other vertebrates have a cardiovascular system for circulation. The system has a network of vessels through which a heart pumps blood that carries nutrients and oxygen to body tissues and waste away from them.

- Blood contains cells. Red blood cells transport oxygen. White blood cells help to defend against pathogens and foreign substances. Cell fragments called platelets help to repair damaged blood vessels by clotting.

- The cardiovascular systems of vertebrates and of some invertebrates such as the earthworm are closed; blood vessels connect to one another in a circuit. Many invertebrates such as mollusks and insects have an open circulatory system in which blood vessels lead to a fluid-filled cavity.

- Blood flow through the heart creates sounds, the lubb and the dupp. The lubb sound is made by the valves closing between the atria and the ventricles. The dupp sound is made by the valves closing between the ventricles and the large arteries.

- Special branches of muscle tissue in the heart send an electrical impulse to the other muscle cells in the heart, making the heart muscles contract.

- An EKG records the activity of the electrical conduction system as a series of waves.

- Each heartbeat is a single cardiac cycle. In a cardiac cycle, the atria contract and the ventricles relax, then the ventricles contract and the atria relax.

- The force exerted on the blood vessel walls is blood pressure. Blood pressure for a young adult is about 120/80 mm Hg. The first number is the systolic pressure (the pressure when the ventricles contract), and the second number is the diastolic pressure (the pressure when the ventricles are at rest).

■ In the capillaries, blood and the fluids surrounding the body cells exchange nutrients, oxygen, CO_2, amino acids, and other substances by diffusion or vesicle transport.

■ The kidneys' nephrons filter blood and create urine, a fluid that holds dissolved body wastes that will be excreted from the body.

CHAPTER 12 ASSESSMENT

Concept Review

See TRB, Chapter 12 for answers.

1 Why does blood from an artery fill sample tubes faster than blood from a vein?

2 Using the accompanying figure, show the path of blood flow through the heart.

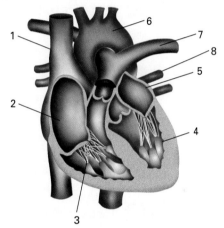

3 If glucose and oxygen can diffuse across capillaries, why can't red blood cells do the same?

4 Leaky vein valves cause varicose veins. Which veins are less likely to become varicose, leg veins or neck veins? Give a reason for your answer.

5 Why is it important that the walls of arteries are thicker than the walls of veins?

6 Imagine that you are a red blood cell leaving the right atrium. Are you carrying oxygen molecules? Give a reason for your answer.

7 When you squeeze the tip of your finger and release it, the color changes from a pale pink to a deeper pink color immediately. What causes this color change?

8 Why is venous blood easier to use than arterial blood for blood analysis?

9 Hiccups are a muscle spasm. What muscle is being rapidly contracted, causing the hiccup?

10 What are the main functions of blood?

11 After a one-on-one game of basketball, your face is red and sweaty. What causes the red coloring and why?

12 What is the function of the heart valves?

13 What is the function of the kidneys?

14 Water passing over the gills of a fish exchanges gases with the fish's blood. How must the oxygen concentration in water be related to that in the fish's blood in order for the fish to benefit?

15 Diuretics, a type of medication, lower the amount of water in blood. How would a diuretic help a person with high blood pressure?

16 A patient's history reveals that his lifestyle decisions included little exercise, smoking, and a high-fat diet. After several diagnostic tests, he is told that he has severe high blood pressure. If the blood pressure is left untreated, what might happen? Speculate why.

17 Why do people lie down with their feet slightly elevated when they feel faint?

18 Discuss the structural differences between arteries, capillaries, and veins.

Think and Discuss

19 In a small city in Texas a large number of children were found to have high levels of lead in their blood. Health officials were able to determine that the children had been playing at a playground downwind from a lead smelting plant. Explain how the lead entered the children's bloodstream.

20 The brain produces a hormone, antidiuretic hormone (ADH), that helps to control how much water you excrete in your urine. ADH causes the kidney to reabsorb water into the blood stream. Caffeine and alcohol keep the brain from producing ADH. Knowing this, predict how the specific gravity of urine would be changed after someone drinks coffee or alcohol.

21 Carbon monoxide's ability to bind with the protein that carries oxygen in the blood is 200 times stronger than oxygen's. How do low levels of carbon monoxide gas in the bloodstream affect the circulatory system?

22 If a fetus doesn't have working lungs as it develops inside the mother's uterus, how do the fetal tissues receive the oxygen needed for growth and development?

In the Workplace

23 What type of respiratory protection would you recommend to lawn-care professionals? Assume that their work includes starting new lawns from bare lots as well as mowing and maintaining existing lots. Give a reason for your answer.

24 Why do emergency medical technicians restore a patient's airway before making sure that the patient has a heartbeat?

25 You are a hemodialysis technician. One of your patients asks why her blood pressure is high before treatment and much lower after treatment. What would you tell her?

26 Why do top athletes tend to have a lower resting heart rate than the general population does?

Investigations

The following questions are based on Investigation 12A:

27 What were the most abundant types of blood cells you observed in the blood sample? What was the least abundant type of blood cell?

28 Heparin is a chemical substance that acts as an anticoagulant, preventing blood from clotting. How might this substance be used to help treat certain medical conditions?

The following questions are based on Investigation 12B:

29 In your experiment you were able to determine that you exhaled a much greater volume of air than what you measured in terms of carbon dioxide exhaled. Explain why your results do not reflect the total volume of exhaled air.

30 If you compared the amount of carbon dioxide exhaled by a conditioned athlete and an inactive person doing the same amount of exercise, who would exhale the most carbon dioxide? Why?

CHAPTER 13

WHY SHOULD I LEARN THIS?

What do you like to eat? When your stomach growls, do you look for the closest vending machine, stop at a fast-food stand, or head for the grocery store? All of these options supply food, but are they the best choices? Will the foods you select give you the nutrients you need for good health? Do the foods have too much carbohydrate, fat, and protein, which your body may store as fat? In this chapter you will learn how the food you eat is disassembled so that the nutrients can be reassembled into the things your body needs.

Chapter Contents

DIGESTION

WHAT WILL I LEARN?

1. the characteristics of different types of digestive systems
2. where enzymes break down foods into nutrients and how nutrients are absorbed
3. what foods supply the six major nutrients that are needed for good health
4. how disease and lack of certain nutrients affect health

What you eat affects your health, so making smart food choices will affect you today and in the future. Magazine articles proclaim that some vegetables prevent cancer, certain vitamins help the immune system, and high-fat diets contribute to heart disease. However, selecting food that is nutritious isn't easy. If it were, dietitians wouldn't be needed to plan diets for hospital patients and help people to choose healthy foods.

Discuss where students get their nutrition information. Is it reliable? How do they know?

Television news reports and infomercials often make claims that vegetables, vitamins, or herbs are the key to healthier living. It seems that every week someone is making a new discovery about how coffee, squash, or popcorn affects our health. Even at the grocery store checkout line you can read about the new diets of movie stars that claim to give you more energy or make you live longer. It is hard to determine what is true and what is not.

Selecting food that is nutritious is not easy. But you can make it easier by learning more about what is in the food you eat and how it is used by your body.

LESSON 13.1 DIGESTIVE SYSTEMS

Animals spend a lot of time and energy getting food. They have developed efficient hunting practices and digestive tracts to maximize the nutrients that they absorb for energy, growth, and development. Getting food into your mouth, the first stop on its way to the stomach, is just the beginning of the process of digestion.

ACTIVITY 13-1 DIFFUSION: IT'S IN THE BAG

SAFETY NOTE

See TRB for additional information on Activity 13-1. The starch in the bag gradually turns blue-black, and the iodine solution may get lighter because of diffusion. The plastic bag is a semipermeable membrane. The iodine molecules are so small that they move from the area of high concentration outside the bag into the area of low concentration inside the bag. Inside, the iodine reacts with the starch, forming a blue-black compound. The starch and the new compound are too large to travel through the membrane. This model represents digested nutrients (the iodine) passing through a cell membrane (the plastic bag) into the cytoplasm (the starch solution), where they create new compounds.

For animals to survive, they have to ingest food, often in the form of complex molecules. These break down into smaller units that the body can absorb. In this activity you'll make a model of how a cell absorbs nutrients.

Wear your safety goggles during this activity.

Working with a partner, pour about 50 mL of starch solution into a plastic bag. Twist the top before sealing the bag with a twist tie. Put the bag in a beaker about one fourth full of iodine solution. Record in your logbook the color of the starch solution and the color of the iodine solution. Look at the bag contents every 5 minutes for the next 15 minutes, recording your observations. Answer the following questions in your logbook:

- What changed in each solution? What caused this change?

- How does this model represent the permeability of a cell membrane?

ZOO ENRICHMENT

BIOLOGY IN CONTEXT

For many years, zoo diets were based on livestock diets. Livestock diets provide energy and concentrated nutrients for rapid growth so that the rancher can get the animal to market as fast as possible. Most livestock are slaughtered fairly young, and any long-term dietary problems are not an issue. For zoo animals the goal is to provide the nutrition the animal needs to live a long and healthy life. Over the years, more and more health and behavior problems have been found to be linked to nutrition.

Zoos face two challenges in feeding their animals: First is preparing diets that are good for the animals; second is getting the animals to eat the food. Researchers study the diets of wild animals to determine their nutrient requirements and to formulate feeds that approximate the animals' natural diets. Many zoos grow some foods; they raise bamboo, use hydroponics to grow grasses, and raise pinks and fluffies (newborn and young mice) for carnivores.

Once zookeepers have the proper feed, they still need to get animals to eat it. Animals grow tired of the same routines and foods. Zoos provide enrichment activities to encourage behavior that is used in

the wild. Joe G., the mammal curator for a Texas zoo, describes an enrichment activity for the zoo's elephants: "In the wild, elephants may travel many miles, grazing all day long. The keepers used to drop the elephants daily hay ration in the same spot every day. The elephants would come out of their stalls, go to the hay and eat, and then go back to their stalls. Now we drop the hay at a different spot in their enclosure each day. The elephant will return to the spot where they were fed the day before, and when they can't find their food, they have to go look for it. They still aren't traveling the great distances they do in the wild, but it does make them forage."

Keepers in other zoos hide raisins, a favorite food of gorillas, under bark and rocks so that the animals have to search to find them just as they would search for food in the wild. These tactics create variety and encourage behaviors that the animal would use in the wild.

DIFFERENT TYPES OF DIGESTIVE SYSTEMS

Additional Activity, Comparing Digestive Systems, can be found in Chapter 13 of the TRB.

Just like the iodine going through the plastic bag, nutrients from food must go through the cell membranes of an organism. Some nutrients diffuse from a high concentration to a lower concentration through the selectively permeable membrane of cells that line the digestive tract. Others need active transport through the cell membrane, which requires energy. To maximize the ability to get food, break it down, and absorb the nutrients, different digestive systems have evolved to fit the animal's source of food and its nutritional needs.

Extremely small organisms such as protozoans can absorb nutrients directly from their watery environment. Larger organisms, such as the cnidarians and flatworms, have more specialized ways to get and absorb food. *Hydra* has a hollow body made of two tissue layers (Figure 13.1A). A mouth takes in food particles and leads to a saclike cavity, where digestion occurs. The cells that line the body cavity absorb the nutrients. Flatworms also have a hollow body (Figure 13.1B).

A **B**
Figure 13.1: *(A)* Hydra; *(B) flatworm*

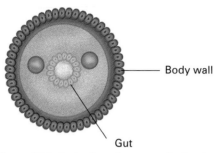

Body wall

Gut

Figure 13.2: *Body plan of a nematode showing tube in a tube design*

Their mouth takes in food and leads to a series of branched tubes that are distributed throughout the body. Food enters the "mouth," and waste leaves through the same opening.

A tube inside a tubular body is a more complex digestive system. Nematodes have a food tube that begins at the mouth and stretches the entire length of their body (Figure 13.2). Waste leaves through an opening called an anus, at the end opposite the mouth. Food and waste move in one direction. The food tube is surrounded by fluid that gives the worms their balloonlike body structure.

Animals such as segmented worms, insects, mollusks, and vertebrates have more complex digestive systems (Figure 13.3). Specialized organs such as the liver, pancreas, and gallbladder assist with digestion. The digestive tract is lined with **epithelial cells,** densely packed cells that cover or line the body and its organs and limit the substances that can pass through, protecting the body's internal environment. The tube is surrounded by two sheets of smooth muscle that push food through the digestive tract as it is digested. The coordinated action of the two sheets of muscles is called **peristalsis.**

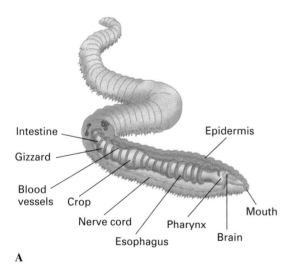

Intestine

Gizzard

Blood vessels   Crop

Nerve cord   Pharynx

Esophagus   Brain

Epidermis

Mouth

A

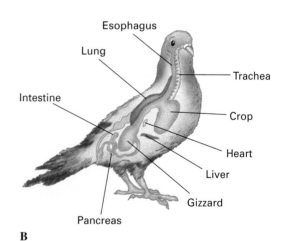

Esophagus

Lung

Intestine

Trachea

Crop

Heart

Liver

Gizzard

Pancreas

B

Figure 13.3: *Digestive system of (A) a segmented worm and (B) a bird*

Cows, sheep, and goats are **ruminants.** Their stomachs have four compartments (Figure 13.4). The **rumen,** the first compartment, contains bacteria that mix with the food. Through the process of fermentation, these microorganisms turn complex carbohydrates and cellulose into fatty acids. The cow gets energy from the fatty acids. The microorganisms, many of which are present in the human digestive track, also synthesize vitamins and amino acids that the cow uses. Cells absorb these nutrients and use them to create proteins for growth and development.

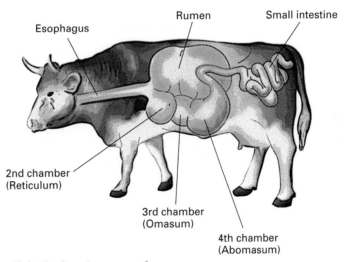

Figure 13.4: *The digestive system of a cow*

The rumen has a large population of bacteria that break down pasture grasses and hay. When cows arrive at the feed lot, the feed lot operator gradually switches their diet from grasses to grain and silage, which are used to fatten the animal. During the dietary switch, grain-consumer bacteria increase, and the hay-and-grass bacteria decrease.

The second chamber is similar to the rumen. Strong muscles in the second chamber periodically return cud to the mouth. The cow chews its cud, which is made of partially digested food that was stored for a while in the rumen and the second chamber. The chewed food passes to the third compartment to be digested by mechanical action and then to the fourth compartment, which works a lot like the human stomach.

RENNIN

Additional Activity, Say Cheese, can be found in Chapter 13 of the TRB.

Calves and other nursing ruminants have a unique enzyme, rennin. Their intestines produce rennin, which separates milk solids from liquids. Commercial cheese makers use rennin to make curds by coagulating a milk protein called casein (Figure 13.5). Other manufacturing processes use casein to make glue, paints, and plastics.

Figure 13.5: *Cheesemaker making Parmesan cheese*

LESSON 13.1 ASSESSMENT

1. The iodine would move through the semipermeable bag to react with the cornstarch in the beaker. The blue-black compound would be outside the bag.

2. A flatworm has one opening for food to enter and waste to exit. Roundworms and segmented worms have a tube that food enters at one end and waste exits at the other.

3. These cells limit what passes through and protects the body's internal environment.

4. Peristalsis, the coordinated contraction of two sheets of smooth muscle, pushes food along. It does not depend on gravity.

5. The foods have some proteins, and the bacteria in the rumen synthesize others.

1 Predict what would happen in Activity 13-1 if the iodine were on the inside of the bag and the cornstarch were on the outside.

2 Distinguish the characteristics that make the digestive plan of flatworms different from that of roundworms and of segmented worms, such as earthworms.

3 Explain why having epithelial cells lining the entire digestive tract is important.

4 Speculate why you can swallow while standing on your head.

5 If a cow eats only grasses and grains, how does it get the proteins it needs?

LESSON 13.2 DIGESTION

The digestive system is like a factory but in reverse. Instead of putting together a product, it takes a product and breaks it into small subunits. In this lesson you will discover how the food you eat is broken apart and passes through your digestive tract.

PLEASE PASS THE POTATOES

BIOLOGY IN CONTEXT

You might like potatoes mashed, baked, or served with cheese sauce. Some potatoes now have an additional ingredient: a vaccine to prevent infection by a bacterium, *Escherichia coli.* Volunteers have tested an edible vaccine against a variety of *E. coli* that causes diarrhea. The gene for a bacterial protein is inserted into the potato, causing the potato to manufacture the protein. When you eat the potato, your body's immune system creates antibodies against this protein. The potato becomes a way to deliver immunity.

Millions of children die yearly from *E. coli,* cholera, and other intestinal diseases that cause diarrhea. Cholera and *E. coli* produce similar toxins that change the way the cells lining the intestine work. The poison causes water to leave these cells rapidly and enter the intestine, where it passes out of the body in diarrhea. In less than 4 hours, people can lose 25% of their body fluid. The infected person quickly becomes dehydrated and dies. Potatoes that contain antibody-producing proteins and that are easy to grow could save lives.

ACTIVITY 13-2 COMPARING MOUTHPARTS

Materials: models of human teeth, skulls from various animals, and diagrams of herbivores, carnivores, and omnivores that demonstrate the types of teeth in the jaw.

The front teeth, or incisors, are for cutting. Horses and cows have a row of incisors. The large front teeth of rodents are incisors.

The cuspids or canines are more fanglike and are for tearing. These are just outside the front incisors in animals like dogs and cats.

Molars are in the back of the mouth.

An animal's feeding habits shape its teeth and mouth. Teeth cut, tear, and grind the food into smaller pieces. In this activity you will identify what type of teeth animals have and what their diet includes.

Compare the skulls of various animals and diagrams of their jaws and teeth. Answer the following questions in your logbook:

- Which teeth are used for cutting and where are they?

- Which teeth could grasp prey and rip flesh? Where are they?

- Which teeth are for grinding? Where are they?

- Predict which teeth belong to herbivores, carnivores, or omnivores.

- How do human teeth differ from other animal teeth?

MECHANICAL DIGESTION IN THE MOUTH

Most food is broken apart in the mouth. Birds use beaks and vertebrates use teeth to break food into smaller pieces. A bird's

A   B

Figure 13.6: *The macaw (A) uses its beak to crack seeds open. The bald eagle (B) uses its beak to tear into the flesh of its prey.*

beak matches the bird's diet; some beaks, like the eagle's, capture prey and tear its muscle (Figure 13.6B). Some beak shapes are good for cracking seeds open (Figure 13.6A).

Herbivores, such as horses and cattle, have front teeth called **incisors,** which crop the grass they eat, and flat molars, which grind it. Horse teeth protrude more as the horse ages, providing a way to estimate the age of the animal (Figure 13.7A). A rodent's incisors grow continually, and small **molars** grind their food (Figure 13.7B). Carnivores such as dogs and cats have well-developed canines (also called **cuspids**) which are used to tear and shred food, and they have sharp molars for cutting (Figure 13.7C). Omnivore teeth are like those of both herbivores and carnivores.

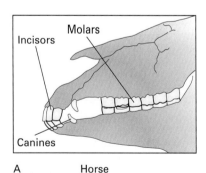
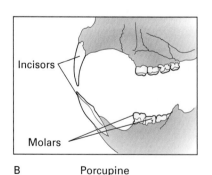
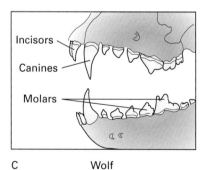

A   Horse   B   Porcupine   C   Wolf

Figure 13.7: *The shape and size of an animal's teeth matches their function. The sharp thin incisors are used to chop while the flat molars grind food in preparation for digestion.*

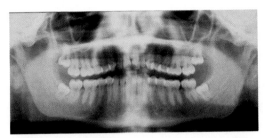

Figure 13.8: *Panoramic dental X ray of adolescent with no cavities*

Human teeth do not grow continually. Primary teeth begin to grow through the gum at 6 months of age. Beginning at age 6, the 20 primary teeth are gradually replaced by 32 permanent teeth that require continuous care to last a lifetime (Figure 13.8). Dental hygienists monitor the development of primary and permanent teeth and check the condition of each tooth during checkups. Hygienists also clean patients' teeth and apply sealant to protect them from the damage that mouth bacteria can cause.

CHEMICAL DIGESTION IN THE MOUTH AND STOMACH

Additional Activity, The Disassembly Line, can be found in Chapter 13 of the TRB.

Starting in the mouth, digestive enzymes break apart the large molecules of food (Figure 13.9). **Saliva** is a liquid which contains enzymes that initiate carbohydrate digestion. Amylase and other enzymes join in, breaking long chains of starch, a carbohydrate, into smaller sugars such as maltose. Recall that enzymes are proteins that speed up a reaction and aren't used up in the process. To understand how important saliva is, swallow three times quickly. Notice that saliva enters your mouth from glands under your tongue. Next, eat several crackers and try to swallow three more times. It's much harder, because your saliva hasn't had a chance to moisten the food. Stroke victims sometimes have

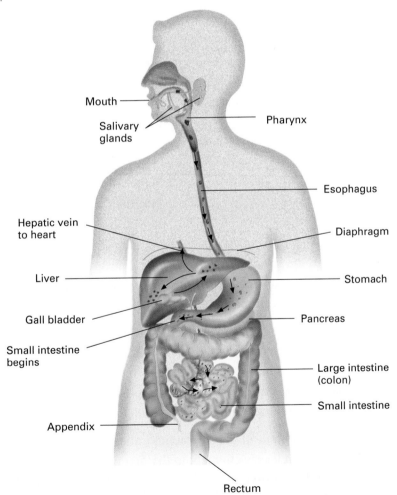

Mouth
Salivary glands
Pharynx
Esophagus
Hepatic vein to heart
Diaphragm
Liver
Stomach
Gall bladder
Pancreas
Small intestine begins
Large intestine (colon)
Small intestine
Appendix
Rectum

Figure 13.9: *Food breaks down in the mouth, stomach, and small intestine. The wall of the intestine absorbs nutrients, which travel to the liver in the bloodstream.*

difficulty swallowing. Speech pathologists work with them to develop a diet that has the right consistency and moisture, making it easier for the client to swallow and not choke.

The chewed food is swallowed. It passes through the pharynx into the esophagus. The **pharynx** is the part of the digestive tract between the mouth and the esophagus and is shared by the respiratory system. A flap of cartilage called the **epiglottis** closes the trachea so that food won't enter the lungs. Peristalsis pushes the chewed food through the esophagus to the stomach.

The stomach is a specialized section of the food-processing tube that runs through your body. It is just under the left lower ribs. It can hold about 1.2 L of food (Figure 13.10). Primarily, it digests protein. Small glands just below the inner surface of this muscular organ produce hydrochloric acid, an enzyme called pepsin, and alkaline mucus (Figure 13.11). The hydrochloric acid unfolds the twisted proteins so that pepsin can cut them into amino acids. Pepsin works only in this acidic environment. Eventually, the amino acids are absorbed in the intestines and transported to the various cells of the body. There, cells reassemble

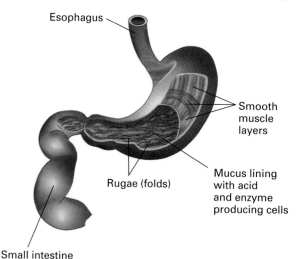

Esophagus

Smooth muscle layers

Mucus lining with acid and enzyme producing cells

Rugae (folds)

Small intestine

Figure 13.10: *Shiny mucus protects the stomach from hydrochloric acid and protein-digesting enzymes made in the gastric pits.*

them into the specific proteins that they need. The mucus coats the lining of the stomach so that the hydrochloric acid and pepsin don't digest the stomach itself.

The acid in the stomach inactivates the enzymes that began digesting carbohydrates in the mouth. It also destroys most bacteria that are eaten with the food and those in the mucus draining from the nasal area. One bacterium, called a helicobacter, can survive in the acid environment, infecting the lining and causing an ulcer.

The liquid content of the stomach is called chyme. Little by little, the stomach releases chyme into the first part of the **small intestine,** a digestive organ that lies just below the stomach.

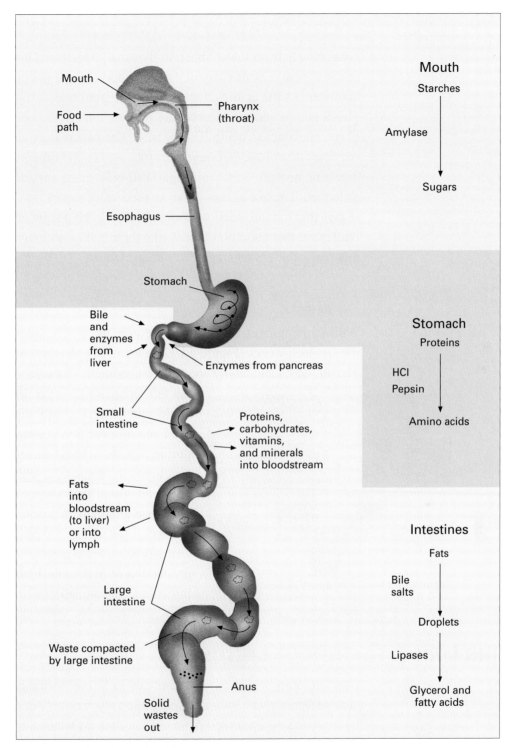

Figure 13.11: *Digestion and absorption of starch, protein, and fat*

ENZYMES IN THE SMALL INTESTINE

Additional Activity, Commercial Enzymes, can be found in Chapter 13 of the TRB.

Food travels from the stomach to the small intestine. This portion of the digestive tract is about 3 m long and about 2–3 cm in diameter. As the acidic chyme enters, the **pancreas,** a digestive gland, pumps sodium bicarbonate into the lumen (the open center of the intestine) to neutralize the acid. The pancreas produces enzymes that break carbohydrates into sugars and others that finish digesting protein. Some intestinal wall cells make enzymes, such as lactase. Lactase breaks down lactose (milk sugar) into simple sugars that the intestinal wall can absorb. Some people are lactose intolerant: this condition means that their bodies no longer make this particular enzyme and can't digest lactose.

A WINDOW ON DIGESTION

CHANGING IDEAS

William Beaumont, a military surgeon in the 1820s, studied how the stomach works (Figure 13.12). One of his patients, Alexis St. Martin, was accidentally shot in the abdomen. The injury healed

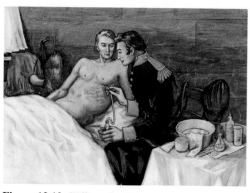

Figure 13.12: *William Beaumont examines a patient.*

but left an opening, so the doctor could sample the partially digested food and observe how it changed during digestion. He observed the development of gastritis, which is inflammation of the stomach. This condition gradually deteriorates the surface of the stomach. Today, gastritis is linked to alcoholism, over production of hydrochloric acid, infections, and sometimes stress.

LEARNING LINK LIFE WITHOUT LACTOSE

Some key words for Internet searches include lactose intolerance, dietitian, food allergies, milk.

Use the Internet and other references to further investigate lactose intolerance. What ethnic groups are most often affected by this missing enzyme? What side effects do lactose-intolerant people experience when they eat dairy foods? Some cheese and yogurts can be digested without the side effects. Why? What foods contain hidden lactose? What foods supply you with calcium when you don't eat milk products?

INVESTIGATION 13A   HOW DO DIGESTIVE ENZYMES WORK?

See Investigation 13A in the TRB.

Digestive enzymes unlock complex food molecules, creating smaller units that the intestinal wall can absorb. When and where do these enzymes work, and why don't they work in other parts of the digestive system? In this investigation you will digest proteins and starches in test tubes, mimicking the enzymes in the environment of the stomach and small intestine.

ABSORPTION IN THE INTESTINES

Now that the small intestine has neutralized the hydrochloric acid and added all the digestive juices, the nutrients are small enough to be absorbed. The small intestine is lined with fingerlike projections called **villi,** which increase the absorption area (Figure 13.13). Here is a demonstration of how to increase surface area: Make a fist and let the area across the knuckles represent an area of the intestinal surface about 3 inches long that absorbs nutrients. Open up your hand, and notice how your fingers provide much more surface area across the front of each finger, the sides, the back, and the top for nutrients to diffuse into the cells. The same is true for the villi in the small intestine. If the villi in one person could be stretched flat, they would cover about half of a tennis court.

Some nutrients, such as amino acids, diffuse through the villi cells much as the iodine passed through the plastic bag in Activity 13-1. Others, including iron and vitamin B_{12}, are actively transported into the villi cells. They pass through the cells and enter the extensive capillary network underneath the villi. Once they enter the bloodstream, their first stop is the liver on their way to the heart.

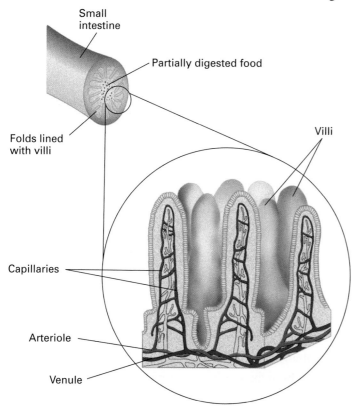

Figure 13.13: *Villi line the small intestine, increasing the surface area for food absorption. (Inset) Capillaries carry away the nutrients.*

Food that is not absorbed passes into the **large intestine** (also called colon), a digestive organ that is responsible for the removal of water from undigested material. The cells that line the colon reabsorb water and salts that are needed for electrolyte balance. When water isn't properly reabsorbed, diarrhea occurs. Bacteria that live in the colon help to break down food residue and may create gas; other bacteria make vitamin K, and some make B vitamins. These nutrients are absorbed through the colon wall. Bacteria, excess bile, and undigested food exit the body through the anus. The trip through the entire digestive tract takes about 12 hours.

Electrolytes are ions such as sodium, chloride, and potassium dissolved in water. The solution conducts electricity.

OTHER DIGESTIVE ORGANS

The pancreas is a long, flattened gland that makes digestive juices. It is sandwiched into the first curve of the small intestine, as it joins the stomach. The pancreas makes many of the enzymes that break apart carbohydrates and fats, and it produces the sodium bicarbonate that neutralizes stomach acid. (The gland also makes insulin and glucagon, which will be discussed in Chapter 14.) These digestive juices enter the small intestine through a duct.

Skin is the largest organ of the body.

The **liver** is tucked under the lower right ribs, just below the diaphragm. It is the second largest organ in the human body and has many lobes or sections.

The liver processes all the nutrient-rich blood leaving the intestine. When the villi absorb nutrients, active transport and diffusion carry them into a capillary network that surrounds the small intestine. These blood vessels go into veins that lead to the liver. The liver is like a car's oil filter (Figure 13.14). As the blood flows through tiny openings, the cells lining the openings filter out old blood cells, process proteins, and convert sugars into animal starch called **glycogen,** a highly branched polymer of glucose. When the body needs more sugar, the liver converts the glycogen to glucose, which enters the bloodstream. A histotechnician in the pathology lab of a hospital stains liver specimens to map glycogen, a process that helps to diagnose liver disease.

An important liver function is to detoxify incoming substances, such as pesticides, solvents, drugs, and alcohol. The drug rehabilitation

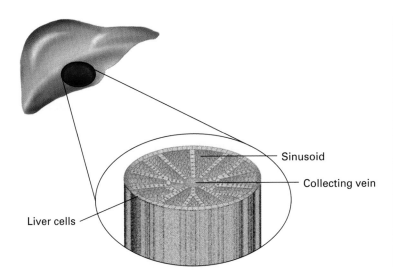

A   **B**

Figure 13.14: *A liver (A) and an oil filter (B) both remove substances from circulation.*

specialist counsels alcoholics and their families about the damage that alcohol does to the stomach lining and the liver (see Figure 13.15). With continued abuse, an alcoholic may destroy this vital filtering system and die. Pharmacists and scientists who develop new drugs must test oral drugs to be sure that they will be effective after they go through the digestive system and the liver.

Figure 13.15: *Liver damage from alcoholism*

After the liver filters nutrients, they travel in the blood to the heart for distribution throughout the body.

LEARNING LINK TEENAGE DRINKING

Some key words for Internet searches include Alcoholics Anonymous, alcoholism, binge drinking, substance abuse, and dependency.

Using library and Internet resources investigate teenage drinking. What percentage of teens drink and what percentage are alcoholics? What is binge drinking? How does the body respond to chronic drinking? How does the alcoholic's nutritional status change, and what damages the liver? Make a poster presentation about your findings.

The liver is an amazing chemical factory. It filters out old red blood cells, recycles the iron in bone marrow to make new red blood cells, and synthesizes bile. This greenish fluid is stored in the **gallbladder,** which is in a hollow area in a lobe of the liver. When you eat fatty foods, smooth muscles contract around the gallbladder, squeezing bile into a duct leading to the small intestine. Bile breaks up the fat in chyme into very small droplets, which are suspended in solution. Then the lipases can split them into glycerol and fatty acids. These molecules can pass through the intestinal wall, for transport to the circulatory system.

LESSON 13.2 ASSESSMENT

1. Meat.

2. Pepsin breaks apart proteins, lipase breaks apart fat, and amylase breaks apart starch.

3. Pepsin works only in an acid environment; the pancreas neutralizes acid when it enters the small intestine.

4. It collects waste food, synthesizes and absorbs some vitamins, and reabsorbs water and electrolytes.

1 A wildlife biologist finds a jaw with several teeth. The molars appear to be pointed, with a prominent cuspid. She would assume that this animal eats primarily what kind of food?

2 Describe the functions of the following enzymes: pepsin, lipase, amylase.

3 Explain why pepsin will not digest proteins in the small intestine.

4 Identify three important jobs of the large intestine.

LESSON 13.3 NUTRITION

If the saying "You are what you eat" is true, the decisions that you make today will influence who you become. Diet has been linked to heart disease, cancer, and longevity. In this lesson you will examine the nutrients your body needs, what foods supply them, and how to use nutritional labels to evaluate foods.

A dietitian explains how to make healthy food choices.

EATING FOR HEALTH

A hospital dietitian plans meals that meet the patients' nutritional needs. The 14-year-old diabetic needs a diet without sugary foods such as soft drinks, jelly, and cake. The burn patient needs a diet that is high in protein, similar to the diet for a pregnant woman. The man with congestive heart failure needs a low-sodium diet. The dietitian plans a daily diet that supplies the six nutrient groups and considers the other nutritional needs of each patient.

Because you choose most of the foods you eat, you are your own dietitian. The National Research Council of the National Academy of Science has developed guidelines for essential nutrients and the **food pyramid,** a diagram that is used as a guide for the development of a nutritional diet (Figure 13.16).

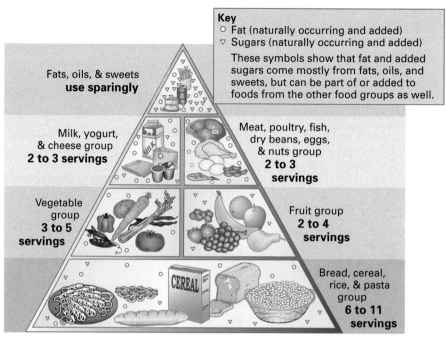

Figure 13.16: *The food pyramid*

Key
○ Fat (naturally occurring and added)
▽ Sugars (naturally occurring and added)
These symbols show that fat and added sugars come mostly from fats, oils, and sweets, but can be part of or added to foods from the other food groups as well.

Fats, oils, & sweets
use sparingly

Milk, yogurt, & cheese group
2 to 3 servings

Meat, poultry, fish, dry beans, eggs, & nuts group
2 to 3 servings

Vegetable group
3 to 5 servings

Fruit group
2 to 4 servings

Bread, cereal, rice, & pasta group
6 to 11 servings

NUTRITIONAL LABELING

Figure 13.17: *A nutritional label*

NUTRITION INFORMATION PER SERVING

SERVING SIZE . . . . ONE CUP
SERVINGS PER PACKAGE 4
CALORIES .200
PROTEIN . . . . 10 GRAMS
CARBOHYDRATE . . . . .34 GRAMS
FAT . . . . .3 GRAMS
SODIUM . . . . 140 mg

PERCENTAGE OF U.S. RECOMMENDED DAILY ALLOWANCE (U.S. RDA)

PROTEIN 20   NIACIN . . . .*
VITAMIN A 2   CALCIUM 40
VITAMIN C . . . . . . *   IRON . . . .*
THIAMINE 8   VITAMIN B12 20
RIBOFLAVIN. . . . 35   PHOSPHORUS. . .30

*CONTAINS LESS THAN 2% OF
U.S. RDA OF THESE NUTRIENTS.   Ⓚ

NO ARTIFICIAL ANYTHING™

Food labels are based on a 2000-Calorie diet. The manufacturer may include other information such as the amount of sodium or cholesterol.

You can check the nutritional value of various foods by reading the label. The label gives the serving size and the number of servings in the package (Figure 13.17). It lists the Calories and the grams of carbohydrates, fats, and proteins that the food supplies. (A calorie with a small "c" is the heat energy needed to raise 1 g of water 1°C. Energy in food is measured in Calories; 1 Calorie = 1000 calories.) The label gives information about vitamins A and C and the minerals calcium and iron. A lot of Americans don't get enough of these vitamins and minerals in their daily diet. Some nutrients are measured in grams, some in milligrams, and others in international units. The National Research Council determines how much of each substance the average person needs and calls that a daily value (DV), equal to 100%. At the end of the day, add up the DV percentage for every food you ate, and the totals should equal about 100% of each nutrient.

Nutritional labels provide important information on critical nutrients, but they don't tell the whole story. Foods contain six major types of nutrients: carbohydrates, fats, proteins, vitamins, minerals, and water. Not all mammals need the same combination of nutrients. Pork producers may supplement the diets of pigs with

B vitamins. Most other livestock get their B vitamins from the microorganisms in their intestines. Animal breeders who raise guinea pigs include food with vitamin C; dog foods don't need vitamin C because dogs produce this nutrient. Dogs have enzymes that effectively break down proteins from meat but not vegetable proteins. Many dog foods include protein from plant sources, which is useless because a dog doesn't have the enzymes to digest it.

ACTIVITY 13-3   WHAT DID YOU EAT YESTERDAY?

Collect empty food containers for reference, including a small milk carton and a soda can. Students will need $\frac{1}{4}$-, $\frac{1}{3}$-, $\frac{1}{2}$- and 1-cup measures, a juice glass, and a tumbler. Food models from the National Dairy Council will help students to visualize the serving sizes.

One of the first things a dietitian does is to analyze a client's existing diet. Besides foods consumed and the nutrients supplied, where the client eats is also important. In this activity you'll analyze your own diet.

In your logbook, write down every food that you can remember eating yesterday, including the amount. Use the food pyramid (Figure 13.16) to identify how many servings you had from each food group. Record the results in your logbook and answer the following questions:

- Report whether you had enough, too little, or too much from each food group.

- Circle the foods that are high in fats. Underline the foods high in sugars. Did you eat these foods sparingly?

- Where did you eat: at home, at school, in the car? Did you eat alone or with others? Does this influence what you ate and how much?

- List the foods that you could have added or eliminated to better match the pyramid recommendations.

Many teenagers do not eat enough fruits and vegetables. They often eat too much protein, fat, and sugar. Eating alone or in the car is common for teens.

CARBOHYDRATES

WORD BANK

mono- = one

di- = two

sacchar(o)- = sugar

A monosaccharide is one simple sugar; disaccharides combine two simple sugars.

Carbohydrates are mainly plant sugars and starches. As you learned in Chapter 1, glucose, fructose, and galactose are simple sugars, or monosaccharides. Each contains the same number of carbon, hydrogen, and oxygen atoms, but the atoms are arranged differently. Two monosaccharides can join to make a disaccharide, such as maltose, sucrose, and lactose. Lactose comes from an animal, not from plants. Fruits, desserts, soft drinks, and candy all contain sugar.

Sugars can be joined in long chains called starches or complex carbohydrates. Plants make starches to store energy. We humans, and a lot of other animals, take advantage of these high-energy stores. The enzyme amylase in saliva begins the digestion of starch, breaking it into sugar units. You can chew a cracker for several minutes and detect that it tastes sweeter; this taste is the work of the enzyme. Once the sugars circulate, cell respiration metabolizes them for energy.

Athletic trainers plan "carboloading" (carbohydrate-loading) diets for long-distance runners and other endurance athletes. Carboloading maximizes the storage of human starch, glycogen, in the muscles, which need energy to run the race. A carboloading diet may include potatoes, peas, pasta, breads, and cereals. Carbohydrates are the body's main source of energy, and the food pyramid recommends 6–11 servings daily. How many servings an individual should have depends on how much she or he is growing, how many calories are used in daily activities, and any special conditions, such as pregnancy, that require additional calories.

FOOD TECHNOLOGIST

CAREER PROFILE

A national restaurant chain makes a famous baked potato soup. Customers expect the soup to taste the same at all the restaurants.

Keiko K.'s job is to develop a dried version of the soup that ensures high quality and uniform taste so that each restaurant lives up to the chain's reputation. She determines the volatile oils that smell so good, the spices and seasonings, and the starch thickeners that will create the perfect soup when mixed with liquids. Sensory panels taste and evaluate the trial samples, helping to refine the mix of ingredients. Using microbiology techniques, Keiko checks the shelf life of the powdered product. Once the recipe is set, the ingredients are weighed and mixed in a blender that handles 1000 pounds of ingredients. Keiko got much of her expertise on the job, preparing food. Her employer has provided additional training. Many food technologists have degrees in food science.

Sensory panel testing food product

FAT

Fat is a very concentrated source of energy. One gram of carbohydrate supplies 4 Calories, while 1 g of fat supplies 9 Calories. Some fats contain fat-soluble vitamins, such as the vitamin A in milk fat. Fats add flavor to food and stay in your stomach a long time, making you feel full.

Cheese, butter, sour cream, and bacon add excessive fat to our diets.

Some stages of growth and development require high fat intake. Pediatricians monitor an infant's diet to ensure that it gets enough fat from formula or breast milk and whole milk for proper brain development. Also, high-fat/low-carbohydrate diets reduce some types of epileptic seizures. However, for many people, fatty products such as chips and sour cream dip taste good but provide too many Calories. The excess Calories are stored as body fat. Look at the food pyramid (Figure 13.16). The small circles indicate that not only salad dressings, butter, sour cream, and oils (at the top of the pyramid) contain fat, but also milk, meat, poultry, eggs, and nuts. Dietitians recommend that people eat these fatty foods sparingly, so many food producers have developed reduced-fat products and leaner animals.

VITAMINS

Vitamins are organic substances that an organism can't make. They are often key components of enzymes. Table 13.1 identifies the major vitamins that humans need. Vitamins A, C, and D are often lacking in the American diet.

Milk is a good source of vitamin A. Beta carotene, a plant pigment that humans convert into vitamin A, is found in dark green leafy vegetables and many yellow and red fruits and vegetables. Vitamin A is essential for good vision. People who don't get enough vitamin A for a long time may develop night blindness: Their eyes don't adjust well to dim light. Vitamin A also contributes to healthy skin. Dermatologists prescribe a form of vitamin A to reduce acne and improve the health of skin.

Vitamin C is a key component of collagen, a connective tissue that wraps muscle cells, surrounds organs, and is found in skin and bones. Sailors during the Age of Discovery and the 49ers of the

Vitamin	Functions	Common Sources
Fat Soluble		
A (Retinol)	Maintains eye and skin health; helps the body resist disease; aids absorption of calcium and phosphorus	Deep yellow and green leafy vegetables, liver, fortified milk
D	Aids absorption of calcium and phosphorus; promotes normal growth through carbohydrate use	Salmon, egg yolk, sun-cured hays; a cholesterol found in the skin changes to vitamin D when exposed to sunlight
E	Prevents cell damage from oxidation; helps keep red blood cells and muscle cells healthy	Vegetables, vegetable oils, cereal grains
K	Needed for blood clotting	Green leafy vegetables, potatoes, tomatoes, liver; produced by bacteria in human intestine
Water Soluble		
B_1 (Thiamine)	Used in carbohydrate metabolism; promotes normal appetite	Pork, liver, legumes, nuts, cereals
B_2 (Riboflavin)	Used in release of energy from carbohydrate, fat, and protein	Milk, liver, eggs, meat, green leafy vegetables, legumes
B_3 (Niacin)	Used for carbohydrate, protein and fat use; maintains body tissues	Lean meat, fish, poultry, liver, cereal, pasta
B_6 (Pyridoxine)	Used in protein and fatty acid metabolism; needed for red blood cell formation	Meat, liver, fish, bananas, green leafy vegetables, whole grains
B_{12} (Cyanocobalamin)	Used in red blood cell formation and protein use. Needed to absorb folacin.	Meat, liver, eggs, milk, cheese, fermented products
Folacin	Related to energy use, red blood formation, and production of genetic material	Liver, legumes, green vegetables, nuts, whole grains
Biotin	Used in energy release from carbohydrates; fatty acid synthesis	Meat, egg yolk, green beans
Pantothenic acid	Essential for enzyme production involved in fatty acid, cholesterol, and amino acid synthesis	Liver, eggs, whole grains, nuts, green vegetables
Choline	Essential for enzymes used in fat metabolism; part of cell structure	Meat, vegetables, whole-grain products, grains
C (Ascorbic Acid)	Strengthens body cell structure and resistance to disease; promotes wound healing; increases iron in intestines absorption	Citrus fruits and juices, tomatoes, strawberries, watermelon, cantaloupe, potatoes, cabbage, broccoli

Table 13.1: Functions and Sources of Vitamins

Gold Rush didn't get enough vitamin C. Gradually, their teeth would loosen and fall out, sores on their skin wouldn't heal, and eventually the people would die. This extreme deficiency is called scurvy. Potatoes, broccoli, strawberries, and citrus fruits are all good sources of vitamin C. Vitamin C is in the watery pulp of fruits and vegetables. It is easily oxidized, so the more the food is exposed to air and heat, the more vitamin C deteriorates. How you prepare foods containing the vitamin influences how much of it gets to the body.

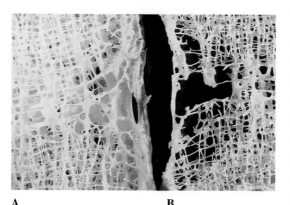

A   B

Figure 13.18: *Healthy bone (A) and bone damaged by osteoporosis (B)*

Vitamin D is crucial for absorption of calcium and phosphorus through the intestinal wall. It is added to milk so that vitamin D and the minerals are together in the digestive tract. Older women and others who have osteoporosis may take vitamin D and calcium supplements to maintain bone strength (Figure 13.18). Skin that is exposed to sunlight produces vitamin D.

The B vitamins mainly regulate the energy that is released from carbohydrates, protein, and fat. Meat, eggs, and grain products, such as breads and noodles, contain B vitamins.

MINERALS

Minerals are inorganic substances; they don't contain carbon. Plants provide most of the minerals that humans need. Table 13.2 identifies many essential minerals. Iron and calcium are lacking in many diets. Iron, supplied by red meat, dried beans, and many leafy green vegetables, is part of the hemoglobin molecule that carries oxygen in red blood cells. People who do not have enough iron are anemic and feel tired much of the time. Milk and some vegetables, such as broccoli, supply calcium. Calcium is important in bone formation, muscle contraction, and blood clotting. Sodium, potassium, and chloride are all electrolytes that help to maintain the water balance of the body. After a hard workout, sports drinks may help to replace quickly electrolytes that are lost in sweating.

Minerals	Function	Sources
Calcium (Ca)	Bone and tooth formation; important in muscle contraction and blood clotting	Milk and milk products, green leafy vegetables, legumes
Chlorine (Cl)	Combines with sodium to regulate osmotic pressure and acid-base balance	Table salt, cured meats, cheese, pickled products
Cobalt (Co)	Component of vitamin B; involved in synthesis of several enzymes	Meat, liver, poultry, fish, milk
Copper (Cu)	Needed for formation of bone, nerve tissue, and hemoglobin	Liver, oysters, crab, nuts, whole-grain products, legumes
Iodine (I)	Component of thyroid hormones, regulates rate of body reactions	Iodized salt
Iron (Fe)	Part of the red blood cells and many enzymes; needed for oxygen transport in red blood cells	Red meat, liver, molasses, green leafy vegetables, legumes
Magnesium (Mg)	Enzyme activator in metabolic reactions, especially carbodydrate use, muscle relaxation	Milk and milk products, nuts, green leafy vegetables, bananas, oranges
Manganese (Mn)	Needed to make enzymes for use of fatty acids	Whole-grain products, nuts, green leafy vegetables
Phosphorus (P)	Combines with calcium to give bones and teeth strength and hardness	Meat, poultry, fish, milk and milk products, whole grains
Potassium (K)	Used in regulation of acid-base balance and osmotic pressure	Meats, potatoes, nuts, avocados, dried apricots
Sodium (Na)	Regulates acid-base and water balance; controls nerve and muscle activity	Table salt, cured meats, cheese, pickled products
Sulfur (S)	Component of some amino acids, vitamins, and insulin	Meat, milk and milk products, eggs, legumes
Zinc (Zn)	Part of several enzymes involved in digestion, respiration, energy and bone metabolism; needed for wound healing, genetic material, sense of taste	Seafood, meat, vegetables, whole grains, legumes

Table 13.2: Functions and Sources of Minerals

Plants supply many essential vitamins and minerals. The food pyramid recommends 3–5 servings of vegetables and 2–4 servings of fruit daily.

PROTEINS

Mixing certain plant sources such as rice and beans that are rich sources of different amino acids can supply many of the essential amino acids.

Proteins are found in every living tissue; they form part of cell membranes, genetic material, hormones, and enzymes, among other things. Proteins are long chains of amino acids. Out of the 20 amino acids, eight are not made in the human body. These must be supplied by food. Foods that contain the eight essential amino acids in the proportion in which they are used are complete proteins and most often come from animal sources. Many cultures in the world are vegetarian. Some do not eat any animal products and receive all the complete proteins they need from plant sources.

Protein is used for energy when fat and carbohydrates aren't present. The American diet is high in protein and fat, but children in some developing countries don't get enough protein. They lose weight, are listless, and catch more infections because of a condition called kwashiorkor (Figure 13.19A). The child consumes a chronically low-calorie diet, using the little protein that is eaten as an energy source, not for growth and development. A similar condition is marasmus. The symptoms of marasmus are inadequate growth, weakness, edema (swelling), and loss of appetite (Figure 13.19B). Both kwashiorkor and marasmus can be fatal. The food pyramid recommends 2–3 servings of meat, dried beans, or eggs and 2–3 servings of dairy products daily. Both of these categories supply complete protein.

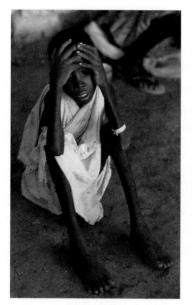
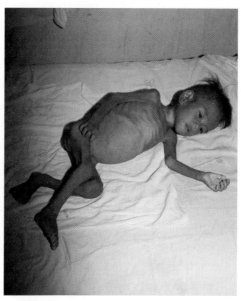

A **B**

Figure 13.19: *Examples of people suffering from protein deficiencies: (A) kwashiorkor and (B) marasmus*

WATER

Last but not least, water is an essential nutrient. About two thirds of the body's weight is water. Water is in every cell. The circulation system depends on watery plasma to carry nutrients, gases, and other molecules throughout the body. Water is not only in what you drink but also in many foods. Watermelon, tomatoes, and soups contain lots of water.

INVESTIGATION 13B NUTRIENT IDENTIFICATION

See Investigation 13B in the TRB.

Who figures out the nutritional information that goes on a food label? Biotechnology firms use sophisticated equipment, such as gas chromatography, to run food assays. They determine which nutrients are in a food and in what amounts. In this investigation you will use simpler procedures to test for protein, complex carbohydrates, glucose, and vitamin C in common foods.

CALORIE CONSIDERATIONS

See TRB for diet analysis activity suggestion.

When you plan a healthy diet, you consider calories as well as the six nutrient groups. Different stages of growth require different amounts of calories. Puppies get different feed from adult dogs. Babies have special formulas and foods that meet their nutritional needs. A pregnant woman nearly doubles her percentages of nutrients while eating only 600 more Calories each day. Women and teenage girls generally need fewer Calories (2000–3000) than do men and teenage boys (2500–3500) because of different muscle-to-fat ratios and body sizes. Calorie requirements also vary from person to person because individuals' activities vary. A water polo player may eat 6000 Calories a day because of demanding practices and games. The food service for the Olympics provides balanced diets and extra calories for the athletes.

FIBER

Plants supply the human diet with carbohydrates, minerals, and vitamins. They also supply another necessary substance: fiber. Fiber is indigestible matter, such as the strings in celery. Peristalsis pushes this waste through the small and large intestines. Why is this important? Cancer researchers link increased fiber consumption with decreased colon cancer. Some fats and other foods break down quickly, creating a burst of energy from oxidation. This burst of energy can damage surrounding intestinal cells, and repeated bursts may cause cancerous changes in the epithelial cells. Fiber moves the chyme along so that exposure to the oxidizing molecules is shorter. Antioxidant vitamins, such as A and C, also help to absorb the energy burst.

The American Cancer Society and the American Heart Association recommend diets with not more than one third of the calories from fats and plenty of fruits and vegetables to increase fiber. Food technologists have developed many high-fiber food products in response to these health concerns, including high-fiber cereals.

LESSON 13.3 ASSESSMENT

1. Carbohydrate, protein, fat, vitamins, minerals, and water.

2. Labels provide easy-to-use information about key nutrients. Daily values should equal 100% for each nutrient at the end of the day.

3. Answers should include dark green leafy vegetables (spinach, kale) and yellow and red fruits and vegetables (e.g., carrots, tomatoes, sweet potatoes, apricots, peaches, cantaloupes). These foods would supply vitamin A.

4. Women need more calories when their activities burn more calories than the men's activities do. For example, a woman who runs track would need more calories than would a man who does little exercise. Pregnancy or nursing would also increase the caloric requirement.

1 Name the six nutrient types that humans need for health.

2 Discuss how nutritional labels help the consumer.

3 Make a list of fruits and vegetables that supply at least 10% of the daily value for vitamin A. Explain why this list is important for people who don't get vitamin A from milk and milk products.

4 Women generally need fewer calories than men. When would women need more calories than men?

CAREER APPLICATIONS

The applications that follow are like the ones you will encounter in many workplaces. Use the biology you learned in this chapter to complete the activities. Share your work with the class.

AGRICULTURE & AGRIBUSINESS

Nutritional Deficiencies

Gather information about animals that have deficiency diseases. You can get this information from agriculture texts, county agricultural extension agents, and/or veterinarians who treat livestock. What problems do specific deficiencies cause? Make a display of photographs or illustrations that show signs of the most common nutritional deficiencies in one or more species of domestic animals. Make your display into a quick reference chart for farmers, ranchers, or pet owners. Include suggestions about how to correct the deficiencies.

Animal Rations

Devise rations for an animal at different stages of life: newborn, juvenile, adult, pregnant or lactating female, old age. Choose a pet, farm animal, or zoo animal and find information on its dietary requirements. Consult with an agricultural extension agent, veterinarian, local producer, or zoo keeper, or use the Internet, to develop balanced rations for growth and maintenance. Base the rations on the energy and protein requirements of the animal at each stage of its life.

BUSINESS & MARKETING

Labeling Requirements

Research nutritional labeling requirements. Describe the requirements and answer the following questions: Do the requirements create special problems for the food industry? How does the food industry handle these problems? Collect food labels that have minimal information and some that include additional information. When would it benefit a company to include additional information? Write up your findings in your logbook.

Selling Lower Cholesterol

Choose a meat industry (e.g., beef, pork, lamb, chicken, turkey, fish, or seafood) and design and deliver a sales presentation on how one or more of the industry's products can contribute to a diet with low saturated fat and low cholesterol. Define the FDA's label use of the words "low," "lean," "reduced," "light," and "free." Use as many of the following as you like in your presentation: pictures of live animals, pictures and names (labels) of low-fat meat cuts, appropriate menus that include the animal product, recipe handouts, food samples, and brochures emphasizing the low saturated fat and low cholesterol in the animal product.

FAMILY & CONSUMER SCIENCE

Digestive Aids

Read labels (or other product information) on antacids, fiber supplements, supplemental enzymes, and other digestive aids. Share your findings with the class in a table of digestive products, comparing main (active) ingredient, how the product is supposed to work, advertising claims, warnings of dangers or side effects, and the type of person or condition each product might benefit.

Hold onto Vitamins

Water-soluble vitamins are easily destroyed during some types of cooking. More vitamins disappear when the cooking water is discarded. Design a brochure or a video that demonstrates food preparation that helps to retain vitamins in fruits and vegetables. Be sure to include information in your brochure or video about vitamins, their sources, and why they are needed in our diet.

Modeling Digestion

Allow students to work in small groups. You might want to have some groups modify this assignment to talk about digestive system disorders. For example, one group of students could use their model to explain heartburn and how different treatments work, while another group might show diverticulitis, ulcers, or diabetes.

Create a visual display depicting what happens to a cheeseburger with lettuce and tomatoes as it passes through your digestive system. Be as creative as possible. Show where the food experiences physical, and/or chemical, breakdown and where the different nutrients are absorbed into the body. Be sure to label the parts of the digestive system. Create a lesson on digestion that uses your display, and use your model to teach the lesson to your class.

HEALTH CAREERS

Put Teeth in Your Diet

Talk to a dental hygienist or dentist about the connection between poor dietary habits and dental problems such as tooth decay. Ask how spinach, chocolate, sugar, vinegar, carbonic acid, hard candy, sticky candy, salt, and fluoridated water affect teeth. Ask the expert to discuss the causes, prevention, and treatment of tooth decay and other tooth and gum problems, including letting babies fall asleep with a bottle. Try to get brochures or other literature to show the progression of tooth decay and treatment, from filling the first cavity through caps and root canals to false teeth or implants.

Diets for Special Needs

Hospital dietitians must plan diets for patients who have special needs. Select a special need and research suitable foods and recipes: pregnancy, lactation, infancy, old age, liquids only, soft foods, bland foods, kosher, vegetarian, high-calorie, burn patient, low-calorie, diabetes, fat control, high-protein, sodium restrictions, allergy, ketogenic, tube feeding, and modified residues. Use the Internet, cookbooks published by organizations for people with special needs, interviews with a nutritionist or dietitian, or other means. Prepare a day's menu for your special needs patient. Summarize the special needs of your patient, the number of calories, the RDA percentage for each nutrient, and how your diet takes care of the special needs. Prepare a class booklet of menus and recipes.

Calcium and Osteoporosis

Create a pamphlet or computer presentation on osteoporosis. Research foods that are naturally rich in calcium and include those in your presentation. Because many people don't get enough calcium in their diets, they take supplements. Include information on the various forms of calcium that are included in supplements and which forms are most easily used by the body. Include other information on ways to prevent osteoporosis.

INDUSTRIAL TECHNOLOGY

Rennin and Casein

Use encyclopedias, cookbooks, the Internet, or other sources to research rennin and casein. Where does rennin come from? How is rennin used to separate casein from milk? How is casein used in food and in manufacturing? Create a poster that shows how to separate casein from milk and casein's use in a variety of products.

Food Processing

Select a type of processed food (e.g., pasta, potato chips, cereal, pickles, salad dressings, or canned meats). Consult a home economics extension agent, books on food processing, the Internet, or other sources for information about the processing method, the steps involved, and the equipment that is required. Share the information with the class. As a class, try to determine what kind of background knowledge and experience would be ideal for a food production supervisor. How much would the supervisor need to know about the electrical, mechanical, and electronic equipment that is used in production? About food safety and spoilage, growth of microorganisms, and sterile technique? About the nutrients in the food and how to preserve them? What courses should the future food production supervisor take in school? If possible, check with a food production supervisor to see how close you came to the knowledge and experience that companies look for when hiring these supervisors.

CHAPTER 13 SUMMARY

- Some small animals such as flatworms absorb nutrients directly through the cell membrane. Larger animals such as roundworms, segmented worms, and vertebrates have digestive systems that are a tube within a tubular body.

- Food and nutrients move through the vertebrate digestive tract because of peristalsis.

- Teeth and other mouth parts mechanically break apart food. Incisors mainly cut food, and cuspids tear. Herbivores have flat molars for grinding; carnivores have pointed molars for cutting.

- The saliva in the mouth begins the chemical digestion, adding enzymes that start to break down carbohydrates.

- Digestive juices containing enzymes in the stomach and small intestine break food down so that the villi lining the small intestine can absorb nutrients.

- The liver, a digestive gland, produces bile, which is stored in the gallbladder. Bile breaks apart fat globules so that lipase, an enzyme, can break them down further for absorption. The liver also stores glycogen, a starch. Glycogen is readily converted to glucose when the body needs it.

- The pancreas is a gland that produces enzymes that digest carbohydrate and fat and sodium bicarbonate that neutralizes stomach acid in the small intestine.

- The large intestine absorbs water and some vitamins before passing waste food, bacteria, and bile out of the body.

- Animals need a variety of foods that supply the six nutrient groups: carbohydrates, fats, proteins, minerals, vitamins, and water.

- The food pyramid assists in selecting foods that supply needed nutrients and encourage good health.

- The bacteria in the rumen of a cow ferment hay and grasses. Their by-products provide energy and nutrients for the ruminant animal's growth and development.

CHAPTER 13 ASSESSMENT

Concept Review

See TRB, Chapter 13 for answers.

1 Explain the differences between the digestive system of a planaria and that of a bird.

2 What are the building blocks of the nutrients that help muscles to contract and increases disease resistance?

3 Suppose you eat a leafy vegetable salad with large amounts of creamy dressing; a sandwich made of peanut butter, jelly, and white bread; a bag of potato chips; a soft drink; and hot apple pie with ice cream. Why isn't this meal good for you?

4 Thinking back to Activity 13-1, why wouldn't a person want to drink saltwater if he or she were stranded in a lifeboat?

5 How do the nutrients get from the center of the small intestine into the capillaries that surround the intestine?

6 What is the benefit of animals converting excess protein, fat, and carbohydrates into body fat?

7 Life processes require highly specific enzymes that function as catalysts; therefore our cells must produce thousands of different enzymes. What three nutrients provide the necessary building blocks for producing enzymes?

8 Discuss the nutritional value of the following daily intake of a teenager: one cup of yogurt, two cups of milk, three ounces of cheese, one-half cup of ice cream, two tablespoons of peanut butter, one egg, three ounces of fish, one cup of juice, one apple, two cups of vegetables, two slices of bread, two ounces of dried cereal, and one hamburger bun.

9 What are the building blocks for starches and for fats?

10 What two important digestive functions does saliva have?

11 Why is the production of mucus important in the stomach?

12 What information does a nutritional food label provide?

13 How is kwashiorkor like marasmus?

14 What happens in each of the four parts of a ruminant's stomach?

Questions 15–18 refer to Figure 13.20 on page 496.

15 Give the names and functions of the organs that are labeled A–G.

16 What nutrient(s) is (are) broken down in organ C?

17 Which organ secretes a chemical that breaks fats into droplets?

18 Which organ reabsorbs water into the bloodstream?

Think and Discuss

19 Your doctor has put you on a long treatment of antibiotics. How might that affect your nutrition?

20 Your small intestine does not have villi. How does this affect the absorption of nutrients?

21 All the nutrient-rich blood leaves the small intestine and travels to the liver. What might happen if the nutrients entered the bloodstream for immediate distribution to the body's cells?

22 Molasses and other sweeteners are frequently added to livestock feed. Why can't the sugary waste from cheese making be fed to adult cows?

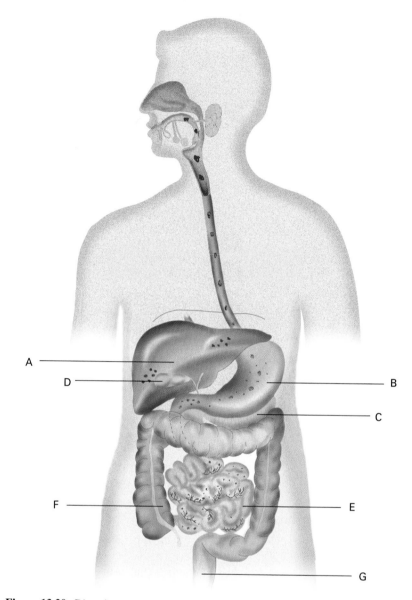

A

D

B

C

F

E

G

Figure 13.20: *Digestive system*

23 What problems must drug researchers overcome to develop a new drug that can be taken orally?

24 When making cheese using rennin, should the milk be cold, warm, or hot? Give a reason for your answer.

25 A dietitian has been asked to evaluate what a client ate one day.

Breakfast: apple juice, two pieces of toast with butter, coffee

Lunch: ham sandwich, apple, carrot sticks, glass of milk

Snack: chocolate milk, slice of cake

Dinner: roast beef, potato, sour cream, broccoli, roll, apple pie with ice cream

How does this diet compare to the food pyramid? Are there any concerns?

26 Imagine that a biotechnology company is designing a machine that will temporarily replace the liver as a dialysis machine replaces the kidney. What functions would it need to perform?

Investigations

The following questions refer to Investigation 13A:

27 What environment is best for stomach enzymes? For small-intestine enzymes?

28 At which temperature did the digestive enzymes work best? Why didn't they work as well at the other temperatures?

The following questions refer to Investigation 13B:

29 When you add orange juice to indophenol, the solution immediately turns clear. It takes a lot more potato water to achieve the same color change. Why?

30 Is fat more commonly found in meat or vegetables? Why isn't starch found in meat?

31 Using the same procedures as in Investigation 13B, you test apples. They test low for glucose. Does that mean that apples don't contain sugar?

CHAPTER 14

WHY SHOULD I LEARN THIS?

Picture yourself washing equipment after a successful lab experiment. Have you ever wondered how your hand can jerk back from scalding water before you have a chance to think about it? Do you know where that painful feeling comes from? What about those good feelings you get when a job is done? What parts of your brain are responsible for those? The answers lie within your nervous system. Your nervous system is responsible for your senses, your thoughts, and all of your actions. And you're using it right now as you think about the answers to these questions.

INTERACTION AND CONTROL

WHAT WILL I LEARN?

1. how a nerve cell functions
2. differences among animal nervous systems
3. how our senses function
4. the glands and hormones of the endocrine system
5. how the endocrine system interacts with the nervous system

---

Legend tells of a Chinese warrior who suffered great pain when shot by an arrow. His pain was relieved when a second arrow found its mark in a special spot that is known to acupuncturists. Although this story is just a legend, acupuncture is now widely used to treat pain and illness. It appears to work by stimulating certain cells of the nervous system, a procedure that interrupts messages about pain sent by other cells. No one knows exactly how acupuncture accomplishes this. But scientists do know how

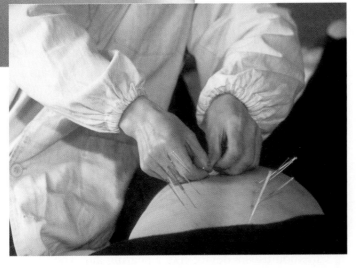

cells of the nervous system communicate with one another. This communication involves the sending of electrical and chemical messages through a network of cells that includes the special network called the brain. In this chapter you will learn how cells of the nervous system send and receive messages and how they work with another communication system, the endocrine system, to maintain homeostasis.

LESSON 14.1 THE STRUCTURE AND FUNCTION OF A NERVE CELL

WORD BANK

neuro- = pertaining to the nervous system

A neural disorder is a problem in the nervous system.

Pain is one of the most obvious indications that some type of communication is taking place between cells in the body. Your response to a sudden change in your environment, as when you jump at the sound of a backfiring car, is another indication. The speed of your response tells you that cells far apart in the body are communicating by some very rapid method. This communication involves highly specialized cells called **neurons.** Neurons, the basic units of the nervous system, communicate messages very rapidly because they can conduct an electrical current.

ACTIVITY 14-1   REACTION TIME

The procedure for this activity can be found in the TRB.

The success of any task to some degree depends on a person's ability to react quickly to a change in the environment. Reaction time, such as the time it takes to apply a car's brakes when an obstacle comes into view, is routinely tested by product safety experts. In this activity you will carry out a simple version of such a test.

TINNITUS
BIOLOGY IN CONTEXT

Imagine how unpleasant it would be to hear a faint siren all day long. This is what life is like for someone afflicted with a condition known as **tinnitus,** or ringing in one or both ears. The ringing is not a true sound but usually results from damage to the inner part of the ear. The problem can last for hours, days, or a lifetime. Not only is tinnitus annoying, it can also interfere with normal hearing. It becomes difficult to hear clearly in places where there is a lot of background noise, such as crowded restaurants.

It is important to have a complete medical evaluation before any treatment for tinnitus is given.

Unfortunately, it is difficult to treat tinnitus. One method uses a device like a hearing aid. The patient puts the device in the afflicted ear and hears soothing sounds. The device drowns out, or masks, the ringing sounds. Sometimes, a low-level background noise, such as the static from a radio, helps people with tinnitus to ignore the ringing in their ears.

NEURONS, SYNAPSES, AND POTENTIALS

When you reach out to catch a firefly in your hand, the gesture is the result of a series of messages passed along neurons in your nervous system. To understand how neurons convey messages from one part of the body to another, it is important to look at the structure of a neuron

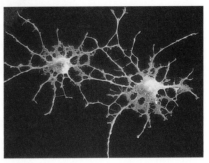

Figure 14.1: *A pair of neurons with several synapses*

(Figure 14.1). A neuron consists of a cell body, which contains the cell nucleus, and several threadlike extensions of cytoplasm. The longest of these threads of cytoplasm is called an **axon.** An axon is the part of the neuron that conducts an electrical current along its length. At the other end of the cell from the axon are many smaller branching threads, each called a **dendrite.** Axons and dendrites communicate with cells adjacent to them (Figure 14.2). This communication takes place across a narrow gap called a **synapse.** Electrical current does not flow across a synapse. Instead, substances called **neurotransmitters** relay messages across the synapse.

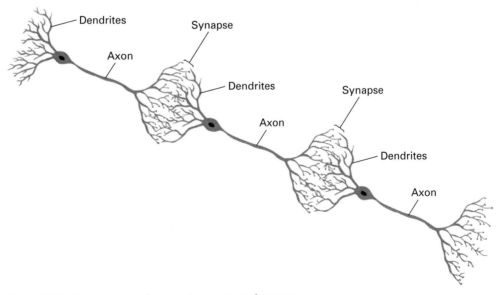

Figure 14.2: *Communication between interconnected neurons*

Figure 14.3: *There is a voltage across the two poles of a battery. A neuron also maintains a voltage across the two poles, similar to a battery.*

You know that a battery stores energy in the form of electrical voltage. This voltage is due to a difference in the charges on the two poles of the battery. One pole is positive, and the other is negative (Figure 14.3). You might be surprised to learn that a neuron also maintains a voltage. The voltage, though many times smaller than that of a battery, is due to differences in charge on either side of the neuron's cell membrane. The cytoplasm of a neuron contains an excess of negative ions; the fluid outside the neuron contains an excess of positive ions, mostly sodium ions (Figure 14.4). The voltage that is measured across the neuron's membrane is therefore said to be negative. This negative voltage, called the neuron's **resting potential,** is present as long as there is no stimulus acting on the neuron. It is maintained by pumps in the membrane that actively transport sodium ions out of the neuron and potassium ions into it.

When a neuron receives a stimulus from another cell or from the environment, changes take place in its membrane (Figure 14.5). The stimulus may make the membrane more permeable to sodium ions outside the cell. Sodium ions then enter the cell at the point of stimulus, disturbing the resting potential. Adding sodium ions to the neuron makes its cytoplasm more positive than the outside

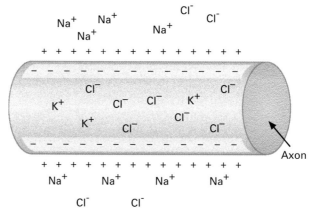

Figure 14.4: *Distribution of ions on either side of a neuron's cell membrane*

fluid, reversing the voltage. A shift in membrane potential from negative to positive voltage is called an **action potential.** The action potential at one spot on a neuron's membrane is very brief. The membrane quickly becomes impermeable to sodium and more permeable to potassium. As potassium diffuses out of the cell, the cytoplasm of the neuron again becomes negative. The resting potential is restored at that spot on the neuron's membrane.

If an action potential took place at only one spot on a neuron's membrane, it would have no importance. Fortunately, an action potential at one spot on a membrane triggers action potentials all along the membrane (Figure 14.5). In this way the neuron conducts an electrical current from the point of stimulus to a synapse.

Axons are wrapped in an insulated sheath called myelin. Myelin acts like the coating on a copper wire. It insulates the axons in a bundle from one another. At intervals along an axon there are breaks in the myelin. An action potential that is initiated at one location is conducted along an axon by jumping from break to break in the myelin. As a result, electrical current is conducted quite rapidly along an axon. In a disease called multiple sclerosis, myelin at various locations in the nervous system is destroyed and replaced with scar tissues. This leads to numerous neurological problems, including lack of coordination, partial paralysis, and double vision.

When electrical current reaches the end of the axon, it stimulates the cell to release a neurotransmitter into the gap of the synapse. This neurotransmitter affects connecting cells, which may be other neurons or muscles or gland cells, which become active when stimulated in this way.

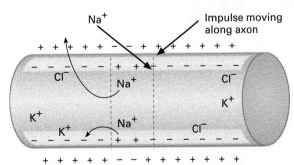

Figure 14.5: *Stimulation of a neuron's cell membrane triggers action potentials all along the axon.*

EARLY NERVE THEORY

Thousands of years ago, the image of the nervous system was more like the plumbing of your home than like its electrical wiring. The Greeks thought that nerves were hollow tubes carrying fluid and that some tubes carried fluid from the sensory organs (the eyes, for example) to the brain, allowing people and other animals to see. Other tubes carried fluid from the spinal cord to the muscles, allowing people and other animals to move around. The true nature of nerves had not yet been discovered.

The invention of the microscope improved the situation, but even in the 1800s, nerves were thought to be made of strange masses called globules. Nerves were viewed while covered with water, and we now know that the water was partly responsible for the early confusion about nerve structure. The recognition that nerves are not fluid-filled pipes but are bundles of axons had to wait until the announcement of the cell theory in 1839. And when the electrical activity of nerve cells was first studied, scientists believed that the signal normally traveled in both directions along the cell. Scientists didn't realize until the early twentieth century that neuron signals normally travel in only one direction.

INTEGRATING MESSAGES

Voltage is electrical potential that causes current to flow.

One neuron may connect through synapses with tens of thousands of other neurons. This is typical of the network of neurons in the human brain. As a message is conveyed from one neuron to another, the original message may be altered or may even be ignored. The fate of a message as it crosses a synapse depends on the types of neurotransmitter in the synapse. It also depends on the membrane receptors on the next neuron or effector to which these neurotransmitters bind. Normally, a neurotransmitter acts by either exciting or inhibiting the connecting neuron.

LESSON 14.1 ASSESSMENT

1. This discussion should include a description of the role of chemical and electrical signals in the pathway.

2. Yes, the primary function of neurotransmitters are to carry a signal across the synapse from the axon of one neuron to the dendrite of another neuron. A boat carries passengers or cargo across a body of water from one shore to the next.

3. Blocking the pumps would allow normal diffusion of ions to proceed indefinitely. This would change the value of the resting potential and interfere with normal conduction.

4. Most of the length of the copper wire is analogous to the axon. The splayed end represents a network of dendrites. The insulation around the wire represents the myelin sheath.

1 As you are walking barefoot through the grass, you step on a bee, and it stings you. Immediately, your foot lifts off the ground. Discuss how the signals traveled from the brain to the muscles in your leg to move your foot.

2 Do you agree that neurotransmitters function like a ferry boat? Give a reason for your answer.

3 You are exposed to a toxin that blocks the activity of sodium/potassium pumps in membranes of the nervous system. How would this affect the resting potential of neurons in your nervous system?

4 Compare and contrast a neuron with a piece of braided copper wire that has had the end stripped of insulation and the individual copper strands at one end separated and spread out.

LESSON 14.2 NERVOUS SYSTEMS FROM SIMPLE TO COMPLEX

When an electrician jerks her hand away from an uninsulated electrical wire, her nervous system performs a very simple operation involving only a few neurons. When that electrician, relieved to find that she is unhurt, sits down to determine why the wire shocked her, her brain performs a more complex operation involving thousands of neurons. In this lesson you'll explore some simple neural pathways and some more complex circuits.

ACTIVITY 14-2 THE KNEE JERK

Students need a reflex hammer, a chair, a flashlight, and a piece of black construction paper.

To maintain homeostasis, your nervous system must sometimes respond very rapidly to an outside stimulus. The response is so rapid that you are not consciously aware of what has happened. Such involuntary responses are called reflexes. Doctors often check reflexes during a routine physical exam or when a neurological problem is suspected.

Have your partner sit in a chair and cross his or her legs at the knee. Using a reflex hammer and standing to one side of your partner, gently tap the exposed leg just below the knee. If the reflex is normal, your partner's lower leg should swing forward almost instantaneously. If this does not occur, try again, striking an area just below or just above your first tap. After recording your observations in your logbook, have your partner perform the test on you. Then answer the following questions in your logbook:

The reaction was a very rapid response to the stimulus. The subject was not conscious of his or her response.

The knee jerk reflex can help to protect the knee by causing the leg to kick away an animal or object that can injure it.

Students should realize that the knee jerk response occurs too rapidly to involve sending nervous signals to and from the brain.

- How do you know your partner has displayed a reflex action?

- What basic function does the knee jerk serve?

- What role do you think the brain played in the knee jerk reflex?

BIOLOGY IN CONTEXT

DISCONNECTED BRAIN

The connections between the many types of neurons in your brain are much more complex than those in the simple reflexes you have studied. When a tumor disconnects the left side of the brain from the right, the strangest things can happen. In one case the patient was asked to close his eyes while a series of objects were handed to him. When he held a ring in his hand, he said that it was an eraser. When given a watch, he said that it was a balloon. A padlock was a box of matches, and a screwdriver was a piece of paper. With his eyes still closed, he received a hammer. When asked how he would use it, he made motions as though pounding a nail but said, "I would use this to comb my hair." When he opened his eyes, he was able to identify all of the objects properly. Clearly, the connections between the two sides of the brain are more complex than the two neurons of the knee jerk reflex.

REFLEXES

The stimulus and response observed in Activity 14-2 require a very simple neural pathway in which the neural impulse does not travel to the brain. This neural pathway is called the **reflex arc.** A reflex arc is a neural pathway that allows for a quick response to a stimulus (Figure 14.6). This reflex arc requires only two neurons: a **sensory neuron** to detect the stimulus provided by the hammer and a **motor neuron,** which sends a signal to a muscle. The reflex arc also requires a special sensory receptor in the muscle that responds to the hammer by stretching. When stretched, this receptor triggers an action potential in the sensory neuron. When the action potential reaches the axon terminal, a neurotransmitter crosses the synapse and acts on the motor neuron. The motor neuron responds with an action potential. When the action potential reaches the axon terminal, a neurotransmitter is released that stimulates the muscle cells to contract.

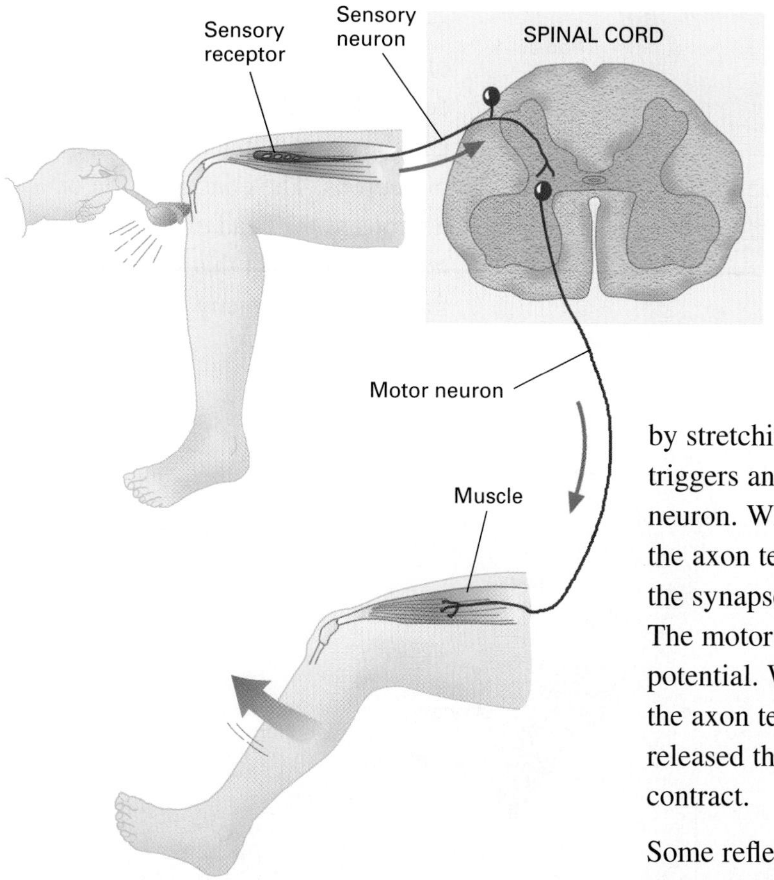

Some reflex arcs are slightly more complex than the knee jerk reflex, requiring the actions of three or more neurons.

Figure 14.6: *Knee-jerk reflex*

The nervous systems of some lower animals resemble the simple reflex arc you have just studied. One of the simplest types of nervous system is the **nerve net** found in *Hydra* (Figure 14.7A). A nerve net does not involve a brain or nerve cord. Nerve cells are loosely organized in a netlike pattern, and action potentials travel in both directions along a nerve cell, unlike the action potentials in our own bodies. As a result, when *Hydra* is touched in one part of its body, the stimulus spreads along the entire nerve net, and the entire organism responds. Sea stars have a more complex type of nerve net with a central ring of nerve cells (Figure 14.7B).

More complex animals such as grasshoppers, flatworms, and leeches, on the other hand, have more complex nervous systems (Figures 14.7C–14.7E) that include a long nerve cord and a concentration of neurons at the head end of the body. This concentration of neurons is referred to as a set of cerebral ganglia. A **ganglion** (plural: ganglia) is a visible concentration of neurons. The concentration of neurons at the head end of an animal relates to the fact that it moves head first and so has many of its sense organs at this end.

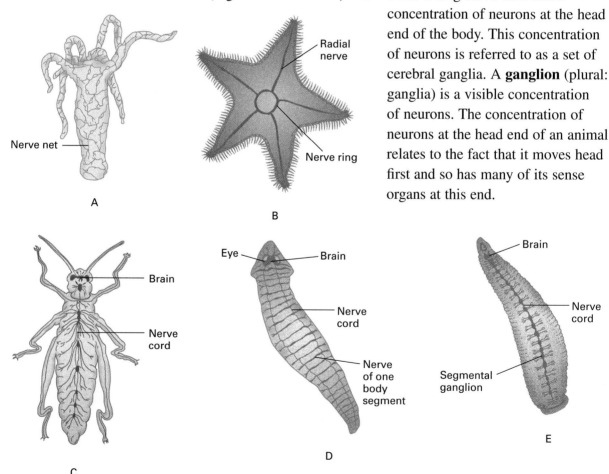

Figure 14.7: *Five simple nervous systems: (A) hydra; (B) sea star; (C) grasshopper; (D) flatworm; and (E) leech*

The flatworm's central nervous system has a ladder-type arrangement of nerve cords. Because of its ladder-type arrangement, the flatworm's brain doesn't need to coordinate body movements. The leech is a segmented worm. Each segment has ganglia that cause the muscles of that segment to contract. In these animals the brain is often responsible for inhibiting rather than promoting muscle activity. If a large portion of an annelid's brain is cut away, it moves continuously. A grasshopper without a brain can walk, jump, and even fly.

THE HUMAN CENTRAL NERVOUS SYSTEM

The nervous system of a small animal might have only a few hundred or a few thousand neurons, connected in simple ways. Animals that are larger and more complex have millions of neurons that are connected in highly complex ways. These connections support their more complex activities and mental abilities.

Humans and other vertebrates have a well-developed **central nervous system** (CNS) that consists of a brain and spinal cord running from the brain to just below the middle of the back. The spinal cord carries long bundles of axons to and from the brain. This is the so-called **white matter** of the CNS; the white color is due to the myelin surrounding the fibers. The cell bodies from these axons are joined in very dense clusters at points along the spinal cord. These clusters are the **gray matter** of the CNS, indicating that they are not covered in myelin.

The spinal cord of a vertebrate is surrounded and protected by the vertebrae of the backbone. Fluid in a central canal of the spinal cord absorbs shock generated by impacts that the body receives during activity. Nevertheless, injury does occur, as when the spinal cord is severed, usually resulting in paralysis.

Neurologists use a medical technique called magnetic resonance imaging (MRI) to detect disorders of the backbone and spinal cord. An MRI shows that at each segment of the backbone a pair of nerves connect to the spinal cord. These nerves carry messages between the CNS and all areas of the body.

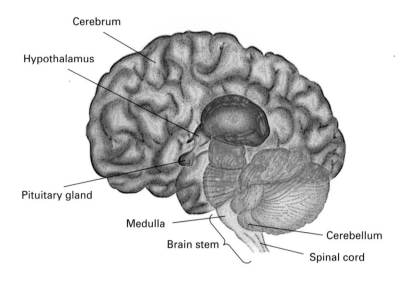

Cerebrum

Hypothalamus

Pituitary gland

Medulla

Brain stem

Cerebellum

Spinal cord

Figure 14.8: *Major components of the human central nervous system*

The human brain has three major regions: the brain stem, the cerebellum, and the cerebrum, which is divided into left and right halves (Figure 14.8).

The **brain stem,** continuous with the uppermost part of the spinal cord, has important roles in maintaining homeostasis. Areas of the brain stem control heart rate, breathing rate, and the diameter of blood vessels. A part of the brain stem called the **medulla** is a pathway for information between the cerebrum and motor fibers.

The **cerebellum,** which lies below the cerebrum at the rear of the brain, is responsible for muscle coordination and maintaining posture and balance. Information from the muscles is constantly relayed to the cerebellum, indicating how body parts are positioned and what state the muscles are in. An athlete such as a gymnast relies on use of this information in carrying out a complex routine (Figure 14.9).

The **cerebrum** is the largest part of the brain. It is responsible for sense perception, mental activity, and control of voluntary movement. A plumber's cerebrum is hard at work when he chooses the proper tools for his job. An actress's cerebrum is hard at work when she switches from a sad scene to a happy scene.

Figure 14.9: *Gymnasts rely on information relayed to the cerebellum to perform complex routines.*

Near the center of the brain is the **hypothalamus,** a region that coordinates the activities of the nervous system with those of the endocrine system. You will learn more about this coordination later in this chapter.

LEARNING LINK BRAIN POWER

The cerebellum of the bird is largest in comparison to the rest of its brain. The muscular coordination required for flight might explain how this came about.

Use the library or the Internet to find illustrations of the entire brain of a fish, a bird, and a human. For each, compare the size of the cerebellum to the rest of the brain. Which has the largest cerebellum compared to the rest of its brain? Now consider the habits of each and try to explain your finding.

LESSON 14.2 ASSESSMENT

1. Loss of muscle coordination and balance.

2. The cerebrum integrates the visual stimuli of the words and interprets them as thought. It also directs your movements as you turn the pages and eat the popcorn. The cerebellum allows you to coordinate the activities of turning pages, handling the popcorn and chewing it.

3. Accept any reasonable answer including: Sensory neurons receive information from the environment and transmit it to the central nervous system; motor neurons transmit the response to a change in environment to the skeletal muscle system.

4. Reflexes have a protective function. The response to a stimulus is faster if the signal does not have to travel to the brain and back.

1 Give the possible complications that may develop if the medulla is severely damaged.

2 What role does each of the major regions of the brain play while you are eating popcorn and reading a mystery novel?

3 Explain the relationship between sensory neurons and motor neurons.

4 Considering the function of a reflex and the distance from the knee to the brain, why is it advantageous to keep the brain out of the circuit?

LESSON 14.3 THE SENSES

No matter what job you do, you rely heavily on your senses. Fashion designers use their sense of sight when selecting colors for a new line of clothes. They use their sense of touch to make sure that the fabric has just the right feel. A waiter must be able to hear the customer's order, and the cook knows the aroma of each spice that goes into a recipe. The customer judges that recipe with her sense of taste, as does the professional taster, who samples coffee, tea, or chocolates and decides whether or not to make a purchase. These are the five senses in action. They are an important part of your complex nervous system. They are your windows on the world around you.

ACTIVITY 14-3   OP ART AND OPTICAL ILLUSIONS

Blackline masters of Figures 1–5 and questions for this activity are in the TRB.

This activity explores your sense of sight. You will study examples of a modern art form, known as op art, and you will discover how your vision can trick you.

Examine each of the five figures in the handouts provided by your teacher. Hold the figure about 12 inches away from you. In your logbook, describe each figure's appearance and its effects on you. Note any perceptions of motion. Answer the questions that are provided by your teacher.

CAREER PROFILE

INDEPENDENT LIVING REHABILITATION COUNSELOR

Cindy S. is a counselor for the visually impaired. "My job," she explains, "is to work with people who are legally blind, to help them develop independent living skills. That might be cooking, writing their monthly checks, or learning Braille, a system of writing and reading for the blind."

Often, Cindy helps people get adaptive equipment that helps them function more independently. "A lot of talking devices are available now—talking watches, talking scales, and very sophisticated talking computers. We have signature guides to help people write checks. But adaptive living is not based on fancy gadgets. Sometimes you can make very simple modifications, such as making a notch in the shampoo bottle to help you tell it from the conditioner."

Cindy says, "A lot of adaptive living is a matter of attitude. Attitude guides you to ask, 'How can I do this?' My biggest challenge in this job is motivating people to think that way."

Cindy has a college degree in rehabilitation and held various summer jobs and school internships in the rehabilitation field.

SENSATIONS DETECTED BY THE SKIN

Animals detect sensations of touch, pressure, vibration, cold, hot, and pain as different receptors in the skin are stimulated (Figure 14.10). Receptors that are capable of transmitting information from an environmental stimulus to an animal nervous system are called sensory receptors. The simplest type of sensory receptor is the **free nerve ending.** Free nerve endings are the dendrites of a neuron. One type responds to changes in skin temperature. When these nerve endings detect something dangerously hot or cold, they signal intense pain.

One type of sensory receptor in the skin detects pressure. These onionlike structures are scattered over the skin surface. The folded cell membrane of the receptor alternates with layers of fluid and surrounds the dendrites of a sensory neuron.

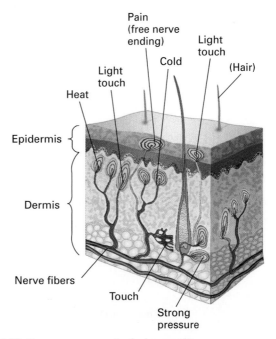

Figure 14.10: *Sensory receptors in the human skin*

Stretch receptors in muscles and joints give us information about our physical position. When the muscle stretches, the stretch receptors relay information to the CNS that helps you do things like keep your balance or precisely manipulate objects with your hands.

INVESTIGATION 14A A TOUCHY POINT

See Investigation 14A in the TRB.

Some parts of the skin of humans and other animals are more sensitive to touch than others. Is this because there are more sensory receptors in these areas? In this investigation you will try to answer this question by mapping the relative density of touch receptors on different areas of the skin of one or more test subjects.

SIGHT

If you are familiar with photography, you know that light enters a camera through an opening and forms an image on film. The amount of light is adjusted with a shutter that makes the opening smaller or larger. Now compare this to what happens when light stimulates your eye (Figure 14.11). The light passes through the transparent outer layer of the eye, called the **cornea.** Directly behind the cornea is the **iris.** The iris is the colored part of your eye. The amount of light that is admitted changes with the size of the **pupil,** an opening in the iris. Muscles in the iris control pupil size. Inside the eye the light is focused by a **lens,** forming an

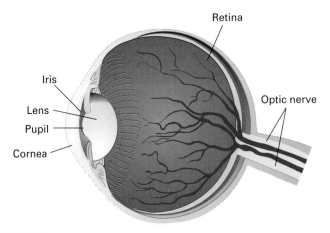

Figure 14.11: *The human eye*

image on the **retina** at the back of the eye. Unlike a camera the human eye never has to be reloaded with film. If you protect your retina it will last a lifetime.

WORD BANK

photo- = light

-receptor = receiver

Photoreceptors are sensory cells in the eye that contain light-sensitive pigment molecules.

Light focused on the retina stimulates sensory neurons called **photoreceptors.** The two types of photoreceptors are named for their shapes. Most of your 131 million photoreceptors are rod-shaped cells called rods. Rods are highly sensitive to light. They provide black-and-white vision only and allow us to see at night. Rods contain a pigment that is easily bleached by light. This bleaching caused you to see an after-image of the black goblet in Activity 14-3. Other photoreceptors, called cones, provide for color vision. There are three types of cones; one type has a pigment that absorbs red light, another absorbs green light, and the third absorbs blue light.

When a pigment in a rod or cone absorbs light, the shape of the pigment molecule is momentarily changed. This absorption causes an action potential in neurons leading to the optic nerve that enters your brain. Visual messages from the retina are processed in a special region at the back of the brain.

ACTIVITY 14-4   THE LOCATION OF THE OPTIC NERVE

Materials: a ruler, a pencil, and a sheet of paper.

Why is it sometimes impossible to see an object that is only inches from your face? In this activity you'll answer this question by studying the internal anatomy of your eye.

Draw a spot 5 cm to the left of the center of a piece of paper. The spot should be about 6 mm in diameter. Draw a second spot the same size as the first, this one 5 cm to the right of the center, so that the two spots are 10 cm apart. Hold the paper at arm's length in front of your face. Close your right eye and look at the right-hand spot with your left eye. Slowly move the paper closer to your face, watching the spot as the paper approaches. Answer the following questions in your logbook:

The spot should disappear when the paper is about 30 cm from the eyes. This happens because the image of the spot is being projected onto a part of the retina that has no photoreceptors, an area called the blind spot.

The blind spot has no photoreceptors because it's where the optic nerve attaches to the retina.

- What happens to the spot? At what distance from your face does it disappear?

- Study Figure 14.11. What feature of the retina might account for your observations?

Another stimulus in the environment that we can detect is sound. Sound is the vibration of air molecules. Sound is detected by the ear, which consists of the outer ear, the middle ear, and the inner ear (Figure 14.12). The outer ear collects sound and channels it toward the eardrum. When sound waves strike the eardrum, that membrane vibrates. The vibrations are relayed in turn to three small bones in the middle ear. The last of these bones relays the

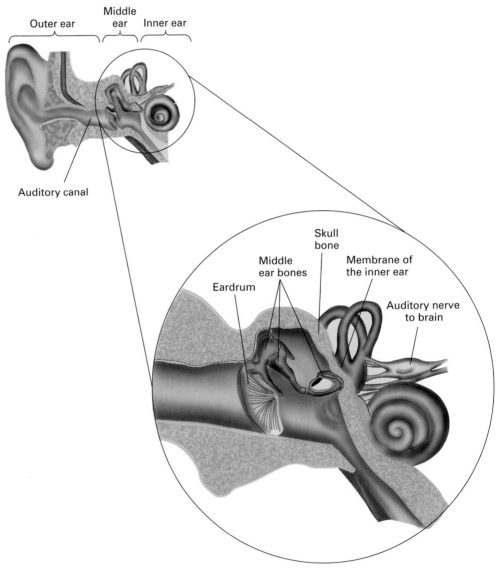

Figure 14.12: *The human ear. The vibrations of the middle ear bones are transmitted to the fluid of the inner ear by the membrane of the inner ear. The vibrating fluid will cause hair cells in the inner ear to produce an electrical current that will stimulate the auditory nerve.*

vibrations to a fluid in the inner ear. A membrane containing sensory receptors called **hair cells** is suspended in this vibrating fluid. As the cilia of a hair cell respond to the vibrations by bending, the cell produces an electrical current. This triggers an action potential in a connecting neuron leading to the auditory nerve. The auditory nerve routes directly to that area of the brain that interprets sound information.

The greater the size of the wave in the inner ear, the more vigorously the hair cells move. The brain interprets the amount of hair cell movement as loudness. Prolonged exposure to loud sounds, such as engine noise, can damage the hair cells of the inner ear and lead to permanent hearing loss (Figure 14.13).

The pitch of a sound, for instance of a note of music, is perceived according to which hair cells are stimulated. Humans with good hearing can detect pitches of 20–20,000 vibrations per second. A dog can hear pitches that are twice as high as the maximum pitch detected by humans. That's why dog whistles are silent to us.

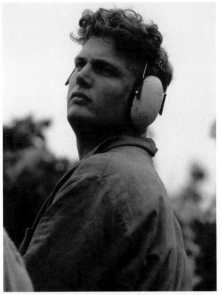

Figure 14.13: *Earplugs or earmuffs are used at certain worksites to protect workers from permanent hearing loss.*

TASTE AND SMELL

The senses of taste and smell, or olfaction, are the means that humans use to detect the presence of chemicals in the environment. The sense of smell depends on sensory receptors in the roof of the nasal cavity (Figure 14.14). Each receptor is stimulated by one or several kinds of molecules that have a shape that matches a site on the receptor. If an airborne molecule binds to a site on a smell receptor, the receptor conducts an electrical current. This triggers an action potential in an olfactory neuron. Olfactory neurons lead to the olfactory nerve, which routes them to the area of the brain that interprets smells.

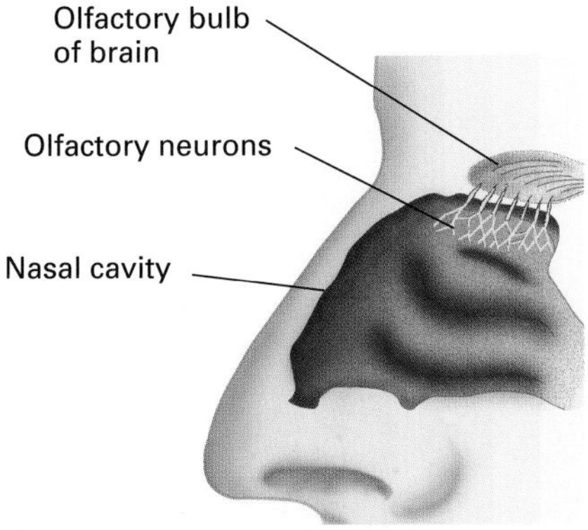

Olfactory bulb of brain

Olfactory neurons

Nasal cavity

Figure 14.14: *Sensory receptors of the human nose*

The sensory receptors that are responsible for the sense of taste act like those for smell. They are on the upper surface of the tongue. The sensory receptors are in structures called taste buds. These receptors detect chemicals as being one of four basic tastes: sweet, sour, salty, or bitter. The variety of taste sensations is due to the variety of odors in combination with the four basic taste sensations.

INVESTIGATION 14B CHOOSING OUR FOOD

See Investigation 14B in the TRB.

Consumer research shows that a person's like or dislike of a food is influenced by many factors: how food is displayed on the package, how food smells when prepared, how food looks on the plate, how food tastes, and how it feels during handling and during chewing. Modifying food to produce dietary specialties may affect one or all of these factors. What are your own biases for or against new types of food? This investigation will help you to find out.

LESSON 14.3 ASSESSMENT

1. Muscles in the iris will decrease pupil size.
2. Free nerve endings in the skin that respond to temperature changes.
3. Cone cells in the retina.
4. Since many foods give off vapors, both the sense of taste and the sense of smell are important.

1 You are sitting in a chair and decide to read, but the room is too dark. Describe what happens to your eyes when you turn on the light.

2 What type of sensory receptor does a mother use in feeling a child's forehead for fever?

3 Color blindness is the inability to distinguish certain colors. What part of the eye does color blindness affect?

4 Do you sense the flavor of your food with your taste buds or your olfactory receptors? Give a reason for your answer.

LESSON 14.4 HORMONES AND THE ENDOCRINE SYSTEM

Almost every day, there is news of someone whose life has improved by taking hormones: an athlete, a woman with a bone problem, an elderly couple trying to stay fit, a traveling salesperson with jetlag. These hormones, although synthetic, are similar or identical to the hormones that are produced by the body's system of endocrine glands (Figure 14.15). These glands help to maintain homeostasis by producing hormones that control normal growth, development, and metabolism.

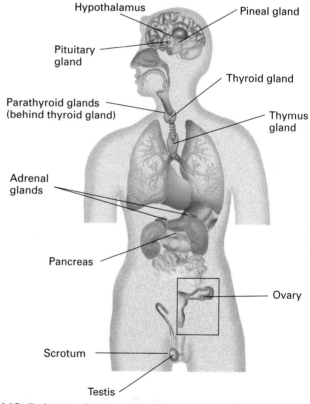

Figure 14.15: *Endocrine glands*

SHIFT WORK

BIOLOGY IN CONTEXT

Shift work is common in manufacturing, public safety, and health care to maintain operations around the clock. It involves working evenings, nights, rotating shifts, or extended shifts.

The night shift at a busy assembly plant had been experiencing low output and an unacceptable rate of accidents. The plant's human resources manager hired an occupational consultant to look into the problem.

The consultant was aware that shift workers commonly suffer from sleep disruption and fatigue, domestic disturbances, and health

problems such as gastrointestinal disorders and increased risk of cardiovascular disease. Night workers often perform poorly and are more accident prone than day workers. Many problems arise because shift workers are often out of sync with environmental and social cues.

To make matters worse, the night workers at the plant were drinking a lot of coffee to stay alert through the night. This was making it difficult for them to get to sleep when they got off work.

There are some health care professionals who work several shifts in one day, which may cause them to perform poorly at work.

The consultant recommended training and counseling sessions for all shift workers. The training would focus on how to sleep during the day, exercising before work and during breaks, eating properly, and dealing with stress. The consultant recommended that the company provide a nap room for employees who felt tired during a shift.

ACTIVITY 14-5 HOW STRESSED ARE YOU?

Provide each student with a self-assessment form. A blackline master of the self-assessment can be found in the TRB. Assure students that this is confidential and will not be collected or shared. To ensure confidentiality, students should not put their name on the self-assessment.

Answers will vary.

When your body can't maintain homeostasis, it is under stress. The fluctuation of hormone levels and the abnormal function of certain organ systems are caused by stress.

Disturbance in normal activity and eating patterns isn't the only cause of stress. In this activity you will assess the amount of stress you have experienced over the past year.

Read the list that your teacher will provide and circle the points for each situation that has happened to you in the past year. Add up the points and compare your total to the scoring guide to find out how much stress you live with. Then answer the following questions in your logbook:

- Do you think you have a stress-related problem? Identify the situations that cause you the most stress.

- Do your scores indicate a pattern of situations that cause you stress?

- What coping behaviors might you develop to manage your stress?

ENDOCRINE GLANDS AND EXOCRINE GLANDS

You have already learned that the nervous system provides very rapid communication between different areas of the body. Another system of communication, the **endocrine system,** operates much more slowly. This system is made up of more than a dozen glands that secrete hormones directly into the bloodstream. These hormones exert their effects when they reach a target cell. Say you missed lunch and by late afternoon your blood sugar level is dropping. In response, the pancreas, an endocrine gland, begins to secrete the hormone glucagon. When glucagon reaches certain liver cells, those cells release glucose into the blood, returning your blood sugar level to normal.

Compare the operation of endocrine glands to those of other glands (called **exocrine glands**) that you learned about earlier: sweat glands, gastric glands, and salivary glands. These glands secrete products into ducts that connect directly to a target location. For instance, salivary glands produce the enzyme amylase, which travels through ducts to your mouth, where it mixes with the food that you chew.

The liver cells that release glucose are the target cells of glucagon. They have receptors on their cell membranes that bind to glucagon diffusing toward the cells from the blood (Figure 14.16). A target cell responds to a hormone by doing something that restores homeostasis, such as maintaining the blood sugar level. Some hormones have target cells in several tissues or organs of the body. For instance, your **adrenal gland,** an endocrine gland that secretes adrenaline, signals cells of the liver to release glucose for a burst of quick energy. Adrenaline also makes your heart beat faster; increases blood flow to the brain, skeletal muscles, and lungs; and decreases blood flow to the digestive organs and skin. Adrenaline prepares you, in the event of trouble, to escape or fight.

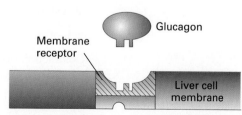

Figure 14.16: *Liver cells have receptors in their membrane for the hormone glucagon.*

THE PITUITARY GLAND

The **pituitary gland** is an endocrine gland that controls the activity of other endocrine glands. It does this by secreting hormones and other substances into the blood to reach target cells in those glands (Table 14.1). The pituitary gland is about in the center of your head and connects to the **hypothalamus,** the portion of the brain that controls the secretions of the pituitary glands. The pituitary gland has two lobes. The front lobe secretes hormones that control the **thyroid gland,** the adrenal glands, and the ovaries or testes. The front lobe also produces growth hormone, which stimulates growth of body tissues and the bone growth plates, and the hormone prolactin, which stimulates the breast to produce milk.

The pituitary gland also interacts with the nervous system to maintain homeostasis. The hypothalamus releases hormonelike substances, which enter capillaries connecting to the front lobe of the pituitary. The pituitary alters its secretion of hormones, depending on the signal from the hypothalamus. For instance, if

Hormone	Abbreviation	Function	Acts on
Front of pituitary gland			
Growth hormone	GH	Stimulates cell division for growth, increases the rate at which carbohydrates and fats convert to glucose for energy	Bones and soft tissues
Prolactin	PRL	Stimulates milk production	Mammary glands
Thyroid-stimulating hormone	TSH	Causes release of thyroxin into blood	Thyroid glands
Adrenocorticotropic hormone	ACTH	Causes release of adrenal cortex hormones into blood	Adrenal cortex
Follicle-stimulating hormone	FSH	Egg and sperm development, production of estrogen	Ovary or testes
Luteinizing hormone	LH	Triggers ovulation or production of testosterone	Ovary or testes
Back of pituitary gland			
Oxytocin	OT	Contraction of smooth muscles of uterus, release of milk	Uterus and breast
Antidiuretic hormone	ADH	Regulates body fluid	Kidneys

Table 14.1: Pituitary Hormones

A   B

The overproduction of growth hormones can cause giantism (A); underproduction can cause pituitary dwarfism (B).

the hypothalamus detects too low a level of thyroid hormone in the blood, it signals the pituitary. The pituitary responds by releasing a hormone that signals the thyroid gland to release thyroid hormone. The hypothalamus also communicates with the pituitary by sending action potentials along neurons that connect to the rear lobe. In response, the pituitary alters secretion of hormones such as antidiuretic hormone (ADH), which helps the body to retain water.

THE THYROID GLAND

The thyroid gland is in the neck in front of the trachea just below the larynx (Figure 14.17). It produces thyroxin, which helps to govern a person's basal metabolic rate, that is, how fast the body burns food for energy. When you get cold, thyroxin increases the metabolic rate, raising your body temperature. To make this happen, the hypothalamus makes thyroid-releasing factor, which travels to the pituitary gland. This gland releases thyroid-stimulating hormone, which travels through the blood to the thyroid. The thyroid secretes thyroxin, which makes muscles and other tissues increase the rate at which they metabolize food for energy and warmth.

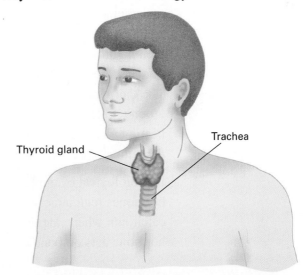

Figure 14.17: *The thyroid gland is located on the trachea just below the larynx.*

hypo- = less than normal

hyper- = more than normal

A person with hypothyroidism has a thyroid gland that is less active than normal. A person with hyperthyroidism has a thyroid gland that is more active than normal.

The mineral iodine is necessary in the diet to synthesize thyroxin. If a person's diet lacks iodine, a condition called hypothyroidism, in which there is too little thyroxin, develops. The most obvious symptom of this condition is a goiter, an enlargement of the thyroid gland (Figure 14.18). People with hypothyroidism also have lower-than-normal metabolic rates. The thyroid gland can also make too much thyroxin, a condition called hyperthyroidism. This results in an abnormally high metabolic rate accompanied by restlessness and anxiety.

Figure 14.18: *Low levels of thyroxin will cause the enlargement of the thyroid gland.*

NEGATIVE FEEDBACK

A constant level of calcium in the blood is critical for muscle and nerve function. Without it, severe cramping and seizures would occur. To maintain the proper calcium level, the endocrine system uses **negative feedback,** a control mechanism that prevents the body from producing too much of a particular substance. To understand this, think of cooking a roast in an oven. You set the oven to 450°F. The oven heats until this temperature is reached, then it shuts off. When the oven thermostat senses that the oven temperature is dropping, it signals the heating elements to turn on. The thermostat keeps the temperature seesawing in a narrow range around 450°. Because the thermostat is designed to turn the heating element on when the temperature falls and turn it off when the temperature rises, the oven operates on negative feedback.

In a biological example of negative feedback, two opposing hormones regulate calcium level. The thyroid gland secretes calcitonin when calcium levels in the blood are too high. This hormone acts on bone cells to prevent the release of calcium and

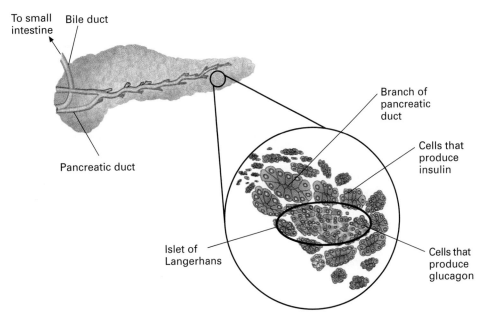

Figure 14.20: *Pancreas and islet of Langerhans*

cells. A second cell type makes glucagon. Its effect is opposite that of insulin; it stimulates the liver cells to convert glycogen to glucose, which enters the bloodstream. Every time you eat carbohydrates, like an ice cream cone, this negative feedback system is activated. As the glucose level in the blood rises, the pancreas releases insulin. This stimulates glucose to leave the bloodstream and enter cells, decreasing the blood sugar level. Eventually, low blood sugar triggers the release of glucagon to raise the blood sugar level to normal.

Diabetes mellitus is a disorder of the insulin-producing cells. This disorder hinders the liver and fat cells from converting glucose into glycogen and fat. A newly diagnosed diabetic may lose weight as his or her body burns fat for energy and may urinate frequently as the kidneys use water to eliminate excess blood sugar. A person with this form of diabetes must have injections of insulin (Figure 14.21), a hormone that is now manufactured by using recombinant DNA technology.

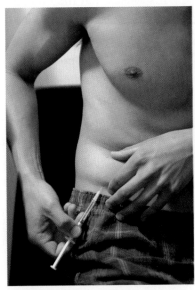

Figure 14.21: *Some diabetics must have an insulin injection at least once a day.*

THE PINEAL GLAND

The **pineal gland** is a small endocrine gland that is located in the brain near the hypothalamus. It secretes the hormone melatonin (Table 14.2), which controls our sleep cycles. Melatonin is sold as a nutritional supplement to encourage sleep in travelers who have jet lag. There are concerns about the use and safety of over-the-counter melatonin.

WORD BANK

circa = about

-dian = day

Circadian rhythms are physiological and behavioral characteristics that follow a daily pattern.

In animals and possibly in humans the pineal gland may be the site of the so-called biological clock. This clock is responsible for the daily rhythms, called **circadian rhythms,** of a variety of bodily functions. Body temperature, sleep/awake periods, and blood pressure are examples of variables that show a repeated daily cycle. Many hormones, including melatonin, show circadian rhythms. The pineal gland has neural connections to the eyes, and daylight has been shown to stimulate the pineal gland to slow down its production of melatonin. A night shift worker has higher melatonin levels than a day shift worker does. This could be a reason for the shift work problems that you read in an earlier part of this lesson.

Figure 14.22: *Changing light levels trigger the pineal gland to initiate estrus, the reproductive cycle in sheep.*

In birds, reptiles, and many mammals the pineal gland may be responsible for responses that they make to seasonal changes. The regulation of reproductive cycles, hibernation cycles, and migration ensures that animals will be in step with nature's seasonal cycles. For example, sheep come into "heat," or **estrus,** the beginning of the reproductive cycle, at a time that will ensure abundant food for the offspring when they are born (Figure 14.22). Insects don't have pineal glands. Their cyclic changes of metamorphosis and reproduction are triggered by special neurons that produce hormones.

LOCAL HORMONES

A variety of cells all over the body secrete hormones that act on specific target cells. For example, certain cells in the duodenum release the hormone secretin in the presence of food. Secretin travels in the blood to the pancreas to stimulate the release of sodium bicarbonate to adjust the pH in the duodenum. Gastrin works similarly in the stomach, stimulating production of hydrochloric acid.

SYNTHETIC HORMONES

Synthetic hormones may have the beneficial effects of naturally secreted hormones but may also cause unwanted side effects. The synthetic hormone tamoxifen is being field-tested for its ability to reduce breast cancer in high-risk women (Figure 14.23). It binds to the receptors that would normally bind estrogen, a hormone that triggers breast cancer in some women. However, tamoxifen may slightly increase a woman's chances of developing uterine cancer. DES, another synthetic estrogen that pregnant women used from the 1940s to the 1970s to reduce miscarriages, was linked to reproductive cancers in their offspring.

DES = diethylstilbestrol

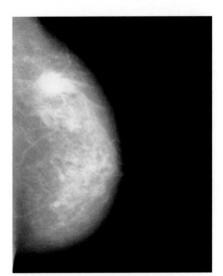

Figure 14.23: *Breast X ray showing a calcified area that indicates breast cancer.*

Synthetic hormones called **anabolic steroids** are used by some bodybuilders to build muscle mass quickly. Anabolic steroids have many side effects, including acne, headaches, fatigue, sterility, stunted growth, and aggressive personality changes, and may lead to cancer and death.

LESSON 14.4 ASSESSMENT

1. Accept any reasonable answer, which may include the description of thermostat for a heating unit, air conditioner, or refrigerator.

2. When the calcium level rises, the thyroid gland secretes calcitonin, which acts on bone and kidneys to reduce blood calcium.

3. If a diabetic has had a recent insulin injection he or she would test high for insulin in the blood. Without this treatment, however, the diabetic would test negative for insulin, since in diabetes mellitus the pancreas does not make insulin.

4. Anabolic steroids have serious side effects if they are taken routinely. They may also be addictive. You should weigh the long-term effects of these steroids with whatever advantages they may have on your performance.

1 In your own words give an example of negative feedback using something that should be familiar to you and your classmates.

2 When there is a low level of calcium in the blood, the parathyroid gland secretes thyroid hormone to increase the calcium level. What prevents the calcium level from getting too high?

3 Would a patient who is being treated for diabetes mellitus test high or low for insulin? Give a reason for your answer.

4 While training to make a sports team at your school, a classmate offers you some capsules and tells you that they contain a steroid hormone that will help you to build muscle. What should you consider before deciding to take or not take the capsules?

CAREER APPLICATIONS

The applications that follow are like the ones you will encounter in many workplaces. Use the biology you learned in this chapter to complete the activities. Share your work with the class.

AGRICULTURE & AGRIBUSINESS

Growth Enhancers

Investigate additives in livestock feed that are used to enhance animal growth and weight gain. Look particularly for additives that are hormones, act as hormones, or stimulate hormone production. How common is their use? How do they work? If you are using the Internet, use keywords such as "feed additive." Write up your findings in your logbook. Discuss in class the question "What are the concerns that consumers have about these additives?" Do you think they are valid or not? What additional information, if any, would you want before making a decision?

Egg Production

Research environmental conditions in chicken houses used for commercial egg production. What conditions have been shown to increase egg production? Which senses are involved? Write up your findings in your logbook and share them with the class.

BUSINESS & MARKETING

Senses and Sales

Talk to someone who works in advertising. Find out how advertising appeals to the different senses. Make a chart listing the senses and what would appeal to each sense. Different media play to your senses in different ways. Pay attention to commercials and ads in the media. Make a list of at least five commercials or ads from each of the following sources: television, radio, slick full-color magazines, and newspapers. To which senses do the ads appeal? How? Share your information in class.

Products That Appeal to the Senses

Walk through a supermarket or discount store. Make a list of all of the products that are designed to make you or something you own more appealing. For example, cosmetics are designed to enhance visual appearance, and paint or stains enhance the appearance of walls or furniture. Compare your list with the lists of others in your class.

Adaptive Living

Contact organizations for visually impaired and hearing-impaired people and find out what kinds of electronic equipment have been developed to extend the abilities of people who have these impairments. What simple items make everyday life easier? If possible, talk with someone who has a visual or hearing impairment. Develop a plan for modifying your home and equipping it for someone who has a visual or hearing impairment.

Infant Development

The first year of infant development is called the sensorimotor stage because the senses are key to interpreting the world because babies are learning to use their senses and motor abilities. Arrange to visit the infant room of a daycare center or to spend time with an infant under a year old. Choose an awake baby to observe closely for a 10-minute period. During the 10 minutes, list everything the baby does (for example, "moves rattle from right hand to left hand," "cries at loud laughter," or "pushes toy with right hand across floor, under high chair." On the basis of your observations, decide what senses the baby is using most in his or her explorations. Research the course of sensory development in infants to see how well research findings correlate with your observations.

Endorphins

Using the Internet or other resources, find out about endorphins. What triggers their production in the body? What purpose do they serve? How long do they remain in the body? What benefits do they provide? Where in the body are they produced? Prepare a report, or work with several other students in the class to make a 1- to 2-minute public service documentary.

Substance Abuse

Select and research a substance that may be abused: alcohol, tobacco, cocaine, heroin, inhalants, methamphetamines, barbiturates, tranquilizers, morphine, LSD, mescaline, psilocybin, ecstasy, marijuana, or other drug. Write a one-page fact sheet that includes the substance, any street names it goes by, whether its use is legal or illegal, what it looks like, what it is made from, and how it is used. What are its effects on the body, particularly the central nervous and endocrine systems? Are the effects reversible? Share your information in class.

Contact Lenses or Glasses?

Visit an ophthalmologist, optometrist, or optician and find out the differences among single-vision, bifocal, and trifocal glasses and among hard, rigid gas-permeable, and soft contact lenses. Pick the kind of lens you would prefer if you needed corrective lenses. Identify conditions that would exclude certain options for you.

Head and Spinal Cord Injuries

Write to the National Organization on Head and Spinal Cord Injuries to answer these questions. What are the leading causes of head and spinal cord injuries, and how can they be prevented? Prepare a poster that shows ways of preventing one type of head and spinal cord injury. What can be done to help victims of head injuries or spinal cord injuries? What are the newest treatments that are being researched?

Audiology

Use career references to find out about careers that deal with ears and hearing. What is the work like? What level of training is needed for each? What kinds of ear or hearing problems can each address? If possible, talk to someone in this field about the most common hearing losses and how they are detected. What can people do to avoid hearing problems? Make a brochure on these careers.

Biofeedback

Research how biofeedback works. Prepare an informational brochure that describes the process, what conditions it can be used for, how long it takes, what the sensors measure, and so forth. Find out whether anyone in your area provides biofeedback services. How successful is biofeedback for different conditions?

Workplace Noise

Interview people who work in noisy environments to find out what steps they take to protect their hearing. Some examples are airport personnel, power plant technicians, jackhammer operators, some factory workers, and crane operators. What are employers required to do under Occupational Safety and Health Administration (OSHA) standards if noise levels in the workplace can't be maintained at safe levels? What hearing protection devices are available and how are they rated?

The loudness of sound is measured in decibels. Research the decibel scale and determine what kind of scale it is—linear or logarithmic—and what that means. At what levels and duration is hearing impaired? Is the impairment temporary or permanent? Make a poster or chart of common sounds and the decibel level of each. If your school has sound meters, you might want to take readings in different settings, such as in the library, a classroom, the halls during class changes, the cafeteria, and the parking lot, and plot the background noise in each of these locations on a decibel scale.

CHAPTER 14 SUMMARY

- The major cell type in our nervous system is the neuron.

- Neurons are highly specialized for their functions and typically produce an electrical signal called an action potential. This signal stimulates the cell to release chemicals that stimulate neighboring neurons.

- Small, simple animals such as *Hydra* have no central nervous system. They have no brain and no major nerve cord. Insects, many worms, and humans have a central nervous system.

- The vertebrate central nervous system has a complex brain and a spinal cord. The spinal cord handles simple reflexes and communicates with the brain.

- The major parts of the brain are the cerebrum, the cerebellum, and the medulla.

- The cerebrum is the site of sensory perception, memory and voluntary muscle movement.

- The cerebellum is responsible for muscular coordination.

- The part of the brain stem called the medulla deals with basic life-support functions.

- The five senses of the human body are touch, sight, hearing, smell, and taste.

- Sight and hearing are complex senses. Our eyes function like cameras, though we have a retina with sensory receptors called rods and cones. The sensory receptors of hearing are in the inner ear and are called hair cells.

- The nervous system communicates with the endocrine system via the hypothalamus.

- Stimulation of endocrine glands makes them alter their secretion of hormones.

- Hormones are chemical messengers that circulate in the body and cause a response to the original stimulus.

- The pituitary gland controls the activities of the other endocrine glands.

CHAPTER 14 ASSESSMENT

Concept Review

See TRB, Chapter 14 for answers.

1 Define neurotransmitters and describe their function.

2 Where are the receptors for equilibrium?

3 Describe the basic structure of a neuron.

4 Why is glucagon often called the "anti-insulin" hormone?

5 Discuss how myelin sheaths make the neuron more efficient.

6 Describe the voltage across the cell membrane when a neuron is not conducting an impulse.

7 Explain how the resting potential is restored after an action potential.

8 What initiates the transmission of an electrical signal from one neuron to another?

9 How are neurons electrically insulated from one another?

10 Explain what an action potential is.

11 List some of the basic functions of hormones in the human endocrine system.

12 Explain the basic role of the pituitary gland in homeostasis.

13 Do you agree with the following explanation of how sound is transmitted within the vertebrate ear? If not, explain sound transmission yourself. "The sound vibrates the eardrum, which makes tiny bones in the ear vibrate. Nerves attached to these bones transmit neuron impulses to the brain."

14 Complete the following description of how light is reflected from an object, converted into an image, and converted into an impulse transmitted to the brain: "Light reflected from an object enters the eye through the pupil."

15 In humans, where are the sensory receptors that respond to chemical stimuli?

16 Explain the roles that the dendrites, cell body, and axon play in the electrical communication of signals through the body.

17 What is the basic difference between the endocrine system and the exocrine system?

Think and Discuss

18 When you hold your nose while eating a peppermint, you can't taste the peppermint. Why?

19 Explain why a signal travels in only one direction across a synapse.

20 Why is the pancreas considered both an endocrine and an exocrine gland?

21 Do you agree or disagree with the following statement: "Light striking any part of the retina will stimulate either rods or cones." If you disagree, give your reason(s).

In the Workplace

22 When traveling on a business trip, you experience jet lag. The company health office suggests a specific schedule of bright lights and darkness to help you readjust. Which gland involved in sleep and waking cycles is this meant to affect?

23 People who work around airplanes and in other noisy places must wear ear protection. Is the outer ear, middle ear, or inner ear least likely to be damaged by frequent exposure to loud noise?

24 You are the supervisor for the night shift at a factory. You want to help maintain high levels of productivity and minimize accidents. Describe some practices that you might enact to help you reach these goals.

25 You are visiting a chicken ranch. The chicken houses have controlled temperature and piped-in music. The rancher says that these comforts reduce stress. Why would it be important to reduce stress in the hens?

Investigations

The following questions relate to Investigation 14A:

26 Based on the accompanying graph, which of the parts of the arm and hand tested probably contains the most nerve endings per square cm?

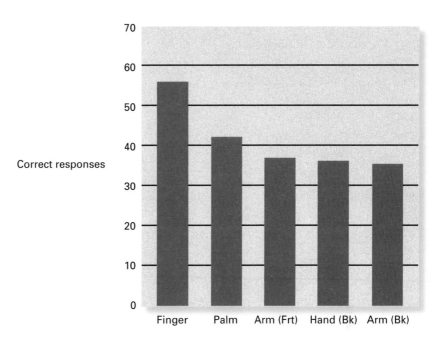

27 Design and describe an experiment to measure the closest distance that an individual can distinguish between one, two, and three probes on the back of the hand.

The following questions relate to Investigation 14B:

28 In Investigation 14B, how did you test the role that smell plays in food choices?

29 List three sensory factors that influence food preference.

CHAPTER 15

WHY SHOULD I LEARN THIS?

Do you know how your body fights and destroys the unseen microscopic invaders that you inhale with nearly every breath? Do you understand how a soldier survives an infection after being wounded on the battlefield? If you were a health inspector, what kinds of diseases and poisons would you worry about in a public restaurant? This chapter answers these questions of life and death. You, the soldier, the health inspector, and the restaurant customer are fortunate in having several lines of defense against dangerous agents in the environment. You will learn about these defenses and how agents that penetrate our outer barriers are destroyed by the immune system waiting inside.

Chapter Contents

PROTECTION AND DEFENSE

WHAT WILL I LEARN?

1. which microscopic organisms cause infection and disease
2. how toxins cause illness and death
3. how the skin functions as a first line of defense
4. the variety of chemical defenses that organisms produce
5. how cells fight invading organisms and poisons
6. how the body recognizes its own cells
7. what happens when the body doesn't recognize its own cells

---

Have you ever seen on television or at a major accident someone in an oversized astronaut suit? This person might be a member of the hazardous materials unit. The hazardous materials unit is called upon to clean up major chemical spills and gas leaks. The oversized astronaut suit is called a biohazard suit. This type of suit is used by the members of the hazardous materials unit and sometimes by industrial workers to protect them against chemicals, pathogens, and radiation. Workers wear special clothing to handle these environmental agents, but the human body alone has some defenses against many of them. We have a variety of built-in protections, and in this chapter you will discover what these defenses are and how they work.

LESSON 15.1 OVERVIEW OF ENEMY AGENTS

Since the 1920s, farmers and agricultural workers in the southeastern United States have had to deal with an uninvited guest from South America: the fire ant. The opening pages of this chapter show one of these insects stinging a person and injecting venom that causes immediate burning pain. Within a day a pus-filled swelling develops. For some people that's the end of the problem. For people who are allergic to the venom, fire ant stings are life-threatening.

INVESTIGATION 15A ANTIBIOTIC ACTIVITIES

See Investigation 15A in the TRB.

Did you know that your skin, mucous membranes, nasal secretions, and tears contain special enzymes that act as antibiotics? These enzymes are called lysozymes. Lysozymes destroy gram positive bacteria that enter into your body. In this investigation you will isolate lysozyme from egg whites and measure its ability to inhibit bacterial growth.

BIOLOGY IN CONTEXT

BROWN RECLUSE SPIDER

Two office workers of a state agency went to an emergency room, complaining of a prickling sensation at the site of a small bruise that became insensitive and hard. The emergency room doctors knew that the workers' immune systems were reacting to something. The doctors recognized that the bruised area was a bite wound and ordered blood tests. The tests showed that the bite was from the brown recluse spider. The bites of these spiders get worse without treatment. Within 24–35 hours the bite forms a bluish-purple ring around blisters. Within 48 hours the blisters begin to break. The blisters scab over, but the tissue under the scab continues to die. The doctors prescribed antibiotics, antihistamines, and steroids to prevent further damage. The workers were fortunate that they got treatment right away. Cases that aren't reported as quickly can require skin grafts or lead to loss of a limb.

The brown recluse has a violin-shaped pattern on its back.

Because the two workers were bitten at work, the rest of the office staff left the building. They were forced to work out of their homes or at another agency. This invasion of brown recluse spiders wasn't the first at this office. The summer before, exterminators had taken nearly a month to get rid of all the spiders. The staff hoped to move back into their offices sooner this time.

PATHOGENS OF ALL KINDS

Organisms must defend themselves against a long list of pathogens that can cause sickness or death. Some of these pathogens are living, some are nonliving, and some are hard to classify. Humans are threatened by pathogens in all five kingdoms of life. From the kingdom Monera comes a bacterium that causes bubonic plague, also known as the Black Death. Bubonic plague is transmitted to humans by fleas, which usually live on rodents, and coughing transmits the disease from one human to another. Hundreds of years ago, this infectious single-celled organism wiped out whole towns. The bacterium kills by destroying lung tissues.

From Fungi comes the cause of athlete's foot. This enemy isn't dangerous, though the itching, scaling, and blistering that it causes can be agonizing (Figure 15.1). Even when the condition seems to be cured, the organism is living below the skin and may erupt in moist conditions such as those of a locker room floor.

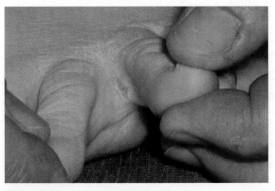

Figure 15.1: *A serious case of athlete's foot*

One dangerous type of dysentery is caused by a protist, and from Animalia comes the 30-foot tapeworm that lives in humans' small intestine. Dysentery can create holes or ulcers in the large intestine, and tapeworms can form holes in the brain, heart, and muscles, though infection usually does little more than disturb the digestive tract. Sewage treatment plants use chlorine, heat, alcohol, acids, and soaps to destroy infectious agents such as the dysentery protozoan. Examples from the plant kingdom are discussed later in the chapter.

Modern entomology books will provide many examples of insect biocontrol. For example, the bacterium *Bacillus thuringiensis* is a popular agent. A tiny predator fly of the family Phoridae is being released in the United States in an effort to control the imported red fire ant.

Insects can't always defend themselves against fungi, bacteria, and parasitic worms. Some of these infectious agents are so effective that farmers use them for biocontrol of insect pests that damage crops. Use the library or the Internet to discover specific examples of insect pests and the biocontrol agents that are used against them.

FOREST RANGER

CAREER PROFILE

An out-of-breath hiker rushes into the ranger station at the park and says, "Help! My friend has been bitten by a rattlesnake about a half mile up the trail." Amber R., a park ranger, grabs the first-aid case and gives instructions to send for an ambulance and to alert the hospital. As they drive to the scene, Amber gets information about the victim. She reassures the hiker that rangers have been trained in first aid and CPR. "We also have special training in dealing with bites and stings from local snakes, scorpions, spiders, bees, wasps, and even nasty plants."

At the scene the victim complains of a strange metallic taste in his mouth. The swelling around the area of the bite doesn't affect the entire leg. These symptoms indicate to Amber that the snake injected a moderate amount of venom when it struck. Moving the patient as little as possible, the ranger administers first aid to the victim, and then loads him into the vehicle. They return to the ranger station, where an ambulance is just arriving.

"Giving first aid to snakebite victims is not a routine part of the job. Much of my time is spent dealing with the park visitors, checking campsites, monitoring trail conditions, and such. In fact about 80% of the job is public relations, with about 10% each for law enforcement and paperwork. In the summer we have a regular program of evening education programs about a variety of topics including such things as outdoor safety, native plants and animals, stars, and folklore.

I first started working for the state parks department when I was still in college as a seasonal aide in the summers. While the pay and the living conditions aren't the greatest, you can't beat the fringe benefits of working outdoors around all of this natural beauty."

MORE ON PATHOGENS

Even the plant kingdom holds dangers for some of us. Hay fever is an allergic reaction to the pollen of flowering plants. Symptoms of hay fever include sneezing, coughing, and a runny nose. Reactions to pollen are a lot like reactions to injected bee or ant venom, though in this case the sufferer's own body produces a different set of unpleasant symptoms. Some people who want to pursue a career in agriculture have been forced to find other employment because of allergies to plants or insects.

Viruses cause diseases as ordinary as the common cold and as deadly as AIDS (acquired immunodeficiency syndrome). The AIDS virus does its damage by attacking a particular type of white blood cell known as the helper T cell. The resulting loss of helper T cells cripples the immune system's ability to fight off organisms that are normally kept under control. T cells are discussed in Lesson 15.3.

Nonliving dangers include poison, venom, and radiation. The source of poison and venom is often biological. For example, puffer fish have poisonous organs, which can kill diners if the fish isn't prepared by an expert chef (Figure 15.2). The fish's poison blocks the action of the nervous system in animals that attempt to prey upon the fish.

Figure 15.2: *The internal organs of the puffer can cause death by poisoning if they are not properly prepared before eating.*

Beekeepers who have venom allergies must beware of their bees, for the insects inject venom with their stingers. A bee's toxin is called "venom" because it is injected. Proteins in the venom bind to molecules on the surface of white blood cells, making the cells release chemicals such as **histamine.** Histamine causes swelling. Some of the chemicals in venom make respiratory tubes constrict, so breathing becomes difficult. This, combined with abnormally rapid heartbeat and abnormally low blood pressure, are the real dangers of a bee sting. The condition of the allergic victim is known as anaphylactic shock.

Poisons don't always come from living things. Coal tars produced by industry and compounds present in cigarette smoke can be inhaled and chemically modified inside the body until they assume a form that damages DNA. This alteration of DNA is thought to be a cause of cancer. Mercury is a metal that is deadly in some forms. Symptoms of mercury poisoning include bleeding of the gums, deafness, trembling, and digestive tract problems. Mercury poisoning is a concern for people who produce the metal from ores or who produce some electrical and mechanical devices, such as pumps, thermometers, and thermostats.

Ultraviolet (UV) light from the sun is natural radiation that can cause skin cancer by mutating DNA. It also damages eyes. Excessive exposure to X rays from medical equipment has been linked to other forms of cancer and to sterility (Figure 15.3). High-energy forms of light such as gamma rays and X rays are classified as ionizing radiation because they can remove electrons from atoms, giving normally neutral atoms a positive charge. The positively charged atoms take part in chemical reactions that can be harmful. Young people are generally more susceptible to X-ray damage than older people because young people's cells are dividing more often as they grow. One of the most common causes of death from excessive X-ray exposure is damage to the immune system cells, causing the victim to die from an infection. Health care workers in hospitals and dentists' offices must be particularly careful to protect themselves from excessive exposure to X-ray machines.

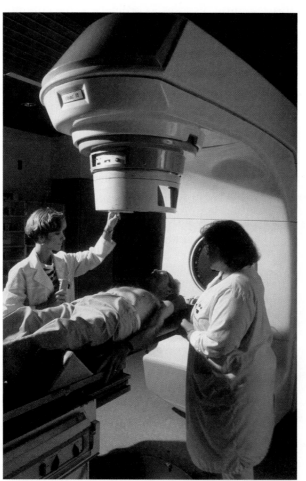

Figure 15.3: *X rays can be useful to the dentist or physician, but excessive exposure can cause illness or death.*

LESSON 15.1 ASSESSMENT

1. To reduce swelling.
2. Accept any reasonable answer, such as making sure the chlorine content is at its proper level.
3. Count the number of living cells and compare it to the number of living cells before infection with HIV.
4. Accept any reasonable answer, such as keeping hands and area clean, making sure food is completely cooked, and not using spoiled food.

1 You are having an allergic reaction to a bee sting, and your doctor prescribes an antihistamine. State the purpose of the antihistamine in this situation.

2 In Atlanta, several children became infected with a virulent strain of the bacterium *Escherichia coli* at a local water park. Recommend a measure that park officials could take to ensure the safety of their customers.

3 You are a research scientist. You have treated HIV-infected cells with an antiviral drug. Using only a microscope, how would you determine the activity of the antiviral agent?

4 Suggest several ways to prevent the spread of pathogens in food preparation.

LESSON 15.2 NONSPECIFIC DEFENSES

Lifeguards, roofers, and landscapers spend a lot of time exposed to the sun. People in these professions risk developing cancer from excessive exposure to UV radiation. Their skin provides some protection against this threat, and it is one of several nonspecific defenses of the human body. The skin layer is called a nonspecific defense because it protects us against many environmental hazards, including bacteria and viruses.

ACTIVITY 15-1 TESTING SUNSCREENS

The procedure for this activity can be found in Chapter 15 of the TRB.

UV light from the sun can damage unprotected skin and may lead to melanoma, a type of cancer. If you are active in the sun, you should shield your skin from exposure to sunlight. One protection is sunscreen. Many types are available. Each is rated with a sun protection factor (SPF). In this activity you'll test the ability of various sunscreens to shield you from UV light. You'll use a test strip that is sensitive to UV light.

TATTOOING

BIOLOGY IN CONTEXT

Tattooing, a technique of marking the skin with colors, is an ancient practice that appears to be making a comeback. A 1990 survey of 10,000 U.S. households revealed that 3% of the population as a whole, and 5% of men surveyed, have tattoos.

Permanent tattoos are applied by using a small electric machine with a needle bar that holds 1–14 needles, each in its own tube. The tattooing machine operates like a miniature sewing machine: The needle bar moves up and down as it penetrates the superficial layer (**epidermis**) and middle layer (**dermis**) of the skin. The tattooist holds the machine steady while guiding it along the skin. A foot switch controls the electric current. The needles protrude only a couple of millimeters from the tubes, so they don't go deep into the skin.

Getting a tattoo can be painful. The severity of the pain depends on the site of the tattoo and the person's tolerance for pain.

Temporary inflammation around the tattoo is common for a day or so. As part of the healing process, the tattooed skin crusts slightly and peels in the first week. Some people have an adverse reaction to a particular pigment that is used in the tattoo, which may result in swelling or itching.

Hepatitis B, a viral infection, can be transmitted from one customer to another if the tattoo needles aren't sterilized. According to Kris Sperry, M.D., a forensic pathologist in Atlanta, "If a tattooist follows appropriate cleanliness procedures, and the person who receives a tattoo takes proper care of it, the risk of infection at the site of the tattoo is minimal, and the risk of picking up any type of bloodborne pathogen is virtually nil."

Source: This article is adapted from the FDA/CFSCAN Web site from an article by Marilynn Larkin entitled "Tattooing in the 90's: Ancient Art Requires Care and Caution," reprinted from *FDA Consumer*, October 1993.

SKIN AND DEFENSE

The nonspecific defenses of the human body include the skin (our largest organ) and the moist mucous membranes lining our internal cavities and organs. Skin forms a barrier against a wide variety of agents, both living and nonliving, that would otherwise cause damage. It doesn't protect against just a single species of infectious agent or type of radiation. (Grasshoppers and other arthropods have an outer surface that provides nonspecific protection. But their surface is actually their skeleton, not a skin.)

The anatomy of human skin reveals its protective abilities (Figure 15.4). The outermost layer consists of dead cells filled with the waterproof protein keratin, which helps to protect the living cells that lie beneath it. The epidermis of the skin consists of the outer dead cells as well as several living layers beneath, including one with melanocytes. Melanocytes are cells that produce a brown pigment called melanin. Different amounts of melanin account for the different skin colors among humans. Melanin has a protective function. UV light from the sun makes melanin move into the upper skin layers, where it protects against further exposure to the radiation by darkening the skin.

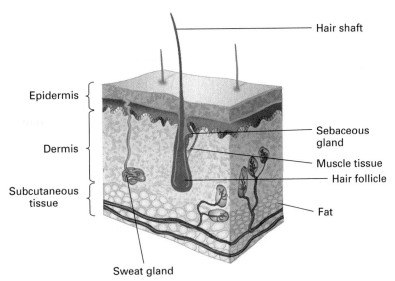

Figure 15.4: *Cross section of skin showing sweat and sebaceous glands*

Hairs are also part of the epidermis. They clearly protect many mammals from cold, wind, and skin damage, but the defensive function of human hairs largely disappeared during our evolution. Below the melanocytes and surrounding the bases of the hairs is a region of the skin called the dermis. It contains sweat glands and oil glands. Sweat glands defend us against excessive heat by excreting sweat that evaporates and removes heat from the body. Oil glands lubricate the skin and produce antibiotics that kill bacteria. The skin secretes acidic substances that make it difficult for bacteria to survive, adding to the simple protection provided by the physical barriers. If you sell reptiles in a pet shop or if you manage a poultry farm, you come into contact with the skin of these animals when you handle them. Scales and feathers are protective skin elements that arise from the epidermis much like the hairs of mammals. Hair and feathers evolved as modifications of scales.

OTHER NONSPECIFIC DEFENSES

Mucous layers line our internal surfaces just as skin covers our outer surface. The mucous lining of the upper part of the respiratory tract (the trachea and bronchi) is protected by cilia. Cilia are cell outgrowths that sweep dust and microbes upward to be expelled when we cough. Cilia are vital protectors of the lungs.

The nonspecific defenders include enzymes secreted with saliva, digestive fluids, and even tears. For example, the enzyme lysozyme

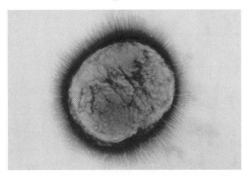

Figure 15.5: *This bacterium is a normal resident of every human intestine.*

helps to destroy the cell walls of bacteria. Saliva contains a slippery protein that protects the mouth lining from physical damage. Much of the food that we eat is naturally contaminated with bacteria. The acids and enzymes of the stomach can kill many of them. Even our own intestinal bacteria (Figure 15.5) provide a nonspecific defense. Invading bacteria must compete with these normal residents.

A less familiar nonspecific defense is provided by the protein **interferon.** Vertebrate cells secrete this substance when they become infected by viruses. Interferon is our most rapid defense against viruses. It makes other cells prepare defenses against the infective agent by producing proteins that slow the invaders' reproduction. There are three types of interferon. One type is released in response to viruses, another is released in response to viruses and bacteria, and the third is released in response to foreign substances of various types. Interferon is used to treat rabies, hepatitis, and eye infections. It is produced as a defense by many species. Most if not all vertebrates make interferon, and plants have a similar defense against viruses.

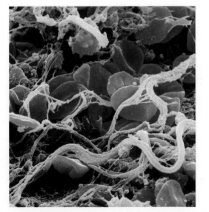

Figure 15.6: *These two red blood cells and the protein threads are part of a blood clot.*

Injury to the skin or mucous membranes is often accompanied by a nonspecific defense known as **inflammation,** swelling caused by body fluid and white blood cells leaking from blood vessels into tissue. The injury could be caused by an insect bite or sting, a wood splinter, or even sunburn. Even a vaccination needle, which is intended to help rather than hurt, can introduce infective bacteria into the body if the needle isn't sterile. Damage to cells caused by the needle itself stimulates the release of histamine and **prostaglandins.** These substances increase blood flow through the capillaries and make them more permeable. This enables phagocytic white blood cells and blood-clotting proteins (Figure 15.6) to more easily leave the capillaries and approach the site of the wound. Clotting has two functions: It begins repair of the wound, and it helps to prevent bacteria from spreading beyond the area of the wound.

The pus found in an infected wound is a collection of body fluids and living and dead white blood cells and bacteria.

LEARNING LINK PYROGENS TO THE RESCUE

Pyrogens are substances that help to produce fevers. Fevers are high body temperatures that can destroy pathogenic bacteria and viruses.

The inflammatory response is a localized reaction to skin or mucous membrane damage and is nonspecific because it responds to a wide variety of damaging agents. Inflammation can be reinforced by a more widespread response from the body. Use the library or the Internet to learn about pyrogens and their role in the inflammatory response. What kinds of cells release pyrogens? What does a pyrogen do?

LESSON 15.2 ASSESSMENT

1. Protecting the patient from infection. When the skin is broken, its defensive function diminishes.

2. Fever is a general response against a variety of pathogens. A higher body temperature kills them.

3. Coughing removes mucus from your lungs. Repressing coughs would leave your lungs congested.

4. Fever or diarrhea or vomiting.

1 You are an emergency room physician. A burn victim comes into the emergency room. His vital signs are stable, and he is conscious, but you notice that the skin in the burned area is broken. What would be one of your major concerns? Give a reason for your answer.

2 When you get sick with a cold, your body temperature rises, causing a fever. Discuss why this is a nonspecific defense and why it works in your favor.

3 You are feeling really sick, and you have been coughing up mucus all morning. When you tried to take a cough suppressant, your mother told you that it will only make you sicker. Why might taking cough suppressant make you worse?

4 A microbe is irritating the lining of your gastrointestinal tract. What nonspecific response would help get rid of this microbe?

LESSON 15.3 SPECIFIC DEFENSES

Humans and other vertebrates can mount very specific defenses against invaders. For example, a particular type of white blood cell will produce antibody molecules that recognize a specific invader and help to attack it. This lesson explores these specific defenses, a part of the immune system's response to invasion by foreign agents.

ACTIVITY 15-2   MODELING AN ANTIBODY

The pattern for the model and the procedure are in the TRB.

Each student will need a sheet of construction paper, a ruler, scissors, a roll of masking tape (3/4" wide), a stapler, white glue, and a pen or pencil.

Antibodies are proteins that defend our bodies against certain invaders. They have a unique structure that allows them to be identified at a glance from their pictures. In this activity you will come to know this characteristic shape by assembling a model of an antibody molecule from paper. Your finished model should look like Figure 15.7. Your teacher will provide a handout for this activity.

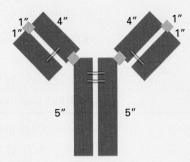

Figure 15.7: *The Y-shaped antibody molecule has two heavy amino-acid chains and two light amino-acid chains. A hinge region is at the bend of each heavy chain. The small squares represent invading molecules bound to the antibody.*

Answer the following questions in your logbook:

Each light chain is bound to its heavy chain partner by one bond.

The heavy chains are bound to one another by two bonds.

Lysozyme and antibodies are both proteins.

Two invaders can be bound simultaneously.

A light chain and a heavy chain bind to each invader.

- How many bonds bind each light chain to a heavy chain?

- How many bonds bind the heavy chains to one another?

- What do lysozyme and antibodies have in common?

- A 1" square in your model represents the invader attacked by the antibody. How many invaders can the antibody attack at once?

- Which antibody chain (or chains) binds to the invader when the antibody attacks?

In 1984 a 12-year-old boy named David (his last name was never made public) died after a brave struggle against a rare genetic disorder: severe combined immunodeficiency. David survived for 12 years without an immune system. An older brother, born with the same disease, died in infancy. Doctors anticipated that David might also have the disorder, so he was placed in a sterile environment right after birth.

Also known as "the boy in the plastic bubble," David progressed from infancy to boyhood inside a series of sterile "bubbles." School came to his bubble through the telephone, and he was an A student. Sterilized food was slipped through a series of airlocks.

The only way for David to begin a normal life was to transplant bone marrow (a source of cells that would rapidly multiply in his body to ward off invaders) from a family member. The best bone marrow donor would be a sibling. David had one sister, whose marrow was not an exact match. However, the doctors decided to transplant 50 grams of the sister's bone marrow.

Sadly, the bone marrow transplant was not successful. After surgery, David had constant fever, diarrhea, and vomiting. He was taken out of his sterile environment to be treated. He grew sicker and finally died of heart failure because of fluid around his heart.

David wanted only to walk barefoot on the grass and some day to live normally in the world. He was able to briefly explore the world in 1977, when NASA built a special suit that allowed him to be safe outside the bubble. The whole country was so interested in David's case that a movie was made about his life.

THE HUMAN IMMUNE SYSTEM

The immune system of humans responds to foreign agents in two ways. In **humoral immunity,** antibody molecules in the blood provide a specific defense. **Cell-mediated immunity** involves interactions between white blood cells; one type of cell regulates the behavior of others. The two types of immune response interact with each other.

Leukocytes
(white blood cells)

Eosinophil

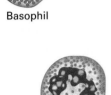

Basophil

Lymphocyte

Neutrophil

Monocyte

Figure 15.8: *The variety of white blood cells*

Both responses involve white blood cells, or **leukocytes** (Figure 15.8).

Lymphocytes, one kind of leukocyte, exist in two varieties: **T lymphocytes** and **B lymphocytes** (also known as T cells and B cells). T lymphocytes have a variety of functions and are important in cell-mediated immunity. B lymphocytes are the basis of humoral immunity. They secrete **antibodies,** proteins that circulate in body fluids. Antibodies attack bacteria, viruses, and even venom that has been injected by animals and poisons that have been swallowed. Antibodies recognize and bind to foreign molecules called **antigens,** which stimulate an immune response. Antigens are usually proteins, but some are not. They are on the surfaces of viruses, bacteria, fungi, protozoans, plant pollen, and parasitic worms. They are also in skin, organs, and other tissues and in venom injected by bees, wasps, ants, and snakes.

The production of antibodies associated with the first exposure to an antigen is called the **primary immune response.** Each B lymphocyte is specialized to fight only one type of invader, and these particular cells increase in number when the antibodies on their surfaces bind to the antigens of their special pathogen. The many identical B cells that are produced by the primary response are called a clone, cells that are genetically identical to the parent cell. Some of the cells, called **plasma cells,** circulate immediately in the blood. Others don't become active until the second infection by the same antigen-bearing agent. The cells that lie in wait are called **memory cells,** and their activation is the **secondary immune response,** the release of antibodies in response to the reappearance of a pathogen.

Activating B cells so that antibodies can spread through the blood vessels requires cells to bind to foreign antigens. It usually also requires input from other body cells in a process called cell-mediated immunity. Macrophages and T lymphocytes called helper T cells are important in the early stages of cell-mediated immunity. It begins when a macrophage engulfs the invader, leaving part of the invader exposed at the surface. A helper T cell also binds to that exposed antigen, so the macrophage and helper T cell form a pair connected by antigen. This encounter activates the T cell, and it divides many times to form active cells. The active cells encounter B cells that have antigen on their surfaces, and the two lymphocytes bind, prompting the B cell to secrete antibodies. Here you see the connection between the B cells of humoral immunity and the T cells of cell-mediated immunity.

How do circulating antibodies help to destroy the infectious agents and poisons? This can happen in several ways. First, they can bind in large numbers to large groups of antigens, creating a clump that phagocytic cells can engulf and destroy. Second, they can bind to an antigen that the invader needs for some vital function (Figure 15.9). For example, when an antibody binds to certain virus proteins, the virus can't invade a host cell because that requires a protein that is now covered up by the antibody. Third, antibodies can work with another type of protein called **complement** to destroy infectious cells directly. Complement proteins make the invading cell burst and die.

The secondary immune response provides natural immunity if the antigen attacks a second time. Vaccinations provide a form of acquired immunity, a defense that is put in place by injecting an individual with weakened or dead infectious agents. Because these harmless forms have the normal antigens, they stimulate the production of memory cells that will immediately respond

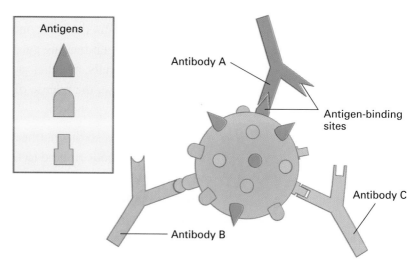

Figure 15.9: *Antibodies binding to antigens on a foreign cell*

to a future attack by an active and therefore dangerous form of the agent. Vaccination is an active form of acquired immunity because it prepares the immune system for the production of antibodies. Passive immunity is different. A mother passes some of her antibodies to her fetus during its development, putting a defense in place without requiring the fetus to produce the antibodies itself.

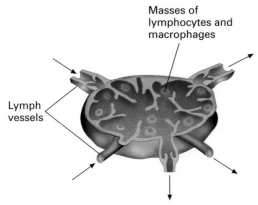

Masses of lymphocytes and macrophages

Lymph vessels

Figure 15.10: *Lymph nodes remove bacteria and viruses from circulation.*

The **lymphatic system** has an important role in defense. Somewhat like the circulatory system of arteries and veins, the lymphatic system consists of vessels that carry fluid. This fluid originates from circulatory system leakage, and it returns to the circulatory system via the lymphatic system. Unlike blood vessels, lymph vessels connect to swellings called lymph nodes (Figure 15.10). The fluid moving through the system is called **lymph.** Lymph is a yellowish, watery fluid. White blood cells in the nodes remove bacteria and viruses from the lymph. The presence of high numbers of white blood cells explains the swollen, tender lymph nodes that often accompany an illness.

A THORNY PROBLEM

CHANGING IDEAS

The discovery that certain cells protect animals from invaders was made by the Russian-born French zoologist Ilya Mechnikov in the 1880s. According to Mechnikov, he had a revelation while watching the tiny, transparent larvae of sea stars. Inside the larvae were large, amebalike cells that crawled about and seemed poised for a defensive function. The scientist tested his idea with one of the simplest experiments imaginable: He stuck a rose thorn into the sea star and waited to see what would happen. The next day, the tip of the thorn was surrounded by the crawling cells, cells that are known today as macrophages. In the human body, macrophages truly behave like feeding amebas, because they ingest bacteria and other small foreign bodies. Macrophages are part of our nonspecific defenses. They destroy many kinds of invaders.

Mechnikov later showed that macrophages can be vital for survival. When an animal's macrophages failed to destroy the spores of an infective fungus, the animal died. Mechnikov emphasized the role of these phagocytic cells in defense, while other scientists emphasized the role of antibody molecules, which circulate in the blood. Modern immunologists know that cells and antibodies cooperate in defense.

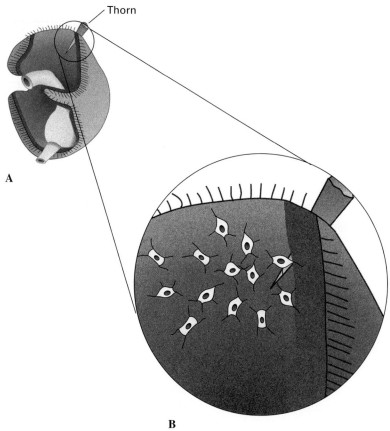

A diagram of Mechnikov's experiment. (A) A sea star stuck by a rose thorn. (B) Macrophage cells swarming around the thorn.

INVESTIGATION 15B ANTIGEN-ANTIBODY INTERACTION

The objective of this experiment is to introduce the principles of antigen-antibody interactions by using the Ouchterlony procedure. Antigens, antibody, and reagents are provided to observe this binding interaction, which results in the formation of a white precipitate after diffusion in agarose.

See Investigation 15B in the TRB.

The binding of an antibody to an antigen is a fundamental reaction of immunology. Antibodies and antigens frequently form complexes that drop out of the solution in which they were dissolved, making it possible to analyze antibody-antigen systems. This dropping out is called "precipitation." In this investigation you'll learn about antibody interactions by studying precipitation.

LESSON 15.3 ASSESSMENT

1. Yes, if T and B cells aren't functioning, antibodies aren't produced to fight off pathogens that invade the body.
2. The yeast must produce antibodies that will destroy the target virus.
3. Memory cells give the same antibodies the ability to fight off the same pathogen after the first exposure.
4. A flu vaccine will produce both a specific defense and a nonspecific defense.

1 An ill person has gone for medical tests. The results indicate that T cells and B cells aren't performing their normal functions. Would this problem be serious if the T and B cells remained inactive? Give a reason for your answer.

2 You are a genetic engineer. You are trying to design a vaccine by inserting human genes in yeast. Summarize what genetically altered yeast have to do to provide an effective vaccine.

3 Animals often produce antibodies in response to the first exposure to a pathogen such as a virus. Explain how the body defends itself from several invasions of the same pathogen.

4 When getting a flu vaccine, you are warned that you may develop fever. Would a flu vaccine produce a specific defense, a nonspecific defense, or both?

LESSON 15.4 DISTINGUISHING SELF AND NONSELF

If you were working on the computer security in a corporation, you might be responsible for providing passwords for yourself and the other employees. You and your coworkers would have access to privileged information forbidden to outsiders. In fact, outsiders might be prosecuted if they were caught attempting to hack into your computer system without the proper password. In other words, your job would require you to distinguish employees from outsiders. That is the subject of this lesson: how your body recognizes its own cells and those of foreign agents.

ARTIFICIAL SKIN

BIOLOGY IN CONTEXT

If you've seen cartoon characters such as Sylvester the cat or Donald Duck get severely burned, you know that in the next scene their skin has grown back. The cartoon character looks as if nothing had happened. It seems as though a group of scientists in Massachusetts have been watching Saturday morning cartoons, because they have found a way for burn patients to regrow skin.

One of the major concerns for burn patients is how their immune system will react to skin grafts. The immune system fights off invaders, and transplanted organs seem like invaders. Kidney, heart, liver, and skin transplant patients require immunosupressant drugs to prevent the immune system from attacking the transplanted organ. If you grow your own skin, however, the immune system will not consider it an invader.

This was good news to Mark W., an electrician who had serious burns over 80% of his body. His doctors used microscopic fibers made from cow tendons, shark cartilage, and a biological polymer to regrow his skin. When implanted, the microscopic fibers became a frame for Mark's skin cells to grow on. Most artificial parts are rejected by the immune system, but because of the polymer's biological properties, Mark's immune system did not reject the fibers. After 30 days, Mark's skin had grown back, and enzymes had dissolved the implanted fibers. His skin looked almost as if nothing had happened.

THE MAJOR HISTOCOMPATIBILITY COMPLEX

It shouldn't be surprising at this point to discover that your body recognizes its own cells by the proteins on their surfaces. These glycoproteins are the passwords of self. Part of their structure is a sugar. The genes that code for these self-recognition proteins make up the **major histocompatibility complex, (MHC)** (Figure 15.11). The group of genes was discovered while scientists were searching for a way to avoid tissue rejection when tissues are transplanted from one individual to another.

There are at least 20 different MHC genes, and each gene has many alleles. This variability means that no two people (except for identical twins) will have the same set of MHC alleles. Your MHC genes are as individual as your fingerprints. The proteins for which these genes code, "password" proteins appearing on the surfaces of your cells, are known as antigens even though they aren't foreign to you. The two types of MHC antigens are known as Class I and Class II. Class I molecules are on all nucleated cells. Class II molecules are restricted to macrophage and B cells and have a critical role in the cell-mediated immunity that we discussed earlier.

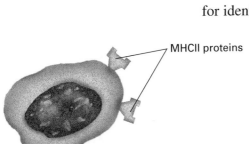

MHCII proteins

Figure 15.11: *MHC proteins are on the surface of cells.*

To see how the MHC II proteins help the immune system to do its jobs, let's reexamine the interplay of macrophage, helper T cell, and B cell. The macrophage uses its MHC II protein to bind with the foreign antigen, possibly a bacterium. The helper T cell becomes active when it binds with this combination of self (MHC) and nonself (antigen) protein. Then B cells become activated when their combination of self and nonself binds to the helper T cell.

Differences among people in their MHC proteins cause the problems and frustrations of tissue transplantation. Skin transplants, or grafts, help to heal severe burns, and the need for compatible liver, heart, and bone marrow tissues is a frequent news item. Unless two individuals are very closely related, the recipient's immune system will reject and destroy the transplanted tissue. Here's how it works: If the donor's MHC I antigens are different from those of the recipient, cytotoxic T cells, a type of lymphocyte, identify the tissue as nonself and attack with a form of cell-mediated response. Someone who needed an organ transplant would be in an ideal position if he or she had a willing identical twin. Otherwise, brothers, sisters, or parents are the best candidates for donation. Drugs that suppress the immune system response are often used so that the recipient's system doesn't reject the donor's tissues. However, the drugs make the recipient more susceptible to infection. **Cyclosporine** is the drug of choice because it suppresses the cell-mediated response by cytotoxic T cells without shutting down the humoral response.

Normal differences among individuals' MHC proteins cause problems for transplants; disorders or mutations in the MHC play a role in **autoimmune diseases.** In this case the problem has nothing to do with a donor. In an autoimmune disease an individual's antibodies attack his or her own tissues. For example, mutant MHC II genes are common in people who have multiple sclerosis and rheumatoid arthritis. Let's look at some examples of autoimmune disease.

WORD BANK

cyto- = cell

-toxic = poison

Cytotoxic T cells kill other cells.

WORD BANK

sclero- = hard

-sis = condition

In sclerosis, something becomes hardened.

Activity 15-3   Understanding Autoimmune Diseases

A procedure for this activity is in Chapter 15 of the TRB.

Each student will need two blocks of modeling clay (each 4" long and about 1" high and 1" wide, though these last two dimensions are not too important), two 4" pieces of solid, insulated wire (14-gauge or smaller; white insulation is best), a wire stripper, a knife, a 1" square of rough sandpaper, double-stick tape, and a new pencil with an unused eraser.

The insulation represents myelin.

The action potential travels from right to left.

What's it like to have an autoimmune disease? In this activity you'll use simple materials to understand what goes wrong in three autoimmune diseases. You will learn about defects at the cellular, molecular, and tissue levels. Your teacher will provide a handout with the procedure for this activity.

Answer the following questions in your logbook:

- In the central nervous system model, what does the white insulation represent?

- In the model of the neuromuscular junction, in which direction does the action potential travel?

When a Body Self-Destructs

Multiple sclerosis (MS) is an autoimmune disease that destroys the myelin sheath (Figure 15.12) of the central nervous system. It typically develops in young adults and afflicts about twice as many women as men. Destroying the myelin sheath has effects comparable

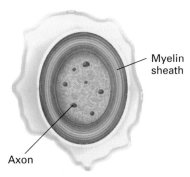

Figure 15.12: *Some nerve cells are normally protected by a layer called the myelin sheath.*

to removing insulation from an electric wire. The bare neuron does not conduct action potentials well. Symptoms include blindness in one eye because of improper signal conduction in the optic nerve, weakness on one side of the body, lack of coordination, and tingling and numbness in the arms or legs. The condition gets worse until walking or even speaking becomes difficult. The victim's own antibodies destroy the myelin sheath. Although the disease runs in families, virus infection at an early age seems to trigger MS. Hormones and the drug cyclosporine can slow the progress of the disease. Although people with MS usually become handicapped in some way, few die from the disease.

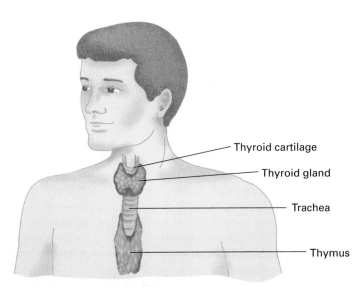

Figure 15.13: *The thymus gland helps white blood cells develop but might also produce disease-causing antibodies in myasthenia gravis.*

Myasthenia gravis is an autoimmune disease in which neurotransmitters can't bind to the receptors of muscle cells. The body's antibodies bind to the receptors instead, as if the receptors were foreign agents or invaders. The problem-causing antibodies may come from the thymus gland (Figure 15.13) in the upper chest, which helps white blood cells to develop. The disease tends to strike young women more often than young men. Muscles become weaker over time, and the condition can kill when it affects the muscles that are needed for breathing. Patients take drugs that slow the normal breakdown of neurotransmitters in the synapse.

Rheumatoid arthritis deformed these hands.

The precise cause of rheumatoid arthritis is unknown, but it may be an autoimmune disease. The joints between bones swell, deform, and eventually fail to work. It usually strikes young or middle-aged adults. The lubricating cartilage between the joints is destroyed, and even the exposed bone erodes when the protective cartilage is gone (Figure 5.14).

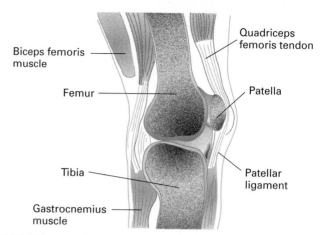

Biceps femoris muscle

Femur

Tibia

Gastrocnemius muscle

Quadriceps femoris tendon

Patella

Patellar ligament

Figure 15.14: *This normal knee joint shows a band of cartilage between the upper and lower leg bones. Rheumatoid arthritis destroys the cartilage.*

Lupus erythematosus is an autoimmune disease caused by antibodies that destroy the body's tissues. A strange symptom of the disorder is a butterfly-shaped rash across the nose and cheeks (Figure 15.15). Lupus usually strikes young or middle-aged women. Mild cases don't require treatment, but severe lupus can damage the kidneys. Treatment in these cases includes aspirin and steroid hormones, which reduce the inflammation in internal organs.

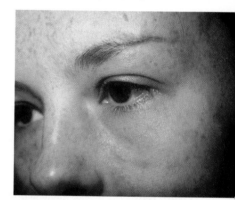

Figure 15.15: *Those who suffer from lupus erythematosus often have red rashes on their hands and face.*

Another autoimmune disorder is insulin-dependent diabetes, also known as juvenile-onset diabetes or type I diabetes. It occurs when the body's immune system destroys the pancreas cells that produce insulin, a protein that regulates sugar in the blood. Daily insulin injections prevent disturbances of the body's acid/base balance and accumulation of waste molecules in the tissues. This form of diabetes (there are several others) affects one out of 600 children. At one time, diabetics had to rely on insulin extracted from cattle or pigs. Biologists have altered a species of bacterium that lives in our intestine to produce human insulin in the laboratory.

LESSON 15.4 ASSESSMENT

1. The immune system will treat the donor skin, but not your own skin, as a pathogen.
2. Damaged cells.
3. Take antibiotics to fight off infection.
4. No.

1 Explain why grafts of your own skin are more successful than skin grafts from donor skin.

2 To monitor the progress of a heart transplant patient, small pieces of the transplanted heart are removed and examined. What would the doctors look for?

3 Immunosuppressant drugs prevent the body from rejecting a transplanted organ but increase the possibility of infection. What can be done to prevent the invasion of pathogens?

4 Without drugs, will the immune system ever stop treating a transplanted organ as a pathogen?

CAREER APPLICATIONS

The applications that follow are like the ones you will encounter in many workplaces. Use the biology you learned in this chapter to complete the activities. Share your work with the class.

AGRICULTURE & AGRIBUSINESS

Animal Health Care Careers

Use the Internet and career references, or contact veterinarians, veterinary hospitals, farms, ranches, zoos, pet shops, animal shelters, and similar places to research employment opportunities in animal health care careers. Where are the jobs? What are the educational requirements for the jobs you identified? Present the information to the class in a chart, graph, table, or poster. If possible, invite people who are involved in different types of animal health care to come to class for a panel discussion.

Animal Health

Select an animal and find out what diseases it can catch. What is done to prevent the disease? What lowers the animal's overall resistance to disease? What increases the animal's overall resistance to disease? Prepare a brochure or fact sheet that discusses the animal and its diseases. Present your findings to the class.

BUSINESS & MARKETING

Allergies and Airplanes

Many people fly frequently for their jobs. Long flights can be a problem for people who have allergies. The air in an airplane recirculates, so allergens can build up. Imagine the problems that someone with a peanut allergy might have when most of the passengers open their bags of peanuts within minutes of each other. If the allergy is severe, it could trigger a life-threatening reaction. Certain fragrances can also trigger an allergic response. Talk to an allergist, an airline representative, a travel agent, and/or someone you know who has allergies and must fly often. Find out what passengers with allergies can do before, during, and after a flight to minimize allergic reactions. Create a brochure for people with allergies that will tell them how to fly with minimal allergy problems.

Marketing Good Health

Work in a small group. Prepare a five-minute or shorter skit that addresses some aspect of the immune system. The skit should include facts about the immune system and facts and fallacies about foods, supplements, home remedies, and over-the-counter drugs that are supposed to boost the immune system's function. If possible, videotape your skit.

**FAMILY &
CONSUMER
SCIENCE**

Chemicals in Cosmetology

Invite a beautician or cosmetologist to speak to your class about protection from the chemicals in beauty products. Ask questions such as the following: How do you protect your skin from the chemicals at work? What special substances do you use to sanitize skin and nails before starting manicures? What level of knowledge about chemical reactions is necessary or helpful in your job?

Skin Cancer and the Sun

Contact a dermatologist or the American Cancer Society for pictures and written information that help to distinguish melanoma from other skin conditions. Develop a list of warning signs for skin cancer. Visit the American Association of Dermatology home page on the Internet to get information on the safety of tanning, using either natural sunlight or tanning beds. What misconceptions do sunbathers have about the practice of tanning? Make a poster of guidelines for people to follow to protect themselves from the effects of harmful UV radiation.

**HEALTH
CAREERS**

Shots for Foreign Countries

Use a globe or an atlas to pick a foreign country at random. If a classmate has already picked that country, choose another one. Use the Internet to visit the World Health Organization's Web site, call a travel agent, or use other resources to find out what health advisories, if any, are in effect for that country. What immunizations are required? Why do people often become ill when traveling outside their country while their native tour guides do not? If you needed to stay a long time in the country you chose, what could you do to lower the risk of becoming ill? Present your findings to the class, using a visual aid.

Immunization Schedule

Check with your county health department, family physician, pediatrician, school nurse, or other health professional to find out immunization requirements and recommendations. Try to find immunization records for yourself and other members of your family. If someone in your family is not up to date with their immunizations, try to convince him or her to get the required immunizations. New vaccines are always under development. Did your parents or grandparents have any of the diseases that you are immunized against?

Filters for Faster Recovery

One problem with heart transplant surgery is that the patient's blood is rerouted through a "heart-lung machine" that acts as an artificial heart while the patient's defective heart is removed and the donor heart is implanted. During that time the blood goes through plastic and metal tubing and pumps. The immune system mounts defenses against these materials. Research the filter system that has been proposed and is being tested that reduces this problem. How does it work? What difference does it make in recovery time? Write a short report in your logbook and present your findings to the class.

INDUSTRIAL TECHNOLOGY

Producing Sterile Products

As a class, make a list of products that you know are sterile and those that you think might be. Using the Internet or other resources, research how sterile products are produced and packaged. How do workers maintain the sterility of the product in its manufacture and packaging? What types of materials are used for the packaging? What special precautions must workers take to keep from introducing contaminants into the product? What clothing or equipment is used? Why are all of these safeguards put in place? What might happen if the product you researched was contaminated? Write a short report in your logbook and present your findings in class.

Personal Protective Equipment

The human body has its various defenses for normal types of antigens, pathogens, and poisons, but some workplaces expose workers to materials against which these defenses are insufficient. Obtain catalogs of personal protective equipment or use the keywords "personal protective equipment" to search the Internet. List the types of equipment and the types of antigens, pathogens, and poisons that each protects against. Each student should select a specific industry or workplace and prepare a brochure or poster of the personal protective equipment that can be used in that environment to protect workers against exposure to harmful materials. Use illustrations to show the proper way to wear the equipment.

CHAPTER 15 SUMMARY

- Organisms defend themselves against a wide variety of infectious agents, toxins, and radiation.

- Humans have enemies in all five kingdoms of life. Some of these are infectious, some cause damage by injecting venom, and some produce a poison that can be accidentally consumed. Bacteria and viruses are the most common infectious agents. Ants, bees, wasps, spiders, and snakes inject venom.

- Ultraviolet light and X rays are forms of radiation that can cause cancer and death.

- The nonspecific defenses protect against a wide variety of hazards.

- Four important nonspecific defenses are the barriers that the skin and mucous membranes provide, the inflammation response within the skin and the membranes, enzymes on and in the organs, and interferon secreted by infected cells.

- The specific defenses specialize against particular species or types of foreign cells or agents.

- Animals have a specific defense known as the immune system.

- The two types of immune response are the humoral response and the cell-mediated response. They are not entirely independent.

- B lymphocytes provide the humoral response by secreting antibodies that circulate in the body fluids.

- T lymphocytes and macrophages initiate the cell-mediated response by binding to one another or to other cells.

- The binding sites of cell-mediated immunity are combinations of antigen and MHC protein.

- The primary immune response is the immediate production of antibodies. The secondary immune response occurs when a second infection or invasion stimulates memory cells.

- Vaccination and the passage of antibodies from mother to fetus are forms of acquired immunity. Vaccination is an active form of acquired immunity, and passage of antibodies from one individual to another is a passive form.

- The lymphatic system plays an important role in defense. Bacteria and viruses are removed in the lymph nodes.

- The immune system is normally able to tell an organism's own cells apart from foreign cells and substances by the proteins that are coded for by the major histocompatibility complex of genes.

- Cytotoxic T cells recognize foreign MHC I proteins on cell surfaces and begin the cell-mediated response, which rejects or destroys the nonself material.

- B cells and macrophages use MHC II proteins to bind with antigen and to bind with T cells during the cell-mediated response.

- Autoimmune diseases occur when the body can't distinguish its own cells from the cells of other organisms. The immune system attacks the body's own cells. Examples include multiple sclerosis, myasthenia gravis, rheumatoid arthritis, lupus erythematosus, and insulin-dependent diabetes.

Concept Review

See TRB, Chapter 15 for answers.

1 What parts of the body are the first line of defense against infection?

2 Compare autoimmune diseases and AIDS.

3 Several kinds of pathogens can invade the human body. Which ones can't reproduce on their own and must live within the host cell?

4 Explain what causes an autoimmune disease.

5 Which types of blood cells are involved in inflammatory response?

6 Why does infection develop as a result of a cut or burn?

7 The eye is exposed directly to the environment. What protects the eye from pathogens?

8 Penicillin is an antibiotic that is produced by a fungus. Some people react violently to penicillin injections. Why?

9 How do antibodies burst foreign cells?

10 Identify the following as a nonspecific or specific defense:

a. inflammation

b. release of interferon

c. watery eyes

d. increased number of white blood cells

11 How do white blood cells fight bacteria at the site of an infection?

12 Pathogens enter our bodies in several ways. If you eat improperly refrigerated or stored food, what kind of pathogen is most likely to enter?

13 Explain why immunosupressant drugs are used to treat some autoimmune diseases.

14 Explain the differences between humoral immunity and cell-mediated immunity.

15 Is the secondary immune response the immediate production of antibodies? Give a reason for your answer.

16 Describe two conditions that would trigger an inflammatory response.

Think and Discuss

17 Summarize how the body recognizes the transformation of normal cells to cancerous cells.

18 Why are transplant patients who take immunosupressant drugs more likely to get virus-associated cancer?

19 Discuss the problems that could result from a mutation in the MHC.

20 Why won't the immune system of an identical twin reject the organs of the other twin?

In the Workplace

21 You have the grounds maintenance contract for a large daycare center. Your region is currently experiencing an outbreak of army worms (a type of caterpillar) that are destroying lawns. You have found them in the center's lawn. Because small children play on the lawn, you don't want to use a chemical pesticide. What might you use instead?

22 The highway department has hired you to talk to its road repair crew on steps they can take to minimize skin damage from being out in the sun and elements. What advice would you give them?

23 You are a public health nurse, and part of your job is to vaccinate people against various diseases. A young couple brings in their baby for the second round of vaccinations. They report that after the first shots, the baby was fussy, had a large knot where the injection was made, and ran a fever. How would you explain to the parents the reason for the reaction?

24 You work for a large medical center. Part of your job is to work with people who need tissue and organ transplants that can be obtained from living donors. These include things like bone marrow or a kidney transplant. A family brings in a list of family members, friends, and acquaintances who have volunteered to act as donors. Everyone on the list will be tested to find the closest match, but you want to start testing with the people who are most likely to be good candidates. List the relationships in priority order that will most likely match.

These questions refer to Investigation 15A:

25 How did the egg white affect bacterial growth? What does this indicate?

26 Explain why this would be important for an egg.

These questions refer to Investigation 15B:

27 What is the purpose of row 1 (PBS) and row 2 (positive control) of your microtiter plate?

28 Why is it important to screen all donated blood for viral antibodies?

CHAPTER 16

WHY SHOULD I LEARN THIS?

Advertisers use sex appeal to sell all sorts of products. Why does this work? Because the sex drive is strongly motivated by hormones and by marketing strategies. A constant stream of ads, songs, and movies portrays sexual activity without dealing with the possible biological consequences: pregnancy, sexually transmitted disease, and emotional impact. Learning more about the biology of reproduction can help you to make intelligent decisions about sex issues and reproduction.

ANIMAL REPRODUCTION

WHAT WILL I LEARN?

1. the different types of asexual reproduction
2. what influences sexual reproduction
3. the difference between external and internal fertilization
4. the organs and hormones in the human reproductive system
5. how a new human develops

When you were younger did you ever ask your mother or father where babies came from? Depending on your age, they may have told you that specially trained storks pick up babies from the baby factory and deliver them to their parents. Now that you are older you know that storks and baby factories are not a part of human reproduction.

People, like most other animals, reproduce sexually. A male sex cell and a female sex cell join, creating a single cell that will grow into a new individual. But some animals can reproduce without mating—they reproduce asexually. In this chapter you'll learn about several types of animal reproduction.

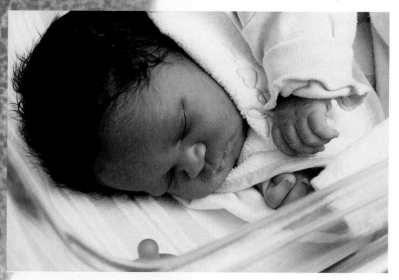

LESSON 16.1 ASEXUAL REPRODUCTION

In asexual reproduction an organism makes a copy of itself. The offspring's genetic material is identical to the parent's. Both the aphid and the sea sponge can reproduce this way. This lesson describes several types of asexual reproduction.

ACTIVITY 16-1   STYLES OF SELF-REPRODUCTION

Each student or pair of students will need prepared slides of *Hydra* budding, *Daphnia* with eggs, and a microscope (live cultures may be substituted for prepared slides).

A bud forms on *Hydra* as on budding yeast. It will grow larger and eventually break off. *Daphnia* can form eggs without mating at certain times of the year (parthenogenesis). These eggs develop into females. Both of these organisms reproduce asexually when food is abundant and the environment is fairly constant. Each can reproduce sexually as well.

Many small animals that are sessile (that can move very little or not at all), that can't find mates easily, or that live in a stable environment reproduce asexually. They do not need a male and a female to create eggs and sperm. In this activity you'll see some of the ways in which animals reproduce asexually.

Look at each prepared slide of an organism. In your logbook, draw the new organism that is beginning to grow on *Hydra*. Look at other students' slides and compare the different stages of development of the new organism. Record your findings in your logbook.

Find the developing eggs in *Daphnia* and hypothesize how they formed through asexual reproduction. Discuss with the members of your group ways to test your hypothesis.

HIGH-TECH CLONING

BIOLOGY IN CONTEXT

A sheep named Dolly captured the headlines in 1997. With the help of technology, scientists removed the genetic material from an egg cell and replaced it with the genetic material from an adult cell. They implanted the newly formed egg into a surrogate mother sheep. For the first time a mammal was produced without the union of male and female gametes.

Dolly is a **clone** (exact genetic copy) of the adult sheep whose genetic material was put into the egg cell. The scientists were trying to develop sheep whose milk would contain special proteins that are used in the drug industry. The sheep's cells would become factories that produced the proteins. Making an exact copy of the original sheep would increase the protein production.

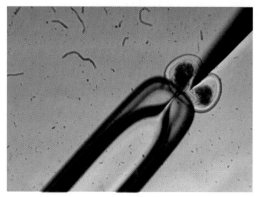

Using a glass scalpel and a microscope scientists can split a developing egg to produce a twin.

Scientists have for many years been able to use the same cloning technique that occurs in nature to create identical twins. They have been able to separate cells in newly formed embryos that are dividing by mitosis. The separated cells develop into two genetically identical offspring, and twins are born.

CHARACTERISTICS OF ASEXUAL REPRODUCTION

Animals that reproduce asexually often share several characteristics. Like *Hydra,* they may be sessile. Other sessile animals, such as sea sponges and coral, have limited or no movement, so they cannot search for mates. Asexual reproduction cycles are often related to the seasons. Animals may produce many identical offspring when food is abundant and when the young have good chances for survival. For example, *Daphnia* may produce many eggs that mature in the spring and summer.

Some animals, such as sponges, hydras, bees, and aphids, have the ability to reproduce both sexually and asexually. The worker bee is from a fertilized egg. *Daphnia* and aphids produce males to fertilize the eggs that survive the winter. These hatch as females in the spring.

Asexual reproduction quickly produces many offspring. This type of reproduction is successful when the environment is stable and changes very little. It is also successful for small populations of animals that can't easily find mates. The drawback to asexual reproduction is that there is little genetic diversity among the offspring. Because each offspring receives the genetic material only from the mother, fewer genetic combinations are available to adapt to changing environments.

BUDDING

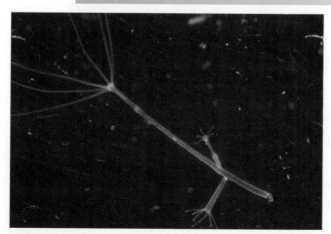

Figure 16.1: Hydra *reproduces asexually.*

Hydra, a cnidarian, often reproduces asexually. The offspring begins as a small bump or bud on the parent's tubular body (Figure 16.1). The bud cells divide and increase in number until tentacles develop. Once the emerging *Hydra* can feed itself, it breaks away to form an independent organism. Other organisms that bud are sponges. The buds form internally until the parent's body breaks down, releasing the newly budded sponges.

REGENERATION

Some animals can regenerate or regrow parts that were lost or injured. Some lizards can regrow tails, and crawfish can regenerate pincers. A few animals, when their bodies are broken or cut into fragments, can regenerate missing cells, forming new individuals. This type of regeneration, found in less complex animals such as planarians, is a method of asexual reproduction. When a planarian population is very small, the planarians can reproduce by transverse fission. **Transverse fission** is a process in which an organism reproduces by splitting in half and regenerates the missing parts.

dictionary

WORD BANK

trans- = across

fission = split into parts

The planarian's body splits crosswise during transverse fission.

INVESTIGATION 16A PLENTY OF PLANARIANS

See Investigation 16A in the TRB.

Have you ever lifted a rock in a stream and seen a planarian on the underside? These flatworms reproduce asexually by regeneration (Figure 16.2). This type of reproduction occurs when the population is very small. In this investigation you will observe this interesting form of reproduction.

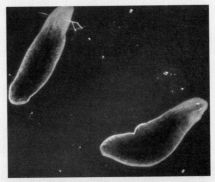

Figure 16.2: *Planarians reproduce by transverse fission.*

PARTHENOGENESIS

Several species of whip tail lizards, living in the American Southwest, are only female. These lizards, along with bees, wasps, and several other animals, can reproduce asexually by a process called **parthenogenesis.** In this method the female produces an egg that develops into a fully formed organism without being fertilized. In a beehive the queen bee produces eggs that aren't

Figure 16.3: *Aphids reproduce asexually through parthenogenesis. They can also reproduce sexually.*

fertilized by a male. The haploid bees that develop from these eggs are drones (males). Aphids, small insects that live on plants, produce many daughter offspring parthenogenetically in the spring and summer when food is abundant (Figure 16.3). In the fall, these females produce males and females. These mate and produce eggs that hatch the next spring.

LESSON 16.1 ASSESSMENT

1. Lack of movement prevents them from finding a mate, needed for sexual reproduction.
2. Stable environment with abundant food, such as in the spring and summer when the plants they feed on are growing.
3. Mitosis.
4. Male.

1 Hypothesize why sessile animals are likely to reproduce asexually.

2 What environmental factors promote asexual reproduction in an organism such as an aphid?

3 When a *Hydra* bud grows, is the cell division meiosis or mitosis?

4 A beekeeper has a beehive that has unfertilized eggs. What sex will the bees that develop from the unfertilized eggs be?

LESSON 16.2 SEXUAL REPRODUCTION

Male peacocks fan their dramatic tail feathers to attract females for mating. Lightning bugs flash signals to future mates. Moths release odors that signal their readiness to mate. Once a mate has been selected, the male animal releases sperm to fertilize the female's ova, creating a new generation of offspring. Breeders, farmers, and pest control researchers use their knowledge of mating behaviors and sexual reproduction either to increase a herd or population or to decrease an insect swarm. In this lesson you'll learn what happens in sexual reproduction.

ACTIVITY 16-2   CHICKEN EGGS

Each group will need a fresh, uncooked chicken egg and a small plate. **Caution students that some uncooked eggs have *Salmonella* bacteria; they shouldn't eat any of the raw egg and must wash their hands carefully after the lab.** One option is to offer gloves to students to use while handling the eggs. Duplicate the drawing of the fertilized and unfertilized egg. The egg white (albumin) is added after the egg is fertilized.

Students might not be able to determine correctly what each part of the egg does. Review the correct answers with the class. The yolk has nutrients and is part of the ovum. The white supplies nutrients and cushions the embryo. The chalazae are two twists of protein that anchor the embryo in the center of the shell.

A chicken's newly laid, fertilized egg has a small cluster of dividing cells called an embryo, which will grow into a chick. The rest of the egg is made up of the nutrients that make growth possible. In this activity you will examine the parts of an unfertilized egg (more correctly called an ovum) and compare it to a diagram of a newly fertilized bird's egg (Figure 16.4).

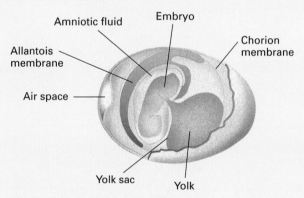

Figure 16.4: *A newly fertilized egg*

Carefully examine the egg shell. It is the last part to form before the egg leaves the mother's body. Crack the egg open and pour the contents on a clean plate. **Caution: Do not eat any of the raw egg.** Like the eggs you buy at the store, this egg is unfertilized. Find a small white dot on the yolk. This germ is where the chick embryo would form if the egg were fertilized. Look inside the shell and notice the thin membrane that lines it. Draw the parts of the egg in your logbook and label as many parts as you can.

A membrane just under the shell forms an air space at the blunt end. Gases are exchanged through the shell.

In the fertilized egg, stored food and nutrients are metabolized as new cells are produced. New structures form that assist the chick's development.

Respond to the following in your logbook:

- Hypothesize what each part of the egg provides a growing chick.

- Compare the fertilized egg in the diagram and the unfertilized egg you cracked. How are they alike? How are they different?

WILD FISH RUNS

BIOLOGY IN CONTEXT

Wild salmon that struggle to swim upstream against the river currents to mate and lay eggs are getting scarcer. Salmon hatch from eggs laid in the gravel beds of freshwater streams. In the spring they swim downstream to the Pacific Ocean. After five or six years, the fish swim back upstream to the same spawning grounds where they hatched. There, the female lays her eggs, and the male fertilizes them with sperm. Both adults then die. Each generation repeats the cycle.

In the late 1880s, over 43 million pounds of salmon were harvested from the Columbia River. Today, few salmon make the run upstream. Dams that generate electricity interrupt the river's flow. The newly hatched fish die in the turbines of the dams. To preserve the declining salmon population, hatcheries were built to breed salmon. Studies determined the best places to build the hatcheries and the best times to release the young salmon into the river. Hatchery workers supervised the fertilizing of salmon eggs, their incubation at the correct temperature, and the care and feeding of the young fish until the salmon could be released into the wild stream. Scientists found that the first year's hatchery fish crowded out the already small populations of wild salmon. In the second year the remaining hatchery fish had poor survival rates, further reducing the number of salmon. The chinook and coho salmon are now listed as endangered species. Unfortunately, what seemed like a good plan to increase fish reproduction led to population reduction.

SPERM AND OVA

Remember from Chapter 4 that during meiosis the cell undergoes two divisions, creating haploid gametes (ova or sperm. *Ovum* is singular; *ova* is plural). When a sperm cell fuses with an ovum, fertilization occurs. This new diploid cell is called a zygote. It can develop into an individual organism.

More sperm are produced than ova. The human male makes 10 billion to 30 billion sperm each month in the testes. Human sperm are tiny, about 0.05 mm long, with little room for anything except DNA. Sperm are the only human cells that have flagella, which they use to propel themselves (Figure 16.5). The human female is born with about 500,000 cells that can develop into ova. These cells, called **oocytes,** are in the ovaries. In these cells the division process has stopped temporarily in the first division of meiosis. During puberty, female hormones cause six to ten oocytes to begin maturing each month. As they mature, meiosis produces haploid gametes. Although all four gametes have the same number of chromosomes, only one oocyte gets most of the cytoplasm. This oocyte will become the ovum that is released that month. The others, called polar bodies, will disintegrate. Ova are larger than sperm cells and have more cytoplasm. The ovum may have stored nutrition to nourish the offspring as it develops. For example, the yolk in a bird's egg nourishes the developing chick.

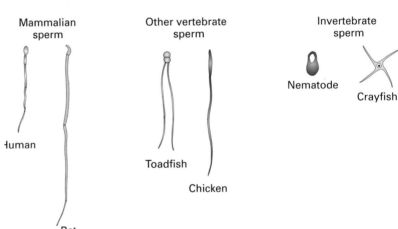

Figure 16.5: *Various types of sperm*

EXTERNAL FERTILIZATION

Ocean coral release sperm and ova simultaneously, to float briefly in the current (Figure 16.6). The synchronized release of gametes into a water environment is called **spawning.** The sperm

Figure 16.6: *The coral's sperm and ova are released at the same time. The fertilized eggs become larvae that drift in the water as they mature.*

immediately use flagella to move toward the ova to fertilize them and create the first diploid cells of the next generation.

Spawning is an example of **external fertilization,** fertilization that takes place outside the parents' bodies. It occurs in water, allowing the sperm to propel themselves toward the floating ova. The wet environment prevents the gametes from drying out. Spawning is a typical method of reproduction for coral, sponges, and other sessile animals. But other types of animals, such as fish, also spawn. Animals produce large numbers of gametes during spawning. A female cod produces more than 4 million eggs. Because of predators and the environment, few fertilized eggs develop and live long enough to reproduce.

Changes in day length and temperature are the trigger for hormone production in many animals. Hormones regulate reproductive cycles and courting behaviors. Many fish, amphibians, and other aquatic animals court mates. Courtship and mating often involve a complex set of sounds, colors, behaviors, and scents.

LEARNING LINK BREEDING PATTERNS

Wood frogs are explosive breeders. Their endocrine system responds to sudden changes in the environment. Puddles attract adult wood frogs that court and mate within a few hours to a few days. They lay many eggs, which develop rapidly and tolerate a wide range of temperature, oxygen supply, and water level. Bullfrogs have a prolonged pattern. Males defend territory near a permanent water source, and the eggs mature slowly into large adults.

Frogs, like other animals, have one of two breeding patterns: prolonged or explosive. Prolonged breeders produce few offspring at a time in an environment that is rather predictable. Explosive breeders produce many offspring in a short time in an unpredictable environment. Constantly changing temperature, moisture, and other factors reduce the chances that the offspring will become adults. Research different animals' reproductive patterns, identifying the animal as an explosive or a prolonged breeder. Describe the conditions that trigger the breeding cycles. Discuss behaviors that attract mates and how adults care for eggs and young.

INTERNAL FERTILIZATION

Land animals such as reptiles, birds, mammals, worms, and many insects reproduce sexually with **internal fertilization.** The male deposits sperm into the female's reproductive tract, fertilizing the ovum inside the female's body. Internal fertilization reduces the need to produce vast numbers of ova, since the ova are protected. The sperm has to travel a long way, in comparison to its size, to reach an ovum. Out of 100 million human sperm released during intercourse, fewer than 100 will reach the ovum. When these sperm reach the ovum, their enzymes begin to break through the ovum's membrane. Only one sperm enters and fertilizes the ovum.

As in external fertilization, the ovum and sperm have to be in the right place at the right time for fertilization to occur. Seasonal cycles, hormonal changes, and courtship behaviors contribute to the synchronization of animals' reproductive cycles. **Pheromones** are key for some animals. These chemical messengers, produced by one organism, influence the behavior of another organism in the same species. For example, the female gypsy moth releases pheromones that indicate her readiness to mate. The male follows the scent upwind to find her. Chemists have copied this pheromone. Fruit growers use it to bait traps that destroy crop pests.

Most female mammals have a breeding cycle called **estrus.** The estrous female breeds only at a particular time of the year when food is plentiful and the survival of the offspring is more likely. The animal's uterine lining prepares for pregnancy; if pregnancy doesn't occur, her body reabsorbs the lining. Primates don't have estrous cycles; their monthly menstrual cycle is described in Lesson 16.3.

Different species care for fertilized eggs differently. In many land animals, such as reptiles and birds, the fertilized egg and yolk are

Snapping turtles abandon their eggs after they have hidden them on a beach.

enclosed in a clear layer of albumin. Just before the egg leaves the mother's body, a shell encloses and protects it. Egg-laying reptiles produce a leathery shell. They often cover their eggs with sand or soil to keep them warm and then abandon them. Birds have a calcified shell; they incubate their eggs and care for the chicks until they develop adult feathers and food-gathering skills. Mammals give birth to live young.

The ovum in humans and most other animals is fertilized internally and is enclosed in a **uterus,** a muscular organ that holds the developing offspring within the female.

HERMAPHRODITES

Some animals that are isolated from other members of their population or are sessile have a unique type of sexual reproduction. They have both female and male sex organs, making them **hermaphrodites.** When two hermaphrodites mate, each one fertilizes the other's ova. Planarians are hermaphrodites. When they mate, two flatworms line up so that the opening of the female reproductive organ in one aligns with the male organ in the other. Each fertilizes the other's ova. They encase the eggs in a cocoon and attach it to a log or rock in water where the eggs will hatch. Earthworms, which are annelids, are also hermaphrodites. The light band of tissue that circles an earthworm's body makes a cocoon for the newly fertilized eggs (Figure 16.7). Other hermaphrodites, such as sponges and coral, release the ova and sperm in spawning.

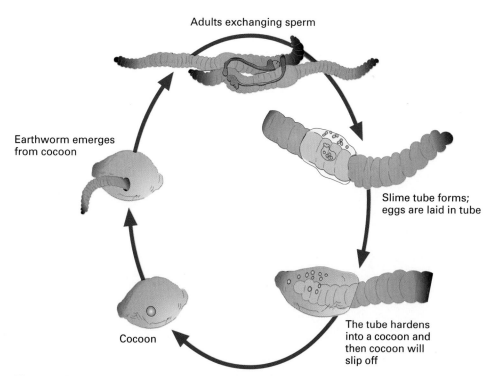

Adults exchanging sperm

Earthworm emerges from cocoon

Slime tube forms; eggs are laid in tube

The tube hardens into a cocoon and then cocoon will slip off

Cocoon

Figure 16.7: *Earthworms' mating and reproduction cycle*

LESSON 16.2 ASSESSMENT

1. The large number of sperm cells increases the odds of sperm reaching the ovum and supplying enough enzyme to allow fertilization.

2. External fertilization involves gametes being released into a wet environment; many more gametes are produced than in internal fertilization, where the sperm are deposited inside the female's reproductive tract.

3. Changing day length is an important trigger for hormones to initiate the breeding cycle. This lab doesn't change much and so won't encourage breeding.

4. Pheromones of one species do not affect another species.

1 Explain why a technician who breeds cows would insert at least 15 million living sperm cells into the cow's reproductive system to fertilize one ovum.

2 Contrast external fertilization with internal fertilization.

3 What are your chances of increasing the population of an endangered frog species if you keep the frogs in a lab with uniform heating and light? Give a reason for your answer.

4 In a stream, male trout were exposed to the pheromones of a female goldfish. Scientists couldn't detect any behavioral response to the pheromones. Explain the results of this experiment.

LESSON 16.3 THE MAMMALIAN REPRODUCTIVE SYSTEM

A zoo announces the birth of a new baby rhinoceros. A marina captures on videotape the birth of a baby whale. Proud parents film the birth of their new daughter. Mammals give birth to live babies, but where do the babies come from? In this lesson you'll learn the basics of the male and female reproductive systems in mammals.

CAREER PROFILE

QUARTER HORSE BREEDER

John B. uses artificial insemination (AI) to breed quarter horses. AI is a technique in which the breeder or an AI technician inserts semen (sperm and fluids) collected from a male horse into a female that is in estrus. As a breeder, John selects a mare and a stud that will produce offspring with certain traits, such as strong hips and muscles, a short back, a long neck, and large eyes.

John collects semen and puts it into test tubes. He keeps the tubes at the correct temperature until insemination. John injects the semen in the mare's reproductive tract. The sperm in the semen swim to the ovum and fertilize it. John's responsibilities do not end after the mare has conceived. He also manages the care of the mare during her 11 months of pregnancy and the care of the foal after it is born.

You don't have to have a college degree to become a horse breeder, but it is very helpful. John recommends that aspiring breeders work around farms for several years to get to know the different breeds and to learn how to read a pedigree.

ACTIVITY 16-3   EMPATHY BELLY

The Empathy Belly with its weighted water bladder, padding, and weights allows both male and female

Pregnancy causes many emotional and physical changes in the mother-to-be. She may experience mood swings, morning sickness, weight gain, and frequent urination. In this activity you'll experience for a short time a few of a pregnant woman's physical changes.

students to try simple activities while temporarily "pregnant." The Empathy Belly may be borrowed from agencies that teach prenatal education or focus on women's health issues. Follow the safety precautions that come with the device, including obtaining parental permission to participate. **Students with back problems or other physical problems should not participate in this simulation.**

Answers will vary.

While you're wearing the Empathy Belly, observe all safety precautions listed in the manufacturer's instructions. Do these tasks: Tie your shoe, lie on the floor and then get up, get up from a chair. Describe in your logbook how you feel doing these tasks.

Record in your logbook how your life would change if you were pregnant.

- How would you look and feel?

- How would your daily activities change?

THE HUMAN MALE

Follicle-stimulating hormone (FSH) and luteinizing hormone (LH) have names that reflect the hormones' functions in females. Their effect in males is equally important. LH is sometimes called interstitial cell–stimulating hormone. It stimulates these cells to produce testosterone.

Between the ages of 10 and 14, a boy's pituitary gland starts releasing hormones that trigger the onset of puberty. Luteinizing hormone (LH) makes the **testes** (the male sex organ) produce testosterone, the male sex hormone. With the onset of testosterone production, the voice deepens, muscles get bulkier, baby fat disappears, body hair grows, and the penis grows. Another pituitary hormone, FSH (follicle-stimulating hormone), makes the testes create sperm.

The testes are in the **scrotum,** a muscular sac (Figure 16.8). Sperm develop normally at a few degrees below body temperature. The muscles in the scrotum help to regulate the

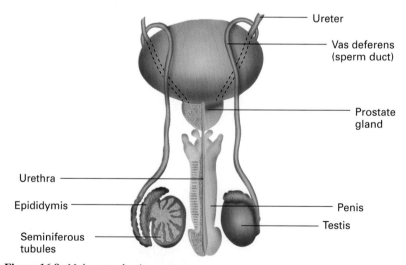

Figure 16.8: *Male reproductive system*

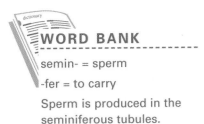
temperature by holding the testes near to or farther from the body. Sperm form in the **seminiferous tubules** in the testes and move into the **epididymis,** which is a coiled tube about 6 m long in the back of each testis. In these tubes the sperm mature for about three weeks and become motile. Peristaltic muscle contractions push the sperm through tubes to the penis, where they leave the body. This process is called ejaculation. The muscles around the male reproductive system force the sperm out of the epididymis and into the **vas deferens** (sperm duct). The sperm are carried over the top of the bladder, past the opening of the gland called the seminal vesicle, through the prostate gland, and into the urethra, which leads to the tip of the penis. The seminal vesicle adds sugar to fuel the sperm, and the prostate adds an alkaline fluid to neutralize the woman's acidic reproductive tract. The sperm, the prostatic fluid, and additional fluids are called **semen.** Sperm that aren't released disintegrate and are reabsorbed in the epididymis.

THE HUMAN FEMALE

The human female reproductive system parallels the male system in many ways (Figure 16.9). The same hormones, LH and FSH, initiate puberty; girls grow taller, develop breasts and wider hips, and start to **menstruate,** the shedding of the uterine wall lining in nonpregnant women. Each month during the menstrual cycle, an

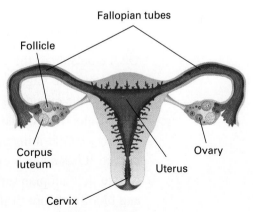

Figure 16.9: *Female reproductive organs*

oocyte matures and the uterine lining thickens, preparing for a possible pregnancy. If pregnancy doesn't occur, the oocyte and lining break down and pass from the body. At the beginning of the cycle, FSH affects several oocytes in the ovary. They begin to mature by completing meiosis to form haploid gametes (Figure 16.10). Each oocyte is surrounded by clusters of supporting follicle cells. The follicle cells produce the hormone **estrogen.** Estrogen stimulates the thin uterine lining to start growing blood vessels and glands that could nourish a fertilized egg. Midcycle, about 14 days after the oocytes begin to mature, the LH level increases, making one mature ovum break out of the ovary. The ovum's release from the ovary is called ovulation. The mature ovum moves into a fallopian tube (also called the oviduct) on its two- to three-day journey to the uterus. The remaining follicle cells in the ovary become the **corpus luteum.** It produces the hormones called **progesterone** and estrogen. These hormones help to maintain the thickened uterine lining for a possible pregnancy. If a sperm fertilizes the ovum in the fallopian tube, pregnancy begins. Otherwise, the ovum, corpus luteum, and uterine lining break down. Levels of LH and FSH decrease. Within a few days, menstruation begins. This monthly loss of the lining and blood lasts three to five days. The entire cycle lasts about 28 days, with the oocyte beginning to mature on day 1 and ovulation on about day 14. Menstrual cycles may be very irregular when they first start. Toward the end of a woman's reproductive years, called **menopause,** the cycle of ovulation and menstruation gradually ceases. Stress, excessive exercise, illness, and severe weight loss may also interrupt the menstrual cycle.

On the outside of the female body, protective folds of skin called labia cover the **vagina** (the tubular passageway that connects the uterus to the outside of the body) and the urethral opening. Toward the front of the folds is the clitoris. At the base of the uterus, where it joins the vagina, is a ring of thick tissue called the **cervix.** It has a small opening through which sperm pass on their way to the fallopian tubes. During menstruation the uterine lining and blood pass out of the uterus through the cervix.

WORD BANK

corpus = body

luteum = yellow

The corpus luteum is a yellowish area in the ovary.

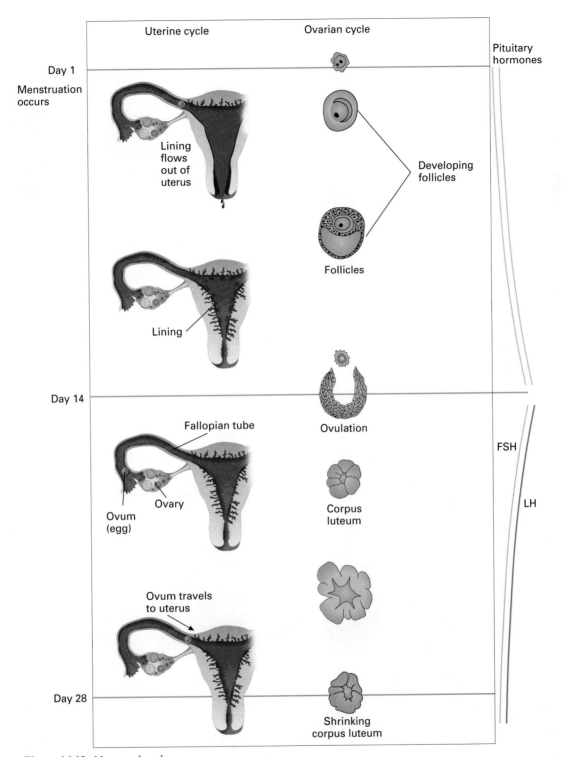

Figure 16.10: *Menstrual cycle*

INVESTIGATION 16B ARTIFICIAL INSEMINATION

See Investigation 16B in the TRB.

Animal breeders and dairy farmers often artificially inseminate the female animal. They purchase sperm produced by a male that has the characteristics that they want in the offspring. Using special equipment, the breeder puts sperm into the female's reproductive tract. In this investigation you'll explore the conditions that keep a sperm viable, that is, able to swim and fertilize an ovum.

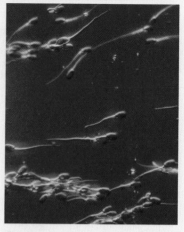

Sperm cells

MULTIPLE BIRTHS

You might know people who are identical twins and others who are fraternal twins (Figure 16.11). Identical twins occur when the fertilized egg begins to divide and the first two cells separate from each other. Each develops into a fully formed person. Both individuals have the same genetic material because they developed from one fertilized egg. Fraternal twins occur when the ovary releases two ova during ovulation and each ovum is fertilized by a different sperm cell.

A   B

Figure 16.11: *Identical twins (A) share the same genetic material. Fraternal twins (B) result from two ova, each fertilized by different sperm.*

Many animals have multiple births. For example, a mouse may have more than 15 litters of five to six offspring in a single year. Animals such as a mouse, a dog, and a cat have a uterus that is divided into two long branches. Each fertilized ovum develops in a different location in the uterus.

SEXUALLY TRANSMITTED DISEASE

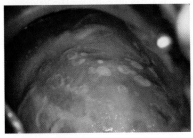

Figure 16.12: *Genital herpes causes periodic outbreaks of painful blisters.*

Figure 16.13: *Pubic lice cause itching.*

Sexually transmitted diseases, commonly called STDs, are a group of infections that are passed through sexual contact. Two STDs, chlamydia and gonorrhea, are bacterial infections that often occur together. Both males and females may experience painful urination or discharge of pus. These infections can cause pelvic inflammatory disease, which may lead to infertility or to birth defects in a developing child.

Penicillin and other antibiotics cure many STDs, but they cannot repair the damage that has already been done. New strains of disease can develop resistance to available treatment. Some STDs are not curable, especially those caused by viruses. These include AIDS, genital warts caused by the human papillomavirus, and genital herpes (Figure 16.12).

Other STDs include *Trichomonas vaginalis* (a protist) and "crabs," insects that are more correctly called pubic lice. Pubic lice often spread through intimate contact, but they also survive on and can spread on clothing and damp towels (Figure 16.13). These organisms cause intense itching. Medication can kill the lice and their nits (egg sacs). Yeast infections may be passed through sexual contact, but often they intensify as a result of taking antibiotics. The medication disrupts the bacteria in the reproductive system, which normally keep naturally occurring yeast populations in check.

LEARNING LINK WHAT'S IN YOUR NEIGHBORHOOD

Use the Internet or the library to research STD transmission, symptoms, damage, and treatment. Make a table in your logbook of the information you obtain on STDs. Contact the county health department to identify the most common STDs in your area.

CHANGING IDEAS

CONTRACEPTION

A nurse at a women's clinic counsels a young married couple about the method of contraception they want to use. Contraception means using any of a variety of methods to prevent pregnancy. This young couple believes that it is important to finish their education before having children.

In the past, when infant mortality rates were high, most people wanted a large family, and contraception was not usually an issue. Nor were there safe, reliable methods of contraception. Now that more children reach adulthood, personal choice and responsibility are major factors in deciding to use contraception. Abstinence, or refraining from intercourse, is the only completely effective contraceptive method. Other methods not only are less effective, but also may have unwanted side effects. Natural family planning tracks the menstrual cycle, using changes in body temperature and vaginal mucus to identify when ovulation occurs. The couple practices abstinence during the days around ovulation. Barrier methods that attempt to prevent sperm from reaching and fertilizing an ovum include male and female condoms and the diaphragm (a soft cap that covers the cervix). Using spermicide with these measures increases their effectiveness.

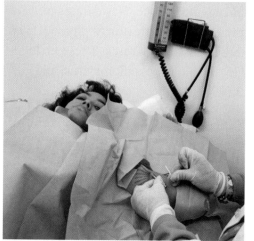

Norplant® birth control capsules are implanted under a woman's skin to provide continuous protection

The greatest scientific advancement in contraception was the birth control pill, developed in the 1950s. It continually provides prevention, not just at the time of intercourse. The pill prevents pregnancy by changing the sequence of hormones that regulate ovulation. Norplant® capsules, which are implanted under a woman's skin, and Depo-Provera® injections work in a similar way for longer periods of time.

Some individuals choose sterilization, which permanently prevents sperm from leaving the epididymis or ova from reaching the fallopian tubes. Male sterilization involves a vasectomy, which severs the vas deferens. Female sterilization is accomplished with tubal ligation, which surgically closes the fallopian tubes.

PREGNANCY

For a normal pregnancy a sperm cell must reach an ovum while it is in a fallopian tube and fertilize it (Figure 16.14). The fertilized egg in which the nuclei of the sperm and the ovum have fused is called a zygote. The zygote begins to divide by mitosis into a ball of cells called the **embryo.** The embryo travels toward the uterus for a couple of days. The corpus luteum makes progesterone and estrogen, which maintain the thickened uterine lining. The embryo implants in the

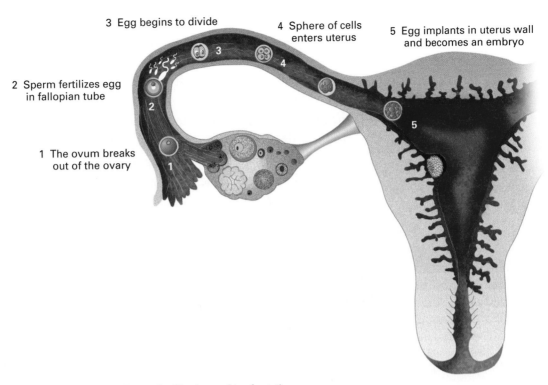

3 Egg begins to divide

4 Sphere of cells enters uterus

5 Egg implants in uterus wall and becomes an embryo

2 Sperm fertilizes egg in fallopian tube

1 The ovum breaks out of the ovary

Figure 16.14: *Events during fertilization and implantation*

lining, about eight days after fertilization. By this time it is a hollow ball of cells with an indented fold in one side. The innermost layer of the ball of cells becomes the digestive and respiratory systems. The middle layer forms bones, muscles, and the circulatory and excretory systems. The fold will develop into the brain and nervous system. The remaining outer cells become the skin.

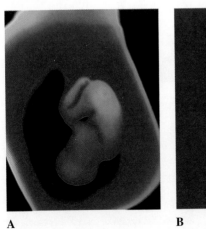

While the fold is forming, three membranes form around the embryo. The inner sac, the amnion, fills with fluid that is a shock absorber for the developing embryo. The middle membrane, the allantois, becomes the umbilical cord. The outer membrane, the chorion, grows into the uterine lining and becomes the **placenta** (Figure 16.15a and 16.15b). This temporary organ produces the same hormones that are produced by the corpus luteum: progesterone and estrogen.

A   **B**

Figure 16.15: *(A) At 5 weeks the organs and the arm and leg buds have begun to form. (B) Within $3\frac{1}{2}$ more weeks the amniotic sac surrounds the fetus. The placenta exchanges food and waste during this period of rapid growth.*

By the end of two months, the organs and the placenta are formed. The embryo has become a fetus. The mother's uterine blood vessels are right next to the selectively permeable membrane of the placenta. The maternal blood flows out of small open arteries and pools around the membrane separating it from the fetal capillaries. Nutrients and oxygen from the mother's circulatory system pass through the placenta's membrane into this capillary network in the placenta. The umbilical cord's arteries carry fetal blood to the fetal circulatory system. Blood carrying the fetus's cellular wastes and carbon dioxide travel back through veins in the cord to the placenta. There, these wastes diffuse out to the mother's veins, returning to her circulatory system. Her kidneys and lungs then filter and remove the waste by-products.

Figure 16.16: *This boy suffers from the effects of fetal alcohol syndrome, which includes physical problems and mild to severe mental retardation.*

Blood cells are too large to pass through the placenta, but certain drugs and alcohol can diffuse through it into the fetal circulatory system. These substances can damage the developing nervous system and organs. Alcohol consumption during pregnancy causes many physical abnormalities (Figure 16.16) and is the most common cause of mental retardation. The impairment may be mild to severe, depending on when and how much alcohol the mother drinks.

The eighth week marks the beginning of the fetal period (Figure 16.17). By now the placenta is nourishing the fetus,

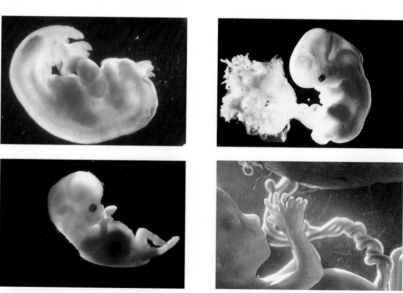

Figure 16.17: *The hand develops from a bud (6 weeks) into finger ridges (7 weeks), separate fingers (8 weeks), and a nearly complete hand (13 weeks).*

and all the fetal organs are formed. During the next six to seven months the fetus continues to grow. So does the mother. She gains about two to three pounds every month. The smooth muscles of her uterus lengthen to enclose a full-term baby, the amniotic fluid, and the placenta. As the uterus grows, it compresses the bladder, causing more frequent urination. Milk glands in the breasts enlarge.

An Additional Activity, Sonograms, can be found in Chapter 16 of the TRB.

FETAL TESTING AND BIRTH DEFECTS

WORD BANK

amnio- = refers to the amniotic sac and fluid

-centesis = puncture

In amniocentesis a doctor inserts a needle through the uterine wall into the amniotic sac and withdraws cells.

Many tests help to monitor embryonic and fetal development. A procedure called amniocentesis collects fetal cells that float in the amniotic sac. A technician prepares chromosomes from these cells for genetic study (Figure 16.18). A geneticist may detect diseases such as Down syndrome or Tay-Sachs disease, a disorder in which fat is improperly stored in the nervous system.

A sonogram uses sound waves to produce an image of the fetus. Technicians study fetal size, the development of the organs, and activity to monitor fetal development.

Down syndrome is a trisomy 21 that causes mental retardation and distinctive physical features. Tay-Sachs is a progressive fatal disease.

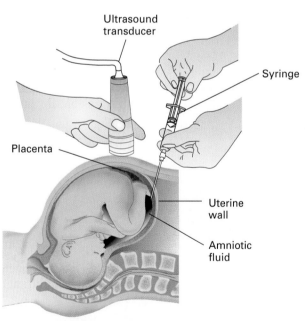

Figure 16.18: *In amniocentesis, cells extracted from the amniotic fluid are studied for genetic problems.*

All organs are formed by the seventh month, and the fetus could survive outside the uterus, though with difficulty. The last two months of pregnancy see a skinny, wrinkled fetus change to a full-term baby, as fat develops under the skin and the weight more than doubles. Surfactant, a substance that lets the lungs expand with the first breath, is first made during the seventh month of development. At the end of the 40-week pregnancy, oxytocin, a pituitary hormone, makes the uterus contract. This is the beginning of labor. The contractions increase in intensity and frequency.

The cervix dilates; it gets thinner, and its opening enlarges. The amniotic sac breaks, releasing the fluid. With the baby's head at the cervical opening, the contractions of the uterus push the baby through the cervix, vagina, and labia and into the world. This passageway is also called the birth canal. A doctor, a nurse, or a midwife immediately assesses the baby's health and reflexes. The uterus continues to contract, pushing out the placenta, which has pulled away from the uterus. The placenta is sometimes called the afterbirth. Humans cut and clamp or tie off the umbilical cord. Other mammals, such as the horse, bite the cord in half to separate the newborn from the placenta.

Figure 16.19: *Mother nursing her child*

During pregnancy, ducts in the mammary glands get ready to produce milk. The pituitary hormones, prolactin and oxytocin, are responsible for **lactation,** the making and release of breast milk (Figure 16.19). The first day or so after the birth, the mother produces a fluid called colostrum. This fluid contains some fat, lactose, and antibodies from the mother that protect the child from infections. Within a few days she produces milk. This milk has a lot of fat and supplies the energy that is needed to triple an infant's size in its first year. Milk fat is also crucial for the brain development that continues after birth. Immune cells migrate to the milk glands and secrete the mother's antibodies into the milk, providing passive immunity for the infant as long as the child nurses.

INFANT DEVELOPMENT

A newborn's reflexes include rooting (turning the head toward its mother's breast when she touches its cheek), attempting to "walk" when held with its feet touching a surface, and being startled by

An infant automatically tries to walk when its feet touch a surface.

loud noises. As time goes on, the infant develops a larger, more complex range of behaviors.

A child's coordination develops systematically from the head down its torso and from the torso to the limbs. An infant turns its head before lifting it. In time, the infant learns to coordinate back muscles, enabling the child to turn over. The ability to sit upright is the result of more muscle control and strength. A child learns to make large movements with its arm before being able to manage fine movements such as holding a crayon and scribbling on paper.

Each behavior is a developmental task that happens in more or less the same sequence for every child. The child must have the physical and mental development to accomplish each new task. Most children have learned to crawl and are beginning to toddle and talk by the end of the first year. The sequence of tasks is the same for most children, but the time at which they accomplish the developmental tasks isn't. Healthy, cared-for children all learn to communicate, walk, and act as independent individuals.

LESSON 16.3 ASSESSMENT

1. Sixteen days is better. By that time, ovulation has occurred.

2. Itching, discharge of pus, rash, painful urination, wartlike growths, blisters.

3. Ovulation.

4. Premature infants are not fully developed and are not ready to live outside of their mothers.

1 A married couple want to have a child. Which is a better time for the couple to try to conceive: three days or sixteen days after the wife begins to menstruate? Give a reason for your answer.

2 Identify several STD symptoms.

3 An endocrinologist, someone who studies hormones, tracked the blood level of LH in a woman. On the tenth day after her menstrual cycle began, the LH levels rose dramatically and then dropped off. What probably happened?

4 The delivery of premature babies carries certain risks. These infants have low birth weights and can have respiratory and cardiovascular problems. Why are premature infants more likely to develop these problems?

CAREER APPLICATIONS

The applications that follow are like the ones you will encounter in many workplaces. Use the biology you learned in this chapter to complete the activities. Share your work with the class.

AGRICULTURE & AGRIBUSINESS

Estrous Cycles on the Farm

Request information from an agricultural extension agent or a livestock farmer on the estrous cycle of a farm animal. Create a poster that describes how estrus differs from the human menstrual cycle. What can farmers/ranchers do to encourage estrus so that the birth season for their livestock isn't spread over many months?

Length of Estrous Cycles

List mammals (other than primates) bred for profit in this country. Find out and describe in a table the estrous cycle of each animal you listed. Are there common traits among species with long estrous cycles that distinguish them from species with short estrous cycles? Write a paragraph in your logbook describing your findings and conclusions.

Egg Quality

Contact an egg producer, a chicken hatchery, or your local agricultural extension agent or use the Internet to answer these questions about chicken (or other) eggs that are sold as food. How are eggs examined and graded for freshness and quality? What characteristics lower the grade of an egg? What are the causes of each characteristic? Should egg producers be concerned about the frequent appearance of these characteristics in their stock? Write a report on your findings and present it to the class.

Gestation

Interview a veterinarian, a farmer, a rancher, or a county agricultural agent about mammal care during gestation (the period between fertilization and birth) and the birth processes of different domestic animals. Ask about special enclosures, restraints, or isolation; nutritional needs; and treatment of discomfort, labor, and delivery. Share your findings with the class in a table that compares domestic animals such as horses, cows, sheep, and swine.

BUSINESS & MARKETING

As an alternative to television, you could bring in various magazines or newspapers and have students analyze print advertising.

Does Sex Sell?

Watch television for two hours. If possible, videotape all the commercials during that time. Otherwise, take notes on each commercial. List each commercial, what it was selling, whether it used sex appeal to sell a product, whether it targeted one or both sexes, when it aired, the network that showed it, and the show(s) during which it aired. Make a class list of all of the commercials that were observed. From the class data, can you draw any conclusions about types of products that are marketed by using sex appeal? Do you think the approach is appropriate for the products being marketed?

FAMILY & CONSUMER SCIENCE

Ethical Issues

Human in vitro fertilization has raised numerous ethical issues. Consider how such issues should be handled. If more than one of a woman's ova are fertilized, what should be done with embryos that aren't implanted? Is genetic screening of embryos a good idea? Who should get custody of a frozen embryo in case of divorce? When do the benefits of in vitro fertilization outweigh potential ethical problems?

Let's Have a Talk

Pretend you are a parent planning your child's sex education. At what age should sex education begin? What should you tell a child about sex at each stage of development? Select an approach for communicating the information that would be the most comfortable for you and a child. Research materials in bookstores (including toy stores) and the children's section of the library, and select the ones that you think are the most effective, accurate, and appropriate for each age level. Report the information in a table.

Planning to Overcome Infertility

Explore the following options for infertile couples: adoption agencies, foster care agencies, artificial insemination, in vitro fertilization, and surrogate mothers. Create a table that shows advantages and disadvantages, what options best suit which couples, and the expenses.

Exercises for Pregnant Women

Contact local hospitals or obstetricians' offices and check the library for information about special exercise programs for pregnant women. Demonstrate some of the most important exercises for the class or invite someone who teaches these exercises to do a demonstration. Focus on exercises that include the expectant father or other members of the family. Explain the purpose of each exercise.

Getting the Family Ready

Use bookstores, toy stores, the children's section of the library, or the Internet for information on preparing a small child to accept a new baby, to be involved as a sibling in the care of the new baby, and to understand as a sibling about the stages of fetal development. Compose a book list, including name, author, publishing company, and date. Evaluate each book for effectiveness, accuracy, sensitivity, appropriateness for age level of the reader, and clarity of illustrations. Prepare a brochure with the information. Ask a local obstetrician to distribute the brochure to expectant parents who already have one or more children.

HEALTH CAREERS

In the Maternity Ward

Ask the maternity unit of a hospital or a nursing school program for a list of jobs in a maternity unit. Use the library or the Internet to find out the role of each job. On the basis of your research, identify and describe how a person in a specific position routinely interacts with others.

Natural Childbirth Classes

Ask to visit a natural childbirth class. Observe the activities in the class. Ask about the type and amount of education and training required to conduct the classes. Report your observations to the class or invite a nurse/clinician to speak to the class on neonatal care, postnatal care, or infant care.

Extraordinary Births

Contact emergency medical technicians, labor/delivery nurses, or neonatal staff. Ask them to compare procedures used during normal deliveries with those used to handle problems such as premature birth, very quick delivery, induced labor, cesarean birth, breech presentation, and multiple births. Ask them to explain what an Apgar score is and to describe the Apgar score for infants at one minute and five minutes after birth.

Drugs and Fetal Development

Investigate the effects of illegal drugs, alcohol, and smoking on the developing fetus. Describe how the substance passes to the developing fetus and what the critical periods are in fetal development and the resulting problems.

INDUSTRIAL TECHNOLOGY

Shipping and Handling

Research the equipment necessary for storing and transporting live sperm, ova, or embryos. Look specifically for descriptions and diagrams of sperm bank systems and any problems associated with those systems. What will the future bring? Diagram a sperm bank system on poster board to display in class.

Pregnancy in the Workplace

Interview someone you know who is working outside the home during her pregnancy and ask the following questions: During which month of her pregnancy will she begin maternity leave? How long does she expect to be away from her job? What, if any, physical discomfort or inconvenience has she experienced as a result of being pregnant while working? Compare your findings with those of other students, especially the effects of different work environments. What are the leave policies for the mother and the father after the birth?

Workplace Hazards

Use the library or the Internet to find articles on chemical hazards, office equipment, and radiation that might contribute to or cause miscarriage or birth defects. Report to the class any solutions you find and ask the class to contribute other possible solutions.

High-Tech Childbirth

Investigate the latest medical equipment used by obstetrics personnel. Include equipment for prenatal examinations and for care of premature babies.

CHAPTER 16 SUMMARY

- Asexual reproduction is one type of reproduction that occurs in some animals that are sessile, can't find mates easily, or live in a stable environment.

- Asexual reproduction in animals includes transverse fission, budding, parthenogenesis, and regeneration.

- Many organisms, such as the sponges and aphids, may have more than one method of reproduction, including both sexual and asexual reproduction.

- Each gamete, the sperm or the ovum, has a haploid set of chromosomes.

- Sexual reproduction requires coordinated sperm and ova production and often involves special organs, courtship behaviors, mechanisms for feeding the embryo, and parenting behaviors.

- External fertilization occurs outside the body and usually involves large numbers of ova and sperm. Internal fertilization occurs inside the body and doesn't involve usually as many ova as external fertilization.

- Hermaphrodites have both male and female sex organs.

- Only one sperm fertilizes one ovum, creating a fertilized egg.

- Pheromones are chemical messages that influence the behavior of another member of the same species.

- Mammals produce a placenta that selectively allows nutrients, gases, and waste products to pass between the developing organism and the mother.

- The human male reproductive organs are the testes enclosed in the scrotum, the epididymis, the vas deferens, the seminal vesicles, the prostate, and the urethra and penis.

- The human female reproductive organs are the ovaries, the fallopian tubes, the uterus, the vagina, the labia, the clitoris, and the breasts.

- Most mammals have an estrous, or breeding cycle, while primates have a monthly menstrual cycle.

- The pituitary hormones LH and FSH control a male's production of testosterone and a female's production of estrogen and progesterone.

- Some sexually transmitted diseases have no obvious symptoms. Many cannot be cured. Some can damage organs and lead to sterility or death.

- Fertilization occurs when the sperm penetrates the mature ovum, fusing the two nuclei and initiating cell division. In humans this occurs in the fallopian tubes.

- In sexual animals a zygote develops into an embryo in which the major organs take shape.

- In placental mammals the longest stage of prenatal development is that involving the fetus. Rapid growth occurs during this stage.

- Human birth has three stages. At the beginning of labor the uterine contractions dilate the cervix, and the baby's head enters the birth canal. The contractions grow in intensity and push the baby out, usually head first. Finally, the placenta separates from the uterine wall and is delivered.

- The infant is born with reflex actions such as rooting. It develops from the head downward and from the center of the body out toward the limbs.

CHAPTER 16 ASSESSMENT

Concept Review

See TRB, Chapter 16 for answers.

1 Why do animals with external fertilization produce more ova than those with internal fertilization?

2 What process produces the gametes? How do these cells differ from the rest of the body's cells?

3 What triggers the release of an ovum in humans?

4 Can a young woman become pregnant before she has her first menstrual flow? Give a reason for your answer.

5 Why do pregnant women need to urinate more frequently than other women?

6 What organ develops during the early stages of pregnancy and transfers materials between mother and embryo? Explain how this exchange of materials occurs.

7 Toads are amphibians that rely on external fertilization. In Europe, many roads that border wetlands have tunnels built underneath for the toads to move from the wetlands to the surrounding area. How do these tunnels encourage reproduction in toads that often live outside the wetland areas?

8 Where are ova fertilized in a human female's reproductive tract?

9 Describe the initial effects of LH and FSH on young teenagers.

10 Explain how breast-feeding provides immunity to newborn babies.

11 Discuss how fertilization occurs in annelids, such as earthworms.

12 Predict how the aphid's reproduction method might change if the weather suddenly got colder for several weeks.

13 Compare budding to regeneration.

14 Wasps have drones and workers. Is the drone genetically identical to its parent? Give a reason for your answer.

15 Why can fraternal twins be different sexes but identical twins can't?

16 Explain how a male dog can tell that a female dog is in estrus.

17 Why is the variety of chromosome combinations produced in sexually produced offspring important to the species?

18 What environmental conditions would allow spawning to occur in a meadow or field?

19 Hiking in the mountains, you see a pair of grouse, a type of bird. One bird's chest is expanded, and it fans its wings and tail as it struts about. The other has fewer distinctive features. Why might the first grouse behave this way?

Think and Discuss

20 The male and female human reproductive systems have many similarities. What female organ corresponds to the male testes? To the vas deferens?

21 Describe how human males regulate the buildup of sperm in the epididymis.

22 When a coral trout spawns, the male and female swim together toward the ocean's surface. The trout release ova and sperm into the water as they turn to swim back toward the ocean floor. Explain the swimming pattern of the coral trout during spawning.

In the Workplace

23 Urine tests reveal the presence of alcohol and illegal drugs in a mother. How might these drugs reach the developing fetus inside her uterus?

24 A doctor assesses an anorexic patient and finds that she no longer has menstrual periods. Explain why the menstrual flow would stop.

25 Many zoos have breeding programs for endangered species. Why is artificial insemination a preferred method for breeding many animals?

Investigations

The following questions relate to Investigation 16A:

26 Can planarians be cut into pieces too small to regenerate?

27 A planarian was cut in half. Which part of the planarian will develop faster—the portion with the head or the portion with the tail?

The following questions relate to Investigation 16B:

28 What does a lab technician look for in a semen analysis to determine male fertility?

29 Why freeze sperm in liquid nitrogen instead of a household refrigerator freezer?

CHAPTER 17

WHY SHOULD I LEARN THIS?

Do you take care of a pet, raise animals for a living, or go fishing? To be successful at any of these, you rely on your knowledge of behavior. You understand how the activities of animals help them to fulfill their basic needs and survive. Understanding an animal's behavior lets you predict what it will do in a situation. Understanding human behavior will help you to do well at school, on the job, and as a member of a family and a society.

ANIMAL BEHAVIOR

WHAT WILL I LEARN?

1. the wide variety of animal behavior
2. the adaptive value of behaviors
3. how genes determine behavior
4. how learning influences behavior
5. how social animals maintain their society
6. the similarity of human behavior to that of closely related animals

People in occupations that deal with the general public understand a lot about human behavior. For instance, police officers constantly observe and evaluate the behavior of individuals. Their understanding of human behavior is a matter of life or death. They must decide whether a person they are speaking with is lying or afraid or if they are a potential threat.

During domestic disputes a police officer must decide whether someone's actions infringe on the rights of another. In protecting the public, they are on the lookout for situations that encourage criminals to break the law. Knowing the motivations and behaviors of criminals helps them do their job.

Police officers, like the citizens he or she serves, are social animals, and the officers' activities are examples of animal behavior within a society. In this chapter you will learn how social and nonsocial animals behave.

LESSON 17.1 TYPES OF BEHAVIOR

When you visit the beach or a public swimming pool, you dress to stay cool. You take a beach blanket or towel to protect yourself from the hot ground. You put on sunglasses to shade your eyes from glare. You drink cool drinks and take an occasional dip in the water when you start feeling hot. All of these actions are human versions of basic behaviors that most animals perform when choosing a place to live, looking for food, or maintaining correct body temperature. In fact, one definition of animal behavior would be all activities that an animal performs as it tries to survive and reproduce.

ACTIVITY 17-1   OBSERVING ANIMAL BEHAVIOR

Procedures for this activity are found in Chapter 17 of the TRB.

Scientists who study behavior try to explain why an animal acts a certain way. This might involve designing an experiment that gives an animal a choice between two or more conditions. In this activity you observe ants that are given a choice between two types of soil: one that is familiar and one that is foreign. You will try to explain the ants' reactions to the different stimuli. Your teacher will provide you with the procedure for this activity.

LEARNING LINK   DEFENDING A TERRITORY

Select an animal species and use the library or the Internet to learn how it recognizes and defends its territory. Record your findings in your logbook and share them with your class.

PET PSYCHOLOGIST

BIOLOGY IN CONTEXT

Does your dog or cat chew your shoe or wrap itself up in your favorite sweater while you're at school? Maybe your pet needs to see Dr. Nathaniel Johnson. Nathaniel is a veterinarian who specializes in animal behavior. His office looks like any other veterinarian's office, but his patients have behavioral problems.

To diagnose a problem, Nathaniel has to depend on the pet owner to describe the animal's behavior. The owner gives Nathaniel a very detailed account of the pet's activities. To make sure that he recommends the proper action, Nathaniel also examines the owner's behavior.

The Peterson family visited Nathaniel when they had problems with their seven-month-old collie, Jasper. Jasper growled and showed his teeth if someone was standing nearby while he was eating. He later began to growl if anyone tried to pet him or give him a treat. Mr. Peterson would hit Jasper across his mouth whenever he growled at one of the kids or did something wrong. The Petersons finally went to Nathaniel when Jasper bit their daughter when she accidentally rolled over Jasper's tail while playing.

Nathaniel told the Petersons that Jasper's aggressive behavior was an attempt to defend himself and to protect his food. Nathaniel encouraged the Petersons to build Jasper's confidence and show him that hands are for positive petting and not for hitting. Several weeks later, the Petersons reported that Jasper was doing much better. Nathaniel said, "Sometimes the problem is not with the pet, but with the owner."

BASIC BEHAVIORS FOR SURVIVAL

dictionary

WORD BANK

thermo- = temperature

chemo- = chemical

photo- = light

-taxis = movement in response to a stimulus

Thermotaxis is a movement toward or away from heat.

Chemotaxis is a movement toward or away from a chemical stimulus.

Phototaxis is a movement toward or away from light.

The fundamental behaviors required for the survival of an individual animal include taxis, thermoregulation, habitat selection, predator avoidance, aggressive defense, and foraging. A **taxis** is a movement in response to a stimulus in the environment such as heat, light, gravity, or chemicals. The movement, either toward the stimulus (positive taxis) or away from the stimulus (negative taxis), helps to orient the organisms in some beneficial way. For example, moving toward heat helps a bloodsucking bedbug to find its warm-blooded host (Figure 17.1A). An organism that moves away from light, such as a bat, may find a cool, damp spot that keeps it from drying out. Many ant species find food by following a chemical trail secreted onto the ground by a nestmate (Figure 17.1B), which first found the food by its smell or sight.

Although taxis is usually a way in which less complex organisms find suitable living conditions, even complex animals can exhibit taxis. For example, a firefighter exhibits negative thermotaxis in moving away from the searing heat of a fire.

A knowledge of animal taxis has helped researchers to design pest control devices. Many insects that are a nighttime nuisance are positively phototactic. They can easily be lured toward an illuminated electric wire that shocks and kills them (Figure 17.1C).

Figure 17.1: *Thermotaxis, chemotaxis, and phototaxis are used by (A) bed bugs, (B) ants, (C) house flies, and (D) lobsters to find food.*

People who fish use the fact that most aquatic species are chemotactic, meaning that they move toward or away from chemicals in the water. Fishers therefore use strong-smelling bait to catch fish and crabs (Figure 17.1D).

Many species with well-developed nervous systems carry out more complex behaviors than taxis to survive. Maintaining optimal body temperature, or **thermoregulation,** requires different behaviors, depending on whether the animal is an endotherm or an ectotherm. **Endotherms** generate their own body heat through their high metabolic rates. Mammals and birds are endotherms. **Ectotherms** cannot use their metabolism to generate body heat. Ectotherms include almost all animals

A

B

Figure 17.2: *Behaviors involved in thermoregulation may differ among (A) ectotherms and (B) endotherms.*

other than mammals and birds. They get heat from their environment, usually by moving to a warmer location. For example, after a cold night in a desert burrow, a lizard's body temperature has dropped and its metabolism is lowered. To raise its body temperature, it must move to an area where it can gain heat, for instance, a rock heated by the sun (Figure 17.2A). If it gets too hot, the lizard moves into the shade or a burrow. Fish are stuck with whatever temperature a body of water is and can do little to maintain their body temperature. They simply conform to the environmental temperature, which they are adapted to tolerate.

In humans and other endotherms, body temperatures are regulated by internal mechanisms. Some of these generate heat, such as increased metabolism and muscle contraction, and some get rid of heat, such as sweating. But behavior plays an important role for endotherms when they are exposed to intense heat or cold. For instance, humans may take off or put on clothing (Figure 17.2B) and seek cooler or warmer environments, often by turning on their air conditioners or furnaces.

Many animals spend a lot of time exploring their habitat, using their mechanical, chemical, and visual senses to find the "right" place. The right place depends on whether an animal is **foraging** (looking for food), thermoregulating, looking for a mate, or satisfying some other urge.

Exploration is often the way an animal makes its habitat selection. For the parasitic twisted-wing insect the habitat is another insect, its host. The larva leaves its eyeless and nearly legless mother (which protrudes from the body of a host insect) to find its own host. It searches for the type of flower that is visited by that host. Crawling around on temporarily well-developed legs and using temporarily well-developed eyes, it is subject to predation, to dehydration, and especially to chance. Only a few of the insects out of thousands will find the right habitat.

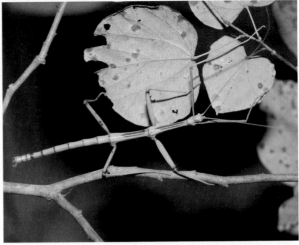

A **B**

Figure 17.3: *Behaviors animals use to defend against predators may be aggressive or passive. Porcupines (A) can be aggressive in defending themselves from predators, while walkingsticks (B) use a more passive defense.*

The most striking behaviors are often the ones that animals use to avoid or fight off predators. The porcupine (Figure 17.3A), a rodent, can be aggressive when provoked. Many of its hairs are long, stiff, sharp spines, called quills. A predator that gets too close will be impaled and sometimes killed by the porcupine's quills, but this happens only if the predator ignores the rodent's warnings. An inquisitive wolf would first hear a whizzing sound as the quills rise up like the fur on an angry cat. Some of the quills rattle like a rattlesnake's tail. The porcupine pounds its feet against the ground and grunts or snarls. A wolf that ignores these defensive signals could become a pincushion for detached quills that can puncture arteries and eventually kill.

Some animals use defenses against predators that are not very striking, which is to their benefit if predators are unaware of them. The walkingstick (Figure 17.3B), so called because of its twiglike shape, is an example of behavior that is linked to an animal's cryptic coloration, or camouflage. In keeping with a color pattern that makes it hard for predators to detect them, walkingsticks move slowly and will remain still when their predators are most active.

INVESTIGATION 17A TAXIS AS AN ANIMAL BEHAVIOR

See Investigation 17A in the TRB.

Animals that lack well-developed nervous systems, such as planaria, use simple movements, called taxis, that take them toward or away from various stimuli. The stimulus might be light, heat, gravity, moisture, or chemicals. A taxis may result in the animal's finding food or mates and avoiding predators or harmful substances.

In this investigation you will examine the responses of planaria to stimulation that might attract, repel, or have no effect on the subject.

REPRODUCTIVE BEHAVIOR

Basic survival behaviors aren't truly useful if the animal doesn't reproduce. Reproductive behavior may include finding a mate, courtship, mating, nest building, and caring for the young. In finding a mate, each sex exhibits behavior, including physical and chemical cues, that stimulate the potential mate. Male bumblebees have large antennae that detect attractive odors produced by unmated queens (Figure 17.4). The flight of a female mosquito causes air vibrations that attract male mosquitoes. They detect these vibrations with their large antennae. Visual stimuli are important for many, if not all, primates. (Primates are an order of mammals that includes humans, apes, and monkeys.) For example, the larger muscles and size of male apes may have evolved because of visual appeal to females, as well as because they give larger males an advantage in fights.

Figure 17.4: *This male bumblebee waits for a mate.*

Female animals tend to be choosier than males when selecting a mate. For example, male insects commonly attempt to mate with many females; a female will often reject male suitors after mating only once. This difference in behavior of males and females of a species is adaptive. A male can increase the number of his offspring every time he mates, whereas

A

B

Figure 17.5: *Reproductive behaviors. (A) Male scorpion flies offer food to their potential mates. (B) Male spiders may court females by plucking lines in the web. (C) The construction of a nest may be a prelude to parental care.*

C

a female might receive enough sperm from a single mating to fertilize all the eggs she will ever produce.

A common event in animal courtship rituals is the presentation of food or other resources from the male to the female (Figure 17.5A). Female spiders sometimes eat males that don't send the right courting signals (Figure 17.5B). Birds and other animals often build a nest for their offspring (Figure 17.5C). A bird's nest in a tree has two advantages: avoidance of predators and protection from wind and rain. The warm nest also confines and contains the developing, defenseless young while one or both parents forage for food.

Parental care occurs more widely than one might guess, even among insects. Mother earwigs lick freshly laid eggs, keep them from scattering, and guard the young after they hatch (Figure 17.6A). However, most insects provide no care to their young beyond selecting a suitable site for egg laying. Parental care also varies among birds. Some birds develop so much in the egg that by hatching time the chick has feathers and spends little time in the nest before learning to fly (Figure 17.6B). Other birds hatch earlier; they lack feathers, are relatively helpless, and must remain in the nest some time before they can fly (Figure 17.6C).

A

B

C

Figure 17.6: *Parental care. (A) The earwig is an insect that protects its young from dangers. (B) Ducklings can enter the water just a day after hatching. (C) Young songbirds are commonly blind and helpless when they hatch.*

The Complexities of Social Life

The exact mechanism by which the ants sense the CO_2 is not fully understood, but it is believed that the sensors are located on the antennae. Vibrations are picked up through their feet.

The behavior of social animals is more complex than that of nonsocial animals. This complexity is due to the need for communication within a group. The members of an ant colony recognize each other by the chemicals in their exoskeletons. Each colony has a different odor, allowing its members to identify each other and interact while excluding or killing ants that intrude from other colonies. Ants stay in touch by relying on chemical cues and on sounds that are produced when body parts rub together. If a nest caves in, ants come to the rescue of their trapped nestmates by tracking the carbon dioxide gradient produced from their respiration and sensing the vibrations set up by their struggles to free themselves.

LESSON 17.1 ASSESSMENT

1. Positive chemotaxis toward the perfume and negative chemotaxis toward the repellent.

2. The mosquito depends on a sense of smell (not only on carbon dioxide) to identify its host.

3. Foraging and habitat selection.

4. Take the animal's body temperature across a range of temperatures to find out whether it continues to conform to the temperature of its environment.

1 On a camping trip, your friend put on perfume instead of mosquito repellent. During the night, several mosquitoes bit her, but they didn't bite those who used mosquito repellent. What type of behavior did the mosquitoes show toward the perfume and the repellent?

2 A scientist was trying to make a mosquito trap using carbon dioxide. His attempts were unsuccessful until he mixed carbon dioxide with a chemical that smells like buffalo breath. Why would this chemical make a difference in his results?

3 The South African small hive beetle has caused a big problem with beekeepers. These beetles love honey, so they enter beehives, lay their eggs, and take over the hive. The beetles' hard shell protects them from bee stings. What types of animal behavior are being shown by the beetle?

4 You trap an animal in its burrow and find that its body temperature is the same as that of the burrow. Describe what you would do to show that the animal is an ectotherm.

LESSON 17.2 WHAT DETERMINES HOW ANIMALS BEHAVE?

Zookeepers, animal control officers, animal breeders, and other people who deal with animals need to understand the basis for an animal's behavior. Is a behavior innate, or inborn, meaning that the animal displays the behavior the first time the stimulus is presented? Or has the behavior been learned, perhaps from other animals of the species or from a trainer? Much research has shown that even innate behaviors are subject to modification as a result of an animal's experiences.

ACTIVITY 17-2   FIERCE FISH

Procedures for this activity are in Chapter 17 of the TRB.

Many animals display aggression toward members of their own and other species. Aggressive behavior may ward off a predator, protect the young, or maintain access to food.

In this activity you will investigate the aggressive behavior of male Siamese

Male bettas sometimes battle to the death over females.

fighting fish (also known as bettas). By conducting some simple experiments, you will decide on the value of the behaviors that are displayed and how the behaviors may be modified. Your teacher will provide you with the procedure.

MIGRATION OF WHOOPING CRANES

BIOLOGY IN CONTEXT

Between 1996 and 1998, only 200 whooping cranes lived in the wild, most of them in a single flock. To build up the flock and relocate the whooping cranes to safer areas, researchers are teaching the birds new migration paths.

Preliminary experiments taught some of the more common sandhill cranes a new migration route from southeast Idaho to central New Mexico. Kent Clegg, a research biologist from southeast Idaho, nurtured and raised newly hatched sandhill cranes. These cranes learned to treat Kent like a parent. In other words, they imprinted on him. These birds followed Kent while he walked, ran, drove a four-wheeler, and flew an ultralight airplane. As they grew, they followed him from Idaho to the Bosque del Apache Wildlife Refuge in central New Mexico. The sandhills found their way back from New Mexico to southeast Idaho in the spring.

After two successful sandhill migrations, researchers expanded their efforts to include the endangered whooping cranes. When the crane eggs hatched in the spring, the baby cranes imprinted on Kent at a research center in Maryland. After a few weeks they were moved to Idaho, where they learned to follow Kent and, after several months, to fly with him. Kent led four whooping cranes and 10 sandhill cranes to New Mexico. He expects that these cranes will return to Idaho in the spring.

GENETICALLY PROGRAMMED BEHAVIOR

Innate behavior is due to the expression of genes and is therefore often referred to as genetically programmed behavior. The knee-jerk reflex that you tested in Chapter 14 is an example of a genetically programmed behavior. Certain of your genes specify how neurons and synapses of a reflex arc get arranged during development of the nervous system. Therefore the knee-jerk response is said to be hard-wired into your nervous system.

Even complex sequences of behaviors may be genetically programmed behaviors. Examples are the daily cycles of activity you experience, such as the sleep/wake cycle and your daily hunger patterns. As you learned in Lesson 14.4, these cycles occur in a circadian rhythm under the influence of your biological clock.

The strong genetic basis of circadian rhythms has been demonstrated in experiments. Animals often maintain their routine of daily activities even when they are confined to the lab, away from normal environmental stimuli. An animal may stick to its normal pattern even when an environmental condition is kept constant, as when it is kept in constant daylight. Mutations have also been found in some of the genes responsible for organisms' internal clocks.

Behaviors that are associated with a biological clock have been studied in white-crowned sparrows. This bird has a biological clock in its hypothalamus. The sparrow's circadian cycle is flexible, changing with the seasons. The clock measures the day as a period of 24 hours, and it measures the amount of light and darkness during each day. These amounts change with the season and so determine the time of year as well as the time of day. The bird's survival and reproductive success depend on its ability to detect seasonal changes, because it migrates twice a year between habitats that are more than a thousand miles apart.

The biological clock of the white-crowned sparrow tells it when to migrate.

LEARNING LINK LIKE CLOCKWORK

Each of us has a biological clock that controls our regular daily variations in sleep and wakefulness, body temperature, metabolic rate, heartbeat, urine excretion, blood cell count, hormone levels, and physical dexterity. Find out through library or Internet research about one or more of these variables. What 24-hour pattern does the variable exhibit? Why should you be aware of these patterns in your daily routine? Record your findings in your logbook and share them with the class.

CHANGING IDEAS

A GENETIC BASIS FOR HUMAN BEHAVIOR

Humans have long considered themselves different from other animals. We are a species with a unique set of behaviors, including unmatched self-awareness and technological abilities. Yet we are also animals, and our behavior should be explainable by the same principles that successfully explain that of other animals. Scientists have long argued about the degree to which human behavior is genetically programmed. Even though traits such as aggressiveness appear to be inborn and adaptive in species such as rats and gorillas, some people have thought that there is little or no genetic basis for human aggression. Their assumption was that aggression is learned and can therefore be overcome by proper training. Some people who try to rehabilitate violent criminals base much of their approach on this assumption.

In the mid-1970s the field of **sociobiology** brought back the notion that human behavior has a significant genetic basis. In his book *Sociobiology,* E. O. Wilson argued that behaviors such as selfishness and aggressiveness had adaptive value in early human societies and therefore were favored by natural selection. Aggressive behavior had value in defending against neighboring groups that were competing for food and territory. Aggressive behavior that today takes the form of cutthroat competition in the marketplace and belligerent driving on the highways can be seen as expressions of a once useful genetic program.

Opposition to sociobiology quickly arose. Social scientists thought that sociobiology implied that genetically programmed behavior was not subject to change. They also thought that sociobiology was dangerous because it seemed to excuse undesirable behaviors such as discrimination. Sociobiologists argued that unless the role of biology in human behavior is studied and understood, it will be impossible to rid society of undesirable behaviors. No doubt the debate will continue for a long time.

LEARNED BEHAVIORS

Figure 17.7: *School is a cultural influence that plays a role in the development of human behavior.*

Learned behaviors include simple types of learning such as habituation, conditioning, and imprinting. More complex behaviors are, in some species, the product of cultural influences. Cultural experiences are particularly important for humans (Figure 17.7).

Habituation is perhaps the simplest form of learning and is widespread among animals (Figure 17.8). It is a loss of sensitivity to an unimportant event or object after repeated exposure to it. For example, a chimpanzee in its natural forest habitat will normally run from a human intruder. However, chimps that live near road crossings see humans frequently and eventually learn to ignore them. They become habituated to the presence of humans.

Conditioning is a somewhat more complex form of learning than habituation. In **conditioning,** learned behavior is associated with a

Figure 17.8: *This panda has become habituated to the presence of humans.*

The film *Fly Away Home* demonstrates the role of imprinting in geese.

stimulus, usually as a result of training. In the 1920s, Ivan Pavlov (Figure 17.9), the great Russian physiologist, showed how conditioning works. He tested the salivary reflexes of dogs by restraining them in harnesses and feeding them by spraying meat powder in their mouths. Before feeding a dog, Pavlov rang a bell. After a number of trials he discovered that a dog would salivate merely upon hearing a bell. The dog became conditioned to expect food every time it heard a bell ring. It associated the bell with food.

Imprinting is a form of learning that can only occur at a specific time in an animal's life. This learning pattern involves

Figure 17.9: *Ivan Pavlov*

recognition and attachment to another object. For example, newly hatched birds learn to follow anything that moves nearby, a behavior that is usually adaptive because that moving object is likely to be their mother. The Austrian behaviorist Konrad Lorenz discovered that goslings would waddle after him if he took their mother's place when they first hatched (Figure 17.10).

Figure 17.10: *Goslings learn to follow a human if their mother is not around.*

The influence of culture on the learning experience is particularly important for humans and highlights their great behavioral flexibility. Sometimes the cultural differences in common behavior seem arbitrary. There are about 4,000 different human languages and at least that many words for the greeting "hello," not to mention the many gestures and facial expressions that humans use to greet one another.

Compare this cultural richness to the limited number of ways in which two beetles "greet" one another. Our species relies more heavily than any other upon instruction and the examples of others.

INVESTIGATION 17B THE BEHAVIORS OF *HYDRA*

See Investigation 17B in the TRB.

Hydras are animals that lack a true nervous system, having only a loosely organized nerve net. Their relatively simple needs and behavior make them easy to study. In this investigation you will observe hydras' hunting behavior, habitat selection, defense, and taxis and determine whether these behaviors can be modified by experience.

ANIMAL TRAINER

CAREER PROFILE

"How do they do that?" guests ask every day at SeaWorld® as they watch majestic marine animals executing mind-boggling maneuvers with attentive trainers.

Dave F., Vice President of Animal Training answers, "It's a matter of preparation, patience, observation, and above all affection for animals. Much of the animals' performance is behavior that they would perform in their daily life in the wild. The trainers observe an animal's behavior, and when it does something that is interesting or could be used in a show, they take note of it. They then discuss how that behavior can be trained."

Dave continues, "Animals don't learn behaviors overnight. SeaWorld trainers break down new behaviors into 'approximations,' which means they teach the animal a brief movement, then add to the original movement. Teaching an animal the signal to swim across a pool starts with getting the animal to turn, then extending this to a movement of two feet. That two feet might then become 10 feet, then 20 feet, until the animal learns the signal means to swim across the entire pool. After the animals learn hand signals for different behaviors, the trainers combine the behaviors into routines that entertain an audience."

Dave explains that at SeaWorld the trainers use positive reinforcement. When an animal performs a desired behavior, it gets a reward. Sometimes the reward is food, but if that were the only reward, the animals would perform only when they were

Orcas performing at SeaWorld®

hungry. Other rewards include playtime, rubdowns, and practicing maneuvers the animals particularly enjoy.

How do you become an animal trainer? It's a hard career to get into because there aren't many positions for animal trainers. Trainers spend many hours each day with their animals, building trust, learning their behaviors, and taking care of their needs. The job is physically demanding. At SeaWorld all candidates must pass a rigorous swim test and a microphone test and have a natural rapport with animals. A trainer's education should include a college degree in psychology or in a natural science such as marine biology. Candidates should also be certified in scuba diving and CPR.

LESSON 17.2 ASSESSMENT

1. A beetle might be stimulated by the similarity of the bottle's surface or color to that of a female beetle.

2. Male bettas raise their fins and expand their gill covers in the presence of another male betta. This may make them appear larger or more massive to the intruding male.

3. Genetic, because it has never had the opportunity to learn it. The first cry clears the lungs and airways.

4. Conditioning.

1 The presence of a brown beer bottle is known to trigger mating behavior in male Australian wood beetles, which appear to attempt to mate with the bottle. Predict what aspect of the bottle might be acting as a stimulus for mating.

2 When a cat is frightened or preparing to fight, it raises its hair and arches its back. This makes it appear larger and more massive than it really is. Did you observe any behavior in a male betta that is similar to the behavior of a threatened cat?

3 Is a newborn baby's first cry a genetic behavior or a learned behavior? Explain.

4 When no one is at home, your dog has the annoying habit of turning over the trash in the kitchen. When you come home and yell at him, he hides under the bed. Lately, he runs and hides as soon as he hears you open the door. What type of learning might best explain the dog's reaction to your opening the door?

LESSON 17.3 SOCIAL BEHAVIOR

Humans, chimpanzees, gorillas, many wasps, many bees, practically all ants, and all termites are social animals. Social animals live in groups in which members are of different generations. The members of the group specialize in different tasks. In this lesson you'll learn about the social behavior that members of groups display.

Wolves are social animals.

ACTIVITY 17-3   DIVISION OF LABOR

See Chapter 17 of the TRB for the procedures for this activity.

One characteristic of a society is that its members carry out different tasks, called division of labor. In this activity you will determine whether seed-eating ants use division of labor when gathering food. Your teacher will provide you with a handout of the procedure.

LEARNING LINK   DIVISION OF LABOR

Use the library or the Internet to find examples of division of labor in an animal society. Some animals you might want to research are lions, wolves, chimpanzees, bees, or the Portuguese Man-of-War. What are some of the roles of the different members of the society? Record your findings in your logbook and share them with the class. Are there similarities in the ways different societies divide the labor?

TERMITE TRAP

BIOLOGY IN CONTEXT

Pest control technicians have a new weapon in their fight against termites. No longer do home owners have to fear termite infestation, because scientists have developed a termite trap.

After detecting termites in a house, the pest control technician places the termite trap outside and away from the infested building. Even though the termites have turned a house into a termite buffet, they still search for their next food source. The technician baits the termite trap to lure one or more termites away from the house. The bait contains enough poison to slowly kill a termite. The technician wants the termite to survive long enough to leave a scent trail. Once the scent trail is in place, other termites will follow the trail to the trap, eat the bait, leave a scent trail, and die. When the trap has attracted termites, the technician sets other traps in the same area. Termites will follow the scent trail until there aren't any termites left.

This trap is just one way in which scientists use their knowledge of animal behavior to produce marketable products. Many technicians consider the termite trap to be one of the most significant advances in termite control in the last 50 years.

SOCIAL BEHAVIOR

Figure 17.11: *The family is the basic unit of human society.*

The basic human social group is the family, consisting of parents, offspring, and often other close relatives (Figure 17.11). Behavioral and physical evidence strongly suggests that in early humans the adult male in the family was responsible for defending the family and for hunting game animals for food. Adult females cared for the offspring, gathered plants for food, and trapped small animals.

Chimpanzees are thought to be the closest relatives of humans, and they exhibit similar social behavior. When exploring new territory, the males of the troop lead the way, and males are more aggressive than females in discovering food.

Insect societies differ greatly from primate societies. Perhaps the most obvious difference is the reproductive division of labor among adults. For example, although the familiar, wingless ant that is seen in kitchens and at picnics is an adult female, it does not reproduce. Instead, it collects food and defends the society. Reproduction is the queen's job. Males are seldom recognized as ants, are seldom seen outside the nest, and don't work or defend the nest. Their only function is to mate with the queen.

Ants are social insects that live in societies of up to several million individuals.

COOPERATION IN SOCIETIES

At times, animals in societies appear to make sacrifices for the good of others, a behavior known as **altruism.** For example, ant workers raise the queen's offspring and defend the colony from attack even though they have no offspring of their own to defend. The queen's offspring are the workers' sisters, and sister ants are more closely related than they would be to their own offspring, if they could have any. Because of this unusually close kinship, the altruism of ants aids the survival and reproduction of individuals that carry many of the same alleles, as would be expected on the basis of evolutionary theory. It is truly rare for any animal to give up its life for distantly related individuals.

African wild dogs provide a striking example of seemingly altruistic behavior (Figure 17.12). When some members of the pack are hunting and eating meat, others stay behind and guard the young. The guards will receive regurgitated meat when the hunters return.

Figure 17.12: *African wild dogs practice a form of altruism.*

Altruism in our own species, though it occurs, is not as readily explained as that in other species. Humans also practice **reciprocal altruism.** Reciprocal altruism occurs when someone behaves in a way that benefits another in hopes of some future benefit. The initial behavior does not benefit the person and may actually require some sacrifice. For instance, a person may donate blood in the hope that, at a later time, some unknown person will repay the favor.

Can you think of any famous or everyday examples of human altruism? Are these examples of truly selfless behavior, or do they have selfish motives? If you believe your examples of human altruism are pure and not merely reciprocal altruism, offer an explanation.

SOCIAL HIERARCHIES

A **social hierarchy** is an organization of groups of animals based on ability, resources, or (for humans) economics or professional standing. The establishment and maintenance of hierarchy and relationships within a society depend on signals. A police officer's badge and uniform are visual signals that remind us of a special authority to enforce the law. An unhappy teacher's facial expression is often enough to restore order and reestablish his or her dominance in the classroom.

The aggressive behavior of this gorilla is used to maintain his social hierarchy among other male gorillas.

In other primate societies, different signals are used to maintain the hierarchy. Male gorillas signal aggression to other males by rising to their full height. Only rarely do they make terrifying sounds and beat on their puffed-out chests as you might have seen in films and cartoons. If you're ever challenged by an aggressive male gorilla, signal back by shaking your head and avoiding eye contact. In gorilla society these signals show nonbelligerence.

Social insects maintain the structure of their societies more by chemical communication than by visual communication. For example, worker ants recognize their queen by her pheromones, chemicals that are present on her body. The queen's pheromones help her to maintain her dominance over the colony. When a foreign ant queen is presented to the workers, she has different pheromones and is likely to be killed.

LESSON 17.3 ASSESSMENT

1. Possible answers are giving blood, contributing to charitable organizations, or rescuing someone from a situation that also puts the rescuer in danger.

2. For most schools the hierarchy would be principal, vice principals, teachers, and students.

3. Accept any reasonable answer that demonstrates a mutual understanding of correction without verbal communication.

4. Altruism, since you are genetically related to your parents; reciprocal altruism, since parents may expect that offspring will take care of them in old age.

1 Give an example of an act of reciprocal altruism that you have performed. What benefit did you expect ultimately to receive from this act?

2 Describe a social hierarchy that you think may exist within your school.

3 What nonvocal signals can a human parent give to a son or daughter to maintain discipline?

4 Debate whether the care and support that you receive from your parent or guardian is an example of altruism or an example of reciprocal altruism.

CAREER APPLICATIONS

The applications that follow are like the ones you will encounter in many workplaces. Use the biology you learned in this chapter to complete the activities. Share your work with the class.

AGRICULTURE & AGRIBUSINESS

Herd Stress

Research the effects of overcrowding on cattle or swine during transportation, in feedlots, and so on. How do the individual animals cope with the crowded conditions? What physical indications show that they are stressed? How does their behavior change? What can be done to alleviate stress in crowded animals? Write a short report and present your findings to the class.

Zoo Behavior

A field trip to a local zoo would be an excellent opportunity for students to observe animal behavior firsthand. When you make arrangements for the visit, tell the staff that you would like a curator to talk to your class about animal behaviors in the wild and how they are accommodated at the zoo.

Use the Internet or the library or visit a zoo and talk to one of the curators to find out about animal behavior in zoos and how it differs from behavior in the natural environment. Some specific behaviors to research include courtship and mating rituals, feeding behaviors, interactions between individuals of the same species, and interspecies interactions in multispecies exhibits. Select an animal and prepare a fact sheet about it that could be given to visitors at a zoo. Include a sketch or photo of the animal, its common and scientific names, where it is from, and what it eats in the wild. Describe some of its behaviors that might be observed at a zoo and what they mean. What behavior problems are encountered in a zoo setting and how are they addressed? Put together a class notebook with everyone's animal fact sheet.

Training Animals

How do animal trainers use knowledge of animal behavior to train animals to perform certain tasks and tricks? Possible sources of information on this topic include books on animal training; animal trainers at zoos, wildlife parks, aquariums, theme parks, colleges (that have live mascots), circuses, and horse farms; and training facilities for canine police, seeing-eye dogs, or companion dogs. Are most tasks or tricks natural behaviors that are performed on cue, or are they learned? From your research, what seem to be the accepted practices for training an animal to perform a certain trick or behave in a certain way? What methods aren't considered productive?

Buying Behaviors

BUSINESS & MARKETING

In small groups or as a class, discuss your buying habits. How often do you buy something on impulse? What was the last item you bought on impulse? What influenced your decision to do this? Are there common behaviors or marketing strategies that influence you and your classmates to buy something? How can retailers use these buying behaviors to sell more merchandise?

Makeup and Fashion

FAMILY & CONSUMER SCIENCE

If there isn't time for all students to survey five people, use one set of pictures and poll only the class members. Ask students whether they think that impressions would be different at different ages.

Clothes, jewelry, and makeup are means of attracting or deflecting attention of other humans. Divide into groups and research the role of fashion in dating or in getting a job. You might want to look at what is considered fashionable in other societies or cultures. What is your explanation for the fairly rapid changes in what is considered fashionable and appealing to society? As a class, gather several sets of pictures (preferably in color) from magazines, newspapers, and other sources. Use the pictures to poll people. Ask them to view each picture and give you their first impression of the person and clothes they are wearing. For example, people might say, "Sexy," "Rich," "Responsible," "Boring," and so on. Each student should conduct a poll of at least five people. Record the age of each respondent. Compile the class results for each picture. Were the impressions similar for each picture? Did the response vary with the age of the respondent?

Abuse and Codependency

If possible, have one or more students visit an organization and take pictures or videotape the facility and the activities that go on there. Be sure to get permission before taking any pictures and have someone tell you what you can and cannot record.

Use the yellow pages of your local phone book to find organizations that work with people with problems such as alcoholism, drug abuse, physical or emotional abuse, anorexia, or bulimia. Divide into groups so that each group can investigate one organization. Your group should interview the director or

Make a documentary or public service announcement for the organization. This might not be possible for some organizations because of issues of confidentiality.

someone who is familiar with the services that the organization offers. Prepare a brochure or video on the organization. Include the services provided, who can obtain the services, special programs and events, costs, if any, and any other information that is pertinent.

Socialization

Research the role of family and daycare in the socialization of young children. What is the role of play in learning behaviors that will be critical in later life? How are good habits, such as washing hands or organization skills, taught and how are bad habits, such as throwing temper tantrums or biting and hitting others, discouraged? Talk to someone who works with young children to find out about some of the games and activities that are appropriate at different ages. What behaviors are being encouraged? Present a game or activity to the class and tell why it would be a good activity for a small child.

HEALTH
CAREERS

Careers in Behavior

As a class, list all of the careers related to human behavior that you can think of. Select one career from the list to research in depth. Each person should take a different career. Contact its professional organization and/or people who work in that career to find out what they do, what education and training they need, and what they find most rewarding about the work. As a class, create a packet of information on careers in behavior. Include an introductory table that lists careers by the amount of education required.

Stress

Everyone experiences some degree of stress, and it is a necessary part of our lives. Excessive stress can be harmful, however, leading to physical and emotional illness. Use a variety of resources to find information on behaviors that cope with stress. As a class, make a list of coping behaviors and describe when they are appropriate and when they aren't. If possible, have a counselor talk with your class and suggest other strategies for dealing with stress. Have the counselor discuss how to differentiate between normal or good stress and excessive or harmful levels of stress. If you know someone who is having trouble dealing with stress, how could you help that person?

INDUSTRIAL TECHNOLOGY

Biofeedback Technology

Use the Internet, phone book, library, and other sources to find information on biofeedback. For what conditions is it commonly used? How does it work? Obtain or draw a picture of the equipment that is used. Explain how the equipment is attached to the patient. What characteristics does it measure? Present your findings to the class.

Management Styles and Worker Behavior

As companies move from top-down-managed, assembly-line work to flexible, self-directed teams, what behavioral changes will workers and management need to make? What programs are available to foster creativity, quality, and motivation? Talk to a supervisor or human resources specialist in an industry and get a description of the work culture and behaviors that are encouraged in the employees. Write up your findings in your logbook and present them to the class.

CHAPTER 17 SUMMARY

- ■ Animals have a variety of behaviors that are necessary for their own survival and for the continuation of their species.

- ■ Basic animal behaviors may include habitat selection, taxis, foraging, thermoregulation, predator avoidance, aggressive defense, and reproductive behaviors.

- ■ Taxis is a response to or away from some physical stimulus in the environment, such as chemicals, heat, moisture, gravity, or light.

- ■ Some behavior is strongly controlled by genes, some is more flexible though significantly influenced by genes, and some is readily altered by experience.

- ■ Cultural influences are particularly important for human behavior.

- Reproductive behaviors that are necessary for species survival vary among animals, and may include finding a mate, courtship, mating, nest building, and parental care.

- Learning is a form of behavior that requires previous experience with a stimulus.

- Depending on the species of animal, habituation, conditioning, imprinting, and cultural influences are involved in the process of learning.

- Social animals live in groups, and the need for communication within societies adds another level of complexity to their behaviors as compared to those of nonsocial animals.

- The social hierarchy of a society is maintained by various types of communication among the society's members.

CHAPTER 17 ASSESSMENT

Concept Review

See TRB, Chapter 17 for answers.

1 Describe how a taxis shown by a pest animal is used to control the pest.

2 An armadillo is a mammal that has a head and body encased in armor of small bony plates. When you approach an armadillo, it curls up into a ball, exposing only the bony plates. What function does the armadillo's behavior serve?

3 Each morning, an animal exits its burrow, moves into daylight, and remains motionless on a rock. After an hour it returns to its burrow. Explain the animal's behavior.

4 List types of behavior that may, depending on the species, be associated with reproduction in some way.

5 There is a species of fish in which the male make a nest for the females to lay eggs in. After the male fertilizes the eggs, both the male and female leave the nest and don't return. What degree of parental care is being provided for the offspring?

6 What does a queen ant do that helps maintain the social hierarchy of the ant nest?

7 Would you consider yawning a learned behavior? Give a reason for your answer.

8 Research has shown that ducks that are raised from birth with another species will not mate with their own species. Offer an explanation for this.

9 Give an example of a cultural influence on your own behavior.

10 You have a dog that runs to the door and barks every time he hears the doorbell ring. What has conditioning taught the dog to expect?

11 Refer to Question 2. If you repeatedly approach the armadillo, it stays motionless when you approach and no longer curls into a ball. Explain the change in the armadillo's behavior.

12 The seasonal migration of animals such as the whooping crane depends on what feature of the animal's biological clock?

13 Identify the role of aggressive behavior of males toward one another in a society of gorillas.

14 Using a real-life situation, explain how your doing someone a favor is an example of reciprocal altruism.

15 How do ducklings learn to recognize their mother?

16 When Koko, a gorilla, learned sign language from her caretaker, what influenced her learning most?

17 Give examples of nonverbal signals that would communicate each of the following feelings: impatience, happiness, displeasure.

18 A turkey hen successfully defends her chicks as soon as the chicks are hatched. Is the behavior of the turkey hen genetically programmed behavior or conditioned behavior?

Think and Discuss

19 The belly of a male stickleback fish is red. When female fish were given a choice of colorless replicas of male sticklebacks and oddly shaped models that were red on the underside, they were attracted to the oddly shaped models. Explain this behavior.

20 A baby chick was placed in a cage with a maternal model that made no sound and with a box that produced chicken sounds.

The baby chick responded to the box rather than to the model. Predict the chick's response if the model were moved within the cage.

21 An infant monkey was removed from its mother and caged with two models. One model was made of wire and produced milk, while the other was a non-milk-producing terrycloth model. The monkey spent most of its time with the terrycloth model. The monkey went to the wire model only when it was hungry. What can you conclude from these results about the value of the two different models as stimuli for the monkey?

22 Scientists have used models of herring gulls and a knitting needle to determine what stimulates the feeding response in young herring gulls. To the scientists' surprise, young gulls responded more often to the tap of knitting needles than to the models. What types of stimuli are more important to young gulls?

In the Workplace

23 Some businesses and organizations require a uniform, and others have dress codes. What behaviors are these businesses trying to encourage or discourage by restricting what people wear?

24 "The customer is always right" is the motto for most customer service departments. Describe what behaviors a customer service representative might have to adopt to deal with a wide range of customer behaviors. List three to five potential customer complaints and behaviors and how you might deal with them.

25 You work as an animal keeper in a zoo. When a new animal comes into the zoo, it is quarantined for one month. During that month the keeper who will be taking care of the animal when it is on exhibit is in charge of observing the animal. In addition to looking for health problems, what behavior traits should you look for?

26 Why is knowing about animal behavior important to pest control operators or animal control wardens?

Investigations

The following questions relate to Investigation 17A:

27 Describe a series of experiments that would determine how thermotaxis, chemotaxis, and phototaxis affect the behavior of daphnia.

28 Explain why Test 1 was used as a control for Investigation 17A.

The following questions relate to Investigation 17B:

29 Why did the hydra eject its nematocysts when it was exposed to acid?

30 How do hydras sense the presence of food?

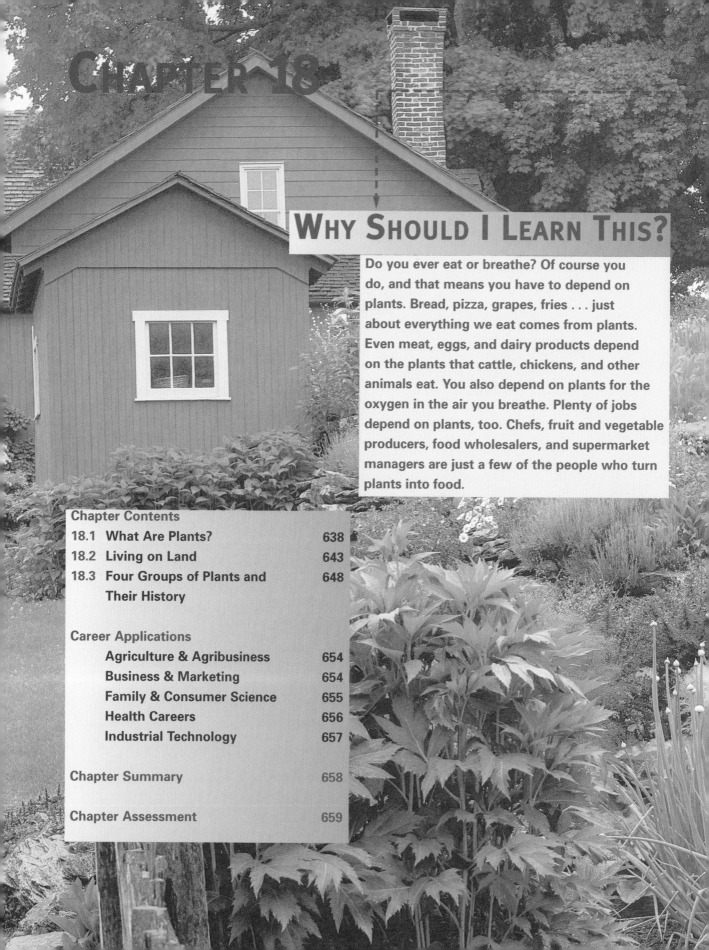

CHAPTER 18

WHY SHOULD I LEARN THIS?

Do you ever eat or breathe? Of course you do, and that means you have to depend on plants. Bread, pizza, grapes, fries . . . just about everything we eat comes from plants. Even meat, eggs, and dairy products depend on the plants that cattle, chickens, and other animals eat. You also depend on plants for the oxygen in the air you breathe. Plenty of jobs depend on plants, too. Chefs, fruit and vegetable producers, food wholesalers, and supermarket managers are just a few of the people who turn plants into food.

INTRODUCTION TO PLANTS

WHAT WILL I LEARN?

1. the major plant organs and their functions
2. how plants have adapted to life on land
3. the similarities and differences among four important groups of plants

---

Your landscaping business has some demanding new clients. One client wants moss on an old stone wall and ferns around the stone benches nearby. These plants need a cool, damp, shady environment. Maybe you can plant some trees around them to provide shade. But what kind of trees: flowering or evergreen?

Another client wants a lawn but doesn't want to mow it. Maybe you should suggest ground ivy instead.

The toughest request is a flower garden that won't attract bees. Bats, not bees, pollinate some tropical plants that would not grow well in this client's garden.

Each plant species represents a different compromise in adapting to life on dry land. Your knowledge of plant biology and your own creativity will help you to please these customers and keep them coming back.

LESSON 18.1 WHAT ARE PLANTS?

How would you define a plant? Most are green organisms, but so are some lizards and other animals. They photosynthesize, but so do many bacteria and protists. Although it is not easy to define plants without including some other organisms, you know one when you see one. In this lesson you'll learn the characteristics that define plants.

ACTIVITY 18-1   PLANT ORGANS

The procedure for this activity can be found as a blackline master in Chapter 18 of the TRB.

It is not hard to distinguish stems, leaves, flowers, and roots from each other. It is harder to figure out what these organs do for a plant. In this activity you'll observe plants to discover what their organs do.

Working with two other students, follow the instructions on the handout that your teacher gives you. Be sure to record all your observations in your logbook. Then discuss the following questions with your group and record your answers in your logbook:

The green color of leaves indicates that they are the primary chlorophyll-containing, photosynthetic organs.

Green leaves, stems, and immature fruits supply the rest of the plant with carbohydrate.

Roots anchor the plant and absorb water and dissolved nutrient minerals from the soil.

The veins and stem transport water and soil nutrients from the roots and carbohydrate from leaves throughout the plant.

- Think about what you learned about plant cells. What is the function of a plant organ that is green?

- Which plant organs are usually green? What substances do those organs probably produce?

- Describe all the ways in which roots might contribute to the life of a plant. Why does a plant need the large surface area of its finely branched roots? What substances do the roots provide to the rest of the plant?

- How could the veins and stem help to distribute important substances throughout a plant?

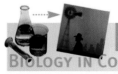

HIGHWAY VEGETATION MANAGEMENT

BIOLOGY IN CONTEXT

Every state has a highway or transportation department. Vegetation managers for state highway departments spend a lot of time making sure that roadside trees and weeds are cut often enough to keep them under control. But they also plant vegetation along highways. Trees don't just block road noise and air pollution. They also release oxygen and help to reduce the problem of global warming. Much of the carbon dioxide that trees absorb from the air becomes part of their tissue. This carbon stays out of the atmosphere during the tree's lifetime, which helps to reduce the greenhouse effect.

Highway vegetation managers need to know the plants that they work with well enough to choose the best ones for planting. They must also know how to control the weeds that crack pavement or cause problems for neighboring farmers or allergy sufferers. For example, some species of pine tend to lose their lower branches as they grow tall. They would not be a good choice for a buffer strip that is meant to shield neighboring homes from the sights, sounds, and smells of the highway.

GETTING TO KNOW PLANTS

Everyone knows that plants have green leaves and stems. You have also learned that the green color of these organs is from the photosynthetic pigment chlorophyll. Plants are autotrophs: Photosynthesis provides them with carbohydrate and a source of energy. You also know from Chapter 6 that plants are eukaryotes that develop from embryos into complex organisms with many types of cells.

Roots anchor plants in the soil and absorb water. The many fine branches of a root system have a large surface area that absorbs water and dissolved nutrients such as mineral ions. Commercial greenhouse operators sometimes raise plants without soil in solutions of water and fertilizer, a system called **hydroponics.** It is used by some tomato growers in Florida. In places where limestone or fossilized coral is exposed, they drill holes in the ground and plant tomatoes in the holes. A fertilizer solution nourishes the plants without soil. Hydroponics will probably provide food and oxygen to astronauts during long-term space flight.

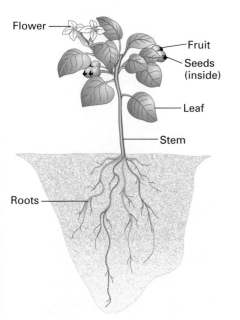

Water absorbed by the roots travels up through stems to the rest of the plant (Figure 18.1). It evaporates and returns to the atmosphere through microscopic pores in the leaves. CO_2, needed for

Figure 18.1: *Major parts of a typical plant*

photosynthesis, and oxygen, needed for respiration, diffuse into the leaf through these pores. Most leaves are flat and have a spongy internal layer with many air spaces between the cells. These adaptations enable the photosynthetic cells to absorb more CO_2. Carbohydrate that is produced in leaves during photosynthesis travels through the veins into the stem. From there it travels to the rest of the plant, where it is consumed in cell respiration and growth.

Flowers are the reproductive structures of most plants. Specialized cells in flowers undergo meiosis to produce haploid gametes—female egg cells or male **pollen** cells (Figure 18.2). Sexual reproduction also occurs in simpler, nonflowering plants, such as ferns and mosses. Some more complex species, such as bananas and certain grasses, have lost the ability to reproduce sexually. They can be grown from cuttings of existing plants.

After fertilization, one or more seeds develop in each flower. Each seed contains a diploid embryo (Figure 6.16, page 223) with a small root, a short stem, and one or two leaves. Other organs, such as fruits and flowers, form later as the plant matures. People who process and sell seeds handle them with care. Seeds that fall on hard surfaces may have damaged embryos that can't grow normally. Starch, fat, or protein that is stored in a seed nourish the embryo until it can maintain itself through photosynthesis. These stored nutrients make seeds a valuable food source for many species, including humans. Most of our important food crops—such as wheat, corn, rice, and beans—are seeds.

A plant organ that contains seeds is a **fruit.** Fruits are specialized to promote survival, wide distribution, and growth of the seeds that they contain (Figure 18.3). Many fruits provide additional nutrients to the embryo as it grows into a young plant.

Figure 18.2: *A tulip flower. The large central tube contains the female egg cells. The long, dark red structures are male organs that contain pollen.*

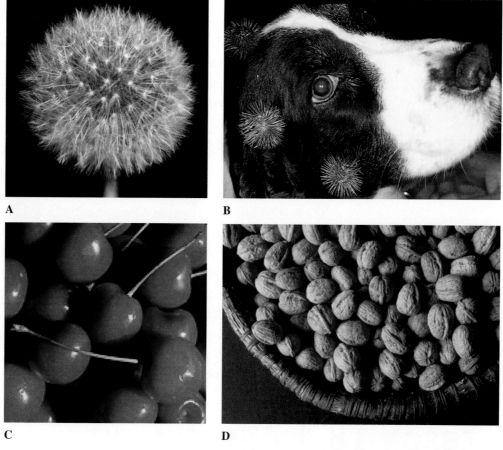

Figure 18.3: *Fruits show a variety of adaptations. (A) "Parachutes" attached to small seeds allow winds to carry dandelion seeds long distances. (B) Burrs spread by sticking to fur and clothing. (C) Cherry seeds sprout in moist, nutrient-rich rotting fruit if they fall to the ground, or in nutrient-rich manure if the fruit is eaten by an animal. (D) Hard fruits protect seeds such as walnuts from being eaten by animals.*

LEARNING LINK SLEEPY VEGETABLES

Many fresh fruits and vegetables are living organs that contain seeds. In nature, seeds often begin to grow and develop into new plants when conditions such as moisture and temperature are favorable. Until then, they remain dormant (inactive). Use the Internet or local resources such as libraries, the cooperative extension service, or interviews with the produce manager of a local supermarket or a plant breeder at a college of agriculture to find out how the seeds in food are kept dormant during storage.

As an embryo develops into a new plant, its cells divide frequently and become specialized. Changes in cell walls are an important part of this process. Enzymes in a ripening peach, for example, partly break down cell walls, making it soft enough for you to eat. Bark-forming cells develop a layer of cork in their walls that keeps moisture inside the tree and harmful insects and fungi out. When tree surgeons and orchard workers prune branches, they often paint a protective layer of tar or other material on the cut surface to take the place of bark.

INVESTIGATION 18A PLANT CELLS AND TISSUES

See Investigation 18A in the TRB.

In Chapter 10 you learned how the different structures of animal cells and tissues suit them to different functions. Plants also have many kinds of cells and tissues. In this investigation you will see how various types of plant cells and tissues produce everything from a soft, juicy peach to a tree trunk whose wood is strong enough to support a house.

Carpenter at work building a wooden frame for a house.

LESSON 18.1 ASSESSMENT

1. The grower will need to purchase vines that produce seedless grapes, since they cannot be grown from seeds.

2. Young fruit tissues perform photosynthesis, contributing to their own growth.

3. Roots absorb water. It then passes through stems and leaves. It evaporates from the surfaces of cells in the leaves, and the water vapor escapes through pores.

4. Carbohydrate and other nutrients are transported to the roots and stored there. The stored food provides resources for regrowth.

1 Consumers prefer seedless grapes to those with seeds. Why is this preference a problem for someone who wants to start growing grapes for supermarkets?

2 Most fruits and seeds are green at first. Some change color as they mature. Color is one way to tell when a crop can be harvested. What does the color of a green fruit tell you about how it gets energy?

3 Describe water's path from soil through a plant into the atmosphere. In what order does it pass through the various plant organs? At what point does it evaporate?

4 Some plants can grow new stems and leaves from the roots that are left in the ground after they are cut. This regrowth reduces the need to replant, but it also makes weeds hard to control. Where does the carbohydrate that feeds these regenerating plants come from? How does it get there?

LESSON 18.2 LIVING ON LAND

Pick the skills that might come in handy when you're working with plants: driving a tractor, cleaning muddy clothes, scuba diving, fixing a pump. You would be right if you rejected scuba diving. Managing the water supply is an important part of working with plants, whether in a greenhouse, on a farm, or in an office, but plants are basically land organisms. The ones that float on a pond are specialized species that have lost the ability to live out of water. In this lesson you'll learn how plants keep from drying out—but leave your wetsuit at home.

ACTIVITY 18-2   CELERY STRAWS

The procedure for this activity can be found in Chapter 18 of the TRB.

Why does water travel from roots to leaves? Several forces are involved, and you can probably figure out some of them pretty easily. The others might not be so obvious, but if you want to raise plants, you'll need to understand how they use water. Your teacher will provide you with the instructions for this activity.

MAINTAINING PLANT WATER BALANCE

BIOLOGY IN CONTEXT

Greenscene® is a company that leases and sells plants. The owners, George S. and Malcolm L., started the business after taking horticulture classes at their local community college. They divide responsibilities for the business: George takes care of the plants, and Malcolm is in charge of finances and the greenhouses.

On hot, sunny days, Malcolm uses evaporative coolers to keep the humidity up in the greenhouses. "Humidity is a key factor in plant health," he explains, "because if the plant intake of water does not keep up with the loss by evaporation, the plant starts to wilt. In our business, every plant delivered has to look in tip-top shape."

What does George do to maintain plant water balance? "If you see a plant in trouble, you can mist it to cool the leaf surface and slow the rate of water loss. You can pull a shade over the top of the greenhouse to cut the radiant heat. If many plants are looking parched, I usually turn on the supplementary coolers and increase the humidity in the entire greenhouse."

HOW PLANTS KEEP FROM DRYING OUT

Adaptations that prevent dehydration were essential for plants' move from water to land. Roots, which absorb water from soil, are an important part of this adaptation. The drawback is that they force plants to be sessile.

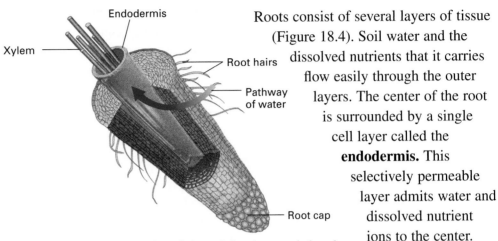

Roots consist of several layers of tissue (Figure 18.4). Soil water and the dissolved nutrients that it carries flow easily through the outer layers. The center of the root is surrounded by a single cell layer called the **endodermis.** This selectively permeable layer admits water and dissolved nutrient ions to the center. The water and dissolved nutrients

Figure 18.4: *Soil water passes through the endodermis to reach the xylem, where it travels up to the rest of the plant.*

WORD BANK

endo- = inside

-dermis = skin

The endodermis is like an internal skin surrounding the inner part of the root.

xylo- = wood

Wood is made mostly of xylem.

pass up through the root and stem in a system of tubes called **xylem.** Xylem tubes consist of the hollow cell walls of dead cells. They extend from the roots into every living part of the plant and are the main constituent of wood.

Several factors cause water to pass up through plants. The most important factor is evaporation from the leaves. Water evaporates from leaves through pores that close when water is scarce and open to admit CO_2 for photosynthesis or to let evaporation cool the leaves on hot days. Xylem in the leaves provides a steady supply of water to replace what evaporates, so water flows steadily from soil, through the plant, and out to the air.

Why doesn't the water evaporate before it reaches the leaves? And why do some people have to work so hard to keep plants and plant parts from drying out? Florists spray or dip blooms to reduce evaporation and to extend the product's freshness. But flowers are the exception: Most plant organs are covered by a cuticle, a thin layer of wax that keeps moisture in and harmful microorganisms out. Produce wholesalers often add a thin layer of wax or oil to fruits and vegetables to reduce evaporation and to protect these expensive products from fungi and bacteria. Bark on woody plants also helps retain moisture.

INVESTIGATION 18B HOW WATER MOVES THROUGH PLANTS

See Investigation 18B in the TRB.

Proper management of the water supply is critical for commercial plant production.

Many greenhouse operations and large irrigated farms use soil moisture sensors and computers to control watering equipment. Big farms in dry regions sometimes even hire consultants to advise when and how much to water the crops. Managers of large office buildings pay for horticultural services to care for the plants in their lobbies and offices. Water isn't the only thing plants need to stay healthy, but it *is* essential. In this investigation you'll explore several forces that make water flow through plants.

MECHANICAL SUPPORT OF PLANTS

An additional activity, How is a Bean Plant Like a Balloon?, can be found in Chapter 18 of the TRB.

Turgor pressure keeps herbs (nonwoody plants) upright and gives them shape, just as inflating a balloon or a tire gives it shape and support. Even woody plants depend on roots for a continuous supply of water to maintain cell turgor while water evaporates from their leaves. Woody tissues also help to support weight and maintain shapes during temporary droughts. Woody plants include large, long-lived trees and some smaller plants that also benefit from the protection that their tough stems provide against plant-eating insects and high winds. Tough, long fibers in the stems of several species make them useful crops. Fibers of the flax plant are used to make linen fabrics and high-quality papers (Figure 18.5). Ramie plants also supply fibers for fabric and paper.

A

B

Figure 18.5: *Fibers from the stem of the flax plant (A) are woven to make linen fabric (B).*

REPRODUCTION ON LAND

An additional activity, How Do Flowers Work?, can be found in Chapter 18 of the TRB.

In Chapter 10 you learned that before animals could thrive on land, their gametes had to adapt to survive out of water. The ancestors of today's plants faced the same barrier. Like frogs and other amphibians that must return to water to breed, the simplest plants must have water to carry gametes. Mosses and ferns, for example, live in moist, well-shaded soils. Their male gametes use flagella to swim through wet soil to the ovaries of another plant of the same species. Other species have a different way to exchange gametes: by being carried on the wind or on the surfaces of animals such as bees and bats that visit flowers. A water-tight layer covers pollen grains, the male gametes of flowering plants so that they don't dry out during their travels (Figure 18.6). Because this layer is so tough and resistant to decay, fossilized pollen provides clues to ancient crops, vegetation, and climates. Police labs often look for pollen on clothing and other objects as evidence of where an object or its owner has been. Because many crops are pollinated by bees, beekeepers often earn less from selling honey than they do from farmers who pay them to bring hives to their fields. Fragrant flowers, a popular item in the florist business, are adapted to attract bees, bats, or other pollinating animals.

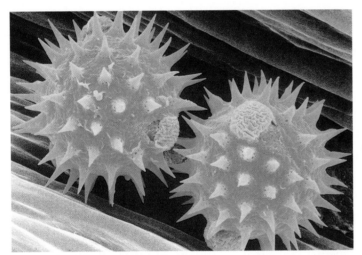

Figure 18.6: *These daisy pollen grains are protected from dehydration by a waterproof surface layer (color added).*

After fertilization, each zygote in the ovary can develop into an embryo. A seed develops to package an embryo with the stored nutrients that it will need to begin growing. The outer layers of the seed coat protect the embryo from cold or dry conditions until it reaches a warm, moist place in the soil where it can begin to develop into a plant. Plant embryos remain **dormant** (inactive) in seeds that are cool and dry. Plant breeders have been able to grow plants from some dormant seeds after storing them for many years. Seeds can deteriorate in poor storage conditions, however, so seed companies put expiration dates on the packets of seed that they sell for use in home gardens.

LESSON 18.2 ASSESSMENT

1. Spraying crops that are pollinated by insects kills the insects, so the plants do not get pollinated. Other crops do not depend on insects for pollination, so spraying them is not harmful.

2. Xylem tissue conducts water in living plants, so it can also be used in the form of paper towels to absorb watery liquids.

3. Ferns need cool, moist, shady environments to reproduce. They are not adapted to desert conditions.

4. Pollen contains only male gametes and does not grow into a plant. Seeds contain embryos that produce whole plants.

1 Farmers often use chemical sprays to kill insects that attack their crops. Why do you think it is illegal to spray insecticides on some crops, but not others, when they are flowering?

2 Paper is made mostly of wood. Explain how xylem tissue makes wood a good material for paper towels used to wipe up spilled liquids.

3 Why would a landscaper working in the deserts of the southwestern United States probably not plant ferns around a customer's house?

4 Pollen grains are small and easy to pack in large numbers. So why don't seed companies sell pollen instead of seeds?

LESSON 18.3 FOUR GROUPS OF PLANTS AND THEIR HISTORY

By now you know that there's no such thing as a generic plant. Some plants surround their seeds with fruit, some leave their seeds naked, and some don't have seeds at all. In this lesson you'll learn about four of the largest groups of plants. You will also learn how plants have evolved in complexity and adapted to particular climates.

ACTIVITY 18-3   SIMPLE PLANTS

The procedure for this activity can be found in Chapter 18 of the TRB.

Most plants are large, complex, diploid organisms. They live as haploid gametes for a short time before fertilization makes them diploid again, and they grow into embryos in seeds. In this activity you'll look at some simple plants and figure out what changed as more complex plants evolved. Your teacher will provide you with the instructions for this activity.

HORTICULTURIST

CAREER PROFILE

Nashville's Opryland Hotel is so big that guests can tour it by boat. Each of its varied environments contains different types of plants.

Hollis M. is in charge of all of the plants at Opryland Hotel, both on the grounds and indoors. Caring for indoor plants is a big job when there are more than eight acres of them growing under glass—and that doesn't include the plants in the public areas, the individual rooms, and displays for special events such as conventions. As you might imagine, Hollis has to deal with a wide variety of environmental conditions.

How does Hollis select plants that will thrive in each of these environments? "You have to match the environmental conditions where the plants would normally grow to the conditions in and around the hotel. You also have to take into account how well a plant can adapt to less than optimum conditions." Hollis explains that most complex plants can adapt to a wider range of conditions than simpler ones can. "Ferns, for example, generally require high humidity or they just fall apart on you. Our guests at the hotel would not tolerate high humidity, so we don't have many ferns.

Plants in the guest rooms must tolerate very low light and dry conditions. We have to rotate them out fairly regularly to keep them looking good."

How do you prepare for a career in indoor horticulture? "Many agriculture programs have at least a course or two that cover indoor plants, if not a specialty." Hollis advises future indoor landscapers to get as much greenhouse experience as possible while they are still in school.

THE ORIGIN OF COMPLEX PLANTS

An additional activity, Cladograms, can be found in Chapter 18 of the TRB. This activity teaches one way that diagrams such as Figure 18.7 can be constructed.

As you saw in Activity 18-3, mosses haven't adapted in ways that have made other plants succeed in diverse habitats. Like all plants, mosses and their relatives, the small, fleshy plants called liverworts and hornworts, are covered with a waxy cuticle and produce gametes in a protective jacket. But these groups are thought to be a side branch on the family tree of Plantae (Figure 18.7), not the ancestors of the larger, more complex groups that are listed in Table 18.1. Green algae and complex plants have similar cell division and similar structure and composition of cell walls and chloroplasts. These similarities suggest that the first plants evolved from this group of protists.

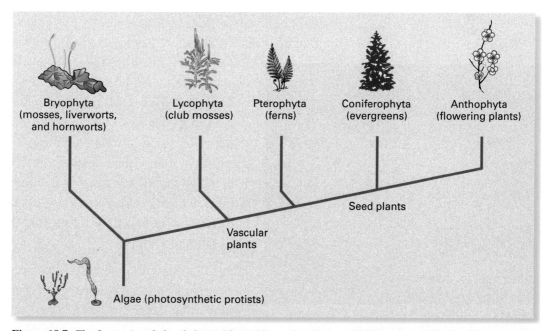

Figure 18.7: *The five major phyla of plants. The red lines show their probable ancestral relationships.*

Scientific Name	Bryophyta	Lycophyta	Pterophyta	Coniferophyta	Anthophyta
Common Name	Mosses, liverworts, and hornworts	Club mosses	Ferns	Conifers (evergreens)	Flowering plants (angiosperms)
Examples	Moss	Ground pine	Bracken fern, beech fern	Pine, fir, spruce	Oak, maple, bean, tulip, wheat, palms, ragweed

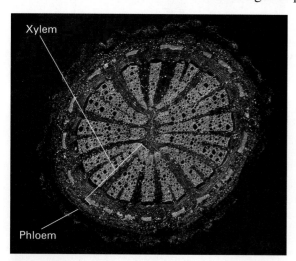

Pygmy moss (*Phascum cuspidatum*)

Shining club moss (*Lycopodium lucidulum*)

Ferns

Engelmann spruce (*Picea Engelmannii*)

Ornamental flowering plants

Note: Many biologists use the term "division" instead of "phylum" when classifying plants and fungi.

Table 18.1: Major Divisions of Plantae

Several important features make possible the large trees, seeds, fruits, and flowers on which we depend for wood, food, clothing fibers, and decorative blooms. One feature is **lignin,** a hard, plasticlike substance that strengthens the walls of cells that support the weight of plants. Some cells produce tough cell walls with lots of lignin; then they die and break down their own cytoplasm. Columns of these dead, hollow cells become xylem. Massive accumulations of lignin-embedded xylem become the tree trunks that give us wood. Another feature is a second transport system, called **phloem** (FLO-um). Unlike xylem, phloem consists of columns of living cells that actively transport substances such as sugars and amino acids throughout the plant body (Figure 18.8).

Plants that have these features can support greater weight and can transport water and nutrients throughout a larger organism. Ferns, for example, are vascular plants. They have vascular, or

Xylem

Phloem

Figure 18.8: *Cross section of birthwort* (Aristolochia) *stem. (Magnification ×11, color added.)*

transport, systems composed of xylem and phloem. These systems permit ferns to grow taller and compete more effectively for light than nonvascular plants such as mosses. Phloem, xylem, and lignin also enable ferns to grow strong roots that force their way through soil, extracting water and dissolved minerals that mosses can't reach and transporting them to their stems and leaves. That's why ferns can live in somewhat drier places than mosses can.

ADVANTAGES OF SEED PLANTS

An additional activity, Big Changes, Big Trees, can be found in Chapter 18 of the TRB.

Changes in reproductive systems are an important part of plant evolution. Mosses spend most of their lives as haploid organisms. They produce diploid structures that live only long enough to make cells that undergo meiosis to form gametes. The reverse is true of simple vascular plants such as ferns. The large, leafy plant is diploid, and the tiny, hard-to-find fern is haploid. Meiosis occurs in the reproductive spore cases that you saw on the leaves in Activity 18-3. The haploid spores drop to the soil, where they grow into the small haploid forms that produce gametes. This cycle of haploid and diploid forms is like the alternation of generations of some Protista that you studied in Chapter 8.

In seed-producing plants the haploid generation has been reduced to a gamete-producing tissue in the reproductive organs of a diploid plant. Let's look at conifers such as pines, which are important trees for loggers, paper and lumber producers, and wildlife managers. The cool, mountainous regions of northern and western North America are almost entirely covered by conifer forests (Figure 18.9). These trees produce most of the wood and paper that industry and consumers use. All trees are diploid. A few specialized cells in the cones undergo meiosis to produce the haploid egg and pollen cells. A surface coat protects pollen from dehydration, so fertilization doesn't have to be in water or wet soil. When pollen from male cones falls onto the female cones, fertilization

Figure 18.9: *Drought-resistant conifers cover many cool, dry mountain slopes where other plants would not survive.*

follows, and diploid embryos develop in the ovaries. Each embryo and its food reserves are protected in a tough, dry seed. Like pollen, seeds remain dormant while dry. Pollen and seeds lessen a plant's dependency on water. Reproduction takes place without surface water, and seed plants can adapt to drier, more diverse environments, such as the Rocky Mountains.

FLOWERS AND FRUITS

About 95% of all plant species, including most edible ones, are seed plants that flower. What makes these species so successful and so useful? How are they different from conifers?

Flowering plants differ from conifers and other seed plants in two important ways. Their xylem is more complex than that of other seed plants, with separate cells specialized for support or transport. And of course they have flowers. Ovaries in flowers nourish the egg cells and develop into fruits, which protect the seeds. Fruit may help to scatter seeds and nourish young plants.

Figure 18.10: *One way to spread pollen. This bee orchid flower resembles the female of a certain species of bee. When the male tries to mate with the flower, pollen sticks to its body. When the insect visits other flowers, it leaves some of this pollen.*

Most flowering plants depend on insects, bats, or other animals to transfer pollen among flowers; these pollinators are more accurate than the wind and gravity that deliver conifer pollen. Nectar or edible flower parts encourage pollinators to visit flowers. Complex symbiotic relationships are common between individual plant and animal species and probably contribute to speciation in plants and animals (Figure 18.10).

The seeds of many species spread by sticking to fur or feathers or by passing through an animal's digestive system after it eats the fruit. You reap the benefits of plants' tastiness whenever you munch on an apple or crunch on sunflower seeds. The farmer reaps the problems when adding manure to a field: many weed seeds pass unharmed through the digestive systems of cattle and other farm animals. These can interfere with crops that are fertilized with manure.

In general, flowering plants devote more resources to reproduction than other plants do. The vascular plants tend to have larger seeds with more nutrients than spores have. Producing fruits and flowers also takes more energy and nutrients. Judging from the number of flowering species, this investment pays off.

LESSON 18.3 ASSESSMENT

1. Mosses do not have the lignin or vascular system that are needed to produce large, treelike forms.

2. Bats help to pollinate some flowers, including those of some fruit trees. (Bats also help to control insect pests.)

3. Bats would not help to pollinate pine trees, which are not flowering plants.

4. There is too much salt in the soil. The salt travels up through the xylem, dissolved in water from the soil. It accumulates on the leaves as the water that carries it evaporates.

1 Could a plant breeder develop a tall, treelike moss to use as an ornamental plant in dimly lit rooms? Explain why this might or might not be possible.

2 Why do some fruit growers build bat houses in their orchards?

3 Tree farmers raise pine trees like corn or other field crops, for paper or lumber production. Would tree farmers encourage bats to settle on their farms for the same reason that orchard owners do? Give a reason for your answer.

4 The plants in your garden are all sick or dead. They also have tiny clusters of salt crystals on their leaves. What's wrong with your soil? Describe the path of the salt as it travels to the leaf surface.

CAREER APPLICATIONS

The applications that follow are like the ones you will encounter in many workplaces. Use the biology you learned in this chapter to complete the activities. Share your work with the class.

AGRICULTURE & AGRIBUSINESS

Thirsty Crops

Use an almanac, atlas, or Internet resources, or talk to an agricultural extension agent to find out what crops grow in your state or region. Which crops need the most water? What is the typical rainfall in your state during the growing season? Is irrigation needed for the crops that are heavy water users? If so, how do the irrigation systems work? What farming practices help to maintain soil moisture?

Hardwood and Softwood

What do the terms "hardwood" and "softwood" mean? How are they related to the differences between conifers and angiosperms (flowering plants)? Use the resources that are available at your library, in cooperative extension publications, and from paper and lumber companies and their organizations to prepare a report on this subject. Explain the meaning of "hardwood" and "softwood" and why most conifers and angiosperms are in separate wood categories. Show how the terms "hardwood" and "softwood" relate to wood structure and use.

BUSINESS & MARKETING

Plant Business

Look in the yellow pages of your local phone book for businesses that deal with plants. In class, select a variety of plant-related businesses. Each student should contact the manager or owner of a different business and find out how this person got into the business. What level of knowledge about plants does he or she need for their job? How did the person learn it? What does the person find most rewarding in the job? What's the biggest drawback? Report to the class on what you found and write a brief description of the business for a class notebook.

Keeping Flowers and Fruits Fresh

Interview a florist or supermarket produce manager about antitranspirants. These waxes, oils, and other substances protect flowers and produce from drying out or being attacked by decay-inducing pathogens, including certain species of bacteria and fungi. How are they applied? At what stage of the marketing process are they used? How much do they add to the useful life of these plant products? Are there any health concerns for the people who apply the antitranspirants or the people who consume the produce? What happens to the antitranspirants when the vegetable peels or flowers are returned to the environment? Report to the class on your findings.

Desert Plants to Save the Whales

Sperm whales got their name from spermaceti, a waxy fat that fills a cavity in their heads. This substance is the source of sperm whale oil and one reason the whales have been hunted and endangered. Use your library and the Internet to research jojoba (*Simmondsia chinensis*), a desert plant that is found in the Southwest. The seeds of jojoba yield an oil that can replace sperm whale oil in cosmetics and be used as a high-pressure engine lubricant. Find out how the seeds are grown and processed. Report to the class on present and potential uses of this plant. Your report should include a marketing brochure describing the benefits of jojoba oil to potential industrial customers and explaining why they should use it rather than other oils in their products. One question to think about as you research jojoba is why almost all the plant oils that we use come from seeds.

Dirt-Free Houseplants

FAMILY & CONSUMER SCIENCE

Go to the library or use the Internet to find out about growing houseplants in materials other than soil. Look at hydroculture and methods that use synthetic gels. Which common houseplants can grow in each type of system? What are the advantages and disadvantages of each method? If possible, choose one of the methods and try it with an appropriate houseplant.

Raising Simpler Plants

Ferns are popular houseplants for dimly lit rooms, and mosses often grow in the shadier, damper parts of a yard or garden where vascular plants are less able to compete. But how do landscapers, homeowners, and companies that supply ferns to retail stores for sale as houseplants raise them where they want? These plants do not produce seeds that can be packaged and sold. How difficult would it be to produce a new generation of plants from ferns in a home or office? Investigate the life cycles of ferns and mosses in your library, and interview a landscaper or horticulturist to find out how you could raise these plants. Report to the class on your findings.

Medicinal Plants

HEALTH CAREERS

Use the Internet to search for information on ethnobotany, medicinal plants, or herbal medicine. Select a plant that has been used as a medicine. Prepare a short report including a drawing of the plant, its classification (family, genus, species), its medicinal qualities, what cultures have used it, any modern uses of the plant or its derivatives, and reports of harmful side effects. Create a class notebook of all reports.

Plant Allergies

Many people are allergic to pollen or to certain fruits, nuts, or vegetables. Are these people allergic only to certain parts of a particular species? What about other plant parts and species? For example, if someone is allergic to peanuts, will he or she also be allergic to peanut pollen? Is this person likely to be allergic to related species, such as peas and beans? Interview an allergist, family doctor, nurse, or other health professional. Public health organizations and patient groups also provide Web sites, e-mail lists, or other Internet resources for the public or publish reading materials on these subjects. Use the information that you collect to prepare a report on patterns in allergies to plant products.

INDUSTRIAL TECHNOLOGY

Living in a Glass House

Research the technologies for maintaining the best growing conditions for plants in a greenhouse. Divide the technologies into those that control light, temperature, or moisture. Also try to find out how pollination and plant pests are managed in greenhouses. Within each category, compare the technologies' cost and efficiency.

Processed Wood Products

Investigate the production of paper or "engineered wood" products such as plywood, chipboard, particleboard, and pressboard. How are these products made? What kinds of wood are used in these processes and why? How does the processing affect the original raw wood? You might find information on these subjects at a public library, in cooperative extension publications, or from companies and other organizations that are involved in this business. Many of these organizations can be found on the Internet. Prepare an oral report to the class on this subject with pictures or charts explaining the process in terms of its effects on plant cells.

Living Air Fresheners

Investigate plants that are used as living air purifiers. NASA has funded considerable research in this area, looking for ways to maintain air quality on space stations and in extended space flight. How are the systems set up? What chemicals do different plants remove from the air? Can plants make a significant improvement in sick-building syndrome?

- Roots anchor plants and absorb water and dissolved nutrients from soil.

- Xylem carries a stream of water and dissolved nutrients from roots to leaves and other above-ground plant parts. The water evaporates through tiny pores in leaves.

- Thin, flat leaves with porous, spongy internal tissues maximize the surface area for photosynthetic gas exchange with the atmosphere.

- Most plants reproduce sexually. Specialized cells in flowers undergo meiosis to produce pollen and egg cells.

- After pollen fertilizes the egg cells in a flower, the zygotes develop into embryos. Each embryo is encased with stored nutrients in a seed that can tolerate unfavorable environmental conditions. Seeds remain dormant until conditions permit growth of the embryo into a new plant.

- Unspecialized cells in embryos grow and divide, eventually producing all the specialized cells of the mature plant.

- Adaptations that help plants to obtain and retain water include the cuticle, roots, bark, and xylem.

- Turgor pressure and the tough, lignin-rich walls of tissues such as xylem are strong enough to give shape to plants and support even the weight of large trees.

- Lignin, xylem, phloem, and roots permit vascular plants to grow tall and to tolerate drier environments than mosses can.

- Seeds, pollen, flowers, and fruits allow plants to reproduce in dry environments and to develop symbiotic relationships with pollen-carrying animals. These adaptations have contributed to the success of seed plants, especially flowering plants.

- Trends in the evolution of vascular plants include a reduction of the haploid phase of the life cycle, more elaborate and specialized reproductive systems, and the development of specialized supporting and water-carrying xylem cells.

CHAPTER 18 ASSESSMENT

Concept Review

See TRB, Chapter 18 for answers.

1 Explain why trees in the northern U.S. lose their leaves in the winter, but many desert plants lose their leaves in the dry summer season.

2 What are the differences between plants and protistan green algae?

3 Plants such as the Venus flytrap, sundews, and pitcher plants capture and digest insects. Does this make their classification in Plantae a problem?

4 Are plant embryos diploid or haploid? Explain your answer in terms of gametes and fertilization.

5 Do all seeds contain embryos?

6 Do plant embryos contain in miniature form all the organs they will have as adult plants?

7 What process is most important in moving water through plants?

8 List at least three adaptations that keep plants from drying out.

9 How do plants wilt?

10 Why doesn't puncturing a plant with a pin make it lose turgor pressure?

11 Do ferns produce pollen? Give a reason for your answer.

12 How does seed dormancy help plants to survive and reproduce?

13 Would you expect to find more lignin in a walnut or in its shell?

14 Are sugars and other organic nutrients more concentrated in xylem fluid or in phloem fluid?

15 Are pinecones fruits? Give a reason for your answer.

16 Refer to Table 18.1. Which divisions (phyla) of Plantae demonstrate alternation of generations? What has happened to the "missing" generation in the other divisions?

17 Some people pick the large flowers of squash plants, fill them with rice or chopped vegetables, and cook them. What is the squash plant then unable to produce?

18 Do mosses, ferns, conifers, or flowering plants devote the most resources to the production and survival of each offspring? List adaptations as evidence to support your answer.

Think and Discuss

19 Why do mosses and ferns often grow in forests beneath large conifers or other trees?

20 What kind of problem would lead you to suspect that a plant's roots had been damaged?

21 What adaptations did plants need to make in becoming land organisms?

22 Why do many species of microscopic parasites attack plants through their flowers?

23 How could infection with bacteria that block xylem tubes make a plant undernourished? Give a reason for your answer.

In the Workplace

24 Cork comes from the bark of a certain species of oak tree. What function could cork serve in bark?

25 Lumberyard managers and builders keep wood covered up or indoors during rainstorms so that it won't swell up with moisture. Could a breeder develop a tree that produces wood that doesn't absorb water? Give a reason for your answer.

26 When bacteria break down the soft tissues of flax stems, the tough fibers that remain can be used to make linen. What does this tell you about the function of lignin, besides that of supporting weight?

27 What kinds of plants would you select when landscaping a house for a client who is allergic to insect stings?

Investigations

The following questions relate to Investigation 18A:

28 What function do plant cells with thick walls usually have?

29 Most cells in the surface layer of leaves do not have chloroplasts. What is their function? Give a reason for your answer.

The following questions relate to Investigation 18B:

30 Would wiping bark with a solvent such as alcohol or ether dehydrate the tree? What does this say about what prevents trees from drying out?

31 A breeze can make people feel cooler on a hot day. Does wind make desert plants more or less likely to be injured by dehydration?

CHAPTER 19

WHY SHOULD I LEARN THIS?

Do you know people who have a green thumb, people who have a knack for taking care of plants? They understand how plants grow and develop. A green thumb could help you to run a landscaping business that creates, improves, and cares for outdoor plantings or office plants. Or you might use it to keep grasses and weeds out of a little-used road or footpath. Most likely, you would use that thumb to care for your own lawn, garden, or houseplants. No one is born with a green thumb. This chapter gives you a chance to start growing one.

PLANT GROWTH AND DEVELOPMENT

WHAT WILL I LEARN?

1. how an embryo in a seed grows and develops into a plant
2. how plants' main vegetative and reproductive organs grow and develop
3. some common types of plant cells and tissues and how they form organs
4. how the structure and composition of some plant parts make them useful

What do you plant when you want a private yard but you're worried about security? You choose a plant whose growth habits are helpful for both needs. Some plants that provide privacy for a home can also conceal burglars and vandals. Holly plants form shrubs with thorn-tipped leaves. These evergreens provide year-round privacy and a terribly uncomfortable place to lurk. So if holly grows well around your home, you may have the perfect plant.

As you learn in this chapter how plants grow and develop, you'll discover the roots of their versatility.

LESSON 19.1 FROM SEED TO PLANT

Although we usually take it for granted, there is something amazing about the way a hard, dry, little seed unfolds into a vigorous, growing plant all by itself. How does it happen? Gardeners, farmers, plant nursery workers, foresters, and landscapers are just a few of the people whose work depends on the successful development of seeds into healthy plants. In this lesson, you'll learn about how that happens.

ACTIVITY 19-1 PLANTS UNDER CONSTRUCTION

The procedure for this activity can be found in Chapter 19 of the TRB.

A seed looks nothing like the plant it will become. How does it change form so completely? It has to grow quickly, too, to outrace bad weather, mold, and seed-eating animals. Sprouts are a popular food, and restaurant owners and their suppliers have learned to grow them in large quantities. In this activity you will examine sprouts and figure out how seed structures become plant structures. Your teacher will provide you with the procedure for this activity.

SEED ANALYST

CAREER PROFILE

Kerry B. works for her state's seed-testing service. She explains how the service samples seed for testing. "This lab is one of three places in our state where any seed advertised for sale must be inspected and tested. We take representative samples of seed from each bag either by hand or using a sampling probe.

"After the seed is sampled, it is checked for purity. To separate seed from 'trash,' or what we call inert matter, a blower is sometimes employed." This machine blows bits of leaves, broken seeds, and other debris up and away from the heavier seeds.

Seed analysts examining seedlings.

Kerry does other tests, too. "We test some of the seeds for viability and for germination rate. Other tests may determine seed moisture content, whether seedlings grown from seed can tolerate stress conditions, and if seeds are under attack from fungi."

Seed analysts don't have a formal degree program, but they need to know about plant biology and classification. After her first two years working in a federal lab, Kerry took a test to become certified by the Association of Official Seed Analysts (AOSA). People working for private companies are certified by the Society of Commercial Seed Technologists (SCST) by a similar process.

GERMINATION AND EARLY SEEDLING GROWTH

Most plant species produce seeds. Seeds are dormant until environmental conditions such as temperature, moisture, and light are right. Then the seed **germinates,** or sprouts. First, the seed absorbs water and swells, which brings the embryo's dehydrated cells back to life. A steady supply of water is critical during germination. Growers often plant when they expect rain soon, to ensure that their germinating seeds do not run out of water and die.

After absorbing water, the seed cells start producing enzymes that break down stored protein, carbohydrate, and fat. Cell growth and division soon pick up speed, and the embryo begins to grow. The first visible sign of growth is in the embryonic root, which lengthens, breaks through the seed coat (Figure 19.1), and begins to absorb water from the soil. This **primary root** turns downward as it grows longer, no matter which end of the seed is up. By the time the embryonic stem begins to grow longer, the primary root has already formed a branched root system of **secondary roots** (Figure 19.2). This system anchors the new plant, or **seedling,** to the soil and absorbs water and minerals.

Seedlings are easily damaged by pests, cold or dry weather, and herbicides. Growers water their seedlings carefully and may apply herbicides early in the season to kill weed seedlings when they are most vulnerable.

Seedlings quickly consume what's left of the seed's food reserves. The shoot (the above-ground part of the plant) soon provides a

Figure 19.1: *The emerging root tips are a sign that this peanut is beginning to germinate.*

Figure 19.2: *Early development of a corn seedling.*

new source of nutrition: photosynthesis. The tiny embryonic leaves expand to their mature size, and the embryonic cells at the tip of the shoot produce more leaves and stem tissue. As each leaf grows, it produces the enzymes and pigments of photosynthesis. The leaves start feeding the seedling as it exhausts the nutrients stored in the seed.

INVESTIGATION 19A | THE CERTIFIED "GERM" TEST

See Investigation 19A in the TRB.

Growers and landscapers go to a great deal of trouble to ensure that most of the seeds they plant germinate and produce healthy seedlings. Successful germination depends on several factors, including how the seeds were stored, their moisture content, age, and any special requirements for ending dormancy. Each species has its own requirements for temperature, moisture, and light or darkness during germination. In this investigation you will discover how germination of one species is affected by light and by nicking the seed coat to allow water absorption.

CHANGING IDEAS

WHAT'S IN A NAME?

At the beginning of the Renaissance (around 1400), scientists were busy naming and grouping organisms. They wanted to develop an organized system to classify all living things and describe their relationships to each other. They started by collecting plants from different countries and comparing them. Botanists from various countries in Europe gave Latin names to plants that grew in their own countries. In the late 1500s, Carolus Clusius of Belgium wanted to get to know "every flowering thing," rather than just giving them names. He described the various forms of plants. He also grew plants and dried them for his collection. Natural history museums began to collect dried plants for scientists and other interested people to study. Such a collection is called a herbarium.

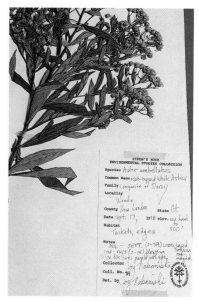

A typical herbarium specimen with its descriptive label

The Swedish naturalist Carolus Linnaeus, working in the eighteenth century, used herbaria in his research. He emphasized the importance of using dried plants as examples for writing very specific descriptions of each species. Linnaeus developed the system of species classification that we use today. His books describe each species he studied in great detail, enabling his readers to identify any plant they found.

PLANT TISSUES

Each plant organ is made up of three main types of tissue (Table 19.1). **Dermal tissue** forms the outer layer of the plant, or **epidermis** (Figure 19.3). The epidermis helps to protect the plant from insects, fungi, and drying out. It is usually only one or a few cells thick. The epidermis produces a waxy cuticle on its outer surface that helps the plant to retain water. People who work for chemical companies to develop new herbicides have found a way around this waterproof barrier. Hydrophobic compounds that diffuse easily through the cuticle wax kill weeds better than do water-soluble compounds.

Ground tissue fills the plant and gives it shape. There are three common types of ground tissue cells. The simplest of these cells

Tissue type	Dermal tissue	Ground tissue	Vascular tissue
Major functions	Reduces evaporation of water; keeps pathogens out	Fills out organ shapes and provides flexible support; stores water, starch, or protein	Transports water and minerals (xylem), sugars and amino acids (phloem)
Examples	Waxy skins of leaves, apples, and cucumbers	Flesh of potato and pear; interior of nonwoody stems; photosynthetic cells in leaves	Wood; central core of carrot and other roots; fibers in corn stalks and bamboo

Apple and section

Lumber

Potato

Table 19.1: Three Main Types of Plant Tissue

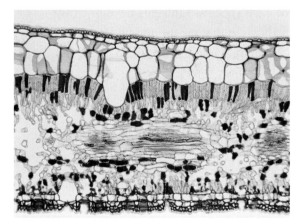

Figure 19.3: *Cross section of a mature rubber tree leaf (*Ficus elastica*), showing upper and lower epidermis (Magnification ×65)*

have thin, flexible walls and a large central vacuole (Figure 19.4A). These cells are important in plants' metabolism. They make and store carbohydrates or other compounds. Examples are the chloroplast-rich cells of leaves and cells that make up the outer parts of roots and the fleshy tissue of fruits.

A second type of ground tissue cell has thick, relatively soft walls (Figure 19.4B). Tissue that is composed of these cells supports growing organs such as celery stalks without limiting their growth.

Some ground tissue cells have thick, rigid walls embedded with lignin (Figure 19.4C). They support mature, nongrowing tissues. When these cells die, they leave their hollow cell walls, which continue to provide support. They include the long, slender cells that make up the fibers of the flax plant, which is used to make linen. Similar fibers of other species are used to make rope and paper. Other cells with lignified walls are shorter and more irregular in shape. They make up hard tissues such as nutshells and seed coats.

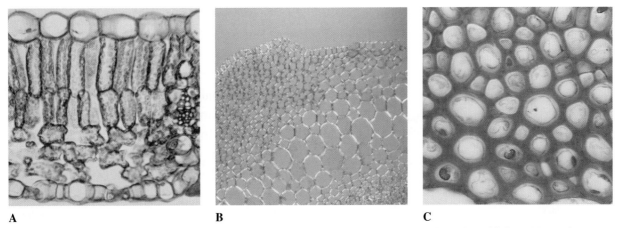

A   B   C

Figure 19.4: *Photomicrographs of typical parenchyma, collenchyma, and sclerenchyma cells (color added). (A) Parenchyma tissue inside a leaf; (B) Collenchyma tissue in a celery stem; (C) Thick-walled sclerenchyma cells in a stem.*

Vascular tissue transports water, nutrients, and other substances throughout the plant body. It includes xylem and phloem. Phloem consists of columns of living cylindrical cells. As these columns develop, part of the cell walls at the cell ends breaks down. The cytoplasm of neighboring phloem cells is connected through these gaps (Figure 19.5). Sugars and other nutrients flow through the openings from cell to cell. That's how plants transport nutrients from leaves and storage organs to growing parts that consume them. Aphids and other insects and pathogens that prey on plants often feed on the nutrients concentrated in phloem. This is one reason that growers spend so much time and money to control these pests.

Figure 19.5: *The structure of phloem. The cytoplasm of neighboring phloem cells is in contact through holes in their cell walls. Sugar and other nutrients flow from each cell to its neighbor without crossing their cell membranes, from one end of the column to the other.*

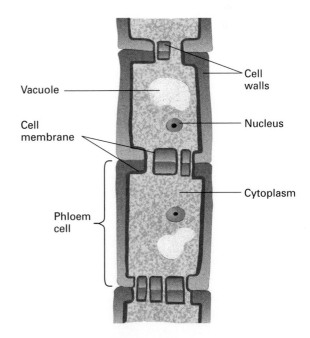

Vacuole

Cell membrane

Phloem cell

Cell walls

Nucleus

Cytoplasm

Xylem transports water from roots to leaves and supports the plant. In conifers and simpler plants, xylem is made mostly of one type of cell. As the xylem develops, these long cells grow thick walls and break down their own cytoplasm. The cells die as they mature, leaving only the thick, strong, hollow cell walls. These xylem cells make up the wood of conifers such as pines, which are used for lumber and papermaking. In these species a single type of xylem cell does double duty, supporting the tree and transporting water. In angiosperms (flowering plants) these jobs are divided between two specialized cell types. Columns of cells with thinner walls make up the "plumbing" that carries water (Figure 19.6). Fibers support more of the weight of these plants. All of these cells are dead and hollow at maturity. Fungi, insects, and other decomposers can destroy much of a tree's dead wood without seriously interfering with its survival, though it may no longer be usable for lumber or paper.

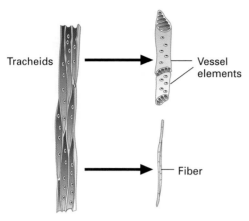

Figure 19.6: *During the evolution of flowering plants, a single type of xylem cell (the tracheid) became two types, one specialized to carry water (the vessel element), and one specialized to support weight (the fiber).*

INVESTIGATION 19B ANALYSIS OF PLANT FIBERS IN PAPER

See Investigation 19B in the TRB.

The fibers in paper come from the xylem tissue of trees. The cells of the xylem tissue are dead. They contain no cytoplasm or nuclei, only cell walls made mostly of cellulose and lignin. Cellulose in the cell walls forms small fibers called fibrils. When wood is reduced to pulp for paper, the fibers and fibrils of the original wood form a network that gives paper its strength. In this investigation you will extract plant fibers from various grades of new and recycled paper and examine their structure. You will investigate the relationship between fiber structure and the type of paper from which the fibers were taken.

Paper mill worker inspecting paper product quality

LESSON 19.1 ASSESSMENT

1. No, the root emerges from the seed before the shoot. Germination may have begun, even though no shoots have grown above the soil surface.

2. The vascular tissue (xylem) has been blocked.

3. The falling sand could have scraped off the cuticle in spots, exposing the plant tissues to fungi.

4. The herbicide would prevent only reproduction, because roots, stems, and leaves are all in the embryo, but flowers are not.

1 You work at a large greenhouse, growing plants for sale. You planted a batch of seeds three days ago, and no seedlings are visible yet. Does this mean that the seeds haven't germinated? Give a reason for your answer.

2 Your bean crop has wilted because of a bacterial infection. Which of the three types of plant tissues has been clogged by the infection, causing the plants to wilt?

3 You are the landscaper for a large, old office building. A few days after the face of the building has been cleaned by sandblasting, the plants in front are suddenly covered with small spots of fungal infection. How could sandblasting lead to these infections?

4 A new herbicide interferes with the formation of plant organs but not with the growth and development of existing ones. If you sprayed it just as the weeds in your field were germinating, would it interfere more with seedling growth or with reproduction? Give a reason for your answer. (*Hint:* Which organs are in the embryo? Which are needed for reproduction?)

LESSON 19.2 VEGETATIVE GROWTH AND DEVELOPMENT

It's spring in northeastern North America. Suddenly, the dead-looking, bare branches are crowded with leaves. A neighbor's lawn is growing enough grass for a herd of cows, and he's mowing it twice a week. Weeds are crowding the vegetables in the community gardens. What makes plants grow . . . and grow and grow? In this lesson you'll learn about vegetative growth—how leaves, stems, and roots grow and develop.

ACTIVITY 19-2   VEGETATIVE GROWTH: LOVE IT OR LEAF IT

Observations might include fertilizing, seeding, pruning, treating with growth hormones (which encourage plant growth) and mowing, trimming, clearing brush, cutting limbs, use of herbicides, and mulching (which discourage plant growth). Some activities may discourage growth of one type of plant while encouraging the growth of another, such as mulching to eliminate weeds that may compete with a desirable garden species.

Around your neighborhood or school, observe how people respond to plant growth. Are they encouraging plant growth or discouraging it? You might need to interview some homeowners, gardeners, or other people who work with plants. If you are doing this activity in the winter, you might investigate how people care for bulbs and cold-sensitive outdoor plants during cold weather. List what you find in your logbook. Discuss the reasons underlying people's horticultural activities.

TREE DAMAGE

BIOLOGY IN CONTEXT

Asplundh® is a company that manages vegetation for utility companies. Its crews keep trees and brush out of power line rights-of-way. The safe distance between branches and wires varies with the line voltage and the temperature and weather ranges of the area. Unchecked, trees can grow onto power lines, exerting enough pressure to snap wires. High winds may break off limbs or uproot shallow-rooted trees, allowing them to fall onto power lines.

Workers trimming trees near power lines

Do Asplundh's crews just cut off the tops of trees? No, they climb trees or use a lift truck to trim branches. They cut branches from large limbs or at the trunk, training the tree to grow parallel to the lines. Sometimes, they cut limbs off at the trunk, so the tree heals with a bull's-eye scar. This kind of cut prevents sprouting from the edges of the scar.

So if you're planting a tree, look up. If the tree is near a power line, it shouldn't crowd the line when it reaches its full height. If you're planting under a power line, select low-growing shrubs and trees that can be pruned to a safe height.

HOW PLANT CELLS GROW

An additional activity, Finding the Growth Centers, can be found in Chapter 19 of the TRB.

Ever since you were a small child, you've probably heard adults telling you to eat well so that you will "grow big and strong." Nutrition is important for both animal and plant growth, and growth quickly stops when plants don't have enough water. Growth can continue only when plant cells expand. As active transport concentrates mineral ions and other solutes in the cells, water follows and the cells swell up. The increase in cell size is responsible for plant growth. That's why drought is so damaging to crops and why growers spend so much money on irrigation systems. As plant cells expand, however, their walls grow thicker and stronger. Eventually, the cell wall becomes so stiff and strong that it resists more expansion. The size and shape of a plant cell depend on the balance between the opposing forces of turgor pressure and cell wall stiffening.

HOW PLANT TISSUES AND ORGANS GROW

WORD BANK

meris- = part, or division into parts

-em or -eme = part of a structure

Meristems are the parts of plant structures that continually divide to produce new cells.

During germination and early seedling growth, nearly every cell divides and grows. In older plants, growth depends on the expansion of new cells produced in specialized zones called **meristems** (Figure 19.7). These zones remain embryonic even in very old trees. Cell division in the meristems of growing organs provides a steady supply of new cells. Most of these cells expand and differentiate to become part of one of the three main tissue types. A few remain permanently in the meristem, producing many generations of daughter cells that contribute to growth.

Apical (tip) meristems, in the tips of roots and stems, produce cells that divide and expand. As these cells grow, they make roots and stems grow longer, pushing the meristematic tips ahead. The stem grows taller, and the elongating root pushes the root tip through the soil.

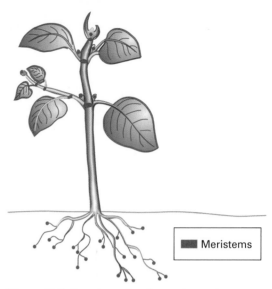

Meristems

Figure 19.7: *Locations of major meristems in a growing plant.*

ROOTS

Behind the root's apical meristem, cells expand to their full length and differentiate. Some epidermal cells elongate to form delicate **root hairs** that increase the surface area for absorption of water and minerals (Figure 19.8). The endodermis, a cylinder of specialized ground tissue, develops around the vascular tissue at the center of the root (Figure 18.4, p. 645). Each endodermis cell develops a band of cork in the cell wall. This waterproof strip encircles the cell, so water reaches the xylem only through the cell's membrane and cytoplasm. A layer of cells just outside the endodermis (not shown in Figure 19.7) remains meristematic; branch roots begin in this layer and grow out through the outer ground tissue. Utility workers who try to keep tree roots from cracking sidewalks or growing into water pipes cut roots to remove the root tip meristem, but the root stump may still produce branches that can grow into the same area.

Other specialized root cells include the **root cap,** tough tissue that protects the root tip as it pushes through the soil. The root ground tissue of some species stores sugars, starch, or other substances.

Figure 19.8: *Root hairs near the tip of a root help absorb water and dissolved nutrients.*

Because roots store substances in their ground tissue, they are often good sources of food and useful chemicals. Useful compounds from plant roots include rotenone (an insecticide), reserpine (a blood pressure medication), and sugar from sugar beets. Find out which plant species are valued for their roots, what useful compounds come from roots, and what these compounds do. Agricultural books from a local library, Internet searches, and your county cooperative extension office can help you to get started.

STEMS

Stems produce many important products. Latex, cinnamon, asparagus, cane sugar, and of course, wood all come from stems. Plant diseases or storms that damage the trunks of rubber trees can cause serious problems for health care workers and painters who depend on a steady supply of latex gloves or paints. Failure of the sugar cane crop can spell economic disaster for entire tropical nations that get most of their income from sugar exports. Cactus stems store water that desert travelers can use in emergencies. Edible underground potato stems, called tubers, store starch.

Like roots, stems start growing at an apical meristem in the tip. Ridges on the growing tip become new leaves. The stem tip and its surrounding cluster of tiny young leaves form a bud. Leaves grow from the stem at each **node** (Figure 19.9), and stems grow taller as cells in the **internodes** (the stem sections between nodes) elongate.

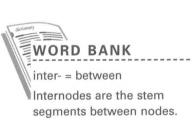

WORD BANK

inter- = between

Internodes are the stem segments between nodes.

During this growth, different types of cells develop. Some stem cells develop into vascular tissue, extending existing columns of xylem and phloem into the new internode. Epidermis cells cover the stem's surface with a cuticle. Cells of the ground tissue in the leaves produce many chloroplasts, where most of the plant's photosynthesis takes place. However, the stem tissue in each node remains meristematic. Cell divisions there add new cells that elongate the internodes. Another small meristem on the stem surface just above each leaf produces a bud that can grow into a new branch or a flower.

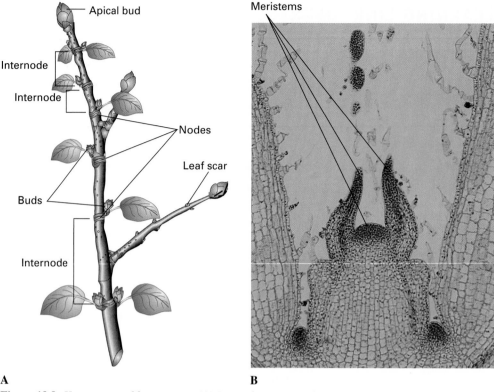

Figure 19.9: *How stems and leaves grow. (A) Parts of a growing plant stem. (B) Vertical section through a stem apex. Note the small, embryonic cells of the apical and nodal meristems and the larger, expanding cells of the internodes.*

LEARNING LINK STEM STORAGE

Find out what plants are cultivated primarily for their stems. What compounds in stems are commercially valuable? What applications do they have? Consult your county cooperative extension office; books about agriculture that are available at your library; the Internet; or encyclopedia articles about bamboo, latex, wax, maple syrup, turpentine, cinnamon, ginger, or taxol.

SECONDARY GROWTH

The growth and development of stems and roots from apical meristems is **primary growth.** This growth increases the length of a plant shoot or root but doesn't add much thickness. **Secondary growth** produces the thick stems (or trunks) and roots of trees on which many industries and occupations depend (Figure 19.10).

A   B

Figure 19.10: *Secondary growth produces huge trees that are important to tourism (A) and to lumber and paper production (B). Trees are also critical for the survival of many species, including humans.*

Remember that branch roots grow from a cylindrical meristem located just outside the endodermis. In secondary growth this meristem acts as a **cambium**—a meristem that produces xylem cells on its inner face and phloem on its outer face (Figure 19.11A).

The accumulated xylem becomes the woody core of large, old roots. Older layers of phloem cells are crushed as the root grows thicker, and surviving cells near the surface produce a tough, corky bark. In most trees and other woody plants, the cambium extends up into the stem. The stem cambium forms as undifferentiated cells between xylem and phloem bundles divide (Figure 19.11B). The rate of cell division soon increases in neighboring cells of the ground tissue until the cambium completely encircles the stem.

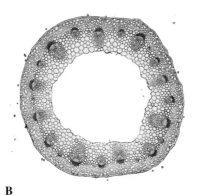

A   B

Figure 19.11: *Secondary growth in (A) a root of willow* (Salix *sp.) and (B) a stem of buttercup* (Ranunculus *sp.). (Magnification ×20, color added.)*

The pattern of secondary growth in stems is like that in roots. Seasonal changes in rates of cell division and growth produce the annual rings that show a tree's age. Hardwood species such as oaks are prized partly for their attractive "ring porous" appearance. These species produce xylem vessels and fibers at different times of the year, so their wood consists of alternating rings of hollow xylem tubes ("pores") and dense, solid fibers. This alternation doesn't happen in pines and other conifers, because their xylem consists of only one cell type.

LESSON 19.2 ASSESSMENT

1. Secondary growth does not occur.

2. No, the height of the trees is the result of growth near the apical meristems, not lower on the trunk.

3. Very dry weather would produce narrow rings because cell growth depends on adequate water to maintain turgor pressure.

4. The blade is a leaf, not a stem. The apical meristem is protected as long as the stem remains short and only leaves grow high enough to be eaten or cut.

1 The vascular tissue of grass stems is arranged in bundles without vascular cambium. How does the absence of vascular cambium affect the growth of grass?

2 You just sold a hammock to a family that plans to hang it from nails driven into the trunks of trees in the yard. One of the kids asks you whether the hammock will rise over the next few years as the trees grow taller. What's your answer and your reason for it?

3 A scientist announces that measuring the rings of very old trees has shown that the climate in your area was very dry in the past. How would the annual rings show this? Give a reason for your answer.

4 Explain how grass can continue to grow after the tip of each blade has been removed by mowing or grazing. (*Hint:* What part of a grass plant is the blade?)

LESSON 19.3 REPRODUCTIVE GROWTH AND DEVELOPMENT

Most flowering plants bloom at a specific time each year. The best time to bloom and reproduce depends on available water and sunlight and the temperature. The appropriate pollinators have to be ready when the flowers are. And florists and beekeepers (who gather honey from bees that have visited flowers) also have to be ready for the harvest. So you could say that flowers rule the lives of florists and beekeepers. They certainly govern the development of flowering plants.

A   B

Florists (A), beekeepers (B), and other workers deal with flowers and other aspects of plant production.

ACTIVITY 19-3 DEVELOPMENT OF REPRODUCTIVE STRUCTURES

Provide students with one or more plants that have reproductive structures at several stages of development (floral buds, flowers, fruits, and seeds). Peppers or legumes such as peas, beans, or alfalfa are ideal. Students will also need hand lenses, dissecting tools, and possibly dissecting microscopes.

Many flowers, seeds, and fruits look like no other plant organs. In this activity you will compare plant reproductive organs at various stages of development and discover which organs develop into which others.

Your teacher will provide you with plants that have reproductive structures at various stages of development, including buds, flowers, fruits, and seeds. Examine each of the reproductive structures carefully with two other students. Sketch and label what you observe in your logbook. You might need to dissect some of the structures to see all their parts.

Compare the various structures you observe. Discuss the following questions and write the answers in your logbook:

- In what order do these structures appear?

- Do your flowers have male parts? Female parts?

- Which parts of each structure become each part of the structure in the next stage? For example, which part becomes the petals? The fruit? The seeds?

Buds appear first, followed in order by flowers and seeds contained within fruits.

Answers will depend on the species of flowers students examine.

In general, the stem apex in the bud becomes the ovary, which in turn becomes the fruit. Cells at the center of the stem apex become the ovules in the ovary that develop into seeds after fertilization. Structures like embryonic leaves in the bud become petals and sepals, which may become part of the fruit.

FLOWER PRODUCTION

Flowers are miniature stem segments with four whorls of specialized leaves: sepals, petals, stamens, and carpels (Figure 19.12). Very short internodes separate these whorls. These specialized leaves grow out from the apical bud or a bud above a leaf, like any other leaf. The meristem in the bud becomes part of the flower. Each part has a role in reproduction. **Sepals** enclose and protect the flower bud. They are the most leaflike flower parts. **Petals** with their bright colors help to attract insect pollinators. **Stamens** produce pollen. The **carpels** fuse together, edge to edge, to form the ovary. In most species, flowers have both male and female parts; but in some, each flower or plant has only one or the other. You'll learn more about plant reproduction in Chapter 21.

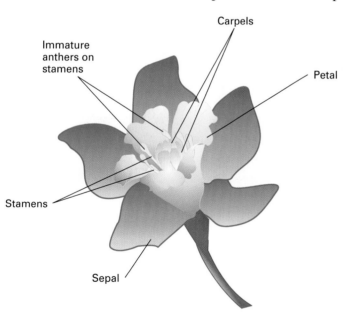

Figure 19.12: *A developing flower. The sepals and petals have been peeled back to show the inside of the flower bud. The carpels fuse to produce an ovary, where meiosis will produce egg cells.*

SEED AND FRUIT PRODUCTION

Fruits and seeds, which include grains such as wheat and corn, nuts, and many vegetables (corn, squash, tomatoes), are harvested for their food value. Fruits also provide commercially important chemicals: Citric acid from oranges and lemons is used in drugs and foods. Pectin extracted from citrus rinds is used in glues, jellies, and jams. Flavorings such as lime and lemon oil are extracted from fruits. The nutrients that are stored in seeds make them valuable foods as well.

After pollen fertilizes the egg cell, the ovule matures into a seed, surrounded by the ovary. The ovary (and other flower parts in some species) develops into a fruit. Cell division and growth contribute to fruit growth. Cell walls may break down partly as fruits soften during ripening, or they may become tough and woody, as in nutshells and pits. Phloem supplies the nutrients that accumulate in the developing seeds and fruits. Ripe fruits may burst and expel seeds, fall off the plant, or be eaten. Wind, water, and animals help to carry seeds away from the parent plant.

LESSON 19.3 ASSESSMENT

1. Stems and leaves.

2. Your tree probably produces only male flowers, while the neighbors' tree produces female flowers.

3. Without big petals or other parts to attract insects or other pollinators, grasses probably rely on wind to spread their pollen.

4. The stem or branch stops growing longer because the apical meristem has differentiated into a flower and can no longer produce new leaves and stem segments.

1 From which organ(s) did flower parts evolve?

2 Your fruit tree looks as healthy as your neighbors' tree of the same species, yet every year, yours produces smaller, different-looking flowers and no fruit, while the neighbors' tree is covered with big, showy flowers that develop into fruit. How can you explain this?

3 Most grasses produce small flowers without large, colorful petals. How does pollen probably travel from one grass flower to another? Give reasons for your answer.

4 What happens to the growth of a stem or branch after its apical meristem produces a flower?

AGRICULTURE & AGRIBUSINESS

Landscape Design, Part 1

Choose a property (public or private) that you think is well landscaped. Contact the owner or manager to get permission to walk the grounds and, if possible, talk to the landscaper or groundskeeper. Sketch the area and take notes on what plants are growing in different locations. How did the landscaper use the different plant characteristics to achieve certain goals? Make a map of the grounds with numbers indicating the various areas. For each numbered area, take a picture or sketch the area, identifying plants and their use in the landscape.

Landscape Design, Part 2

Find a property that needs landscaping. Decide what plants would suit different parts of the site. Use seed or nursery catalogs or talk to a nursery worker to select plants that will accomplish what you want. Make a map of the property with numbers indicating the various areas. Sketch what the lot would look like if your plan were followed. Present your plan to the class.

A Snip in Time

Contact a forester, nursery worker, groundskeeper, orchardist, or someone else who knows how to prune plants. When and why do people prune? What's the proper procedure? What plants benefit from pruning? What are the dangers of pruning? Prepare a brochure on pruning basics. Ask a local nursery or retail plant department to distribute the brochure to its customers.

Wood Works

Talk to someone who works with wood. Find out which woods are good for different purposes and why. What causes the grain in wood? What are knots? How should wood be stored? What causes warping and bowing? What finishes protect and seal wood? Are some woods more resistant than others to insects and fungi? How is lumber treated to resist these decay organisms?

BUSINESS & MARKETING

Native Nursery

A nursery that specializes in native plants has asked you to help plan its marketing. Prepare a sales brochure on plants from your area that could be cultivated. Highlight how to use these plants in various landscapes. Find pictures of the plants or draw them. Describe the benefits of using native plants in a landscape. Include some examples of imported or exotic plants that have become hard-to-control weeds in a new area.

Plant Immigration

You are the import/export manager for a company that wants to make a product using natural fibers from a tropical tree. Interview a U.S. customs agent to find out what regulations you will have to follow to bring in and use exotic plant material. Find out why such regulations exist. Present your findings in the form of an explanatory poster and an oral report.

Dry Land Plantings

In areas such as the southwestern United States, concerns about the size of the water supply have led people to look for alternatives to grass lawns. Use your library, the Internet, or interviews with growers, landscapers, or cooperative extension agents to find out how drought stress affects plant growth and development and to investigate xeriscaping, the practice of landscaping with stones and drought-resistant plant species. Use photos or drawings to create a marketing brochure for a landscaper that compares the benefits and problems of a grass lawn and a xeriscape and explains how xeriscaping is done.

Bonsai

You have been asked to design the interior of a home with a large sunroom. The room will have an Oriental motif, and the client wants a group of bonsai trees at one end of the room. Talk to a bonsai expert to find out how to select the right plants, how they must be cared for, how they are shaped and trained, and how long they take to grow. Find out how much established bonsai trees cost. Draw up plans and care instructions for the plant display.

Plant Organs as Food

Look at the supermarket ads in the newspaper or visit a produce department. For each plant or plant part being sold, determine which organ is eaten. Plan a nutritious dinner using only plants and including all major plant organs (stems, roots, leaves, fruits, and seeds). Write an educational menu that explains how the role of each plant organ in the life of the plant affects its use as food. For example, storage organs are often rich in starch, protein, and other nutrients. Tough, inedible peels may be protective epidermal tissues.

Evaluating Herbal Remedies

Many fruits, seeds, and other plant organs contain oils and other substances that are advertised and sold as home remedies for various illnesses. Are these advertising claims justified? Is there any evidence to support them, or are they just based on customs and the attempt to sell a product? Can any of these products be dangerous to certain people? Select one or two herbal remedies and investigate the science (if any) behind the advertising claims made by the manufacturer or store that sells them. The Internet, your library, and an interview with a pharmacist are good resources to get you started.

Medicinal Plants

Use the library or Internet to find out how plant compounds are extracted for medicines. What economic factors determine the cost of these drugs? What plant parts are used to produce the drugs? What function does the compound have in the plant? Make a poster of medicinal plants. Have a picture of the plant and tell what part of the plant is used, what compound is obtained from the plant, and how it is prepared and used. You might start by researching aspirin, vincristine, taxol, or digitalis on the Internet or at your library.

INDUSTRIAL TECHNOLOGY

Papermaking

Use an encyclopedia, the Internet, or other resource to draw a flowchart of the process of papermaking. What kinds of trees are used most? Why are these species, and not others, the ones most commonly used? How is the process different when recycled paper is used as the material? Write a report that gives the number of trees that are used each year for paper production. Describe what goes on at each stage of the process.

Engineered Wood Products

Research how particleboard, chipboard, and plywood are made. What tree parts are used to make each type of wood product? How do the properties of these materials differ? What is the difference between exterior grade and interior grade? What is veneer and how is it made? Prepare a brochure that describes these materials. List applications for large sheets of each. Which material would be recommended for each application? Information is available from builders and building supply yards and at the library.

CHAPTER 19 SUMMARY

■ Germination begins when a seed absorbs water under suitable environmental conditions. First the embryonic root emerges from the seed coat, followed by the shoot.

■ The structure of a plant depends on the arrangement of vascular, epidermal, and ground tissues.

■ Plant growth depends on the production of new cells by mitosis and cell growth.

- Turgor pressure is responsible for plant cell growth. Strengthening of the cell wall opposes this expansion and determines the cell's final shape and size.

- Meristems produce new cells that enable plant organs to grow. Apical meristems are especially important in the primary growth of stems and roots.

- Primary growth in stems produces stems with alternating nodes and internodes and leaves. Meristems in the nodes add cells to the internodes. Stems grow longer as cells in their internodes elongate.

- A leaf and a bud are attached to the stem at each node. The bud may grow into a branch or a flower.

- During secondary growth the inner surface of a cambium layer produces cells that develop into layers of xylem, and the outer surface produces cells that develop into phloem. Secondary growth produces the thick, woody stems and roots of long-lived plants.

- A meristematic layer just outside the endodermis produces branch roots. It serves as the cambium that produces secondary growth, or thickening of roots.

- Flowers are short, specialized stem segments with four kinds of specialized leaves (sepals, petals, stamens, and carpels). They develop from buds at the stem apex or at nodes.

- Carpels are modified leaves that fuse to form the ovaries of flowers.

- The ovary and sometimes other flower parts develop into fruits.

CHAPTER 19 ASSESSMENT

Concept Review

See TRB, Chapter 19 for answers.

1 Which of the following can be scientifically classified as fruits? Potatoes, onions, green beans, pecans in the shell, tomatoes.

2 Which of the following organs forms only after germination? Stem, root, leaf, flower.

3 Which part of an embryo emerges first during germination?

4 Comment on the following statement. Is it correct? If not, explain what is wrong with it. "Primary growth in both roots and stems results from the expansion of cells produced by apical meristems."

5 What root structure acts most like the buds located at stem nodes? Give a reason for your answer.

6 What kind of vascular tissue transports sugar to the various organs of the plant?

7 How can a plant's stem continue to grow longer after it has produced flowers?

8 Where are the haploid cells in sexually reproducing plants?

9 Do plant organs grow mainly by cell division or by cell growth?

10 Could a large tree develop without secondary growth? Could a hanging vine? Give reasons for your answer.

11 Is primary growth more active in the trunk of a big tree or in its young twigs?

12 During spring and early summer, new stem growth at the tip of a branch on an old tree may be green and soft, while the older part of the branch is stiff and covered with bark. Is this difference a reliable guide to where primary and secondary growth are occurring in the branch? Give a reason for your answer.

13 Sepals are on the outside of many flower buds. They are often green, while petals are not. What's the connection between these two facts?

14–17 For questions 14–17, state the name and function of each structure labeled in the figure.

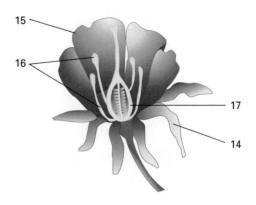

Think and Discuss

18 Compare the growth of a root and a shoot.

19 Some seeds cannot germinate until their seed coats have been scraped with a knife or softened by acid. In nature, many such seeds are swallowed whole by fruit-eating animals. Explain how the scraping or acid treatment makes germination possible and how this probably occurs in nature.

20 What evidence do plant vascular systems provide that flowering plants are more complex and evolved later than conifers?

21 When growers dig up a plant and transplant it, they often cut off many of its leaves. Explain why this is done. (*Hint:* Think about how plants absorb and lose water and how digging up a plant is likely to affect its roots.)

In the Workplace

22 Some grape varieties have flowers with both male and female parts. Other varieties, such as scuppernong, have only female flowers. If growers want to produce scuppernong grapes, what must they do?

23 A church ordered lilies from your nursery as holiday decorations. The church is concerned about pollen because some of the congregation may be allergic to it and because it stains the white flowers yellow. What part of the flower should you tell the customer to remove from each newly opened flower to prevent problems with pollen?

24 Before removing an unwanted tree, a landscaper may "girdle" it by peeling off a ring of bark and the thin, soft tissue between the bark and the wood of the trunk all the way around the tree. How does girdling kill trees?

25 Is it more important to supply water or fertilizer to seeds when planting them?

Investigations

The following questions refer to Investigation 19A:

26 Apple breeders must store their seeds in refrigerators before planting them, to ensure germination. Explain how the seeds of wild apple trees meet the requirement for cold before germinating and how this requirement might help apple seedlings to survive.

27 Why is it important that seed producers conduct their germination tests as close as possible to the time the seeds are sold?

The following questions refer to Investigation 19B:

28 Could spoiled fruit be recycled into paper? Give a reason for your answer.

29 Why can natural cotton rags and discarded clothing be used to make paper?

WHY SHOULD I LEARN THIS?

Fruit and flower growers can control their plants because they know how plants work. Flower growers time their crops to produce blooms just before a holiday. Fruit growers pluck and prune their trees to balance the number and size of the fruits they produce. How do you get a plant to cooperate with your plans? You start by understanding the processes that keep plants alive and thriving.

PLANT LIFE PROCESSES

WHAT WILL I LEARN?

1. how plants balance water conservation and photosynthesis
2. how phloem transports sugars and other substances
3. how plants maintain a steady supply of carbohydrate day and night
4. the importance of plant growth regulators in growth, development, and other plant activities
5. the effect of night length on flowering plants

Every spring in the northeastern United States, squirrels chew through the bark of sugar maple trees and drink the sap that wells up. Maybe they were the first to learn that these trees produce lots of sugary sap. Today, they compete with humans who know how to collect plenty of sap without harming the trees. In other words, maple syrup producers have learned a lot about a tree's life processes.

In this chapter you will learn more about plants as active, living systems. You will also see how people manage these systems for their own purposes.

LESSON 20.1 PLANT TRANSPORT SYSTEMS

When you look at a tree in leaf, it's hard to believe that its bark hides a transport system as busy as an interstate highway. Maple syrup producers depend on this transport for the sap that they collect. And bacteria and fungi can plug xylem or phloem, causing a traffic jam. In this lesson you'll learn what controls this transport system and what it's transporting.

ACTIVITY 20-1 HOLES IN PLANTS

Provide students with geraniums or other plants with thick leaves from which they can easily peel the epidermis. Make a concentrated salt solution by dissolving 20 grams of NaCl in 80 mL of water. They will also need forceps, scalpels or razor blades, microscopes, glass microscope slides, coverslips, paper towels, and eyedroppers.

If time permits, have students peel the upper epidermis and compare the number of stomata on the two surfaces. Encourage them to speculate about why there are more stomata on the lower surface.

A transparent epidermis allows light to reach the photosynthetic cells inside the leaf.

Chloroplast-bearing cells in the epidermis are not a significant part of the photosynthetic capacity of the leaf.

Their chloroplasts allow them to open and close the stomata in response to light.

Water evaporates through the microscopic pores in leaves called **stomata** (plural of *stoma*). How do plants control this process? In this activity you will observe stomata with a microscope and develop a theory of how plants control evaporation from their leaves.

Working with a partner, tear a leaf on an angle to expose the inside of the leaf. Use forceps to peel the transparent epidermis from the underside of the leaf, starting at the torn edge where it sticks out, and lay it flat on a microscope slide. Add a drop of water and a coverslip. Look at the tissue with the microscope, first at low power and then at high power. Note the transparent cells and the pairs of special cells surrounding each stoma. In your logbook, draw and label what you see.

Put a drop of salt solution at one side of the coverslip. Touch a paper towel to the other side of the coverslip to draw the salt solution under the coverslip. Examine the cells around the stomata again. Describe and draw any changes. Answer the following questions in your logbook:

- How does a transparent epidermis benefit a plant?

- Do the epidermal cells with green chloroplasts make up a large percentage of the photosynthetic cells in the leaf?

- If the epidermal cells with chloroplasts are not important in photosynthesis, how does their ability to respond to light benefit the plant?

MAPLE SUGARING IN VERMONT

In Vermont, Quebec, and other parts of northeast North America, buckets and plastic tubes hanging on the trunks of maple trees are a sign of spring. Early spring is harvest time for maple sap, an important natural product in the Northeast.

Sap develops throughout the year. During the summer, photosynthesis in the leaves produces sugar. The sugar travels through the phloem from the leaves to the lower trunk and roots. There, the sugar is stored as starch. Cool fall temperatures signal root cells to change the starch back into sugar. The sugar acts like antifreeze, preventing harmful ice formation by lowering the freezing point of cytoplasm and xylem fluid. During the winter, cell respiration in the roots and lower trunk produces CO_2 gas that builds up in the xylem. In the spring, the rising temperature expands the CO_2. The gas pressure helps to push the sap up in the xylem. The syrup maker drills a hole into the sapwood of the tree, and the sap flows out through the hole and into a bucket or tube. Maple sap is about 98% water and 2% sugar. An evaporator boils away most of that water. When the sap becomes a thick syrup that is 66% sugar, it's ready to be bottled and sold.

CONTROLLING WATER LOSS

Water doesn't flow through plants continuously, evaporating through stomata all day and night. As well-adapted organisms, plants have evolved ways to regulate their absorption and loss of water.

Transparent epidermal cells produce a water-conserving cuticle. Light passes through the epidermis to the cells within the leaf that perform most of the plant's photosynthesis. Nearly all the water that plants lose evaporates through their stomata. In Activity 20-1 you saw that stomata are really gaps in the epidermis between pairs of **guard cells** (Figure 20.1) that contain chloroplasts.

When sunlight and temperature permit photosynthesis and plants have enough water, guard cells break down starch polymers into many small sugar molecules. At the same time they absorb many potassium ions by active transport. The concentrated sugar and

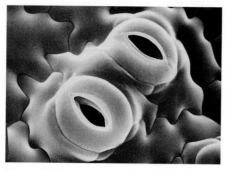

Figure 20.1: *Pairs of guard cells around two stomata in a tobacco leaf (magnification ×485, color added). If the cells lose turgor, they press together. Note the thickened "lip" on each facing guard cell wall that helps to seal the stomatal opening.*

potassium make the guard cells' cytoplasm hypertonic. As a result, the cells absorb water and swell up. The swelling pushes the two members of each pair of guard cells away from each other, opening the stoma between them.

As water vapor escapes through open stomata, it is replaced by mineral-rich water that moves up through the xylem from the roots and soil. CO_2 diffuses into leaves through the open stomata, fueling photosynthesis. At the same time, oxygen gas produced by photosynthesis diffuses out through the stomata. At night or when hot, dry conditions threaten to dehydrate the plant, guard cells convert sugar back to starch and leak potassium and water. Their turgor decreases, and the stomata close.

INVESTIGATION 20A   A MODEL OF STOMATA

See Investigation 20A in the TRB.

Botanists have spent a lot of time studying how guard cells open and close. This process is important for growers, who want their plants to absorb and convert as much CO_2 as possible to sugar for every drop of water that evaporates from their leaves. In this investigation you will build and test models of guard cells.

Water conservation is important to farmers.

HOW PHLOEM WORKS

An additional activity, How Does Phloem Work?, can be found in Chapter 20 of the TRB.

By now, you know that phloem and xylem transport substances in very different ways. Because xylem is composed of the cell walls of dead cells, the flow of water and minerals from the roots is not usually in danger of interruption. The only likely problems are cutting or breaking of the stem, frozen or dry soil, and parasitic microorganisms that block the xylem.

But phloem works differently. The transport of sugars, amino acids, and other vital substances produced in leaves is interrupted when you kill cells in a living stem (for example, with heat). Phloem is made of living cells. If these cells die, the system backs up. Sugar accumulates in leaves, and the chloroplasts convert it to starch. Eventually, the starch grains can get so big that they disrupt the normal activities of the chloroplasts and interfere with photosynthesis.

Because phloem cells are alive, they can use active transport to gather substances for delivery to other parts of the plant. Sugar and other substances diffuse easily from one cell in a column of phloem cells to the next through gaps in the cell walls (Figure 20.2). Growing tissues rapidly absorb these substances from nearby phloem cells and consume them. This reduces the concentration of transported substances in that part of the phloem. In leaves and storage tissues, active transport loads phloem cells with sugars and other substances. The phloem transports these nutrients to the young growing tissues, where their concentration is much lower because they are constantly being used up.

Maple syrup producers aren't the only people who are concerned with sugar transport in plants. Have you ever noticed the sticky goo on cars that are parked under certain trees? It comes from small insects called aphids. They pierce leaves and stems and suck

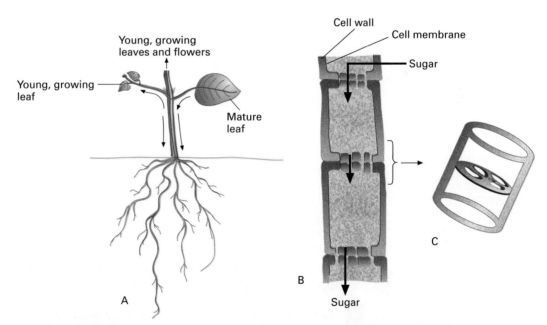

Figure 20.2: *Phloem transport. (A) Sugars and other nutrients travel through phloem from mature leaves to roots and growing tissues. The nutrients move through columns of phloem cells (B) by passing through openings in cell membranes and cell walls (C).*

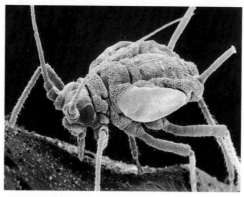

Figure 20.3: *A young rose aphid* (Macrosiphon rosae) *feeding on a rose leaf (color added, magnification ×36).*

the nutritious sap out of phloem tissue (Figure 20.3). The sap contains much more sugar than aphids need; they benefit mostly from its proteins and amino acids. Aphids excrete the excess sugar in a liquid called honeydew, which falls from trees and annoys car owners. Aphids are not just a nuisance. They can multiply so rapidly that they severely weaken or kill plants by consuming the plant's carbohydrates and spreading plant viruses.

LEARNING LINK TREE GROWTH AND LUMBER QUALITY

Wood consists mostly of xylem. Differences between the stresses on the upper and lower surfaces of a heavy branch influence the development of the xylem in those places in ways that affect its use as lumber. Use resources such as your library or the Internet, or contact the U.S. Forest Service or a lumber company to research this process. Start by searching for the keywords "reaction wood," "compression wood," and "tension wood."

LESSON 20.1 ASSESSMENT

1. Guard cells control the opening and closing of stomata. A closed stoma traps more water in the plant, protecting it from dehydration.

2. The CO$_2$ needed for photosynthesis enters leaves through open stomata. Water evaporates through open stomata, pulling nutrient-rich water into the plant from the soil around its roots.

3. The fuzz is fungus that grows on the honeydew, consuming the sugar it contains.

4. If the stomata opened when soil water was frozen, evaporation from leaves could dehydrate the plant and kill it.

1 Plant growth depends on the normal functioning of guard cells. Explain why they are called guard cells and how they protect plants from dehydration.

2 What are two ways in which plants benefit from opening their stomata?

3 Fuzzy dark patches that form over honeydew spots on cars can be difficult to remove. Think about what honeydew consists of. What do you think the fuzz is? Give a reason for your answer.

4 Needles of pines and other trees stay green through the winter. Why do these plants, which live in climates with cold winters, have stomata that close when temperatures drop below freezing?

LESSON 20.2 CONTROL OF PHOTOSYNTHESIS

Plants need water, and their stomata conserve water when they're closed. So why don't growers keep their crops flooded and the plants' stomata closed all the time? Because photosynthesis would never occur and the plants would never grow! In this lesson you will learn how plants and the people who grow them balance photosynthesis with water conservation.

ACTIVITY 20-2   CARBO-LOADING LEAVES

The procedures for this activity can be found in Chapter 20 of the TRB.

On a sunny day, the leaves of a tomato plant are rapidly producing carbohydrates by photosynthesis and shipping them to the rest of the plant through the phloem. What controls how much sugar stays in the leaves and how much is loaded into phloem? The instructions your teacher gives you for this activity will help you gain some clues to answer this question.

GROWING FUEL ON THE FARM

BIOLOGY IN CONTEXT

Coal, oil, and natural gas are the three major energy sources in the United States. Many farmers would like to add corn to this list.

Photosynthesis in the leaves of corn plants produces sugar from CO_2 and water. The energy of sunlight is converted to the chemical energy of sugar. The sugar travels to the corn kernels and is stored as starch. When you eat corn, your body breaks down the starch in the kernels and releases its chemical energy.

So how does a car use the energy of plant carbohydrates? Bacteria and purified enzymes break down starch and ferment it to make ethanol. Ethanol is one of the most environmentally safe fuels. When it burns, the only waste products are water and CO_2. Some engines can use either pure ethanol or a mixture of ethanol and gasoline called gasohol. These engines can power cars, boats, buses, motorcycles, chain saws, and lawnmowers.

Unfortunately, engines don't perform well using gasohol, and producing ethanol is expensive. In spite of the problems, many countries are trying to use gasohol. It could be just a matter of time before ethanol is a major energy resource.

PHOTOSYNTHESIS IN LEAVES

Natural selection has reached a compromise between the need to allow CO_2 into photosynthetic organs and the need to prevent dehydration by evaporation. Leaves and their stomata are part of this compromise. Photosynthesis requires light; it makes sense that light stimulates stomatal opening. But bright sunlight heats up leaves and promotes evaporation, too. Stomata usually close up on hot, dry days. Table 20.1 outlines the effects of the environment on stomata.

Factor	Stomata Open	Stomata Closed
Temperature	Moderate (up to 30°C)	Hot; near or below freezing in some species
Humidity	High	Low
Light	Bright	Dark
Moisture (of plant and soil)	Sufficient	Dry (drought stress)

Table 20.1: Effects of the Environment on Stomata

When light, temperature, and moisture are right for photosynthesis, stomata open. CO_2 from the surrounding air diffuses through the stomata into the air spaces in the leaf to the chloroplast-filled cells within (Figure 20.4). Oxygen and water vapor take the reverse path out to the air. The photosynthetic cells may export carbohydrate to the phloem as sucrose, which exits the leaf, or store it temporarily as starch in the chloroplasts. Many plants store starch in their leaves during the day. They break down the starch into sucrose at night and deliver it to the phloem. This schedule lets the leaves feed the rest of the plant at night, when photosynthesis shuts down.

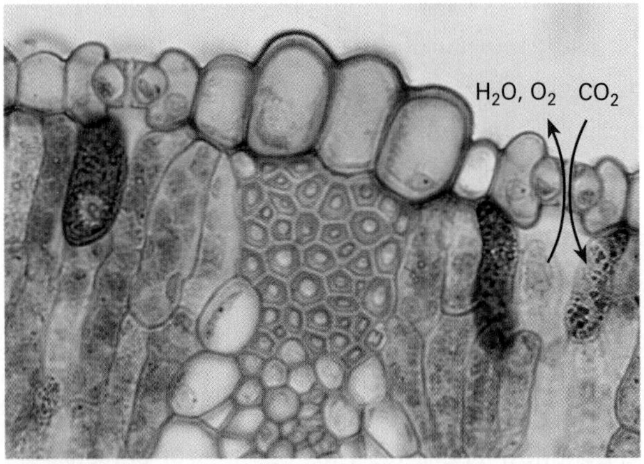

Figure 20.4: *A stained cattail* Typha *leaf section. Pairs of guard cells are visible in the epidermis at the upper left and right. Water vapor and oxygen diffuse out through open stomata, and CO_2 diffuses in.*

Some desert plants (Figure 20.5) have a different way to balance CO_2 absorption and

Figure 20.5: Kalanchoe *is a desert plant that has become a popular house and garden plant. It opens its stomata at night instead of in the day.*

water conservation. Their stomata open during the cool desert nights, and the leaves convert CO_2 into organic acids. During the day the stomata remain closed, and the chloroplasts use solar energy to reduce the organic acids to sugars.

LESSON 20.2 ASSESSMENT

1. Adding CO_2 to the air is more effective during the day, when photosynthesis is active.

2. Adding CO_2 to the air is more effective when the air is humid, because humid air promotes stomatal opening.

3. Hot, dry wind increases the rate of evaporation.

4. Having stomata mostly on the lower surfaces of leaves protects them from the sun and reduces loss of water vapor. In humid climates this loss is less important, so stomata may occur on the upper surfaces as well. Leaves that float on water have stomata only on their upper surfaces because the lower surfaces are in water.

1 Greenhouse operators sometimes add CO_2 to the air in their greenhouses to promote growth. Would the day or the night be a better time to do this?

2 When adding CO_2 to the air in a greenhouse, should growers make the air dry or humid? Give a reason for your answer.

3 An engineer is writing a computer program to control an automatic irrigation system. What assumptions should the engineer make about the effect of hot, dry wind on the rate at which plants lose water?

4 Most plants have more stomata on the lower surfaces of the leaves and fewer on the upper. Some plants that live in humid places have an equal number of stomata on their upper and lower leaf surfaces, and floating water plants often have no stomata on the lower leaf surfaces. Explain these differences.

LESSON 20.3 REGULATION OF GROWTH AND DEVELOPMENT

If a worker is accidentally injured at work and loses a finger, a surgeon may be able to reattach it, but no one can grow a replacement finger. But if a plant loses a branch to high winds or hungry animals, several dormant buds may grow out into new branches. What controls this process? You'll learn in this lesson what controls plant growth and development and what makes it so adaptable.

ACTIVITY 20-3 DOMINANT STEM TIPS

Have you ever noticed that some plants (such as corn) usually grow straight and tall, with few or no branches, while others (such as tomatoes) grow bushy with branches? In this activity you will see how the growing tip of a stem or branch affects the growth of other branches.

Examine two coleus or other plants with your group. Using Figure 20.6 as a model, make a branching stick diagram in your logbook of the growth pattern of each plant. Label the location of buds that haven't grown into branches.

On one plant, cut a section 2 cm long from the tip of each large main branch, as in Figure 20.6. Mark your diagram with dashed lines to show the removed parts. Mark the plant pots with your group's name and set the plants in a sunny spot. Water them as needed.

Examine the plants daily for 10–15 days. Then make a new branching diagram of each plant in your logbook. With your group, compare the two plants. Answer the following questions in your logbook:

- How did cutting off the main growing tips affect growth and branching?

- Suggest how one growing part of a plant could block the growth of other parts. Describe an experiment that could help you to test your idea.

Removing a growing tip allows short branches and buds to grow more quickly because they are no longer inhibited by a substance produced in the growing tip.

This can be confirmed by grafting growing tips onto plants, by blocking transport out of the growing tip, or by applying extracts from growing tips to plants.

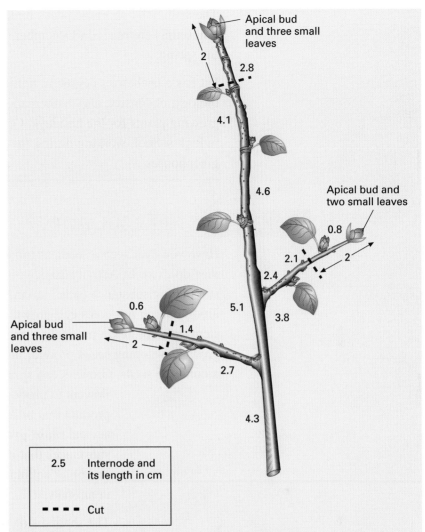

Apical bud and three small leaves

2
2.8
4.1
4.6

Apical bud and two small leaves

0.8
2.1
2
2.4
5.1
3.8

0.6
1.4
Apical bud and three small leaves
2
2.7
4.3

2.5	Internode and its length in cm
▄ ▄ ▄ ▄ Cut	

Figure 20.6: *A branching diagram. The colored marks are bud locations. Dashed lines indicate parts to be cut off.*

GREENHOUSE PRODUCTION MANAGER

Pat L. is the production manager at a large greenhouse. He supervises several growers who produce special holiday plants. Pat's job is to make sure that the plants are healthy and are the right size and color at the right time. "No one wants to buy green poinsettias or Easter lilies that have already bloomed," Pat explains.

Sometimes, Pat tricks plants into growing outside of their natural season. At 5:00 A.M. for eight months, Pat covers the greenhouse with black plastic to block the sunlight for part of the morning.

The artificially shortened days make the plants bloom early. So poinsettias are ready in December, and Easter lilies are ready in the spring.

Pat has a bachelor's degree in horticulture and took courses in biology, chemistry, and mathematics. After college Pat began growing plants for the holidays. Other growers start their careers in high school, working during the summers at nurseries and greenhouses.

PLANT GROWTH REGULATORS

Have you ever seen a seedless tomato? They aren't common, but they do exist. In Activity 20-3, you saw evidence of a growth-stimulating substance called auxin. Growing stem tips produce **auxin,** which promotes cell growth below the tip. Removing the stem tip stops auxin production and lets other parts of the plant grow. Developing seeds produce auxin, too, making a fruit grow around the seeds. Growers can spray auxin on unfertilized tomato flowers to produce seedless fruits, an unusual product that brings a high price. Auxin is one of several **plant growth regulators** (PGRs), substances that control growth, development, and other cellular activities, just as hormones do in animals.

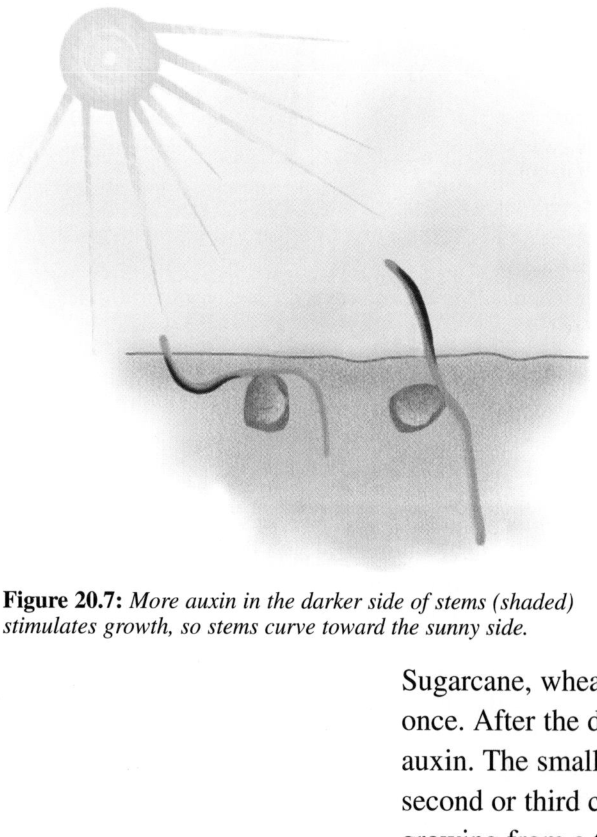

The shaded side of stems tends to have more auxin than the sunny side. The extra auxin makes stems grow toward light and away from the ground (Figure 20.7). Auxin also promotes the downward growth of roots. A synthetic auxin kills weeds because it disrupts the balanced growth of plant organs. Auxin in cells at the tops of some trees favors growth of the main trunk, producing tall, straight, unbranched trees that make good lumber.

Figure 20.7: *More auxin in the darker side of stems (shaded) stimulates growth, so stems curve toward the sunny side.*

Sugarcane, wheat, and some trees can be harvested more than once. After the dominant main stem is cut, it no longer produces auxin. The small buds or branches on the stump grow to produce a second or third crop. You may have seen these small shoots growing from a tree stump.

DISCOVERING AUXIN

In 1880, Charles Darwin and his son Francis described an "influence" from the tip of a grass shoot that makes the lower part of the shoot bend toward light. Covering the tip of the shoot with a light-proof material made it grow straight up. Covering the shoot tip with a clear material made it bend toward the light. When they put a light-proof sleeve around the part of the shoot that bends but left the tip uncovered, the plant still bent toward the light.

In 1910, Peter Boysen-Jensen showed that the "influence" of the tip is a chemical. He sliced the tips from seedling shoots and inserted either a thin gel or an impermeable piece of mica between the tip and the base of the shoot. Shoots with tips that were reset directly onto the base or that were separated by the porous gel bent toward the light. Shoots with tips that were separated by a slice of mica grew straight.

In the 1920s, Fritz Went sliced the shoot tips from seedlings and placed them on agar blocks in the light. He thought that if the tips contained a chemical that influenced growth, the chemical would diffuse into the agar blocks. To test this idea, he placed some of these agar blocks on shoots that had their tips removed and grew them in the dark. The shoots grew straight up. Some blocks he offset so that they covered half of the stem. The cells on that half of the shoot grew longer than the cells on the other half. The shoot bent away from the side with the block. Went named the growth-promoting chemical that diffused into the agar "auxin," meaning "to increase."

OTHER PLANT GROWTH REGULATORS

An additional activity, A Seedy Case, can be found in Chapter 20 of the TRB.

Although auxin was the first PGR discovered, other PGRs are just as important. Many plant compounds act as PGRs. Growers spray some PGRs on plants to influence fruit development, harvesting, and ripening. Table 20.2 outlines the actions and uses of the major PGRs.

If you did the additional activity, A Seedy Case, you had to wash an inhibitor from lettuce seeds before they would germinate. That inhibitor is **abscisic acid,** a PGR that maintains dormancy. The

Name	Major Effects	Commercial Uses
Abscisic acid	Dormancy; opposes growth and stomatal opening	
Auxin	Stem and fruit growth; opposes growth of other branches and buds	Herbicides, seedless fruits
Cytokinin	Flowering and seed germination; opposes aging	Making cut flowers last longer
Ethylene	Aging and ripening	Ripening and harvesting fruit
Gibberellin	Reproductive development (bud growth, flowering, fruit growth, germination)	Seedless fruit, developing more and larger fruits

Table 20.2: Major Plant Growth Regulators

seeds of other species, including some desert plants, also contain abscisic acid that must be washed out before they germinate. This ensures that growth will begin only when plenty of water is available. Cambium and bud tissues also become dormant when they accumulate abscisic acid as winter approaches. Abscisic acid is also a stress signal: It makes stomata close during dry conditions.

A group of PGRs that stimulate growth are the **gibberellins.** There are more than 70 gibberellins, all slight variations on the same basic chemical structure. They are especially important in reproductive growth. Like auxin, they promote rapid stem growth, flowering, and the formation and growth of fruit.

Cut flowers last longer and retain their market value after being dipped in cytokinin solution.

Farmers spray a mixture of gibberellin and auxin on seedless grapes so that they develop without fertilization and seeds. In spring, gibberellin in buds and seeds helps to overcome the dormancy caused by abscisic acid.

Like abscisic acid and gibberellin, cytokinin and ethylene have opposite effects on a plant. **Cytokinin,** produced in roots, prevents aging. It also helps to overcome auxin's repression of branch growth.

In most trees, cytokinin makes the lowest branches, which are closest to the roots, grow the most. Cut flowers that are dipped in cytokinin solution last longer. In the future, grocers may treat fresh produce in the same way to prevent spoilage.

A gas called **ethylene** (C_2H_4) encourages aging, such as ripening fruit and dropping leaves and fruits in autumn. You can put fruit in a paper bag to speed up ripening; the bag keeps some ethylene from escaping. Tomatoes are often picked while they are still green and hard so that they won't bruise. Before produce distributors sell the green tomatoes, they ripen them with ethylene.

Orchard workers sometimes spray a synthetic chemical that produces ethylene on fruit and nut trees. A special harvesting machine then shakes the trees and collects the fruit. Because all the fruit falls at once, it's cheaper to harvest. Apples can be stored for months in an airtight refrigerator filled with CO_2, which flushes out ethylene. That's how wholesalers can supply supermarkets with apples all year.

This orchard worker uses a harvesting machine to collect fruit that had been treated with ethylene so that it falls at one time.

INVESTIGATION 20B LONG DAYS OR SHORT NIGHTS?

See Investigation 20B in the TRB.

Seasonal changes in temperature and day length determine the time of year when many crops flower and produce seeds or fruits. Cooperative extension agents who are helping to introduce a new crop work out planting and harvest dates that allow enough time for these reproductive structures to develop.

Radishes flower and produce seeds only when the days are the right length. How does day length affect flowering? Or is it night length that matters? In this investigation you will examine the effect of light and dark on the flowering of radishes.

Producers of outdoor flower crops must select varieties suited to their specific geographic zones.

PHOTOPERIODISM

Remember the greenhouse production manager on page 701 who darkened the greenhouse to force poinsettias to flower? Poinsettias are **short-day plants;** they flower only when the days are shorter than some critical length. As you might have figured out, what they really need are long nights. Lettuce is a **long-day plant,** or short-night plant. It begins to flower when the nights get shorter than some critical length. The control of flowering by night length is called **photoperiodism.** Because night length depends on the time of year and distance from the equator, crops that are sensitive to night length are bred for specific geographic zones (Figure 20.8).

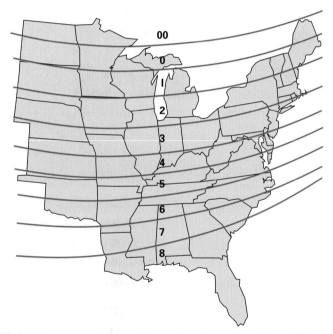

Figure 20.8: *Soybean varieties come in maturity classes from 00 to 8. These classes depend on photoperiodism and temperature sensitivity. The map shows the approximate range of each class in the eastern United States.*

LESSON 20.3 ASSESSMENT

1. Auxin produced at the top of the tree promotes growth of the main trunk. It travels down, repressing growth of branches near the top. Cytokinin travels up from the roots, promoting growth of branches at the bottom of the trunk. As a result, the main trunk grows straight up, and the length of branches increases from top to bottom.

2. No, PGRs affect how resources are divided between plant parts, but they can't create nutrients out of nothing.

3. No, the plants need a certain length of uninterrupted dark. A few minutes is not enough.

4. Apples produce lots of ethylene, which promotes ripening.

1 How do auxin and cytokinin in pine trees interact to produce the cone-shaped trees that are sold as Christmas trees?

2 Can growth-promoting PGRs such as gibberellins take the place of fertilizer, water, or light? Give a reason for your answer.

3 A florist successfully sued the owner of a brightly lit sign for disrupting the development of his flowering plants. The sign was lit all night. Could the florist have avoided the problem by covering the plants with a dark cloth for a few minutes each night? Give a reason for your answer.

4 Green bananas and tomatoes ripen faster in a paper bag with a fresh apple. Explain how the apple speeds up ripening of the other fruits.

CAREER APPLICATIONS

The applications that follow are like the ones you will encounter in many workplaces. Use the biology you learned in this chapter to complete the activities. Share your work with the class.

AGRICULTURE & AGRIBUSINESS

Year-Round Mums

Chrysanthemums bloom naturally in the fall, yet colorful pots of the plant can be bought at virtually any time of the year. Talk to a greenhouse operator, florist, or other knowledgeable person to find out how chrysanthemums are forced to bloom at all times of the year.

Rose Bowl Parade

Find out how many flowers go into a typical float for the Rose Bowl Parade. What kinds of flowers are used most often? Where do float makers get the flowers? Do plants have to be forced to produce enough flowers for this winter event? Are they long-day or short-day species?

Field and Greenhouse Irrigation

Research either field irrigation methods or greenhouse irrigation methods. What methods are used? How do they work? How much do they cost? What are the limitations of each system? Are the systems continuous, timed, or automated? Find out the advantages and disadvantages of each method. As a class, compare methods of providing water to field plants and to greenhouse plants. Make a chart comparing irrigation methods.

BUSINESS & MARKETING

Take a Shine to a Leaf

Florists sometimes spray leaves to make them shiny. Talk to a florist or search plant care books or the Internet and find out what makes the leaves shine. How does the spray affect the operation of the stomata? Is it eventually absorbed by the plant or does it evaporate? What do your sources recommend for long-term maintenance of foliage?

Ethanol-based Fuels

A group of growers has hired you to develop a publicity campaign. They want to promote ethanol-based fuels produced from crops. Research the current use of these fuels in your state. What are the incentives for consumers to use these fuels? How can they be used? How are vehicles or other equipment modified to burn the fuels? What are the environmental benefits? Create a public service announcement and an informational brochure.

Photoperiod-Insensitive Plants

Plant breeders can manipulate or even eliminate the sensitivity of flowering crops to photoperiod. This enables the crops to grow and flower over a wider geographic area. Design an advertisement for a new "day-neutral" (photoperiod-insensitive) variety of chrysanthemum or poinsettia (which are normally short-day plants) or iris (usually long-day plants). Aim it at home gardeners, explaining photoperiodism and the advantages of this new variety.

FAMILY & CONSUMER SCIENCE

Making Syrup and Molasses

Research different types of syrups and the plants that are used to make them. How are these syrups made? Are any plants from your state used for syrup making? If so, make a display of syrups from crops in your state and describe the process that is used to make them. If not, select two or three types of syrup to display. Explain how storing sugar or other carbohydrates benefits the plant and fits into its life cycle. Find recipes that use the syrups, and prepare a brochure that promotes syrups from your state. You might include a comparison of tastes, uses, and facts such as how many plants, fruits, etc., it takes to make a gallon of syrup.

Teach Kids about Transport

Prepare a science lesson for elementary school children that shows how water moves through a plant. Include a demonstration using celery or carnations and colored water. Arrange to teach the lesson to an elementary school class or to the younger brothers and sisters of someone from your class.

Bullying Bulbs

Home gardeners often grow flowers from bulbs that they save from year to year. What does it mean to "force" a bulb? Is it the same as forcing plants to flower by artificially varying the photoperiod? Consult gardeners, gardening books and magazines, or your county cooperative extension agent about forcing bulbs. Do some plants detect the photoperiod as bulbs, through their leaves, or via other parts? Are plants ever more or less sensitive to photoperiod? Report your findings to the class.

Jet Lag

Most living things seem to have internal clocks that are affected by photoperiod and other factors. Research the phenomenon of jet lag, and find out how alike human and plant internal clocks are. Do they respond similarly to changes in photoperiod? What is so disruptive about rapid airplane travel to a different time zone? Experimenters have flown plants and animals to different parts of the world and observed the changes in their internal clocks. Do organisms respond to other factors, such as geographical differences in the earth's magnetic field? Find out as much as you can about "biological clocks" and report your results to the class. Provide practical tips on coping with jet lag, and explain how they work.

Human Hormones in Plants

After menopause, some women are told by their doctors to eat more soy products such as tofu. Soybeans have some of the effects and health benefits of the hormone estrogen. Other plants also contain substances that are similar to animal hormones. For example, sweetclover contains something like estrogen. It can create reproductive problems in cattle and sheep that eat it. Abscisic acid promotes dormancy in both plants and hibernating animals. Are these examples just coincidences? What role do these substances play in the lives of plants? Are they both animal hormones and PGRs? Do they protect wild plants from being eaten by making herbivores sick? Find out what you can about these substances. A library or Internet search with the keywords "plant secondary compounds" or "dietary antimetabolites" would be a good place to start.

INDUSTRIAL TECHNOLOGY

Ethanol Production

Research the production of ethanol. Create a flowchart showing the production of ethanol from crop residues. What equipment is required? How long does the process take? What wastes does it produce and how do they affect the environment? Present your findings to the class.

HVAC Systems

Research heating, ventilating, and air-conditioning systems that have automated systems for bringing air into a building, based on environmental factors such as temperature and humidity. Make a flowchart showing how these systems are controlled. Then do the same for stomata and compare the two kinds of systems.

CHAPTER 20 SUMMARY

■ Guard cells control stomatal opening and closing. Light, moisture, and moderate temperatures make stomata open. Darkness and dry or very hot or cold conditions make them close.

■ Guard cells respond to changing conditions to conserve water while letting CO_2 into leaves and oxygen out so that photosynthesis can continue. They swell when they absorb water and shrink when they release water.

■ Phloem consists of living cells that actively absorb sugars and other substances in leaves and release them in tissues that need them.

■ Many plants store some carbohydrate in leaves during the day and transport it to the rest of the plant through the phloem.

■ If phloem transport breaks down, carbohydrate builds up in leaves and can interfere with photosynthesis.

■ Some plant species have adapted to dry climates by developing a system for opening stomata only at night.

■ Auxin produced in stem tips makes the tip grow and prevents the growth of other branches and buds. It also helps the growth of fruit, of stems toward light and away from gravity, and of roots toward gravity.

■ Plant growth regulators include gibberellins and cytokinin, which stimulate growth and development; ethylene, which promotes aging; and abscisic acid, which slows growth and causes dormancy.

■ Long-day plants flower only when nights are shorter than a certain length, and short-day plants flower only when nights are longer than a certain length.

CHAPTER 20 ASSESSMENT

Concept Review

See TRB, Chapter 20 for answers.

1 Explain the function of leaf cells that do not have chloroplasts.

2 Compare how the chloroplasts of guard cells and other leaf cells contribute to photosynthesis.

3 How do guard cells open stomata?

4 How does sugar pass from one phloem cell to the next?

5 Do short-day plants flower when the day is interrupted by a dark period? Give a reason for your answer.

6 Explain why it does or does not make sense for seeds to produce ethylene.

7 Which would be a better way to stimulate seeds to germinate: soaking with gibberellin or with abscisic acid? Give a reason for your answer.

8 Why do people treat fruits with ethylene in closed containers but spray other PGRs on plants?

9 Which PGR would you expect to accumulate in leaves on a very hot, dry day? What would its effect be there?

10 Which PGR could you spray on a beautiful shade tree to make it keep its leaves longer in autumn? What problems could this cause for the tree in a northern climate with cold winters?

11 Light signals desert plants such as Kalanchoe to close their stomata. Would you expect them also to reverse the usual stomatal response to humidity? Give a reason for your answer.

12 Aphids don't need to suck phloem fluid out of plants. It is under such great pressure that it flows out into their bodies. What is the source of this pressure?

Think and Discuss

13 Suppose you treated a plant stem with a poison that interferes with the ability of cells to use ATP energy. Lots of sugar and starch accumulated in the leaves, and photosynthesis slowed down. Explain these results. (*Hint:* Think about how phloem works.)

14 Which PGR would you use to make cotton plants drop their leaves before harvesting so that you could pick a cleaner crop of cotton?

15 Is the following statement correct? Give a reason for your answer. "The critical night length for short-day plants is longer than the critical night length for long-day plants."

16 Auxin, cytokinin, or abscisic acid purified from one plant species can be used in experiments on another species. Why isn't this true for gibberellin?

In the Workplace

17 On the basis of what you have learned about phloem and xylem, would you expect leaf wilting to be one of the first signs of injury in trees that have been damaged by toxic pollution? Give a reason for your answer.

18 A plant breeder tried to produce seedless strawberries, but berries without seeds wouldn't grow large enough to sell. What was probably missing?

19 Explain how cutting back on watering can speed up fruit ripening.

20 A florist shop started selling fruits and vegetables, too. Suddenly, the flowers weren't lasting as long as they used to. Explain why this might happen. What could be done to solve this problem?

Investigations

The following questions relate to Investigation 20A:

21 How do the thickened walls of guard cells on the side facing the stomatal opening help to control opening and closing?

22 Can models prove that a hypothesis is correct or incorrect? Explain how models can help to test a hypothesis.

The following questions relate to Investigation 20B:

23 Can a dark, cloudy period at midday affect flowering? Give a reason for your answer.

24 Would your conclusions about the effects of the light period and the dark period on flowering have been different if you had used a short-day plant?

CHAPTER 21

WHY SHOULD I LEARN THIS?

You are tired of paying big bucks for your favorite fruit—raspberries. You have decided to grow your own, so you order a dozen plants. But the "plants" that arrive are nothing but a bunch of sticks! Feeling foolish, you poke them into the earth, water them, and wait. In a week they are producing leaves; in a year, a few berries; in two years, enough raspberries to satisfy even *your* craving; and in three years it is hard to keep them from invading the rest of your tiny garden. How do plants perform this miraculous transformation?

PLANT REPRODUCTION

WHAT WILL I LEARN?

1. how plants reproduce asexually
2. how to produce plants asexually
3. how plants reproduce sexually
4. what makes flowers, fruits, and seeds useful

Plants can grow not only from sticks but also from even less likely sources. If you worked as a plant tissue culturist, you could be growing whole plants from a pile of cells that looks like apple sauce. It seems like magic!

Even the traditional way of growing a crop from seed can seem magical. Seeds are part of how flowering plants reproduce. Each seed contains a tiny plant embryo that can grow into a plant similar to its parents. The farmer sows a few seeds in the spring and reaps hundreds in the fall—with proper planning and care.

This chapter will give you some insight into the many ways in which plants reproduce and provide benefit to people. Whether you ever grow plants or prepare their seeds as food, or if you only use cut flowers as a decoration, and all you know about food is how to eat it, you will benefit by understanding how plants reproduce.

LESSON 21.1 ASEXUAL REPRODUCTION

Many plant species reproduce without flowers, seeds, and fruits. How do you control weeds that spread by roots and stems? How is asexual reproduction used to grow strawberries, potatoes, and seedless grapes? In this lesson you will learn about asexual plant reproduction and how people in many occupations use it.

ACTIVITY 21-1   NEW PLANTS FROM CUTTINGS

Plants of many species can produce new individuals from pieces that are broken off. Growers use this form of asexual reproduction to mass-produce plants in commercial greenhouses. Your teacher will provide you with the instructions for an activity that will help you to learn one way to generate genetically identical plants.

CAREER PROFILE

PLANT TISSUE CULTURE TECHNOLOGIST

Maureen B. is a plant tissue culture technologist. Maureen says that her job centers on a plant's ability to grow from a single cell. "When someone is studying tissue culture, most of the time they are learning how to grow a whole plant from a single cell." From one 8-cm section of a plant stem, Maureen can produce up to 12 plants. "The procedure is simple. First I'll treat the stem with alcohol to kill any microorganisms. Then the stem is divided into smaller pieces, called explants, which I will place in culture jars. Within one to two weeks, each explant will have produced a healthy little plantlet."

What will Maureen and the scientists she works with do with the plant cells that she grows? They might decide to give them new genes, maybe even human genes! Not only can Maureen's group reproduce several plants from just one small section of a stem, they can also genetically alter the plant and reproduce it. One group inserted the human insulin gene in a sunflower and collected human insulin from the sunflower seed oil. Most plant tissue culture technologists like Maureen have a master's degree in biotechnology, botany, or plant physiology. They have taken courses in biology, chemistry, botany, and genetics.

ASEXUAL REPRODUCTION IN NATURE

Have you ever been responsible for weeding a lawn or garden? Weed control is a constant concern for gardeners, landscapers, farmers, and groundskeepers. The measures that they take depend on whether the weed is an annual or a perennial. An **annual** plant lives for a year or less. **Perennials** are longer-lived plants, such as trees. Annuals such as crabgrass use nearly all their stored nutrients and energy reserves to reproduce sexually and then die after producing seeds. Chemical herbicides are often less effective against annual species, because herbicides don't affect the dormant seeds.

Many perennials, by contrast, spread or reproduce asexually, with organs that are active year-round. Perennial weeds such as quackgrass are therefore more vulnerable to herbicides. Quackgrass has **rhizomes:** roots that spread horizontally and send up new shoots (Figure 21.1A). Strawberry plants and spider plants also cover a lot of ground; they produce horizontal stems called stolons. **Stolons** send down roots to start new plants.

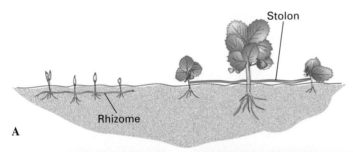

A

Many edible roots, such as sweet potatoes, are asexual reproductive structures. Starch and other nutrients stored in these roots help to produce new plants each spring. White potatoes are

B

C

D

Figure 21.1: *Forms of asexual reproduction: (A) Production of new quackgrass plants from rhizomes and strawberry plants from stolons. (B) Potatoes are cut up and planted like seeds. Each eye is a bud that can grow into a shoot. (C) Garlic cloves separate and grow into new plants. (D) Dandelions produce asexual flowers and seeds.*

underground, starch-storing stems. The "eyes" on a potato are embryonic shoots and roots. You can cut up a potato, with an eye in each piece, and plant the pieces to start a new crop (Figure 21.1B).

Other plants reproduce asexually by breaking into pieces. Garlic bulbs, for example, break into cloves. Each clove can grow roots and shoots, producing a new plant (Figure 21.1C).

Most plant species reproduce sexually as well as asexually. The species that rely only on asexual reproduction have usually lost the ability to reproduce sexually. So dandelions, which reproduce only asexually, still produce flowers and pollen, even though their seeds develop from ovary cells without fertilization (Figure 21.1D).

ARTIFICIAL ASEXUAL REPRODUCTION

Artificial asexual reproduction of plants is thousands of years old. **Grafting** (Figure 21.2) is an ancient technique that joins a branch or stem from one plant to the roots and stem of another. Buds can also be grafted onto stems. People graft plants to combine the benefits of closely related species. For example, branches of orange trees that produce high-quality fruit are grafted onto the roots and stumps of closely related trees that are more resistant to parasitic soil organisms. Although trees that are grown from seeds take years to produce fruit, grafting quickly produces many trees with branches that bear a desirable variety of fruit.

Figure 21.2: *A vigorously growing branch of a productive young olive tree grafted onto the trunk of an established old olive tree*

Figure 21.3: *Plant tissue cultures. (A) Plant cells form a callus in a dish of nutrient-containing agar. (B) Addition of auxin or other PGRs to the agar stimulates the callus to form a miniature plant.*

Plant breeders usually have to wait for each generation of plants to mature before choosing the ones that they want to use for the next round of breeding. One way to speed things up is to grow millions of plant cells in liquid or agar culture (Figure 21.3A), just as bacteria are raised. Plant cells growing on agar may form a disorganized mass, or **callus.** If a breeder can find the callus or individual cells that have the most desirable alleles, he can stimulate them with PGRs that organize the cells into shoots and roots. He can transfer the resulting plantlets (Figure 21.3B) to pots of soil; when they mature, they'll produce seeds for the next step in the breeding program.

INVESTIGATION 21A FRUITS AND SEEDS

See Investigation 21A in the TRB.

What difference does it make whether or not a seed has an endosperm? How does it affect the seed's usefulness? Huge industries, not to mention nearly your whole food supply, depend on seeds with particular characteristics. In this investigation, you will compare the composition of various kinds of seeds with their uses and nutritional value. You will also identify each of the three layers in some fruits and figure out the importance of how each layer functions in the life of the plant and how it affects the fruit's usefulness.

LESSON 21.1 ASSESSMENT

1. Cutting works better for annuals. If they cannot flower and produce seeds, they cannot reproduce.

2. Asexually produced plants are genetically uniform, so a pathogen could kill an entire crop. Sexually produced plants can have more genetic variety, so some plants may be resistant.

3. No, bananas, like many asexual plant species, have lost the ability to reproduce sexually, though they still produce (infertile) fruits and flowers.

4. Callus forms on cut pieces of plant organs in culture just as it would if a whole plant were cut. In cell cultures, the callus continues to grow on the nutrient medium.

1 One of your fellow landscape architects wants to eliminate a weed in a large lawn by cutting the lawn before the weed can flower. Will this method work better with annual or perennial weeds? Give a reason for your answer.

2 Some varieties of a crop may carry alleles that provide resistance to disease-causing microorganisms. Is a grower more likely to have a big problem with diseases in crop plants that are produced sexually or asexually? Give a reason for your answer.

3 Usually, banana plants reproduce only asexually. Does this mean that bananas are more primitive than sexual species such as apples? Give a reason for your answer.

4 Callus sometimes forms on injured plant tissues and provides cells that help heal or scar over an injury. How does the natural process of callus formation make plant cell culture possible?

LESSON 21.2 SEXUAL REPRODUCTION

With the support of governments and international food and agricultural organizations, plant breeders throughout the world have established a network of storage and research centers that collect seeds of millions of varieties of crop plants and their wild relatives. Seeds are a convenient form in which to preserve plants' genetic diversity. Maintaining crop diversity can prevent disasters like the potato famine in Ireland, when thousands of people either starved or emigrated to save their lives. How do plants create these tough packages of life? You'll see in this lesson.

ACTIVITY 21-2   SEEDS COME TO LIFE

The procedures for this activity are in Chapter 21 of the TRB.

Seed quality is so important to our food supply that the law says that any seeds that are marketed must be labeled to indicate the type of seed and the percentages of weed seeds, debris, and live crop seeds. Other required tests include those for resistance to cold and aging, percentage of moisture, and the growth rate of the germinating seeds. In this activity you will perform a viability test to determine the percentage of live seeds in a sample.

A seed company worker checks germinating seeds.

BEEKEEPING

BIOLOGY IN CONTEXT

To many people, bees are just a bother, but other people make a living from them. Beekeepers usually sell honey, but most of their income comes from fees paid by farmers who need the bees to transfer pollen among crop flowers. Some flowers produce a nutritious liquid called nectar. As a bee sucks the nectar, its body picks up pollen. When it visits a different flower, some of the

A beekeeper examining bees.

pollen on its body rubs off, pollinating the new flower. A bee visits only one species of flower per trip from the hive and back, so it only delivers pollen to flowers of the same species. Back at the hive, the bees convert the nectar to honey and eat some of the pollen. Both plants and bees benefit from this relationship.

Farmers who raise cucumbers, melons, apples, sweet cherries, and blueberries pay beekeepers to bring portable hives to their fields when the plants flower. In recent years, however, honeybees have had a rough time. Parasitic mites, competition from aggressive wild Mexican bees, and injury from insecticides have severely reduced their populations and caused big losses to beekeepers and farmers. Entomologists (insect biologists), plant scientists, and beekeepers are struggling to solve the problems that threaten the relationship among honeybees, crop plants, and the people who depend on them.

SEXUAL REPRODUCTION IN SIMPLE PLANTS

An additional activity, Pollen Tubes, can be found in Chapter 21 of the TRB.

In Chapter 18 you learned that sexual reproduction in ferns, mosses, and other simple plants involves alternate haploid and diploid generations. Their sperm swim through moist soil or a film of water on plant surfaces until they reach and fertilize an egg (Figure 21.4). The need for a wet environment restricts where these plants can grow. That's why landscapers and decorators plant ornamental ferns mostly indoors or in the shade of trees or buildings. Because they don't produce seeds or other edible organs containing large reserves of nutrients, ferns, mosses, and their relatives are not important as food crops.

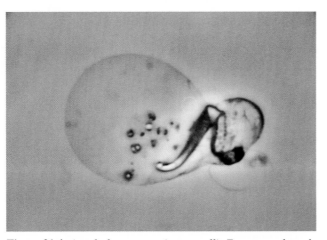

Figure 21.4: *A male fern gamete (sperm cell). Ferns are adapted to reproduce in moist soils where their ribbonlike sperm cells can swim to reach and fertilize the fern's egg cells (magnification ×500, color added).*

SEXUAL REPRODUCTION IN NONFLOWERING SEED PLANTS

WORD BANK

gymno- = naked

angio- = container

sperm = seed

Gymnosperms have exposed seeds. The seeds of angiosperms (flowering plants) are contained inside fruits.

You might wish to refer students to Table 18.1 to remind them of the differences between gymnosperms and angiosperms.

Pine trees, important as a source of lumber, pulp for papermaking, and needles for the manufacture of turpentine, provide a good example of how **gymnosperms** (nonflowering seed plants) such as conifers reproduce. Pines have male and female cones (Figure 21.5). The small male cones produce haploid, single-celled pollen grains that fall from the cones and are spread by wind. Some of the pollen falls on female cones. These are the larger, more familiar cones that you may have seen. Each scale of a female cone has two hollow egg-producing chambers, or **ovules.** Pollen that lands on an ovule begins to grow a long cellular extension called a **pollen tube.** The pollen tube slowly grows into the ovule, breaking through the tissue surrounding the egg-forming cells. At the same time, meiosis occurs in the ovule, producing two or three egg cells. When the pollen tube reaches these eggs, a haploid nucleus inside the pollen tube merges with one of the egg cells.

What happens to all the pollen that doesn't fertilize plant egg cells? Some of it goes into eyes and noses. Allergists often treat people who are sensitive to common types of pollen and who have problems at times of the year when pollen drifts in the wind. The allergist may apply extracts of the pollens of various species to a patient's skin to identify which ones cause an allergic reaction. Some pollen lands on the clothing worn by a suspect or a victim in a crime. It becomes evidence of where these clothes, and presumably their owners, have been. The outer layer of pollen grains has a complex structure that is unique for each species (Figure 21.6). Because this layer is a tough material that resists decay, it shows up in ancient artifacts and soil. The fossilized pollen gives archeologists clues to what vegetation was at a site in ancient times and therefore indicates what the climate was.

The zygote (fertilized egg) develops into an embryo. The embryo, the surrounding nutrient-rich tissue, and the toughened ovule wall make up a seed. The embryo consists of a small root and stem and one or more specialized leaves called **cotyledons.** These structures absorb nutrients from the maternal tissue during seed development and germination, feeding the seedling until photosynthesis takes over (Figure 21.7).

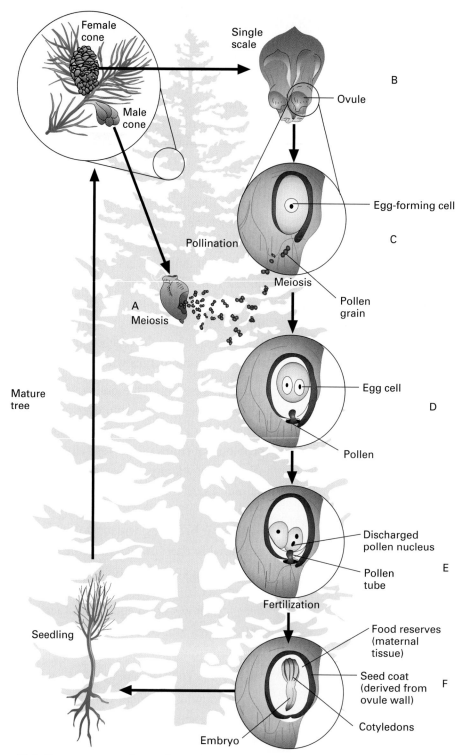

Figure 21.5: *Gymnosperms produce male and female reproductive structures. Male pine cones produce pollen through meiosis (A). Meiosis in female cones (B and C) produces ova in the ovules of each cone scale. Pollination (D and E) leads to fertilization, followed by development of the embryo (F).*

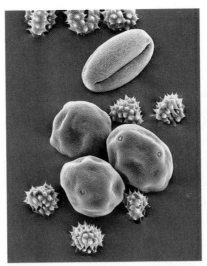

Figure 21.6: *Pollen grains (color added, magnification ×1,100). The various shapes mark these as grains of different species: daisy (prickly), cherry (long), and hornbeam (round).*

Figure 21.7: *These pine seedlings have green cotyledons and needle leaves, indicating that they have begun to make their own food through photosynthesis.*

INVESTIGATION 21B A FEW BUGS IN THE SYSTEM

See Investigation 21B in the TRB.

You may use the following demonstration here. Show the class flowers of one or more of the following species and encourage them to speculate about the possible adaptive value of the flower form and colors. Then darken the room and show them the flowers again with ultraviolet illumination. Species that will work include evening primrose (*Oenothera speciosa*), marsh marigold (*Caltha palustris*), oriental poppy (*Papaver orientale*), and many species of mallows (*Malva*) and daisies. These flowers all have "landing strips": contrasting UV-fluorescent colors at the tips and bases of petals that are visible to pollinating insects and are thought to guide them to the anthers and ovaries at the flowers' centers. Ask the class to speculate and discuss the possible significance of these "invisible" color patterns. You might need to remind them that these flowers are pollinated by insects. An interesting related issue is the difference between the "perceptual worlds" of different species.

About 95% of plant species are flowering plants. Fragrant flowers and edible, nutritious fruits and seeds have helped these species to develop mutualistic relationships with animals, including humans. This is why most

Plant breeders hand pollinating experimental barley plants.

crops are flowering plants—and why many insects and other animals help themselves to farmers' crops. In this investigation you'll use the mutualistic relationship between bees and flowering plants to make genetic crosses between plants.

Sexual Reproduction in Flowering Plants

Refer students to the figure in the handout from the additional activity in Chapter 18, How Do Flowers Work?

Flowering plants, or **angiosperms,** produce ovules in flowers within a protective structure called an **ovary.** After fertilization the ovary develops into a fruit, and the ovules develop into seeds.

Unlike gymnosperms, which the wind pollinates, most angiosperms are pollinated by insects or other animals that deliver pollen directly to the female flower parts. Flowers usually have tasty parts or nectar that the pollinator eats. Pollen that rubs off or falls onto the animal's body travels with it to the female parts of other flowers of the same species. Various adaptations ensure that pollinators mostly visit flowers of the same species. For example, some flowers produce nectar at the bottom of a long, narrow tubular flower that only certain species of insects or hummingbirds can reach with their long mouthparts or beaks. Pollinators may also recognize colors, shapes, or fragrances of specific species of flowers.

Unfortunately, insecticides that help to protect crops from harmful insects can also kill pollinating insects and insects that eat pest insects. Successful farmers watch the populations of harmful and beneficial insects in their fields. They apply insecticides only when necessary and when they will do more good than harm (Figure 21.8).

Figure 21.8: *Careful monitoring of weather and insect populations helps farmers avoid wasted or harmful insecticide applications. This approach is called integrated pest management, or IPM.*

Another important difference between angiosperms and gymnosperms is the way they fertilize and feed the embryo. In gymnosperms, only one pollen nucleus participates in fertilization. In angiosperms, two haploid nuclei from the pollen tube enter the ovule: One pollen nucleus fertilizes a haploid egg cell, and a second one enters a diploid cell in the ovary. No other group of organisms uses **double fertilization,** or fertilization of two cells (Figure 21.9). The result is a diploid zygote and a triploid cell (one with three copies of each chromosome).

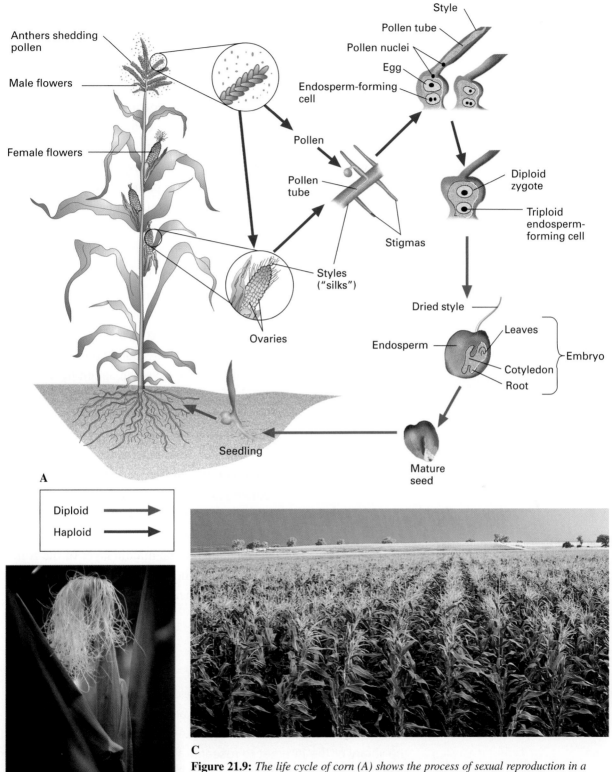

Figure 21.9: *The life cycle of corn (A) shows the process of sexual reproduction in a flowering plant. In some angiosperms, such as corn, the male and female flowers are separate. The female flowers (B) have long styles covered with pollen-binding stigmas. The male flowers (C) shed pollen on the "silks" (styles) of the female flowers. Fertilization by pollen produces a diploid zygote and a triploid endosperm.*

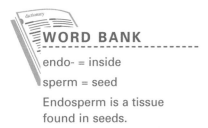
The zygote develops into the embryo, and the triploid cell develops into a tissue called **endosperm.** Endosperm supplies nutrients to the developing embryo. Angiosperm embryos have one or two cotyledons that absorb sugars and amino acids from the endosperm. In some economically important species, such as beans and cotton, the endosperm breaks down, and the embryo completely absorbs its nutrients by the time the seed is mature and dry. The cotyledons store these nutrients until the embryo uses them during germination. In other species, such as the grass family, which includes wheat, rice, corn, barley, and oats, the endosperm persists and nourishes the germinating embryo until it can feed itself. Starch and protein stored in endosperm are what make wheat and other grass seeds such great food sources.

A fruit develops from an ovary, which has three tissue layers, named for their positions in the fruit (Figure 21.10). The characteristics of these fruit layers vary from species to species. In some species, other flower parts also become part of the fruit. A peach has a skin that protects the interior; a fleshy, edible middle layer; and a hard pit that protects the seed inside when an animal eats the outer layers. The outer layer of a coconut consists of fibers that are toughened by lignin. People gather the fibers and sell them for making rope and mats. The hard middle and inner layers are sold with the large seed as food; the white edible part is fatty endosperm. If you like potato chips or other fried snacks, you have probably eaten palm oil. This oil is squeezed from the middle layer of the fruit of the oil palm, a type of palm tree, and used in commercial baking and frying.

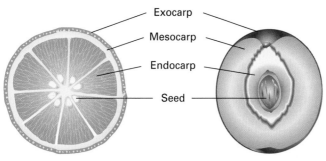

Exocarp

Mesocarp

Endocarp

Seed

Figure 21.10: *The three layers of a fruit: exocarp, mesocarp, and endocarp.*

LEARNING LINK SEEDS IN THE OVEN

Some seeds can be ground into flour for bread. Bread made with wheat flour rises well, but rice flour alone can't be used to make bread. Why not? Why are cornbread and whole-wheat bread coarser than white bread? What makes rye bread chewy? Use the library or the Internet to research the differences among flours and how they affect flour use. Some keywords to search for include gluten, cereal grains, storage proteins, aleurone, and zein.

THE FIRST FARMERS

Before people began planting seeds and growing crops, they spent much of their time foraging for food, and they had no dependable supply to fall back on in times of need. People might have begun to plant crops after they found that some of the seeds they collected and stored to eat later had sprouted.

Ancient seeds discovered by archeologists provided evidence that some of the first farmers lived in the Middle East. For example, 9,000-year-old wheat and barley seeds were found among the bones of goats in what is now Iraq. Many people once thought that agriculture was born only in the Middle East and then carried to other parts of the world. Today, it seems more likely that agriculture was invented more than once. Peas and beans almost as old as those found in Iraq suggest that Southeast Asians discovered the benefits of agriculture without help from other cultures. Likewise, squash, beans, peppers, and corn were grown in Mexico 7,000 years ago.

LESSON 21.2 ASSESSMENT

1. Ferns are not a likely source of allergic irritation. They do not produce pollen, and the spores on the undersides of their leaves fall to the ground; they do not float in the air like pollen.

2. Gymnosperm seeds do not contain endosperm tissue, which is where the starch and protein necessary for bread flour are.

3. This will be difficult but not impossible. Most elaborate flowers depend on insect pollination. The fragrance comes from the nectar that attracts insects. It might be possible to find flowers pollinated by wind, birds, or bats that please the client, but they probably will not be fragrant, and the garden may still attract plant-eating pest insects.

1 Your patient tells you that she is allergic to the pollen from the ferns in her garden. She asks whether you think she should replace them with something else. How would you respond? Give a reason for your answer.

2 Explain why millers make flour from the seeds of flowering plants such as wheat and corn but not from the seeds of gymnosperms such as pines.

3 One of your landscaping clients has a dangerous allergy to bee stings and is afraid of all kinds of insects. He wants you to plant a garden that will produce many beautiful, fragrant flowers throughout the summer but will not attract insects. Is this job likely to be easy, difficult, or impossible? Give reasons for your answer.

4 Food stores often advertise raw sprouts (germinated seeds) as being more nutritious than dry grains, which must be cooked. How does the nutritional composition of a seed change during germination?

4. As the stored nutrients in the endosperm or cotyledons are consumed, the percentage of starch and lipid decline. Stored protein also breaks down, but the seedling uses its amino acids to make new proteins.

CAREER APPLICATIONS

The applications that follow are like the ones you will encounter in many workplaces. Use the biology you learned in this chapter to complete the activities. Share your work with the class.

Growing Seedless Fruits

Watermelon and grapes are summertime treats for many people. But some people think that eating these fruits isn't worth the effort because of the seeds. As a result, growers have developed seedless varieties of watermelon and grapes. Find out how to grow seedless varieties of grapes, watermelons, or other fruits. Write a plan for growing one of these seedless fruits and present it to the class.

Grafting

There are several reasons for grafting trees. The most important ones are to create trees that would otherwise be hard to multiply, to replace the existing root system of a tree with a better one, to reduce the time a tree needs to produce flowers or fruits, to repair damage to older trees, or to rejuvenate them with young and improved material. Use the Internet to find out what plants are often grafted. Select one species and find out when and why it is grafted and the procedure(s) used. Prepare a poster showing the grafting techniques and present it to the class.

The Fastest Lawn

Grass can be grown from seed, from sprigs, or from sod. Talk to someone in the lawn care business. Find out how long it takes to establish a lawn by each method. How do the costs compare? How much time and effort does each method require? Prepare a brochure that discusses the options for people who are interested in putting in a new lawn. Recommend grasses for your area.

Blemish-Free Produce

Talk to a produce manager about what buyers look for when they are selecting fruits and vegetables. What do harvesting, storing, and displaying those fruits and vegetables require? Some fruits and vegetables are picked before they are fully ripe so that they will hold up during shipping. Does that affect the flavor of the fruits and vegetables in the store? Which fruits and vegetables ripen and develop full flavor when picked in an immature state and which don't? How can you select flavorful fruits and vegetables? Write up your findings in your logbook.

Want to Be a Florist?

Talk to a florist about how flowers are selected for cut arrangements. How can flowers that bloom at one time of the year be available all year long? How are flowers shipped across the country? How does one become a member of the FTD network? How can you make a cut-flower arrangement last a long time? What are the busiest seasons for florists? What seasons are slow? What training does a florist need? Where can you get that training? Is continuing education required? Work with several other students to develop a plan for starting a new florist shop in your area. Present your plan to the class.

Plant Politics

Some companies in the United States and other developed countries collect native plants from less developed countries to use for medicines, ornamental plantings, or other purposes. The governments of the less developed countries want to claim royalties from all products that are produced from their native plants, whether or not they were grown in their country. Divide the class into two groups and debate whether the governments should receive royalties and under what circumstances.

Plants from the Kitchen

Prepare a lesson for a child about growing plants found in the kitchen. Choose a plant to grow from a seed such as an avocado, from a cutting such as a carrot top, or from a root such as a sweet potato. Try out your lesson with a younger brother or sister or with an elementary school class.

FAMILY & CONSUMER SCIENCE

Dangerous Plants

Some plants produce substances that are dangerous to people and their pets. Using plant books or information from a poison control center, the local extension service, or the Internet, make a table of common dangerous plants, including the parts that are poisonous, problems that they cause, and antidotes or treatment in case of contact or ingestion. Are most of the poisons in reproductive or in vegetative parts of plants? How does this help plants to survive? Survey your home or school for the problem plants. If you find one, make a small sticker or card that describes the danger and the antidotes. Attach it to the plant or its pot. If small children or pets might be endangered, make recommendations for replacing the plant or for keeping it out of their reach.

HEALTH CAREERS

Pollen Allergies

Many people are allergic to plants and particularly to their pollen. Contact an allergist in your area and find out which local plants cause the most problems. When do they produce pollen? What can people with allergies do to relieve their symptoms during pollen season? Make an informational pamphlet about seasonal pollen allergies in your area.

Plant Cell Culture for Medicines

Research the use of plant cell cultures in drug manufacture. You might use the resources of your library or contact a company that manufactures medicines. How are the cells harvested from plants? How are they cultured? How are useful substances removed from the cells? How are the substances processed before they become part of a drug that the patient takes? Drugs that you might research are digoxin and digitoxin from foxglove (*Digitalis purpurea*), vincristine and vinblastine from the Madagascar or rosy periwinkle (*Catharanthus roseus*), and taxol from the Pacific yew (*Taxus brevifolia*).

INDUSTRIAL TECHNOLOGY

Equipment for Plant Tissue Culture

One of the most common problems in plant tissue culture is contamination by other organisms such as fungi and bacteria. What special equipment prevents contamination in large labs or industries? Describe the operation of a laminar flow hood. Why is it used in plant tissue culture operations? How is the hood disinfected between uses?

Making the World Safe for Tapioca

The starchy granules that are used to make tapioca pudding come from the roots of a tropical plant called cassava or manioc. The roots store nutrients and function in asexual reproduction. Find out how the roots are processed to remove a cyanide-producing poison. What kind of equipment is used? How does it work? Is the processing done near where the plants are grown or near the consumers who will buy the product? What are the practical reasons for this decision?

Taxol Production

Taxol and other substances come from the bark of the Pacific yew. Find out where these trees grow and when the medicinal substances can be harvested. Does harvesting the bark kill the tree? Develop a plan for growing Pacific yews to ensure a supply of taxol and other substances for the future. You will need to consider how the yew reproduces.

CHAPTER 21 SUMMARY

- Most plants reproduce sexually. Some species, especially perennials, can also reproduce asexually.

- Asexual reproductive organs include rhizomes, stolons, nutrient-storing roots, and fragmenting bulbs or other organs.

- Methods of artificial asexual reproduction include grafting, cutting up and planting pieces of roots or stems, and culturing cells, callus, and organs.

- Sexual reproduction in simpler plants requires at least a thin layer of liquid water through which the male gametes can swim. More complex plants have nonmotile gametes. The male gamete, pollen, is dispersed by wind or animals.

- Pollen dispersal in most nonflowering seed plants (gymnosperms) depends on wind.

- Seeds consist of an embryo, stored nutrients that feed the germinating seedling until it makes its own food, and a protective seed coat made of maternal tissue. They can remain dormant until environmental conditions permit germination and growth.

- The ovules of flowering plants (angiosperms) are enclosed in an ovary.

- In most flowering plants, pollen is dispersed by animals. Close mutualistic relationships provide particular species of insects, birds, and bats with food and ensure that these animals deliver pollen to flowers of the same species.

- When pollen lands on the stigma, an extension of the ovary of a flowering plant, it begins to grow a tube down into the ovary.

- The seeds of flowering plants experience double fertilization. One pollen nucleus fertilizes an egg cell, and the zygote develops into the embryo. The other fertilizes a diploid cell, which develops into the triploid endosperm.

- Endosperm is a nutrient-rich tissue that nourishes the embryo during seed development and, in some species, during germination and early seedling growth.

- Fruits develop from the ovary wall and other flower parts. The three tissue layers of the ovary develop into the three layers of the fruit. The composition and characteristics of these layers determine their usefulness.

CHAPTER 21 ASSESSMENT

Concept Review

See TRB, Chapter 21 for answers.

1 Why are asexually reproducing annual plants unusual?

2 Name three foods that come from different annual plants and three from perennials. Don't use examples mentioned in this chapter.

3 Could you graft apples, grapes, and oranges all on the same tree? Give a reason for your answer.

4 What is the difference between a stolon and a rhizome?

5 Wild carrots can live as perennials, though carrots are raised as annuals on farms. What is the most obvious way for a carrot to survive as a perennial?

6 Are insects on a crop a reliable sign that the crop is in danger and needs to be sprayed with insecticide? Give a reason for your answer.

7 Pollen grains are said to germinate, as seeds do. What process involving pollen could this refer to?

8 Many conifer species have sticky surfaces on their female cones that trap any insects that land there. Why doesn't this defense interfere with pollination by insects?

9 "Cat-faced" fruits have large dents where the fruit tissue failed to grow. They result when insects eat part of a flower. Which organ

do the insects probably eat to cause cat-facing? Give a reason for your answer.

10 Plant tissue culturists can grow endosperm tissue in culture, but these tissues don't produce plantlets that can flower and produce seeds. Why has this been impossible?

Think and Discuss

11 How can stolons or rhizomes make a plant population more vulnerable to disease?

12 A group of aspen trees in the western United States is thought to be the largest single organism in the world, although no above-ground connections between the trees are visible. How can a whole grove of trees be considered one organism?

13 Why is a tomato considered a fruit?

14 Large quantities of pollen have been found around the bones in the graves of prehistoric cave dwellers. What does this suggest about the behavior of these people?

15 Why is pollen considered a male, not a female, gamete?

In the Workplace

16 Do plant breeders find it easier to make improvements in varieties of sexually or asexually reproducing species? Give a reason for your answer.

17 Why might a plant breeder treat a plant cell culture with a mutagen? What is the advantage of doing this rather than treating whole plants with a mutagen?

18 Some orchard owners killed lots of bats because they saw them eating fruit. Loss of the bats reduced the size of the following year's harvest much more than their eating fruit would have. Why did the harvest decrease?

Investigations

The following question relates to Investigation 21A:

19 Describe each of the three tissue layers (endocarp, mesocarp, and exocarp) in a watermelon.

The following question relates to Investigation 21B:

20 How could natural selection produce bees with specialized organs on their legs that collect pollen?

CHAPTER 22

WHY SHOULD I LEARN THIS?

Changes in populations affect just about everyone. Marketing specialists watch how quickly human populations grow and how the proportions of people of various ages change. This variation in size and age of a population affects demand for their clients' products, and that demand affects how much they can charge you for the products. Level of demand affects consumption of all types. Are you looking for an apartment? If a lot of people are moving out of town, you'll get a palace in town for a low-price rent. But don't expect to find an inexpensive palace in a crowded neighborhood!

Chapter Contents

POPULATIONS

1. how populations grow in size and density
2. how birthrates, death rates, and migration affect population size
3. what limits the size of populations
4. the connection between population density and the environment
5. the prospects for the human population

"There are plenty more fish in the sea." This common expression of unlimited opportunity no longer rings true. What happened to all those fish? Is there any way for them to recover from overfishing and other disasters? A lot of people are looking into this problem—fishers, policy makers, resource managers, and consumers.

Populations of other species are important, too. Wildlife and fisheries managers watch the size and health of the populations of endangered species and those that are caught for food or hunted. A sudden increase in the death rate of certain desert rodents in the southwest may signal public health officials of an outbreak of plague or other serious diseases that can affect humans.

In this chapter you'll take a close look at some vital statistics and find out how populations grow and shrink.

LESSON 22.1 POPULATION GROWTH

What do school officials and the operators of commercial fishing boats have in common? They are concerned with populations. School officials need to anticipate changes in birthrates and housing patterns, which signal the need for more classrooms and teachers. Fishers are always trying to maximize their catch, but will they leave enough fish behind to reproduce future populations? In this lesson you'll learn about the factors that affect the size of populations.

ACTIVITY 22-1   HOW FAST CAN A POPULATION GROW?

MATH TIP

Because the number of bacteria doubles every half hour, their number is always twice what it was 30 minutes earlier. After 30 minutes the population is equal to 2×1 (or 2^1). After 60 minutes, it is twice that: $2 \times 2 \times 1$ (or 2^2), and after 90 minutes it is 2^3. After any period of time, it is equal to 2^n, where n is the number of 30-minute periods that have passed.

Approximately 281 trillion (2.81×10^{14}) cells would be produced in 24 hours.

In reality, populations do not grow this quickly.

Their growth slows as they run out of space, exhaust their food supply, attract predators and parasites, and accumulate waste.

You are aware of how rapidly human populations can grow. Are the populations of all species growing as quickly as ours? How fast can a population grow if there is nothing around to slow it down? In this activity you'll calculate how quickly a population of bacteria grows, starting from just a single cell.

Some species of bacteria divide as often as every 30 minutes under the right conditions. What happens when a lab technician puts one bacterium in a test tube of nutrient broth? The cell and its descendants divide every 30 minutes, so the population doubles every 30 minutes.

Working with a partner, calculate or estimate how many bacteria this would produce in 24 hours. (You might need a calculator.) When you have an answer, record in your logbook how you found it, and discuss the following questions:

- How many descendants can one bacterial cell produce in 24 hours?

- Does this number seem reasonable? Assess whether bacterial populations really grow this quickly in their own environments.

- Discuss the factors that you think might slow down the rate at which populations grow in nature.

STATISTICIAN

In a high-rise office building behind a door marked Department of Vital Statistics, Rebecca W., age 23, is entering into a computer her state's data on births, deaths, marriages, and divorces. Records of these events are usually filed at the county level and then forwarded to the state for compilation. Some of the information that Rebecca receives is collected by hospitals, nursing homes, funeral homes, and birthing centers. She sends the state totals to the National Center for Health Statistics near Washington, D.C.

Another part of Rebecca's job is answering public requests for information. School districts frequently need birthrates to help them to predict future kindergarten enrollment. Public health departments, concerned with causes of death, might need to know the number of people who died of rabies last year. The local Bureau of Women and Children recently called Rebecca and asked how many unmarried women had babies last year in her county.

Rebecca has a bachelor of science degree from a four-year college. She took courses in math (especially calculus), economics, sociology, and computer programming. She continues her training at week-long workshops sponsored annually by the National Center for Health Statistics and the Applied Statistics Training Institute.

POPULATIONS AND HOW THEY CHANGE

The U.S. Constitution requires the federal government to conduct a census, or population count, every ten years. The number of people living in each state determines how many congressional representatives that state will have. Pollsters, computer programmers, mathematicians, and mapmakers all get involved in this process.

In Chapter 6 a population was defined as a group of individuals of the same species living in a certain area—in this case, all the people living in a state. But a population is not just the number of individuals in an area. City and regional planners, game managers, and foresters are also interested in **population density** (the

A

B

C

Figure 22.1: *Solitary predators such as hawks (A) usually have low population density and even distribution. Social animals such as mountain sheep (B) often occur at high population densities and in uneven, clustered distributions. Plants that compete for water (C) in an arid environment are usually evenly spaced.*

number of individuals per unit of land area) and **population distribution.** Distribution refers to the pattern in which organisms tend to occupy a space, usually either in clumps or spread randomly or evenly throughout the space (Figure 22.1). Other people who study populations closely include managers of nature preserves who protect endangered species and agricultural consultants who advise growers when population densities of crop-eating insects or other pests are large enough to justify applying insecticides.

GROWTH RATE

In Activity 22-1 you saw that, under the right conditions, populations become huge, sometimes very quickly. Every population has a maximum growth rate, which it reaches only rarely and briefly, when conditions are ideal. This maximum growth rate is the population's **biotic potential.** Since the growth rate depends on the number of reproducing individuals, the larger a population gets, the faster it grows. Under ideal conditions (as in Activity 22-1), the growth rate can speed up like a runaway train rolling downhill. This kind of rapid population growth is called **exponential growth,** a phase of population growth in which the population continually doubles in size at fixed intervals. In a population that grows exponentially, the growth depends on the density of the population (Figure 22.2). At low density the growth rate is low; at high density the growth rate is high.

Real growth is almost always much slower than exponential growth. On the simplest level, all changes in population size depend on three factors: births, deaths, and migration. Each of these factors depends on many others. Population changes in most species depend more on the birthrate and death rate than on migration. In many species, especially insects and small aquatic species, mortality among the young is very high. Large numbers of eggs are produced, making it more likely that at least a few of the young will survive and themselves grow to reproductive age.

Even in humans, infant mortality is sometimes high and has been an important factor in the growth of human populations. In North America and Europe, human infant mortality was much higher

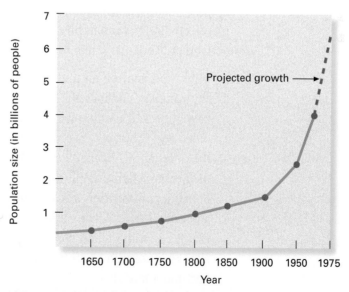

Figure 22.2: *When a population's growth rate is a function of its density, the population grows (or declines) exponentially.*

A   B

Figure 22.3: *(A) High birthrates and falling infant mortality produced many large families in the 1800s. (B) Birthrates and family size decreased significantly in the 1900s.*

200 years ago than it is today. People compensated for this by having many children. Women often had five or more babies (Figure 22.3A). Improved health and sanitation during the nineteenth century reduced infant mortality. Populations then grew more quickly until birthrates fell between 1870 and 1930 (Figure 22.3B). A similar change began in many developing countries in the twentieth century. Social attitudes about the importance of children as workers and support for their aging parents, as well as beliefs about birth control, have had important effects on the human birthrate. Birthrates also tend to decline as women's opportunities for education and employment increase.

Migration is often an important factor in population size when the environment changes. In the last 100 years, new opportunities for education and employment have drawn people from rural areas to cities and suburbs. This movement has, in turn, led to changes in the populations of many wild animals. Deer are drawn to the edges of forests, where they find small shrubs and grasses to eat. When housing construction destroys forests, the lawns and hedges of new suburban homes attract hungry deer. With them come the ticks that feed on them and spread diseases to humans. Raccoons and skunks, in search of an easy meal from garbage cans, soon become a nuisance.

Humans also carry species into new environments, sometimes unintentionally (Figure 22.4). The introduced population often grows rapidly because it has no competitors or predators; it then disrupts the balance among native species. For example, the English gypsy moth was brought to America in the 1700s to produce silk. The silk farming was a failure, but the moths escaped and have been feasting on American trees ever since. Construction of the St. Lawrence Seaway let ships travel all the way from Chicago to the Atlantic Ocean. It also let fish parasites such as lampreys enter the Great Lakes and destroy the profitable fisheries there. Snakes that stowed away in the landing gear and luggage compartments of airplanes traveled to snakeless Hawaii and other Pacific islands. They have been destroying populations of defenseless ground-nesting birds, which evolved in a habitat that had no such predators.

Figure 22.4: *Pacific brown tree snakes endanger ground-nesting Pacific island birds.*

Two important factors that affect birthrate and death rate are the length of a species' generations (how long individuals take to reach reproductive age) and its population structure (the percentage of individuals of each age and sex). Species with short generation times, such as mice, have high birthrates and great biotic potential. Such species are likely to become pests under certain situations. Species with long generations, such as elephants, swordfish, sharks, and tuna, don't recover easily from hunting and fishing. This is why their survival is easily threatened.

People who study populations create useful ways to display their data. Life insurance companies, for example, try to predict how long their customers are likely to live. They do this with a life table (Table 22.1), which is based on observations of the ages at which a large sample of the population dies. Patterns of life and death in a population can also be shown in a survivorship curve. The survivorship curve for a population of oysters (Figure 22.5) reflects the fact that most oyster larvae are eaten by predators before they develop protective shells. Human death rates, by contrast, are generally low until old age.

Current Age (years)	Expected Remaining Life (years)	Number Surviving	Expected Deaths of Original 1,000 People of Specified Age	Expected Deaths per 1,000 People of Specified Age
0 to 1	74.6	1,000	11.15	11.15
10	65.7	986	0.18	0.18
20	56.0	981	1.00	1.02
30	46.6	970	1.09	1.21
40	37.2	956	1.96	2.05
50	28.3	925	5.01	5.42
60	20.3	850	11.33	13.33
70	13.5	697	20.70	29.70
80	8.1	442	29.61	66.98
85 and older	6.1	0	1,000.00	1,000.00

Table 22.1: Life Table for U.S. Population: Average Survival of 1,000 People from Birth

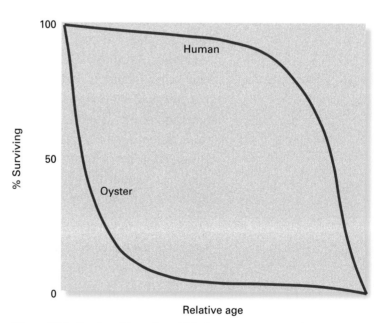

Figure 22.5: *Survivorship varies widely from one species to the next.*

LEARNING LINK CEMETERY DATA

Students who want to obtain cemetery data should ask permission from the cemetery director and carefully observe signs and other regulations about visiting grave sites.

Using library and Internet resources, research how life span has changed for the U.S. population over the last several hundred years. You can use data accumulated by the U.S. Census Bureau, or you can collect data from sources such as headstones in a cemetery. If you use the latter method, be sure to sample randomly. A large random sample should produce data for roughly the same number of individuals born in each decade. Plot your data on a bar graph. Mark the y-axis of the graph "Average Life Span (in years)" and the x-axis "Decade of Birth." Discuss factors that might account for changes in the average life span of individuals over time.

ACTIVITY 22-2 AGE PYRAMIDS

A blackline master of several age pyramids can be found in the TRB.

Real estate developers need to estimate the numbers of houses and schools that an area needs. Social Security planners estimate how many people are likely to retire each year and how long they are likely to live. Such estimates must be based on information about population structure. This information is often displayed in a population's age pyramid. In this activity you will compare age pyramids of populations in different countries and learn to interpret them.

Work with one or two other students. Examine the age pyramids for several human populations. Consider the percentages of people in different age groups: infants, school-age children, people in their child-bearing years, and the elderly people in the population. On the basis of your analysis of the age pyramid, answer the following questions in your logbook:

Mexico has the largest percentage of people of reproductive age, about ages 15–40, and therefore the highest birthrate.

Sweden, with its low percentages of children and people of reproductive age, has achieved zero population growth.

Mexico will need to build the most schools.

Sweden has the largest percentage of elderly people, so it must address the issue of health care for retired people.

- Which country has the highest birthrate?

- Which country has achieved a growth rate of 0?

- In which country will new school construction be a major issue?

- In which country will health care for retired people be a major issue?

TRACKING FISH POPULATIONS

Fisheries managers are always looking for a better way to collect information on fisheries. For years they have depended on the Marine Recreational Fishery Statistics Survey for their data on sports fishing. And for years, fisheries managers have argued about the accuracy of the survey's information.

The National Marine Fisheries Service and the Department of Marine Resources are developing and testing a new survey method for tracking how many fish one person can catch within a given period of time. This study will use fishing charter boats. A 10% random sample of charter boats will be selected each week. Each boat operator will be notified by mail a week in advance that it is his or her turn to participate. Each captain will record for a certain period how many fish of what species were caught and where they were caught.

Charter boat captains will keep data on fish caught by tourists on their fishing charters, as shown here.

If this new survey method provides more reliable data, fisheries managers will be able to set fishing limits accurately. Without reliable data an entire species could be fished to extinction.

Source: This article was adapted from "Charter Captains Can Help in Survey," (published 4/24/98) with the permission of *The Sun Herald*.

LIMITS TO GROWTH

No population can grow forever. Eventually, it would overwhelm its environment. One resource would become the **limiting factor** for the population's growth. A limiting factor is the requirement that is in shortest supply. For animals the limiting factor might be food, water, or nesting sites. For plants it might be space in a crowded forest or meadow or access to sunlight. Most populations do not outstrip their resources; instead, they are held in check by physical factors such as temperature or by competing organisms, predators, and disease.

The maximum population size for a particular species that an environment can support is the **carrying capacity** for that species. Figure 22.6A shows a typical growth pattern as a population approaches its environment's carrying capacity. At first, there are far more births than deaths, and the population grows exponentially, near its biotic potential. As the size of the population approaches the carrying capacity, however, survival becomes more difficult. The death rate rises, and the birthrate falls. Hypothetically, when death rate equals birthrate, the growth rate of a population is zero. At this point population density is stable and **zero population growth** has been achieved.

Species with short generations and high birthrates, such as many insects, may exceed their carrying capacity before the birthrate and death rate become equal. As limiting factors come into play, mortality rises, and the population density falls dramatically (Figure 22.6B). The rapid rise and decline in population density are sometimes called a boom-and-bust cycle. Breeders of tropical fish are aware of this cycle. Fish lay many eggs at once, so a tank can quickly fill with more young fish than can possibly survive in

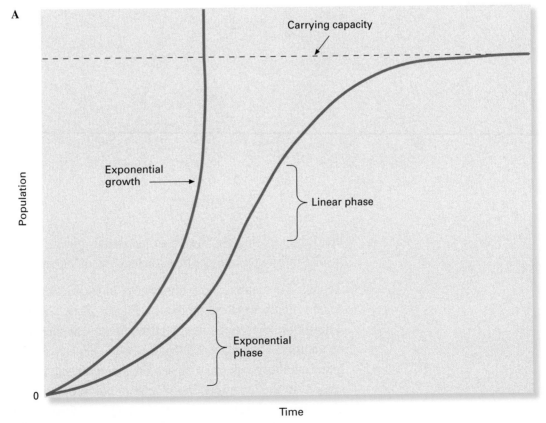

Figure 22.6: *Varieties of population growth curves. (A) An ideal population gradually approaches its carrying capacity (figure continued on next page)*

B

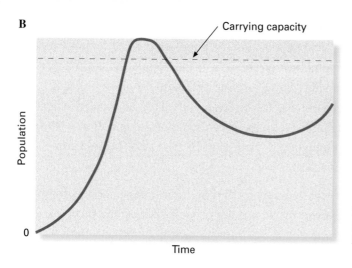

Figure 22.6: *Varieties of population growth curves (continued). (B) A boom-and-bust cycle. (C) A predator-prey cycle.*

C

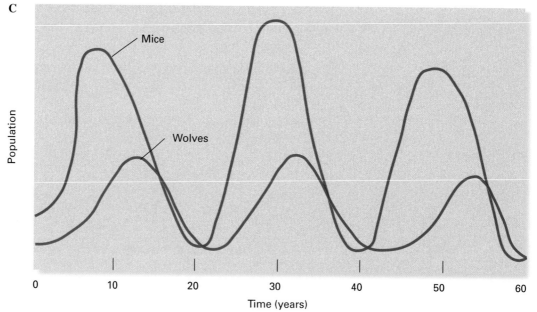

that limited volume of water. In many fish species, overcrowding results in intense competition for food, which keeps fish small.

Regular fluctuations, or cycles, occur in population densities of some rodents and their predators (Figure 22.6C). For a long time, wildlife biologists assumed that the two cycles depended entirely on another. They reasoned that when the density of voles is high, foxes eat more voles and bear more young in the next breeding season. In the meantime the vole population declines because of predation, so foxes have less to eat and bear fewer young the following season. However, evidence is piling up that cycles in rodents are not related directly to predation. Instead, cycles may occur in some rodent species when stress from overcrowding reduces birthrates. In others, populations that are growing exponentially may overshoot the carrying capacity, leading to starvation. What does seem true, however, is that cycles that are observed in predators, such as lynx and fox, are caused by population cycles in their prey.

INVESTIGATION 22A TAKING A CENSUS

See Investigation 22A in the TRB.

Counting a population might sound easy, even boring, but it can be difficult and complicated. Usually, you would count a small sample of the population and use that number to estimate the total population. How accurate are these estimates? In the 1990 census the city of Detroit had just a few more than one million people. If the city's population fell below one million, it would no longer receive certain federal funds. So city officials are concerned about the accuracy of the 2000 census. In this investigation you'll run into some of the challenges of census taking.

Census takers go door to door to collect information for the census.

LESSON 22.1 ASSESSMENT

1. The mouse population will grow from 20 in the current year to 40 in the second year and to 80 in the third year. It will overshoot the carrying capacity in the fourth year.

2. U.S. population growth is now due mostly to immigration.

3. The pattern in Figure 22.6A is the best because it depends on a slow decline in birthrate. The other two involve rapid increases in death rate.

4. As long as there is little or no unemployment, the system works best when the population is growing quickly. In a rapidly growing population the percentage of the population that is of reproductive and working age is high, and the percentage that is retired is low.

1 The carrying capacity of a population of field mice is 120 mice per acre. There are currently 20 mice in the field, and they are expected to double their population each year. How many years from now will the mice overshoot the carrying capacity?

2 The birthrate and death rate in the United States are about equal, yet our population is growing. Explain how this growth can happen.

3 Which pattern in Figure 22.6 represents the most desirable way for the world human population to approach its carrying capacity? Give a reason for your answer.

4 The social security system depends on contributions from people who are working to support retired workers. Does this system work better when the population is growing quickly or slowly? Give a reason for your answer.

LESSON 22.2 ENVIRONMENTAL LIMITS ON POPULATION

How much have you thought about the danger of overpopulation? When does the earth hold too many people? When some go hungry because there's no longer enough land to grow crops? When a city full of people no longer has anywhere to put its trash? Or when other species are becoming extinct because of human activity? In this lesson you will learn about the effects of population growth.

ACTIVITY 22-3 GROWING CITIES . . . SHRINKING POPULATIONS?

As cities and suburbs grow in population, they spread into rural areas. The new construction often destroys environments that supported wildlife. Where do these animals go? Do they leave? Do they die? Or do they stay and become neighborhood pests?

Contact your state's environmental conservation or wildlife management agency to find out what species have been threatened or endangered in your area as a result of urban sprawl. Ask a representative of these agencies, or of others, such as conservation associations, the following questions and record the answer in your logbook:

Answers may vary.

- Which species have been most affected in your area?

- How have they reacted to the new construction? What happened to them?

- How have changes in the populations of these species affected the environment?

- What efforts are being made to protect or restore the species' population, such as regulations on new construction?

Working in a group, develop a plan that will ensure that humans and other species can coexist. Consider not only controls on human activities but also management of species so that they are not threatened with extinction.

The costs are attributable to the need to build new infrastructure such as dams, canals, wastewater treatment facilities, and irrigation systems. The new infrastructure allows lower-quality water to be used and moves water from places of lower needs to places of greater needs.

If China seems exotic to you, then think about this: One out of every five people in the world lives there. At its current rate of growth, China's population will reach 1.5 billion by the year 2050. Food shortages, pollution, waste disposal problems, and declining fisheries are just a few of the problems of this huge population. One of China's major problems is a water shortage of 7 million gallons of water daily. China is spending between $620 million and $1.05 billion a year to build systems to treat more water and move it to water-poor areas.

The Chinese government recognizes that its shortages and environmental problems are related to the growing population. That's why China has implemented policies to slow population growth. In 1979 the government introduced a policy to limit families to one child per family. It also introduced rewards for married couples who have only one child. Families that don't limit births face fines and loss of government benefits. This policy has been successful in the cities, where it is strictly enforced, but not in rural areas, where farm families want many sons to share the work and to support their parents when they are older. As the birthrate falls, those aging parents become a larger percentage of the population. By the year 2020 the number of Chinese people over 65 years old will reach 167 million, nearly twice what it was 30 years earlier.

In rural areas of China large families are valued because there are more hands to do the work.

Human rights groups oppose the one-child policy because it is so severely enforced. Recently, China has tried to encourage smaller families by offering improved health services, guaranteed security for the elderly, and family-planning services. Even with these changes, other efforts will be needed to protect the environment. The fate of China's environment depends on the pace of its population growth and how that growth will be controlled.

How the Environment Limits Population Growth

Many factors affect population growth. Cattle ranchers experience a decline in the growth of their herds when cows and calves are overcrowded. Factors such as stress and competition for food become more intense in a high-density population. Factors that affect population growth and depend on population density are called **density-dependent factors** (Figure 22.7). These include parasites and predators, which take a greater toll when their host or prey populations are crowded. For instance, insects or pathogens can devastate crops in regions that are devoted almost entirely to one crop, such as wheat or corn. Human diseases such as tuberculosis often become epidemics in crowded cities.

Not all influences on population growth are density-dependent. Weather is the most common example of a factor that acts without regard to the size of a population. Such a factor is called a **density-independent factor.** The growth and reproduction of insects, for example, are especially sensitive to temperature. In the northern half of the United States, many insect species live in the soil in winter. Farmers and other growers expect more insect attacks on their crops than usual after a mild winter that allows many of these animals to survive. In some places, severe events such as wildfire and hurricanes

Figure 22.7: *Limited nesting space is available to the bank swallows that built these nests. Nesting and egg-laying are limited by population density.*

Figure 22.8: *Density-independent effects on plant and animal populations. The cycle of fire and regrowth in this forest has an important effect on the populations of many plant and animal species.*

are common enough to affect population growth of various plant species and the animals that depend on them (Figure 22.8).

INVESTIGATION 22B   MAKING A CHANGE

See Investigation 22B in TRB.

You've learned that the environment affects the growth of populations. How do growing populations affect their environments? They may consume resources, provide food or other resources for other species, and affect the chemistry of the soil and water. People contribute to the effects by introducing new populations and promoting their growth. These populations may include humans, plants and animals that are grown for food, ornamental flowering plants and foreign grasses, and pets. In this investigation you'll examine how these populations and the way we cultivate or care for them affect the environment for other species.

EFFECTS OF POPULATION GROWTH

Population growth often has unpredictable effects on the environment and on members of the population. When humans introduce new species of crops or domesticated animals into an

Figure 22.9: *One effect of a growing population. When humans bring foreign plants into an area, such as the kudzu shown here, native populations can shrink dramatically.*

area, these nonnative species, called **exotics,** often escape and grow wild. They may compete with or prey on native species so aggressively that they change the environment or drive local species to extinction. Kudzu, for example, is an Asian vine that was brought to the southeastern U.S. as a feed crop for cattle and to control erosion (Figure 22.9). It soon became a noxious weed, interfering with crops and crowding out or killing native vegetation.

Other effects of population growth include the depletion of environmental resources and the accumulation of toxic wastes. In many lakes and streams the population of algae is limited by the supply of phosphorus, a mineral nutrient. Phosphorus can wash into streams from fields and lawns, where it is applied as fertilizer, or from municipal waste water. This increase in nutrients is called **eutrophication.** With nutrients no longer limiting growth, the algae in a stream or lake multiply quickly, forming an algal bloom (Figure 22.10). When the algae die, they release toxins that kill other organisms. Bacteria that break down the dead algae consume so much oxygen that fish and other organisms suffocate. An overgrowth of algae affects populations of many species and can change the entire character of a body of water.

Some ecologists think that human population growth is the biggest environmental problem facing us in the twenty-first century. They cite the following effects, caused or intensified by overpopulation:

- Burning of fossil fuels and improper disposal of waste materials pollute air, soil, and water.

- Farming and construction on unsuitable land erode soil and contribute to water and air pollution.

- Changes in land use decrease species diversity and can upset the balance between predator and prey species. One example of this is the growth of pest populations.

At high population densities, birthrates fall and death rates rise. As a population approaches its carrying capacity, the possibility of a boom-and-bust

Figure 22.10: *The overgrowth, or bloom, of populations of algae has serious consequences for the rest of life in a body of water.*

Figure 22.11: *The challenge of the twenty-first century: how can we control our population and preserve the environment that supports us?*

cycle increases. Scientists who study human population growth rates worry that the world human population will soon reach carrying capacity and may experience a crash as starvation and infectious diseases cause a rapid rise in our death rate. We are continuing to raise our carrying capacity through advances in medicine, improvements in housing, and better distribution of food. Many regions, however, especially the developing areas of Asia, still experience shortages of food, water, and shelter and are overwhelmed with pollution and waste disposal problems (Figure 22.11). High population densities in urban areas and migrant camps have already increased the spread of infectious diseases and the stresses of overcrowding.

CHANGING IDEAS

FAMILY SIZE AND FAMILY PLANNING

When the European countries and the United States were agricultural societies, large families were important. Children were inexpensive workers on the family farm and could boost its productivity. However, disease, harsh working conditions, and poor sanitation kept death rates high and families small. As these countries became more industrial, advances in medicine and sanitation reduced death rates. Families grew as more children survived and fewer women died in childbirth. But as farm tasks were mechanized and more people moved to cities, big families weren't as important and became a burden to some.

In the early 1800s, Francis Place, a labor leader in England, began advocating family planning. He predicted that working conditions would improve if there weren't so many workers competing for jobs. Others wrote pamphlets urging the use of birth control to promote economic and physical health and describing various contraceptive methods. These pamphlets circulated in England and the United States until the 1870s. The U.S. Congress passed the Comstock Law in 1873 forbidding the mailing of these materials, which they called obscene information. Many states also passed laws forbidding distribution of birth control literature.

In 1912, Margaret Sanger, an American nurse, began to publish articles on contraception. She was prosecuted for her writings, but the charges were dropped. In 1916 she opened the first birth control clinic in Brooklyn. She was arrested, but as a result of her case, laws were changed to permit doctors in New York to prescribe birth control for health reasons. Other states followed, and the last of the Comstock laws was repealed in 1966.

In the twentieth century, birth control has gradually become widely accepted. Medical opposition has disappeared, and few organized religious groups still object to the use of contraceptives by married couples. Still, individual decisions about using these technologies will always depend on people's values and beliefs about what is important in human life and relationships.

LESSON 22.2 ASSESSMENT

1. Possible answers are infectious disease, food shortages, competition for resources, and stress-related illnesses such as heart disease.

2. High birthrates provide plenty of workers and customers.

3. No, because it is the desire of the people living in the household to get rid of all of the pests.

4. Expanding human populations spread into and destroy wild habitats and consume or pollute natural resources on which other species depend.

1 Give an example of a density-dependent factor that affects human population growth.

2 Discuss why a manufacturer in a developed country might be interested in producing and selling its products in a country where birthrates have traditionally been high.

3 Birthrates of many species fall as population density nears its carrying capacity. Could this observation lead to a practical way to control household pests such as roaches and ants? Why?

4 How is human population growth related to the problems of endangered species?

CAREER APPLICATIONS

The applications that follow are like the ones you will encounter in many workplaces. Use the biology you learned in this chapter to complete the activities. Share your work with the class.

AGRICULTURE & AGRIBUSINESS

Predator Release

Using recent issues of a farm journal or publications from the agricultural extension service, find out about predator release and other biological pest control programs that are currently in use in your area. For each program, diagram the food chains that include the pest, the released predator, or both. Present your diagram to the class.

Calculating Carrying Capacity on the Range

Contact an agricultural extension agent or a rancher, or use library or Internet resources to find out how many cattle can be supported on a given piece of pasture or range land. What factors must be taken into account in the calculation? What must a rancher do to support a herd that is too large for the land? What problems will the rancher face in the short term and in the long term if he or she consistently keeps a herd that is too large for the existing land? Write up a short report and share it with the class.

Fish and Game Limits

Find out who sets hunting and fishing seasons and limits in your state. Contact the responsible agency and find out how it calculates the limits. What methods are used to determine the population of a given game species? What other factors are involved in the calculations? Create a brochure that could be given to hunters that explains why and how bag limits are set.

Broadcast Demographics

Divide the class into groups. Each group should select a different radio or TV station. Find out what the daily and weekend programming is. Contact the marketing or sales director at each station. How does the station determine its market share?

BUSINESS & MARKETING

What population does it target with its programming at different times of day? When the station approaches a business to get it to buy advertising, how does the station determine the best time to run the ads? Have each group present a profile of the audience for the station they investigated. Using the yellow pages, select five businesses at random. If you don't know what product or service the business offers, find out. Then determine which station you think would be the best one to carry advertising for that business.

Demographics

Using the library or the Internet, find out what demographers study, how they get their information, and how their studies help us to understand current issues and make public policy decisions. Find out how demographic information was used to solve a local problem. Report your findings to the class. Look up census data for your city or Zip Code and create an age pyramid of the people living in the area. Compare it to the previous census and try to determine whether the population is growing, declining, or remaining stable.

City Planning

Invite an urban designer, planner, or architect to visit your class. Ask the following questions: What role do population issues play in architecture, urban design, and urban planning? How are special populations such as children, handicapped people, or elderly people considered? How do planners make allowances for future population growth? In what way is the ecology of the city considered?

Population Control in Your Community

FAMILY & CONSUMER SCIENCE

Contact a family-planning specialist in your community. This person could be a social worker, a nurse, or a family-planning administrator. Find out why this person considers family planning important in your community. Ask what methods of distributing family-planning information are considered most effective. Why do some people oppose family planning? What are the greatest obstacles to family planning, according to the specialist? Write a report and present your findings to the class.

Problem: Unwanted Pets

Talk to a local animal warden or animal shelter operator or contact a national organization such as the American Humane Society. Try to find local statistics for the numbers of unwanted animals that are picked up or destroyed each year. What are some of the problems that stray animals cause? Develop an advertising campaign to spread the word about how people can help prevent the increase in stray animal populations, such as spaying and neutering their pets, finding homes for unwanted animals, and enforcing leash and licensing laws. If possible, work with a local animal shelter and/or local veterinarians to develop a special event to raise funds to educate the public about the problem of stray animals and to set up a low-cost neutering program in your area.

HEALTH CAREERS

Infant Mortality Rates in Your County

Contact your local or regional health department and find out the statistics on infant mortality in your county. Compare them with the national statistics. Are they higher or lower? With the public health official who gives you the statistics, discuss the factors that may contribute to infant mortality where you live. Write a report on your conclusions in your logbook.

Immunizations

Talk to a public health worker about the impact of childhood immunizations on mortality rates. Research mortality rates in a less developed country where immunizations are not readily available. You can obtain this information from Internet sites such as the World Health Organization. Compare life expectancies and age pyramids using information from the Central Intelligence Agency's (CIA) on-line World Factbook. Besides immunizations, what health-related issues might account for differences in the age structures and mortality rates between the United States and the country you chose to research?

The Black Death and Other Plagues

Research the cycles of the Black Death in Europe during the Middle Ages. What was the relationship between human population increases and rat population increases? Prepare a graph illustrating this relationship. In addition, research a more modern epidemic such as outbreaks of hantavirus. Compare human populations to populations of the host organism. (In the case of hantavirus, rats are also hosts.)

INDUSTRIAL TECHNOLOGY

Responding to Population Size

As a class, make a list of nonretail industries in your community whose workforce size is very closely related to the size of the community's population. (Examples would be government services, the construction industry, and utilities.) Consider the technical problems that both rapid population growth and sudden population decline would pose for these industries.

Technologies for a More Crowded World

Much of the pollution and other environmental damage due to human population growth is related to burning fuel to produce electricity and other forms of energy. Investigate new technologies that are designed to reduce the use of electric power or fossil fuels, such as solar water heaters, energy-efficient houses, and energy-efficient appliances. You might start by investigating the Environmental Protection Agency's Energy Star Program.

CHAPTER 22 SUMMARY

- Under ideal conditions, populations of many species can grow indefinitely at an accelerating rate.

- Population growth rates depend on birthrate, death rate, and migration.

- Population structure and the length of generations strongly affect birthrate and death rate.

- As a population approaches its environment's carrying capacity, the birthrate falls and the death rate rises.

- Rapidly growing populations may exceed carrying capacity and then fall dramatically in a boom-and-bust cycle.

- Populations of some species cycle around the carrying capacity for reasons that are not fully known. This cycle can make populations of their predators follow similar cycles.

- Density-dependent factors such as competition for scarce resources depress population growth more at higher population densities.

- Density-independent factors such as unfavorable weather conditions depress population growth by reducing the birthrate or raising the death rate by the same proportion, regardless of the population density.

- Growing populations alter the environment by consuming resources and competing with other species. The effects of a growing population on a biological community are complex and often unpredictable.

- Large and growing human populations have reduced the earth's carrying capacity for many species through air and water pollution, soil erosion, and disruptive land use practices.

- Overcrowding and competition for resources as the human population approaches carrying capacity create undesirable and unhealthy living conditions for millions of people. A disastrous population crash is also possible.

CHAPTER 22 ASSESSMENT

Concept Review

See TRB, Chapter 22 for answers.

1. Compare and contrast the density and distribution of an urban population with that of a farming community.

2. Decide which has greater biotic potential—bacteria or alligators—and give a reason for your answer.

3. In the United States, more people choose to have children later in life than was common in the past. How does this affect a population's biotic potential and growth rate?

4. Give an example of a change in a species' carrying capacity.

5. Are rabbits or elephants more likely to experience a boom-and-bust cycle? Give a reason for your answer.

6 What is most responsible for changes in the population size of the United States in the past several decades: changes in birthrates, death rates, or migration rates? Give a reason for your answer.

7 How does population distribution affect methods of census taking?

8 Predict when migration might have more effect than birthrate or death rate on population.

9 Identify one or more species whose population decline is related to an increase in human population.

10 The population density on a small Pacific island has fluctuated greatly in the past 200 years because of frequent eruptions of a large volcano. Is the population on the island being controlled in a density-dependent or density-independent fashion? Give a reason for your answer.

11 Give an example of a species whose population has increased because of an increase in human population.

12 A survivorship curve for a particular population is a horizontal line. What does that mean for an individual in the population?

Think and Discuss

13 A population's birthrate is higher than its death rate, but the population is decreasing in size. Offer an explanation for this and identify the type of population that might experience this effect.

14 How could a fall in human infant mortality lead to a fall in a population's birthrate?

15 Accidental release of plant nutrients into the soil or water can cause populations of a few species to rise and populations of many others to fall. How can you explain this effect?

16 A population in a developing country is growing very rapidly. Decide whether this growth will have more of an effect on birthrates or death rates as the population approaches its carrying capacity. Give a reason for your answer.

In the Workplace

17 How could a historian use old newspapers or gravestones to study changes in population structure over a period of years?

18 A ranger at a national park finds that the age pyramid for a certain species in the park is widest at the top and tapers to the base. What is probably going to happen to this population?

19 How does today's huge and growing human population affect what society asks of farmers?

20 Use what you learned in this chapter to explain why the governments of many countries restrict importation of plants and animals.

Investigations

The following questions are based on Investigation 22A:

21 Is it possible to estimate a city's population by counting the population of only a few randomly selected city blocks?

22 Describe possible important sources of error in the estimate in Question 21.

The following questions are based on Investigation 22B:

23 In general, is the number of species in an area greater when it is wild or when it is farmed?

24 How could a fish farm increase the population of nonfarmed species?

25 A wildlife manager uses a trap to catch raccoons without injuring them. The trap catches 20 raccoons, which the manager marks with identification tags and releases. A week later, the manager puts more traps out in the same area and catches 50 raccoons. Ten of them have tags. Estimate the raccoon population in the area and explain how you arrived at this estimate.

CHAPTER 23

WHY SHOULD I LEARN THIS?

If you could move anywhere in the world, where would you choose to live, work, and play? Before you decide, here is the catch: Your criterion is not where you can find the best job, eat the best food, or catch the perfect wave. It is where you, a human, can fit best into the natural setting. How could you even begin to figure that out? And what is the point, anyway? The point is that humans change the world—mostly but not always in positive ways. Minimizing the negative changes is a worthy goal.

COMMUNITIES AND ECOSYSTEMS

WHAT WILL I LEARN?

1. how species that are adapted to each other and their physical environment form communities
2. why plants are more abundant than animals
3. how ecosystems change over time
4. how people solve problems in managing ecosystems such as farms, lawns, and gardens
5. how common biological communities depend on nonliving factors such as climate and soil type

---

Your neighbor's backyard is an eyesore, filled with weeds in various stages of growth and decay, tree limbs that have fallen and sprouted mushrooms, and a stack of firewood that hasn't been moved in several years. It has patches of moss and lichen instead of a lawn. You, on the other hand, keep your lawn mowed, flower garden trimmed and never seedy, leaves raked up. It's a constant fight to eradicate the weeds that invade from your neighbor's yard! So why are all the goldfinches next door, the chipmunks filling the other yard with motion and interest, the butterflies flitting about the weeds? Why do you feel so lonely? You'll find out in this chapter.

LESSON 23.1 BIOLOGICAL COMMUNITIES

You're walking outdoors after a heavy rain. You hear a slow cracking sound, and a huge tree limb drops to the ground, just missing you. Think of the complex biology that nearly killed you. The tree was probably weakened by a fungus, so when the limb was heavy with water, it dropped. The fungus might have arrived on the feet of a woodpecker searching for a tasty wood-boring insect. Now the dead branch will become food for decomposers. In this lesson you'll look at biological communities and the relationships among species.

ACTIVITY 23-1   WHICH WAY IS UP?

The description of a parasite and hosts and the instructions for this activity can be found in Chapter 23 of the TRB.

Some animals eat other animals that eat plants. So even carnivores depend on plants. In this activity you will analyze relations among three species and how they indirectly affect other species.

Your teacher will provide you with the instructions for this activity.

BIOLOGY IN CONTEXT

ANCHOVIES, THE ECONOMY, AND BIOLOGY

Most of the time, winds from the coast of Peru push warm surface water out to sea (Figure 23.1). Cold, dense water from below rises to replace it, a process known as upwelling. This deep seawater is rich in nutrients that stimulate the growth of algae and cyanobacteria. These organisms provide food for fish, especially for anchovies, and make the Peruvian coastal waters biologically very productive.

Every four to seven years, however, a phenomenon known as El Niño (Spanish for "the child") occurs. Coastal winds blow less hard, and warm water moves in. Less upwelling means less nutrients. Algae and anchovies become scarce. Pelicans and other seabirds that eat anchovies begin to starve.

In the past the fish population had time to recover between El Niño years. But anchovy fishers increased their catch more than

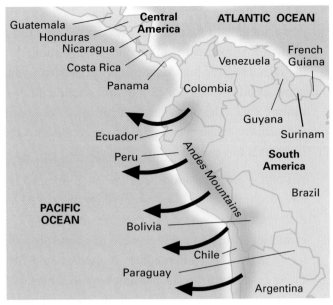

Guatemala
Honduras
Nicaragua
Costa Rica
Panama

Central America

ATLANTIC OCEAN

Venezuela

French Guiana

Colombia

Guyana

Surinam

Ecuador

Peru

South America

PACIFIC OCEAN

Andes Mountains

Brazil

Bolivia

Chile

Paraguay

Argentina

Figure 23.1: *Offshore winds blow warm surface water out into the Pacific. It is replaced by cold, nutrient-rich, deep seawater.*

200-fold between 1955 and the early 1970s. The combination of overfishing and El Niño ruined the anchovy fishery.

Peru lost another major export product as well: guano fertilizer, made from seabirds' droppings. Since the early 1970s, harvest limits for fish and guano have allowed these resources to make a comeback. Fishers have also learned to switch over to fish such as sardines that are attracted to the warmer water during El Niño years. While harvests are smaller than they once were, another population collapse is less likely in the future.

ORGANISMS IN COMMUNITIES

Figure 23.2: *Restoring the environment. These conservation workers are planting tree seedlings to protect against slope erosion.*

Conservation workers who try to reintroduce an endangered species often find that they can't simply truck in plants or animals and plant or release them. They have to restore the whole environment (Figure 23.2). Species survive because they are adapted to use the resources in a specific kind of environment. An organism's environment includes the soil, water, climate, and other organisms in the area. All the interacting species in an area make up a **community.** Together, a community and its nonliving environment—soil, water, and climate—are an **ecosystem.**

Why does one species need so many others? An endangered bird, for example, needs more than just one or two food species. It might need a tree species that provides sites and materials for nests and predators that feed on the bird's parasites.

Other predators help to keep the birds' population healthy by removing less-fit individuals. Alternative prey and nesting sites for similar bird species help to reduce competition. A community includes predator-prey relationships, parasitism, mutualism, commensalism, and competition for food, light, water, or choice nesting sites. The many relationships in a community contribute to the stability of the ecosystem. Some ecosystems have existed for thousands of years with little change.

All of a species' relationships with its living and nonliving environment add up to a "job description" called a **niche.** A niche is a particular way of using resources in the ecosystem. For example, an animal could survive in a forest by eating fruit and living in trees to avoid predators on the ground. This niche is filled by birds in some forests and monkeys in others. A full description of a niche includes a species' needs for water, temperature, space, symbiotic relationships, and other resources, so fruit-eating monkeys and birds fill niches that are similar in some ways, but not in others.

Species also adapt to one another. Tropical juice drinks often include the juice of passionfruit. The fruit grows on a vine (Figure 23.3) whose leaves and stems produce substances that are toxic to many plant-eating insects. *Heliconius* butterflies lay eggs on passionflower leaves, and the larvae eat the leaves and digest the toxins. To defend against *Heliconius* larvae, the vines have evolved small yellow bumps on their leaves that look like *Heliconius* eggs. The butterflies don't lay their eggs on leaves that already have eggs, so the bumps are effective "scarecrows." The leaves also produce nectar that attracts ants and wasps that prey on *Heliconius.* The relationship between the passionflower vine and *Heliconius* is an example of **coevolution,** the evolution of two

Figure 23.3: *Passionflower vines have both predators and mutualistic partners among the insects.*

📖 dictionary
WORD BANK
co- = together
evolution = gradual change
Coevolution lets species that compete or cooperate adapt to each other over time.

species in response to one another. Coevolution may occur between predator and prey or in a mutualistic or competitive relationship.

COMMUNITY STRUCTURE

Two important factors to consider when you are trying to understand a community are its species diversity and its trophic structure. **Species diversity** refers to the number of species in the community and how common each species is. In general, species with identical niches can't exist in the same community. They compete so intensely that the fitter species eventually replaces the other. Differences in the ways two similar species use the environment can allow them to coexist. For example, birds that eat the seeds of the same tree species might feed in different parts of the trees. Variation in the environment also allows similar species to coexist if their needs are different. One seaweed species may dominate on rocks that are rarely exposed by the tide, while another may compete better by surviving daily dry spells during low tide exposure.

WORD BANK

troph- = nutrition

Trophic structure and trophic levels describe how a community's species nourish one another.

The **trophic structure** refers to the transfer of energy and nutrients from one species to another. For example, animals eat plants and other photosynthesizers, predators eat animals, and parasites and decomposers attack all types of other organisms. Communities usually have several levels within a food chain called **trophic levels,** including producers (autotrophs), plant eaters, and predators. Lesson 23.2 discusses energy transfer through these levels in more detail.

INVESTIGATION 23A   WHERE DO THEY GO?

See Investigation 23A in the TRB.

If you were trying to restore a damaged environment to help endangered species survive, would you have to be sure that every species from the original community is there? Or would you have to worry about just a few species? This practical question has deep roots. Ecologists have often debated whether communities are groups of species that occur only together or whether each species tolerates a different range of conditions that roughly matches those of other species in overlapping areas. In this investigation you'll examine the ranges of several species and find out whether or not they always occur together.

CHANGES IN ECOSYSTEMS

The TRB has a blackline master showing succession.

Change is inevitable in all ecosystems. A lawn is a community in which people have tried to prevent change by removing unwanted species. But as anyone who maintains a lawn can tell you, it would quickly revert to a mixture of taller species if people stopped cutting, weeding, and fertilizing it. One biological community tends to replace another in a process called **succession,** a sequence of changes within a community over time.

You can see succession in an abandoned lot. At first, grasses and other weeds appear. Mice, birds, and insects that feed on these plants soon follow. The roots of grasses and other weeds grow into cracks in pavement and begin to break it up. These early arrivals that can live in a fairly lifeless environment are called **pioneer species.** Later, taller grasses and weeds take root, followed by shrubs. Trees that need shade as seedlings begin to grow beneath the pioneer plants and eventually overshadow and replace them. Predators such as snakes, foxes, and hawks begin to feed in the area. If nothing interferes with the process of succession, a community will eventually appear that stays pretty much the same for many years. Whether that community will be a pine forest, a grassy prairie, or something else depends on environmental factors such as temperature, moisture, and soil type. The typical stable community for a particular area is sometimes called a **climax community,** although some change is always going on.

LEARNING LINK   SUCCESSION IN HISTORY

People both cause and are affected by changes in ecosystems. Investigate some historical changes through the resources of your library or the Internet. You might concentrate on archeological evidence that the Sahara desert was more hospitable in the past, how Native American cultures that cut down trees and burned the grass of the Midwestern prairies expanded the range of the bison they hunted, or how some ancient farming practices eroded the soils of the Mediterranean countries that were once forested.

INVESTIGATION 23B OBSERVATIONS OF AN ECOSYSTEM

See Investigation 23B in the TRB.

Most people don't have the chance to observe succession or interactions between species as closely as forestry workers or wildlife specialists do. In this investigation you will take a closer look at these things and try to explain what you see.

LESSON 23.1 ASSESSMENT

1. Fewer trophic levels would be better for the owner. There would be fewer species preying on the trees, so the fruit yield would be higher.

2. The uneven lawn would have more weeds because it provides habitats for more plant species.

3. Accept reasonable answers that show understanding of coevolution. Since the most severe cases kill the patient, who then cannot transmit the disease, natural selection would favor both human alleles that provide some protection and bacterial strains that did not rapidly kill their hosts.

4. Pioneer species are involved in the succession process that would turn homes and farms back into natural communities. People usually try to prevent this process.

1 Would an orchard owner prefer the fruit trees to be part of a community with many or few trophic levels? Give a reason for your answer.

2 Are there likely to be more weeds in a flat, level lawn or in one with many rocks and high and low spots? Give a reason for your answer.

3 Some human diseases caused by bacteria (such as syphilis) are thought to have been more severe in ancient times than they are today. Explain how coevolution could have caused this.

4 Many pest species, such as mice, dandelions, and other weeds, act as pioneer species in nature. Why are they often pest species for humans?

LESSON 23.2 NATURAL ECOSYSTEMS

A maple tree grows, absorbing nutrients from the soil. A mouse eats some of the tree's seeds. Perhaps a wolf will eat the mouse. Worms, fungi, and bacteria decompose the tree's dead leaves and the mouse's wastes. These organisms are part of an ecosystem. Ecosystems exist in your neighborhood, forests, the ocean, and even your aquarium. The whole earth can be seen as a single ecosystem. In this lesson you will learn more about how natural ecosystems work.

ACTIVITY 23-2 TROPHIC PYRAMIDS

Students will need rulers for this activity. You might want to provide calculators and large (8" × 14" or 11" × 17") sheets of unlined paper. The complete instructions for this activity, including a Math Tip, can be found in Chapter 23 of the TRB.

Which is easier to find in your neighborhood: grass and trees or the animals that eat them? How much vegetated land is needed to support a herd of deer or a bear? Estimates of how much space and food an endangered population needs are important to wildlife managers and to landowners whose property is affected by regulations protecting endangered organisms and the resources they need. In this activity you'll use the data in Table 23.1 to construct a "layer cake" diagram like Figure 23.4 to compare two ways of measuring the resources provided by populations at different trophic levels.

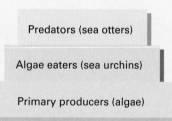

Figure 23.4: *A trophic pyramid is a way to visualize the energy resources of populations at different trophic levels in a community.*

Type of Organism	Population	Total Mass (kilograms)
Edible grasses	600,000,000	6,000,000
Grazers (zebras, gazelles, wildebeests, etc.)	426	55,400
Predator (lion)	1	205

Table 23.1: A Savanna Food Web

A COLD, CLEAN GLASS OF WATER

Every city needs water. In the early 1800s, New York City officials chose the Croton River, which is in the Catskill Mountains. The river provided plenty of clean water and was in an area that was unlikely to become densely settled and polluted.

Since then, residential and industrial development has polluted the Croton's water. Forests, which once absorbed pollutants from groundwater before it reached the river, have been replaced by homes and businesses. The amounts of pesticides, agricultural waste, and sewage in the river have increased. New construction threatens a number of endangered plants and animals.

To control the pollution, New York City, the Environmental Protection Agency (EPA), and the towns near the Croton River have developed a plan to improve the water quality. Under the plan, New York will purchase thousands of acres of forested land surrounding the river and help to pay for environmentally sound economic development near the river's source. The EPA and the towns will monitor the river for pollutants, improve nearby sewage treatment facilities, and find ways to control the pollution of the waters at the river's source.

Remind students that a river's source is its beginnings at the highest elevation or uppermost part.

ENERGY IN THE ECOSYSTEM

Figuring out what is endangering a species and then figuring out what to do about it are often very difficult tasks. Before conservation workers can make recommendations, they have to understand the food webs that include the endangered species. Why is this essential?

The answer has to do with the flow of energy and matter through ecosystems. As in any system, that energy must come from somewhere. The sun fuels most of the Earth's ecosystems. Photosynthesis captures only about 1% of the solar energy reaching the Earth, and almost all life on the Earth depends on that 1%. Through photosynthesis, plants and other autotrophs convert solar energy and materials from the physical environment into food. Animals and other heterotrophs in the ecosystem

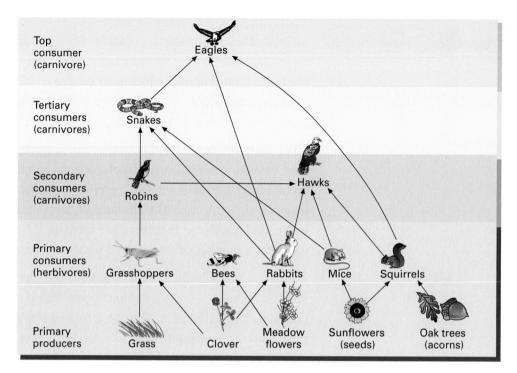

Figure 23.5: *Trophic structure and food web of a meadow ecosystem. Energy flows from producers to top consumers.*

consume the autotrophs and each other, absorbing energy from the food (Figure 23.5). The food made by plants and other producers fuels entire ecosystems. The energy flows in one direction: from producers to consumers.

Tourists and hikers who visit a protected forest may hope to see a moose or a few deer. They expect to see thousands of trees and other plants. But they almost never meet predators such as wolves and wildcats. Why are plants so common and predators so rare? As Figure 23.5 shows, an ecosystem usually includes many more producers than consumers. Herbivores (plant-eating animals) are **primary consumers.** Some of the energy they get from plants is transferred in turn to carnivores (meat eaters) that prey on them. Carnivores, such as robins and hawks that feed on herbivores, are **secondary consumers.** Other carnivores, such as owls or lions, devour the secondary consumers. These third-level predators are **tertiary consumers.** There may be further levels of consumers, but food webs rarely have more than six trophic levels.

Biomass, the total dried weight of the organisms in a population or community, provides a better measurement of the size of each trophic level than population does. In Activity 23-2 you saw that the higher a species' trophic level is, the smaller its population's biomass is. Many prey support each predator. This ratio is why trees are easy to find, deer somewhat harder, and predators of the deer hardest of all.

Natural ecosystems are open. Energy and matter constantly escape from the ecosystem, and the system can't go on without sunlight or some other energy source and a supply of materials such as oxygen and minerals. Most of the energy that is originally captured by producers is eventually lost to the environment as heat.

Alternative Investigation 23C, Niches in a Complex Habitat, can be found in the Laboratory Manual and Chapter 23 of the TRB.

LESSON 23.2 ASSESSMENT

1. The top consumer (or top predator) species should be the last introduced. The top consumer depends on all the species below it.

2. Top predators help to control the populations of other animals, such as deer, and keep them from consuming so many plants that the whole ecosystem collapses.

3. No, energy is lost to the environment at every trophic level. An outside energy input such as sunlight is always needed.

4. Humans are the top consumers in an agricultural ecosystem.

1 In restoring a disturbed ecosystem, which trophic level should conservation workers introduce last?

2 A forest ranger notices that since wolves were wiped out and hunting was banned in the forest, many plant species have become less common and the deer look thin and unhealthy. Explain how predators such as wolves can help to maintain the health of a community.

3 Nutrients are passed from producers to consumers and back through decomposers to producers again. Can an isolated ecosystem maintain itself without an outside energy source? Give a reason for your answer.

4 What is the top consumer species in an agricultural ecosystem?

Lesson 23.3 Managed Ecosystems

Every organism affects the ecosystem in which it lives. Humans are more influential than most species—can you even think of another contender for Most Manipulative Species on the Earth? Your town or city is an ecosystem planned and controlled by humans—a managed ecosystem. Managed ecosystems don't usually work like natural ecosystems, as you'll see in this lesson.

Activity 23-3   How Big Is a Farm?

The procedures for this activity can be found in Chapter 23 of the TRB.

Like natural ecosystems, managed ecosystems interact with the outside world. However, the borders of a managed system are less clear. For example, should farmers and aquarium workers be considered part of the ecosystems they manage? In this activity you will examine a managed ecosystem and decide where its borders are.

CHANGING IDEAS

Fire Ecology

Before Europeans settled in North America, wildfires were more frequent than they are now. Many began from lightning strikes and cooking fires. Native Americans on the Great Plains burned the prairie to remove trees and extend the grazing range of the bison they hunted. European settlers tried to prevent and control fires to protect their property. The federal government began to help fight wildfires in the 1930s. Between then and 1987, the area that was burned decreased from over 50 million acres a year to roughly 2.1 million acres a year.

After years of fire suppression, however, dead wood and leaves accumulated on forest floors, tying up nutrients and leading to sometimes catastrophic wildfires. Unlike earlier fires, which passed quickly through an area, the intense heat from accumulated fuel burns soils and kills seeds, roots, and burrowing animals. Ironically, the fire suppression practices that were meant to protect us have actually increased the hazards from fire.

Since the Yellowstone fires of 1988, the managers of our national parks and national forests have started to set more controlled fires.

Figure 23.6: *Fire releases nutrients to the soil and allows sunlight to reach young plants that nourish wildlife after a forest fire.*

Tree sap from jack pine trees seals shut jack pine cones. Only the intense heat of fire will cause the sap to boil. The boiling sap causes the jack pine cones to pop open, spreading their seeds on the enriched soil.

Periodic fire improves the health of forest and grassland communities and reduces dangerous levels of flammable materials. Some species, such as the jack pine, can't reproduce without fire. Wildlife habitat improves when predator cover shrinks, sunlight can reach young seedlings, and animals can feed on the tender new plant growth that follows a fire (Figure 23.6).

MANAGING AN ECOSYSTEM

Farmers, aquarium managers, and other people who manage artificial ecosystems often have to reinvent the natural ecosystem. Park rangers, for example, try to maintain a healthy mix of producers (plants or algae), primary consumers (elk or moose), and predators (turtles, wolves, or fish) that occupy higher trophic levels.

Often, managers want to maximize the growth of a single species (Figure 23.7). A forester wants to grow wood as quickly as possible and might not worry much about consumers such as birds and spiders. A dairy farmer might raise plants to feed to cattle, caring about plants only to the extent that they contribute to milk production. Other animals that live on grass or cattle are competitors to be controlled or eliminated.

By changing ecosystems to serve human interests, managers create a conflict: They want to maximize the production of one species or natural product. But the more they bend the ecosystem to this goal, the more they undermine it. The Midwestern grain belt has thousands of square

Figure 23.7: *Dairy farmers maximize the primary productivity of dairy cows.*

A   **B**

Figure 23.8: *Regional monocultures such as the American corn and wheat belts (A) can be suddenly wiped out by parasites or plant-eating insects (B).*

miles devoted to a few species (Figure 23.8). These **monocultures** (large areas of a single species) can be quickly destroyed by parasites such as fungi, insects, or bacteria. In naturally diverse ecosystems the parasites' host species are a smaller part of the whole community, so the damage is less severe. Disasters such as the Irish potato famine that you read about in Chapter 6 are rare in natural ecosystems.

One approach to ecosystem management that tries to maintain production while sustaining the system is based on renewability. Dairy farmers have always tried to renew the productivity of their fields by spreading them with the cows' manure. The nutrients that the cattle don't absorb from the plants they eat return to the soil, where they nourish the next year's crop of hay. Solid-waste managers have extended the renewable concept to urban consumers and waterways that receive sewage waste. Applying city sewage sludge to farm fields may return nutrients from the managed food web to the soil while reducing pollution and eutrophication of natural waters.

URBAN ECOSYSTEMS

See Chapter 23 of the TRB for an additional activity, Is a City an Ecosystem?, which can be used here.

Humans can live just about anywhere on the Earth because we change our environment to make it hospitable for us. Our creations are often unbalanced artificial ecosystems, such as cities. About half the world's people live in cities, and that percentage is increasing.

Cities might seem self-sufficient, but they aren't. Most modern cities use resources rather inefficiently. Inputs such as food, clean air and water, and energy come from outside the city, and wastes pass out of the system in sewage and garbage. The decomposers that break down these materials are located in sewage treatment plants, natural waterways, and landfills outside the city, so they do not return nutrients to the system. As cities grow, they consume more water and other resources and produce more waste, putting stress on surrounding and distant ecosystems. City and regional planners are working to balance the immediate needs of city dwellers with the long-term health of the outside ecosystems that support them.

DOES AN URBAN ECOSYSTEM HAVE A FOOD WEB?

People in cities produce little of the food they consume. They are consumers in the food web of agricultural ecosystems outside the urban system. Most plant life in the city and suburbs is found in parks, yards, and narrow rows along streets and highways (Figure 23.9). However, few of these ornamental plants are part of a food web with human consumers. Indeed, landscapers and groundskeepers often try to prevent any species from feeding on them. Insects, birds, rodents, and fish are some primary consumers in the urban food web. Carnivores such as owls, bats, and snakes are among the secondary consumers that feed on them. Food and suitable living areas are often insufficient to support many higher-level predators such as hawks and insect-eating bats and frogs.

Figure 23.9: *Urban plants. What food web is based on these producers?*

Pets are occasionally consumers in the urban food web, for instance, when cats eat mice or birds.

However, most pets eat food from agricultural ecosystems outside the urban environment. Domestic animals, including pets, eat up about 4% of the average family food budget.

Parasites are part of all ecosystems, including cities. In fact, the crowded conditions of cities encourage the spread of parasites such as worms, protists, viruses, and bacteria that infect humans, their pets, and wild urban organisms such as rats. Especially in less developed countries where city water and sewer systems are inadequate, diseases caused by parasites can spread rapidly via contaminated water. Excessive use of antibiotics and rapid disease transmission in cities have contributed to the development and spread of resistant strains of disease-causing bacteria. Public health officers and health care personnel fight hard against disease in densely populated urban neighborhoods.

SOLID-WASTE DISPATCHER

Louis P. has worked for a local waste disposal company for eight years, first as a driver and now as a dispatcher. He directs a fleet of trash collection trucks that serve the city.

Louis's company operates the local landfill. "A lot of people think a landfill has to stink," Louis says, "but it really doesn't. As soon as you dump a load, the compactor compacts it and the scraper covers it with dirt." He explains that the landfill incorporates the latest technology to control groundwater contamination and methane, a gas that is produced during the decay of organic matter.

Louis talks about the problems that are caused when people improperly dispose of hazardous waste. "Sometimes mixing different chemicals will cause fires right in the truck, and you have to dump the load immediately. I've dumped two loads in the street, and once we even dumped a load in front of the fire station!"

When asked about the life expectancy of the landfill, he says, "When I came to work here, it was supposed to last 30 years. It's filling at a faster rate than that, though, because some of the nearby communities are bringing their waste when their landfills close or are filled to capacity." Louis explains that the landfill is closely monitored by the state health department and the air control board. The monitoring is intended to ensure that no one dumps liquid waste or hazardous industrial materials.

LESSON 23.3 ASSESSMENT

1. The farmer should encourage decomposers because they benefit the farm ecosystem by releasing nutrients that can be absorbed by crops.

2. The river can provide water and receive waste such as sewage from the city. The official should monitor use to make sure not too much water is withdrawn from the river and no untreated sewage is added to it.

3. Returning nutrients to the soil reduces farmers' need for fertilizer and city dwellers' need to find a safe place to dump sludge without damaging the environment.

4. Yes, it is. More diverse communities do not have concentrated populations of one species that are vulnerable to a single pathogen. Diverse plantings might not attract or allow the multiplication of pathogens as much as monocultures.

1 Farmers often try to keep animals such as seed-eating birds and mice out of their fields. Should they also try to get rid of insects and worms that act as decomposers? Explain how these species affect the farm ecosystem.

2 How would a regional planner or environmental protection official consider a river that flows through a city in terms of inputs and outputs to the city ecosystem? What concerns would he or she have about limits to the use of the river?

3 How would farmers and urban consumers benefit from closing the loop in the nutrient cycle by applying safely treated city sewage sludge to farm fields?

4 A landscaper claims that planting many kinds of grasses and shrubs and a flower garden, instead of just a large lawn of grass, ensures that customers will have fewer problems with plant diseases. Do you think this claim is based on good science? Give a reason for your answer.

LESSON 23.4 BIOMES

Could you raise bananas in New York or pigs in Arizona? Why do these sound like really bad ideas? You've got to consider water availability, soil conditions, and the overall climate when deciding where to grow a crop. If you're a farmer, a fertilizer seller, or a produce buyer for a supermarket chain, you have to know where crops grow best while doing the least environmental damage. This knowledge will keep you from making a big mistake, such as planting oranges in North Dakota!

ACTIVITY 23-4   IN THE BEGINNING

The procedure for Activity 23-4 is found in Chapter 23 of the TRB.

What did your neighborhood look like before it had so many people in it? What were its original vegetation and wildlife? Knowing the native species helps you to choose what to plant in lawns and gardens and on farms. It helps landowners to predict the weeds they need to identify. It helps potential buyers to know what to expect on abandoned land. In this activity you will find out what the natural community in your region was and how it depends on the local climate. Your teacher will provide you with the procedure.

LIVING LIKE NOAH

BIOLOGY IN CONTEXT

On September 26, 1991, the Biospherians—eight volunteers consisting of four men and four women—entered a large glass enclosure in the Arizona desert to live in isolation from the rest of the world. The Biospherians were in touch with the outside world only by electronic communications. The structure that they shared with more than 3000 species of plants and animals was named the Biosphere 2 (Figure 23.10). The structure and its inhabitants represented a small-scale version of earth's **biosphere,** the parts of the Earth where organisms live.

In the Biosphere's 3.15 acres, plants and animals were grouped (as they are on earth) into regions called biomes. A **biome** is a community of organisms that is adapted to specific climate and soil conditions. Biosphere 2 included several miniature biomes: a tropical rain forest (growing under a 26-meter ceiling), a grassland, a desert, an ocean (7.6 meters deep), and a marsh with both freshwater and saltwater areas. Biosphere 2 also had more

WORD BANK

bio- = living organism

A biome is a major ecological community.

Figure 23.10: *In 1996 Columbia University took over the management of the Biosphere 2 complex, which is now used for environmental science education and research.*

than half an acre of cropland, which the Biospherians cultivated and harvested for their food.

Before two years were up, the Biospherians had to break the guidelines of their experiment. The oxygen level inside the enclosure steadily declined, until the Biospherians had trouble working. Fresh oxygen had to be pumped into Biosphere 2 from the outside, and excess carbon dioxide had to be removed. Despite their inability to remain completely self-sustaining for two years, the Biospherians collected data that they hope will contribute to our understanding of the Earth's biosphere.

CLIMATE, SOIL, AND COMMUNITIES OF ORGANISMS

Figure 23.11 shows the major **terrestrial** (land) biomes. These biomes are climax communities that developed over long periods of time through the interaction of climate, soils, and the organisms of the region. In many places, human activities such as agriculture, logging, and city building have removed the native communities that are typical of a biome. These activities are often most successful when they replace the climax community with a managed system that is somewhat similar. For example, farmers who replaced the grasslands of the Midwest with corn and wheat (also members of the grass family) created the world's most productive grain-producing region.

Biomes are not continuous. Desert communities, for example, occur in several parts of the world. The species that are found in the deserts of Arizona and North Africa are different, but the two communities look somewhat similar. Biomes describe general types of communities, not groups of particular species. The Earth's major land biomes are desert, grassland, chaparral, rain forest, deciduous forest, coniferous forest, and tundra.

Desert plants are short and have small leaves that conserve water. The plants in rain forests are tall and leafy. This enables them to compete with the other plants in the forest for water and sunlight.

Temperatures fall as you travel away from the equator. Figure 23.12 shows how temperature and precipitation (water such as rain or snow that falls from clouds) affect the distribution of biomes. In a dry desert biome, plants are short and have small leaves that

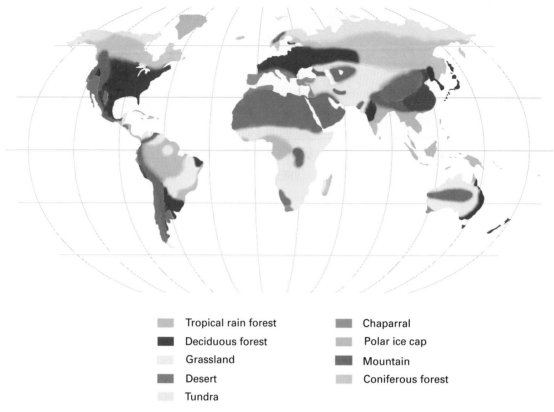

	Tropical rain forest		Chaparral
	Deciduous forest		Polar ice cap
	Grassland		Mountain
	Desert		Coniferous forest
	Tundra		

Figure 23.11: *The major terrestrial biomes of the world*

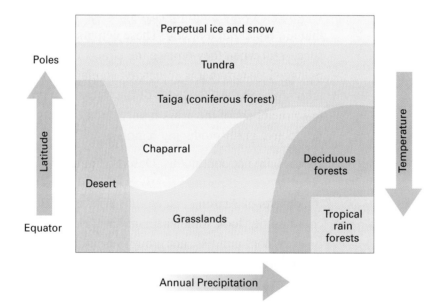

Figure 23.12: *How climate affects biomes*

reduce water loss by evaporation. The ample rainfall in the humid tropics enables trees to grow tall and leafy.

DESERTS

Deserts are many people's idea of useless places where nothing can grow. But the thriving farms of California's Central Valley are a reminder that many desert soils are quite fertile. All they need is water to become productive agricultural regions. Deserts can develop wherever annual rainfall is less than 25 centimeters. Desert organisms have thick coverings that help them to conserve moisture. They also tolerate wide temperature swings. The dry soils cool down quickly at night, just as they heat up quickly during the day. Some plants become dormant during long dry periods, and others complete their entire life cycle quickly after a sudden cloudburst. Common desert plants and animals include sagebrush, cacti, lizards, insects, scorpions, snakes, and birds.

Many Americans who suffer from allergies to the dust, molds, and pollen of moister biomes have moved to the deserts of the Southwest.

GRASSLANDS

Earlier, you read that using the American prairie as grazing land or replacing it with grain crops such as wheat was an ecologically sound practice. Yet plowing up these grasslands to plant grain led to the Dust Bowl of the 1930s, when wind erosion and a long drought caused huge dust storms that buried farms and contributed to economic collapse (Figure 23.13). What went wrong? As Figure 23.12 shows, rainfall is often the main factor that makes the difference between a grassland and a desert. Plowing land that is too dry can destroy a grassland biome, and that is what happened. Grassland biomes are usually in the middle of continents or around deserts.

Figure 23.13: *Dust bowl damage*

The grass family and grazing animals such as camels, zebras, antelope, and cattle are a typical example of coevolution (Figure 23.14). Unlike many plants, grasses tolerate grazing and provide nutritious food for animals adapted to digest them. In the tropical

Figure 23.14: *These savanna grasses and zebras are well adapted to one another.*

grasslands known as savannas, rainfall is seasonal and there is just enough of it to nourish grasses and keep the regions from turning into deserts, but not enough to support many trees. Lightning causes occasional fires that also help to reduce the growth of trees. (Many grasses grow back quickly from their roots after fires.) During a series of dry years beginning in the late 1960s, overgrazing by too many cattle and sheep on the West African savanna contributed to collapse of that community. The result has been starvation for many of the inhabitants.

CHAPARRALS

Figure 23.15: *Typical chaparral country in the mountains overlooking the Owens Valley in California.*

The forest biomes of the world all have tall trees except for the shrubby chaparral or Mediterranean scrub forest biome (Figure 23.15). Chaparral is a transitional zone between dry grasslands or deserts and humid tall-forest biomes. Chaps, the leather leg covering worn by cowboys, are named for the tough, often thorny chaparral vegetation, which can scratch and injure riders' legs. Chaparral covers the foothills of southern California and much of the land bordering the Mediterranean Sea.

TROPICAL RAIN FORESTS

Rain forests are found in the tropics where there is a lot of rain. Early European explorers were impressed with their lush vegetation (Figure 23.16) and took this as a sign of very fertile soil. After cutting down the trees and growing crops for a few years, however, they harvested very little from their fields. Rain forest soils don't hold onto plant nutrients; heavy rains soon wash out these substances when they are not part of living organisms. Decomposers rapidly attack dead materials and plant roots quickly absorb their nutrients before they are washed through the soil into rivers. Interrupting this cycle destroys the ecosystem.

Figure 23.16: *The tropical rain forests may contain over one million species. Most of their nutrients are tied up in living organisms and unavailable to crops.*

Unfortunately, farmers in the rain forests of South America and Africa are still cutting down trees to make room for their crops. Many rain forest trees yield valuable exotic woods such as teak, ebony, and mahogany. Large-scale logging of these species, though profitable for some farmers, is a destructive force in the rain forest biome.

Just as wheat and grazing cattle can sometimes replace natural grasslands, tree crops such as mangos and papayas are often grown in mixtures with other trees in rain forest zones. This mixture produces a product that people can sell while preserving some of the biome's diversity.

Most of the world's species may live in tropical rain forests. A rain forest may have more than 300 species of trees. In contrast, the sort of a forest you might find in Maine or Oregon might have only 10 or 20 species. A rain forest can have 1400 species of birds and uncountable species of moths. Biologists constantly discover new species of plants and animals in the rain forest, often high up in the trees near the sunny top layer, called the canopy.

DECIDUOUS FORESTS

Frequent rains, cold winters, and hot summers are some of the outstanding characteristics of the deciduous forest biome. This biome has moderate temperatures and abundant rainfall evenly spaced throughout the year. The deciduous forest has well-defined seasons, with a growing season that ranges between 139 and 300 days.

A typical deciduous forest consists of a mixture of deciduous trees and some evergreens. The rich vegetation of the forest supports a diversity of animal life. White-tail deer, opossums, rabbits, squirrels, porcupines, snakes, and woodpeckers are just some of the abundant animal life found in the deciduous forest.

CONIFEROUS FORESTS

Coniferous forests are named for the conifers (cone-bearing needle-leaf trees) that dominate this biome. They occur where the climate is too dry or cold for deciduous trees. Because this biome is widespread in northern areas, including much of Siberia and western Canada, it is also known by its Russian name: taiga. Foresters know taiga trees such as pine, spruce, fir, and hemlock as "softwood" species, which are used to make paper and inexpensive furniture and for building construction. Coniferous forests in other cool regions such as the northwestern United States and the Rocky Mountains are also logged extensively. Conifers once grew around the northern Great Lakes before that area was extensively logged.

TUNDRA

When the pipeline that carries oil from Alaska's North Slope was built, the tundra biome presented an unusual problem. Engineers worried that the pipeline would melt the frozen tundra soil. Tundra ecosystems cover 20% of the earth's dry land surface in the Arctic and mountain areas, though most people never see them (Figure 23.17). Only a shallow surface layer of the soil thaws out in the summer, when lichens and shallow-rooted grasses grow and moose and reindeer graze on them. Just below the surface lies a permanently frozen layer up to 500 meters thick, called **permafrost.** This icy barrier prevents drainage of the muddy surface. Tree roots can't grow in the cold, shallow, waterlogged soil. If the Alaska

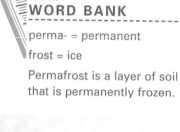

WORD BANK

perma- = permanent

frost = ice

Permafrost is a layer of soil that is permanently frozen.

pipeline had melted the underlying permafrost, it could have disrupted the tundra ecosystem along its path. The pipeline was built on raised supports that protect the soil and permit migrating moose and reindeer to cross underneath.

Above the permafrost, dead plant matter decomposes so slowly that it accumulates and forms peat, partially decayed material. In Canada and other places, peat is dug up from tundra bogs, used for fuel, and shipped to plant nurseries for gardening. Occasionally, someone discovers an undecomposed human body, hundreds of years old, in a peat bog.

Figure 23.17: *Reindeer grazing on lichens in a typical tundra biome*

AQUATIC BIOMES

WORD BANK

zoo- = animal

plankto- = drifting

-on = thing

Zooplankton are microscopic animals and animallike protists that float near the ocean's surface.

Aquatic biomes cover over 70% of the Earth's surface. They include shorelines and the deep ocean, coral reefs, wetlands, streams, and lakes (Figure 23.18A–C). In the ocean and many freshwater biomes, algae and cyanobacteria replace plants as the producers. In fast-moving rivers and streams, living and dead organisms that fall into the water are important, and the food web is based on decomposers. Major primary consumers include a wide variety of protists and microscopic animals known in saltwater systems as **zooplankton.**

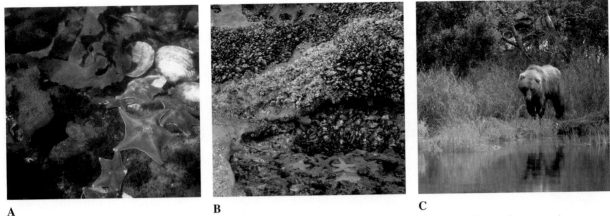

Figure 23.18: *Aquatic biomes are extremely varied. (A) Coral reef community. (B) Mussels, barnacles, and sea stars in an Oregon tide pool. (C) Fish, frogs, insects, water plants, and even bears contribute to a pond community in Alaska.*

Vegetation and animal life depend on a multitude of factors such as water depth, amount of sunlight penetrating the water, currents, temperature, dissolved gases, and minerals.

LESSON 23.4 ASSESSMENT

1. Drier conditions after the dam is built could convert the forest to a grassland or coniferous forest, depending on temperature.

2. No, the savanna is too dry to support a forest.

3. Tree crops make more sense in the rain forest, because they are similar to the native vegetation.

4. No, trees don't grow well in the tundra because they need better drainage and deeper, unfrozen soil.

1 How might a deciduous forest be affected by damming a river upstream from the forest?

2 A grassy meadow is an early stage in the succession that leads to a deciduous forest. If a grassy athletic field in a savanna region is abandoned, will it eventually become forested? Explain why or why not.

3 You are reviewing project proposals from tropical countries for an international agricultural development agency. One group proposes to settle unemployed people on farms in the rain forest that will raise basic food crops such as corn and potatoes. Another wants to have the settlers raise coffee, cocoa, and fruit trees and use money they make from selling their products to buy their own food. Which plan makes more sense? Give a reason for your answer.

4 Figure 23.13 shows that many tundra zones are located next to taiga. Would it make sense to try to expand the lumber industry by planting useful, cold-tolerant trees such as pines in the tundra? Give a reason for your answer.

The applications that follow are like the ones you will encounter in many workplaces. Use the biology you learned in this chapter to complete the activities. Share your work with the class.

AGRICULTURE & AGRIBUSINESS

Monitoring the Tropical Rain Forest

Report on the status of tropical rain forests on different continents. Write to various conservation and scientific organizations to request recent data, or search the Internet. Look for magazine articles and television or radio programs that would be useful for your report. The tropical rain forests in many countries are being burned or clear-cut to make the land available for the production of crops. What effect does this practice have on the local soil? How does this practice affect the global environment? What can be done to correct this situation?

Artificial Reefs

Research artificial reefs. How do they improve the ecosystem? Are they just attractors or are they producers? What policies and legislation are involved in obtaining a reef permit? Report your findings to the class.

Farms as Ecosystems

Talk to a farmer and ask him or her to describe the farm's operation. Consider that person's farm to be an ecosystem. Describe the community of species, their habitats, the nonliving components of the ecosystem, and the disturbances that take place in that ecosystem. Either write your description or use a labeled diagram of the farm to convey your ideas.

Polyculture

Some farmers are experimenting with polyculture, or mixed cropping, to increase the species diversity of their fields and reduce the risk that plant parasites will destroy a crop. Ask your county cooperative extension agent about this practice and how farmers use it in your area. How does it differ from crop rotation? Contact the Land Institute in Salina, Kansas, and find out how its mission relates to native grasses. Report back to your class.

Predator Release and Integrated Pest Management

Using recent issues of a farm journal or publications from the cooperative extension service, find out about biological control programs that are currently used in your area. For each program, diagram the food webs that include the pest and the released predator. Research the practices used in integrated pest management. Present to the class how integrated pest management might be used to reduce pesticide waste and the resulting pollution.

Use It and Lose It?

List the common naturally occurring plant species in your area. Discuss how people and other organisms use each species. Does its use necessarily result in disruption of the ecosystem? Explain how using these species affects your local ecosystem. Use visual aids as appropriate to demonstrate the uses of different species.

Conserving Living Resources

Make a class list of all of the parks, wildlife sanctuaries, grasslands, state and national forests, and other sites in your area that are set aside for recreation and/or conservation of plant and animal life. How many have you visited? Ask a representative from one of these sites to come to the class and discuss environmental management and employment opportunities.

BUSINESS & MARKETING

Ecotourism

Visit or call a travel agency to get information on an ecotour of a tropical rain forest. Have a particular tropical country or region in mind when you call. After you learn the details for a trip, make a marketing plan for it. Consider the following: Would your marketing highlight the chance to experience the rain forest, or would it also emphasize your operation's kindness to the environment? What types of transportation, accommodations, and special services will be provided for ecotourists? How can you reassure ecotourists that they won't adversely affect the environment they have come to see? What kind of scientific information should be provided to tourists before and during their trip? Are any special precautions taken to avoid parasites, predators, and plant toxins in the area they visit or to prevent introduction of foreign species to that environment? Report your findings to the class.

Ecological Recovery from Disasters

Imagine you are a preschool teacher in an area that has been hit by a natural disaster, such as a hurricane or flood. (Make your hypothetical situation specific.) Your class of four-year-olds includes some children who have temporarily or permanently moved out of their homes. As a result, they are anxious and fearful. You want to help the children in your care deal with their fears. (You also have some fears about natural disasters, and you want to do something positive in response to this experience.) Develop an activity for your students, based on the idea of habitat in an ecosystem. Use the displacement of some local species as an example of how living creatures can survive a disturbance to their habitat. Write about your activity in your logbook. Develop supporting materials such as a poster, magazine pictures, bulletin board material, and so on.

Saltwater Pet

Design an aquarium that would be appropriate for a saltwater organism that interests you. Do some library or pet shop research on your selected species. Consider salinity, dissolved oxygen, structural habitat, food, and so on. Describe how you will ensure that each factor is addressed in your aquarium. Develop a maintenance plan for the aquarium that includes a cleaning schedule and diagnostic tests that should be done on a routine basis.

Diet for a Small Biosphere

From articles on Biosphere 2 suggested by your teacher, find out what foods were grown for the crew's consumption. On the basis of this information, design a balanced weekly menu for the crew.

Facts and Fallacies about Organically Grown Food

Research the facts and fallacies about organically grown food. Is there a precise meaning to the claim that a food is "organic"? What health claims are made about organic foods? What benefits can you realistically expect from organically grown foods? What scientific evidence is there for and against the claims made by sellers of organic foods?

An Ounce of Prevention

Make a list of precautions for a crew of Biospherians to follow to reduce the possibility of sickness or injury during a prolonged stay inside an artificial biosphere. Explain to the class how these precautions relate to life in an enclosed environment and how this would be important information for future space explorers.

Music from the Forest

Choose a musical instrument that depends on forest products for its manufacture (e.g., drum, acoustic guitar, violin, clarinet, or oboe). Find out from an instrument builder or repair person what species of plant or animal provides the material for the instrument. Ask how the material is harvested and processed for use in the instrument.

An "Unsoiled" Biosphere

Biosphere 2 was originally supplied with over 30,000 tons of soil. Microorganisms in this large volume of soil may have depleted Biosphere 2's oxygen supply by consuming oxygen gas. Earlier biosphere systems did not rely on soil for agricultural production. Investigate one or two of these earlier systems and identify their advantages and disadvantages. The NASA web site is a good place to begin looking.

CHAPTER 23 SUMMARY

- Species adapt to survive in specific environments that include both living and nonliving components.

- Similar niches may be filled by different species in different ecosystems.

- Coevolution is common between predator and prey and between species that are involved in mutualistic or competitive relationships.

- In general, only one species occupies a particular niche in any community. Similar species can coexist when they use different resources or parts of the environment.

- Producers such as plants and other photosynthesizers support successively smaller masses of consumers on higher trophic levels.

- If left alone, communities tend to undergo succession, culminating in a more or less stable climax community.

- Photosynthesis captures about 1% of the solar energy reaching the Earth. This is the energy source that maintains most ecosystems.

- Food webs rarely have more than six trophic levels.

- Consumers capture about 10% of the energy of the trophic level below them.

- Successfully managed ecosystems compromise between sustainability and productivity. Nutrient recycling and energy flow often make important contributions to renewable managed ecosystems.

- Monocultures are highly vulnerable to predators, parasites, and harsh environmental conditions.

- Biomes are typical climax communities for a particular combination of climate and soil conditions.

- Managed ecosystems that are similar to the native biome are often the most successful.

CHAPTER 23 ASSESSMENT

Concept Review

See TRB, Chapter 23 for answers.

1 Is the water in which fish live part of their community? If not, what is its relationship to the community?

2 Are birds part of a bear's environment or just another species in the same environment?

3 Do thorns that keep animals from eating a plant show that coevolution has occurred? Give a reason for your answer.

4 Do the plants in a community depend on each other in ways that make them always occur together?

5 Does the biological community that forms in an abandoned field change the environment in ways that help that community to survive? Give a reason for your answer.

6 Is your weight a biomass? Give a reason for your answer.

7 What is the trophic level of a praying mantis that feeds on leaf-eating grasshoppers?

8 What percentage of the energy of producers is gained by tertiary consumers in a typical community?

9 What is the trophic level of a vulture that feeds on road-killed rabbits?

10 Explain why chemoautotrophic bacteria should or should not be considered decomposers.

11 When cities compost fallen leaves, grass clippings, and other lawn and garden waste instead of burning them or burying them in landfills, are they creating a renewable system? Give a reason for your answer.

12 How can hunting help to protect the diversity of species in a forest? What else could be done to protect endangered forest diversity?

13 How could recycling nutrients in urban ecosystems help to protect water quality?

14 If Biosphere 2 ever works as intended, what would be its most important input?

15 How would you expect fire suppression efforts to affect a grassland biome?

16 Is succession a change from one biome to another? Give a reason for your answer.

Think and Discuss

17 Can two species of penguins with the same food and temperature requirements but different nesting habits live in the same community? Give a reason for your answer.

18 Explain why animals such as bison and elephants that live in herds must constantly move from one place to another.

19 The major source of nutrients for a stream ecosystem is dead matter such as fallen leaves. How is this system like an urban ecosystem?

20 How do you think limited multiuse policies reduce the stress on ecosystems?

In the Workplace

21 What is the advantage of a cat over poison for controlling mice that attack stored grain?

22 How does the concept of trophic levels help to explain why leather, made from the skins of cattle, is less expensive than the furs of carnivores such as bears and mink?

23 What trophic level do humans occupy?

24 Explain how planting mixed crops can reduce the portion of the crop lost to insect pests.

Investigations

The following questions are based on Investigation 23A:

25 Compare plants to animals in terms of the tendency of different species to be found in interdependent groups.

26 Do small local variations in the environment such as hills and low, wet spots affect the distribution of species in a community? Explain.

The following questions are based on Investigation 23B:

27 Walking out of the woods, you pass through an area of small trees and shrubs and then a grassy meadow. Is it reasonable to say that this is an area undergoing succession? Would your answer be affected if the walk was up a steep slope? Give a reason for your answer.

28 Is a barren desert just a place where succession hasn't taken place yet? Give a reason for your answer.

CHAPTER 24

WHY SHOULD I LEARN THIS?

Everything we produce depends on items in our environment called natural resources. Some people take these things for granted; they can't imagine ever running out of things such as petroleum, clean air and water, soil, plants, and animals. Yet those who work to manage our natural resources are aware that both the quantity and quality of these resources are in danger. They remind us that everybody plays a part in managing and preserving these natural resources.

NATURAL RESOURCES AND THEIR MANAGEMENT

WHAT WILL I LEARN?

1. the variety of natural resources, both nonliving and living
2. the nature of a biogeochemical cycle
3. how a resource can be depleted or degraded
4. ways of reducing resource use
5. ways of sustaining a resource
6. how a renewable resource can be restored

The Ranger uranium mine in northern Australia provides many jobs and an ore that produces energy to run homes and industry. So why are plans for a second mine in the same area setting off a big environmental protest in Australia? Conservation groups worry about radioactive waste from the mine spilling into nearby Kakadu National Park.

Suppose a similar mine were to be built near your community? Do you understand enough about natural resources to make an informed decision about such an issue? In this chapter you'll analyze issues relating to natural resources that will help you make decisions about their use and how they should be managed.

LESSON 24.1 LIVING AND NONLIVING RESOURCES

A **natural resource** is a substance that exists naturally on Earth and that humans can put to use. Natural resources include both living and nonliving things. Park rangers guard living natural resources such as plants and animals. A chef who turns on a gas burner to cook a meal is using a nonliving resource called natural gas. When you drink some water, you are consuming a nonliving resource that might once have been part of a cloud. In this lesson you'll learn about the variety of natural resources.

ACTIVITY 24-1   NATURAL RESOURCES ALL AROUND US

The procedure for this Activity can be found in Chapter 24 of the TRB.

The natural resources of Earth have been used to produce the millions of products that are now a part of our lives. This activity will help you to start thinking about how dependent you are on natural resources.

WOOD SHORTAGE
BIOLOGY IN CONTEXT

In the last 50 years, one fourth of the world's forests have disappeared. The demand for paper is doubling every 20 years. Some companies are worried about the possibility of wood shortages occurring in the near future. So they are attempting to develop alternative ways to produce paper and other products now made from wood.

Home builders are using construction materials made from recycled plastics instead of lumber. Paper companies are looking into using sugarcane waste, bamboo, and fast-growing tropical trees as an alternative to pine trees. Studies have shown that a plantation of fast-growing tropical trees the size of Sweden could satisfy the entire world's need for paper.

Although these measures go in the right direction, we cannot depend on these resources alone to prevent a wood shortage. We must learn how to conserve our use of paper products. By recycling paper, using products made from recycled paper, and planting more trees, we can help replenish one natural resource—wood.

ENERGY RESOURCES

Natural resources that are used as energy include the sun, the wind, tides and rivers, fossil fuels, and nuclear energy. Fossil fuels such as natural gas, gasoline, oil, and coal are derived from the remains of long-dead organisms.

We get energy from fossil fuels by burning them. Industry uses fossil fuels to power factories, airplanes, and ocean liners. People use fossil fuels to heat and cool their homes. Most of the energy that is released from burning fossil fuels is used to generate electricity. The burning of fossil fuels poses problems because it produces wastes. These wastes include carbon dioxide, sulfur dioxide, nitrogen dioxide, and particulates, all of which are pollutants.

Although we use fossil fuels mostly for energy, we also process them to make plastics and other products (Figure 24.1). Fertilizers are also produced from fossil fuels.

Fossil fuels are found trapped deep between rock layers in the ground (Figure 24.2). Coal is mined, usually through a shaft into the ground or by stripping

Figure 24.1: *Petrochemical plants use natural gas to produce components of paints, plastics, and other products.*

Figure 24.2: *This drilling crew is extracting oil from an underground deposit or "trap."*

away overlying layers of rock. Petroleum is extracted by drilling. Because it takes millions of years for fossil fuel deposits to form from decaying organisms, fossil fuels are considered both a limited and a nonrenewable natural resource. A **limited resource** is expected to be entirely exhausted someday. A **nonrenewable resource** can't replace itself or be replaced by other processes.

ALTERNATIVES TO FOSSIL FUELS

Like fossil fuels, nuclear energy is limited and nonrenewable. Because nuclear fuels are more efficient than fossil fuels, they are considered the primary replacement for fossil fuels. However,

The control center of a large nuclear power plant

nuclear reactors pose certain safety problems that fossil fuels do not. If not contained, the radiation that nuclear reactors produce can cause sickness or death. In 1986 a poorly designed nuclear reactor near the city of Chernobyl in Ukraine blew apart because of the release of high-pressure steam. The explosion released more radioactivity than 30 atomic bomb explosions. Cities within a 30-km radius of the accident site were permanently abandoned. Much of the rich farmland of Ukraine was contaminated by radioactive fallout and has been abandoned.

Solar energy may have great promise for replacing fossil fuels. Solar power plants use mirrors to concentrate the sun's rays to heat water and produce steam (Figure 24.3). The steam drives turbines that produce electricity. Some individual homes and businesses use rooftop sunlight collectors that store energy in large batteries to heat water (Figure 24.4). However, solar collecting devices are relatively expensive and are efficient only when the sun is shining. Solar cells contain semiconductors that produce electricity when light strikes them. Solar cells power most of the satellites that circle the Earth.

WORD BANK

semi- = half

-conductor = a material that permits an electrical current to flow easily

Semiconductors conduct electricity at high temperatures but are poor conductors at cold temperatures.

Figure 24.3: *This solar power plant produces electricity.*

Figure 24.4: *Rooftop solar collectors are a common sight in the Sun Belt.*

AIR

Air is a mixture of gases. Figure 24.5 shows the composition of a typical sample of dry air at sea level. Air is an important natural resource because the oxygen in air supports respiration in aerobic organisms and combustion. Air also provides the carbon consumed in photosynthesis and the nitrogen that microorganisms convert to forms that are used to make DNA, amino acids, and other essential compounds.

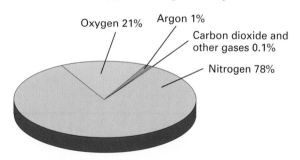

Oxygen 21%

Argon 1%

Carbon dioxide and other gases 0.1%

Nitrogen 78%

Figure 24.5: *Composition of air at sea level*

As you learned in Chapter 7, microorganisms have to convert atmospheric nitrogen to a form that the members of the food chain can use.

WATER AND THE WATER CYCLE

You are familiar with water in oceans, lakes, rivers, and streams. These bodies of water are called **surface waters.** Some water is also found underground, however, in spaces within rocks called **aquifers** (Figure 24.6). Aquifers may serve as the water supply for a city or be tapped by farmers to irrigate crops.

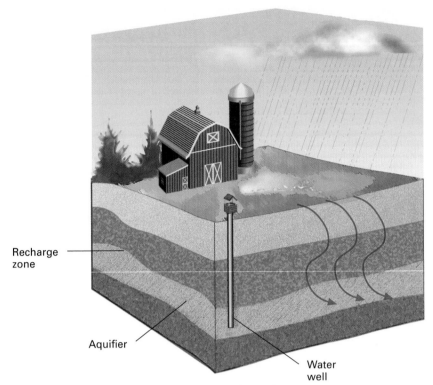

You are also familiar with water as rain, snow, sleet, and hail—forms of precipitation. An easy way to understand water as a natural resource is to look at how it changes form as it moves through the environment. This movement is often displayed in a diagram called the water cycle (Figure 24.7).

Recharge zone

Aquifer

Water well

Figure 24.6: *An aquifer holds water below Earth's surface.*

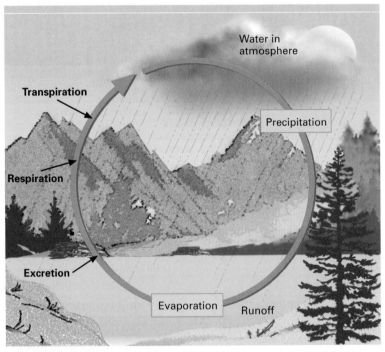

Water in atmosphere

Transpiration

Precipitation

Respiration

Excretion

Evaporation   Runoff

Figure 24.7: *The water cycle. Water falls to Earth as precipitation. Then it runs off into surface waters or seeps into the ground. Along the way plants and animals may use it. Water can then return to the environment through evaporation and excretion.*

SOIL

Soil is the loose part of Earth's crust that gives physical support and nutrition to plants. Figure 24.8 shows how soil forms. Inorganic soil particles are produced when rocks disintegrate. Organic soil particles form when dead organisms decay.

Soil forms in layers called horizons. Most of the organic material in soil is in the A, or surface, horizon. Soils experts often condition soil through **composting,** the mixing of organic matter with soil to encourage the release of nutrients from the organic matter. The B, or subsurface, horizon is rich in minerals. Minerals are often mined and used in many products.

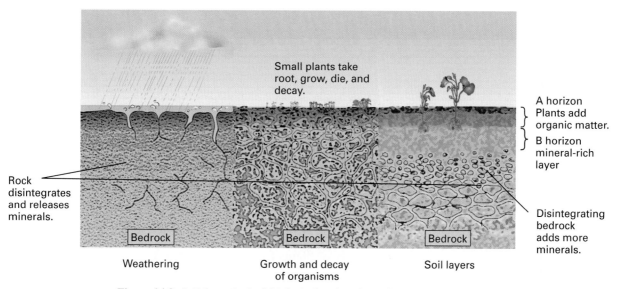

Figure 24.8: *Soil forms by both biological and geological processes.*

BIOGEOCHEMICAL CYCLES

Air, water, and soil are important natural resources involved in a **biogeochemical cycle.** A biogeochemical cycle describes the movement of a key nutrient, such as nitrogen or oxygen, through the environment. The cycle includes various components of the physical environment: the atmosphere, soil layers, groundwater, and bodies of water such as oceans, lakes, and rivers. The key nutrient may combine with elements in these components and be converted from one form to another. An example of a biogeochemical cycle is the nitrogen cycle (Figure 24.9).

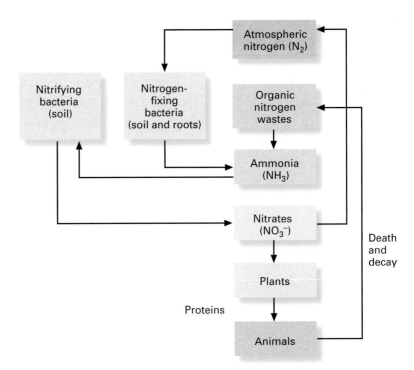

Figure 24.9: *Nitrogen moves through the environment in various forms.*

LIVING RESOURCES

Plants, animals, and other organisms are living resources that we use for food, entertainment, and energy. If you managed the produce section of a grocery store, your job would depend on an ample supply of plants for the human diet. If you owned a nursery, you would try to profit from the beauty that people see in flowers and foliage.

When we think of animals as resources, we tend to think of them as food. But they provide far more than that. Look at racehorses from the perspectives of horse breeders and owners. These animals provide a living for some by providing entertainment for many.

Living resources are considered renewable. **Renewable resources** are resources that are capable of reproducing themselves. Their use, however, may be either consumptive or nonconsumptive. For instance, poplar trees provide wood for furniture and are consumed (used up). Seeds must be planted to renew the supply of trees. On the other hand, maple trees that are tapped for syrup continue to produce sap for years; the trees are not consumed. Proper management of living resources can increase the yield of a product, such as wood, syrup, fruit, or meat.

Although we think of plants mainly as food and decoration, for centuries humans have also used plants for medicinal purposes. Use the library or the Internet to learn how people use plants to treat diseases and other bodily disorders. What chemicals in these plants were later synthesized in the laboratory as pharmaceuticals?

LESSON 24.1 ASSESSMENT

1. Dairy cows are raised for their milk, which is a nonconsumptive use of a resource. Beef cows are slaughtered for their meat, which is a consumptive use.

2. The proposed solution is not effective because it does not mention how the electrical heat will be produced. Most electricity is generated from fossil fuels, so this plan will not have the desired effect.

3. They use products made from natural resources.

4. As the water in the aquifer got lower, the weight of the house in sandy soil could cause the aquifer to collapse.

1 How do dairy farmers and beef ranchers differ in their use of cattle as renewable resources?

2 An environmental activist who wants to reduce pollution caused by fossil fuel burning has suggested banning heating with coal and fuel oil and instead using electrical heat. Discuss the effectiveness of the proposed solution.

3 Explain how the work of a machinist, social worker, bookkeeper, or TV personality is involved with natural resources.

4 Suppose you build a house on sandy soil in an area with an aquifer that is close to the surface. On the basis of your knowledge of biogeochemical cycles, predict what effect a long-term drought in the region would have.

Lesson 24.2 Depleting and Degrading Resources

Cloth or plastic? The types of diapers that are put on babies affects the supply of natural resources. Disposable diapers are made from petroleum (a nonrenewable resource), while cloth diapers come from plants (a renewable resource). Resource-conscious parents use, wash out, and reuse the less convenient cloth diapers, right? Not necessarily. Using cloth diapers would save on petroleum, but what about the water needed to wash the cloth diapers? Decisions about resource use are rarely simple. People must be aware of both obvious and subtle ways in which the use of a natural resource affects both its quantity and quality.

Activity 24-2   Measuring the Half-life of a Resource

The procedure for this activity can be found in Chapter 24 of the TRB.

Uranium and plutonium are unstable, radioactive elements. They break down, or decay, spontaneously, so even without human use they will eventually disappear from the Earth. Each radioactive element has a unique rate of decay, measured by half-life. The **half-life** is how long it takes half of a sample of the element to decay. In this activity you'll determine how rapidly a radioactive element decays.

Career Profile   Forest Technician

Jim S. is a special agent for the U.S. Fish and Wildlife Service in Alaska. This is how he explains his job:

"I'm a police officer for wildlife. I went to the Federal Law Enforcement Training Center, which is a federal police academy. Every year, I attend 40 hours of school to brush up on law enforcement, and I have to pass a target test with a pistol. I have all the training of any other police officer, except I protect wildlife.

"How do I do that? Well, special agents keep a close eye on animals and animal parts coming into and leaving the state. Congress made it against the law to transport protected and illegally obtained wildlife across state lines. That law is called the Lacey Act. Some special agents spot-check airport baggage. We also patrol in the field. Since Alaska is so big and the wildlife refuges are so remote, about half of our agents use small planes for patrols. Sometimes agents conduct stakeouts. A stakeout is when agents stay hidden and wait in a certain area where they think laws are being broken. It's hard catching law-breakers in the act. Sometimes I work undercover, pretending to be just another hunter. That way I see to it that hunters or hunting guides are obeying the laws.

"I've always had a feeling for wildlife. Nearly all the special agents do. Lots of people would like to be special agents like me. I think that's because a lot of people want to do what I do to protect wildlife."

Source: Courtesy of Alaska Department of Education, Office of Adult and Vocational Education, Juneau, Alaska 99811.

RUNNING ON EMPTY

Depleting a natural resource means reducing the amount that exists. A limited, nonrenewable resource such as petroleum may some day be totally depleted.

As a resource is depleted, it can become harder to get. In West Virginia, for example, most of the rich veins of aluminum have been mined. The remaining veins have much less metal, and getting the aluminum out is more expensive.

Figure 24.10: *This tractor is harvesting trees in a tropical rain forest.*

Renewable resources can be depleted, too, when they're used faster than they can replace themselves. For example, trees can be cut down for lumber and other products faster than replacement trees can grow from seeds. Mahogany is a popular furniture wood that grows in tropical forests. Today, these forests are being cut in vast stretches (Figure 24.10). Mahogany trees can't replant themselves when all the trees in a large area are cleared. One reason is that logging activities disturb the soil so much that the seeds can't germinate and grow properly.

Living resources such as plants and animals are renewable resources only if (1) their habitats still provide food and water, shelter, mates, and protection from predators and parasites and (2) they can reproduce at a rate that replaces the population being harvested. When humans clear the land, many wild species lose their habitat and leave or die. Domestic species of plants and animals are rarely in danger of being overharvested. Crops can be renewed from commercial seed, and breeder hens continuously hatch eggs. But if wild animals are harvested faster than reproduction can replace their populations, the species will decline. At some point, a species may become so rare that a catastrophe such as a flood or freezing weather can wipe it out.

A population of organisms that can't renew itself will disappear from a local area or even disappear from Earth entirely. If a species disappears entirely, it is said to be extinct. Approximately 750 species of plants and animals in the United States are currently in danger of extinction. Endangered species have populations so small that they will die out if no steps are taken to save them. Another 200 species in the United States are threatened because their populations are beginning to decline. These species may become endangered if nothing is done to help them increase their numbers.

DEGRADING NATURAL RESOURCES: POLLUTION

dictionary

≡WORD BANK
- - - - - - - - - - - - - - - - - -
de- = remove from

-grade = a standard of quality

Exhaust fumes from cars degrade air quality.

A resource is **degraded** if using it lowers its quality. For example, an automotive worker running tests on a car burns gasoline made from petroleum. Burning gasoline releases carbon dioxide, carbon monoxide, ozone, and other waste gases, all of which degrade the quality of air (Figure 24.11). Carbon monoxide replaces oxygen in our blood by binding to hemoglobin in red blood cells. Ozone, a very unstable and reactive gas, oxidizes tissues in our respiratory passages.

Too much carbon dioxide and other so-called greenhouse gases act like a blanket in the atmosphere. Sunlight that is absorbed at the Earth's surface

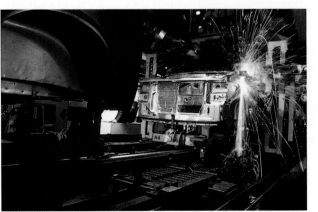

Figure 24.11: *Automobiles, shown here being assembled, degrade air quality.*

reradiates into the lower atmosphere as heat (Figure 24.12). The blanket of greenhouse gases traps this heat close to the Earth. This condition is known as the **greenhouse effect,** because the heat is trapped much as it would be by the glass in a greenhouse. In moderation the greenhouse effect helps to maintain the Earth's temperature within a range tolerable for life. But if the greenhouse gases become too concentrated, they may raise global temperatures, an effect called **global warming.**

Water quality can also be degraded in a number of ways. Dirt and other particles that are too light to sink in water become suspended. The water then becomes cloudy, which makes it difficult for plants to get light and for animals to locate food and mates. Other substances may dissolve in water. These include nutrients, such as nitrogen and phosphorus, which may cause eutrophication, and pesticides, heavy metals, and other substances that may be toxic. Some dissolved substances act as acids or bases, lowering or raising the pH of water. Water quality is also affected by temperature. As the temperature of a body of water increases, the amount of dissolved oxygen in the water decreases. This effect is sometimes called thermal pollution. It can have devastating effects on populations of fish and other aquatic organisms.

Soil quality is degraded when soil nutrients are removed. Too much rain can leach nutrients from the soil into groundwater. Nutrients are also depleted when the same crop grows year after year. Growing crops on land that has been cleared of forest can rapidly deplete soil nutrients.

Burning fossil fuels degrades the quality of air, water, and soil as resources. Fossil fuel burning, like most other types of combustion,

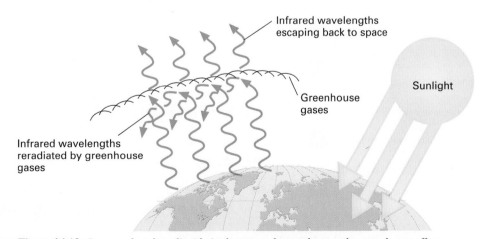

Figure 24.12: *Increased carbon dioxide in the atmosphere enhances the greenhouse effect.*

Figure 24.13: *Acid rain has destroyed living resources in hundreds of lakes in the northeastern United States.*

Equation for formation of sulfuric acid:

$$SO_3 + H_2O \rightarrow H_2SO_4$$

produces nitrogen oxides (NO_x) and sulfur oxides (SO_x). In the air these substances combine with water to form acids. For example, sulfur trioxide, SO_3, combines with water to form sulfuric acid (H_2SO_4). When it rains, these acids fall to Earth, an event called acid rain. Acid rain is a serious problem because it lowers the pH of soil and bodies of water. This leads to die-offs of soil organisms and aquatic organisms (Figure 24.13). Acid rain also directly damages plant leaves, interfering with photosynthesis. Our own health is affected by acid rain. When we breathe polluted air, the SO_x and NO_x in the air react with the moisture in our respiratory passages, irritating their linings.

INVESTIGATION 24A ACID RAIN

See Investigation 24A in the TRB.

When sulfuric acid falls from the atmosphere with precipitation, it may enter soil, lakes, streams, and other water reservoirs. The acid dissolves in the soil or lake water and lowers its pH. These reservoirs, of course, are home for many plants and animals. Changes in pH almost always harm aquatic organisms and lower the species diversity of a lake or stream. In this investigation you will examine how bodies of water are sometimes buffered against the effects of acid rain.

LESSON 24.2 ASSESSMENT

1. A strong case can be made for all of them except sunlight, which is an unlimited resource.

2. Limited and renewable.

3. Students should consider solar power, tidal generation, wind power, nuclear power, geothermal energy, and fusion energy.

4. This conversion will save the school money by using a less expensive fuel.

1 Decide which of these resources a community of people would be most likely to deplete: water, soil nutrients, trees, sunlight.

2 An airplane engine uses alcohol for fuel. The alcohol is produced from corn and other grains. Is the fuel a limited or unlimited resource? Is it a renewable or nonrenewable resource?

3 Choose an alternative energy source that your community might use when fossil fuel supplies run low. Give reasons for your choice.

4 Your school has decided to convert its coal burning furnace so that it will burn trash. Discuss how this might benefit the school.

LESSON 24.3 MANAGING RESOURCES

Personal computers were supposed to create paperless offices—no more letters, books, printed references, and so on. It hasn't happened yet. But imagine if we ran out of paper and were truly paperless. We would have none of these paper products and no paper money, checks, cardboard boxes, or toilet paper. That sounds bad, but things would be much worse if we ran out of drinkable water, a problem in many places today. What can we do to manage natural resources wisely so that we can all have enough? This lesson tries to give some answers to that question.

ACTIVITY 24-3 ALTERNATIVE SOURCES OF ELECTRICITY

The procedure for this activity can be found in Chapter 24 of the TRB.

Managing resources wisely often means finding alternatives to resources that are being depleted or whose use is degrading other resources. This sometimes requires using products in ways far different from the ones for which they were designed. In this activity you'll use a simple galvanometer, a device that measures electrical currents, to detect electrical current produced by oranges and soft drinks.

DIAPERS = SHOES

BIOLOGY IN CONTEXT

Julie L. has a silent but important business partner: nature. Julie is an inventor with a master's degree in nutrition and biochemistry from Oregon State University. She got into the shoe business because she was interested in environmental issues and wondered how she could make an impact. Inspired by huarache sandals made from recycled tires, which were popular in the 1960s, Julie decided to develop environmentally positive shoes. Shoes, she reasoned, are something everyone needs, yet the shoe industry produces them using toxic adhesives and dyes, chemically tanned leather, plastics, and petroleum-based rubber.

Julie had no experience in the footwear industry. But in the summer of 1990 she met Bill Bowerman, a track coach who had worked with his students to develop a running shoe. Bill helped Julie to pull together her concept of a shoe made entirely from recycled materials. He put her in touch with designers and gave her a source book of shoe parts. She called different manufacturers, asking for samples of ecologically sound materials. She ended up having a fabric specially developed from polypropylene from recycled diapers.

Metro, a community service group, gave Julie a $110,000 grant that allowed her to make the first large lot of shoes.

RESOURCE CONSERVATION: THE FOUR RS

It's probably safe to say that no natural resource will last forever, but some are clearly more limited than others. Fossil fuels aren't as reliable a source of energy as sunlight. The future availability of fresh water is also a concern, especially because of its role in producing electricity (Figure 24.14).

We can use ingenuity in maintaining the resources that we already use and in discovering still others. Resources can be sustained or conserved by reducing their use or by increasing their abundance. The search for new petroleum reserves takes the efforts of geologists, petroleum engineers, land surveyors, and many other people. Chemists have developed synthetic fuels such as gasohol, which combines gasoline with alcohol derived from agricultural wastes.

Figure 24.14: *These powerhouses convert the energy of falling water into electrical energy.*

Resource managers often refer to the four Rs of resource conservation: (source) reduction, recycling, reuse, and restoration. **Source reduction** reduces the amount and/or toxic nature of the waste that enters the waste stream. Increasing gas mileage of automobiles, returning beverage bottles for refilling, and packaging products with fewer throwaway materials are examples of source reduction. Building more efficient furnaces and insulating offices and homes are examples of source reduction that reduce fossil fuel consumption (Figure 24.15).

Conserving resources also involves **recycling,** which involves sorting waste into its components for reuse. You can take an object apart and recycle the parts into new products. For example, when you reduce old television sets to plastic, metal, and glass, the components can be recycled. The aluminum of a used food container

Figure 24.15: *Building insulation reduces fossil fuel use.*

Figure 24.16: *These barrels and their contents could be recycled.*

can be reprocessed to make new cans, and its label can be used in new paper products. Recycling extracts useful materials from waste so that they can be used again (Figure 24.16). Reducing waste to its parts can be the first step toward recycling but isn't always necessary. A plain plastic container can be melted down and used in a new plastic part without any sorting. Even buildings can be recycled in a sense. Old warehouses and hotels can be repaired or adapted to new uses as restaurants or condominiums. Recycling saves the energy and resources that are needed to knock down the old structure and build a new one from scratch.

Reuse is the simplest conservation approach, although it isn't always possible. In reuse, a product is used again in its original form or with a little processing. Old lumber can be reused instead of more trees being logged and processed. Used beverage containers can be sterilized and refilled over and over again. Community-based industries often collect used clothing, appliances, furniture, and other items, which are repaired, cleaned, and given to those in need or sold to fund charities (Figure 24.17).

Restoration often involves more effort than reducing, recycling, and reuse. It takes time to restore a forest by planting new pines, but it can be done. Is it possible to bring a species back from extinction? Scientists are trying to restore the quagga, an African horse that is closely related to zebras (Figure 24.18). It has been extinct for more than 100 years. Some scientists believe that they can restore this species to life because its genetic material lives in the genes of its zebra relatives. They select and mate zebras that most closely resemble the extinct quagga. Perhaps the

Figure 24.17: *Before throwing out unwanted items, consider donating them to a thrift shop.*

Figure 24.18: *Scientists hope to restore the quagga, an extinct species of horse.*

offspring from two such zebras will closely resemble a quagga. This project was one of the inspirations for the novel *Jurassic Park*. Other scientists believe that such restoration would only be possible if the complete DNA blueprint of an extinct species is available, a very unlikely situation.

RESOURCE MANAGEMENT

Management of living natural resources has a turbulent and even violent history. Management is rarely a simple issue and often arouses controversy. For example, when it looked as though some African wild animals might become extinct, land was set aside in national parks to protect the animals from hunters who shoot them for hides, meat, tusks, and other products. People who hunt such animals illegally are called poachers. They face severe punishment if they are caught. In the United States as well, unusually rare animals are protected. It is a crime to kill or even possess endangered species.

But does this kind of protection solve the problem? Even some conservationists now realize that private ownership might be the solution, not the problem. If elephant ivory is valuable but elephants can't be owned and bred, poachers will have a reason to capture (and kill) as many elephants as possible. If it were legal to own an endangered species, a breeder might profit from raising it in large numbers. In Australia, ranchers and farmers consider

Figure 24.19: *The emu is a flightless bird, now raised for its meat.*

flightless birds called emus a pest and exterminate great numbers of them. Fortunately, emus are being exported from Australia to the United States, where they are being bred for meat and other products (Figure 24.19). Emu farming ensures that the species will not become extinct.

WASTE STREAM

WORD BANK

bio- = life

de- = remove

-grade = a standard of quality

Bananas are a biodegradable resource.

Reduction, recycling, reuse, and restoration can help to sustain both living and nonliving natural resources. But what happens to the waste material that is left over when these conservation options aren't used? In other words, what happens to our trash? It enters the waste stream. If wastes are biodegradable, they break down (decompose) with the help of bacteria, fungi, and animals. This natural form of recycling takes place without human intervention. It doesn't take these creatures long to recycle table scraps. The decomposed waste enters the ecosystem as food or fertilizer in the biogeochemical cycles. If the waste isn't biodegradable, it won't change much for many years, perhaps thousands or even millions of years. In the meantime, landfills are filling to capacity. In some areas, such as the northeastern United States, there are few places left to build new ones.

INVESTIGATION 24B BIODEGRADATION OF SOAPS

See Investigation 24B in the TRB.

Many makers of synthetic chemicals, including detergent manufacturers, claim that their products are biodegradable. To support such a claim, a manufacturer must have a product tested. Tests for biodegradability involve feeding the product to microorganisms and looking for decomposition. In this investigation you'll look for evidence that microorganisms added to a soap or detergent solution are multiplying, using soap or detergent as an energy source.

WHERE'S THE TRASH?

The waste stream includes food, plastics, textiles, containers, paper, construction and demolition scraps, landscape trimmings, dead animals, abandoned cars, old appliances, tires, and sludge from the treatment of waste water. Most household waste is paper, especially old newspaper. Containers, food scraps, and yard trimmings account for much of the rest.

Modern landfills are lined with materials that prevent groundwater from carrying leachates. **Leachates** are solutions that consist of materials dissolved in water that trickles out into the ground. Pipes collect leachates from the bottom of the landfill so that a truck can

Figure 24.20: *Modern incinerators can generate electricity from waste.*

carry the waste to a water treatment plant. Other pipes deal with the gases that decomposing garbage produces. These gases may escape into the atmosphere or can be collected and used as fuel. So landfills can use at least two active forms of recycling.

One alternative to the landfill is incineration, or the burning of refuse. Many large cities incinerate their garbage because they have no more space for landfills. Although gases and solid particles are released into the atmosphere, modern incineration plants use the heat produced by burning garbage to turn water into steam. The steam is used to generate electricity (Figure 24.20).

LESSON 24.3 ASSESSMENT

1. Water and air used in manufacturing cans and the soil from which aluminum ore must be extracted.

2. Soil nutrients were washed away by the heavy rains.

3. Accept any reasonable answer, which might include recycling of paper, plastic, and glass; well-used public transportation; and use of alternative energy resources.

4. Students should consider improved fossil fuel use, nuclear power, wind power, solar polar, geothermal energy, and so on.

1 Predict which resources, in addition to aluminum, are conserved when aluminum cans are recycled.

2 A developer clears all vegetation from a hillside, and after several years of heavy rains, a homeowner plants a garden. What is the best explanation for why the garden grows poorly?

3 List elements that you would include in a plan to help your community conserve energy.

4 As a future taxpayer, how would you want your tax dollars to be spent on energy research? Select the form of energy you would support and explain why you would support it.

CAREER APPLICATIONS

The applications that follow are like the ones you will encounter in many workplaces. Use the biology you learned in this chapter to complete the activities. Share your work with the class.

AGRICULTURE & AGRIBUSINESS

It's Under Control

A controlled burn is a vegetation fire that is set intentionally. Some people think that the wildfires that swept through Florida in 1998 might have been largely prevented if controlled burns during previous years had kept down the amount of highly flammable dead vegetation. Contact your local fire department or conservation agency to find out what is required to conduct a controlled burn and what precautions must be taken. Research the benefits of periodic burning. Look for information on the Internet, using terms such as "prescribed burning" or "controlled burning vegetation." Develop an informational pamphlet that describes controlled burning and the benefits and drawbacks of this method of ecosystem management.

Dolphin-Safe

Research the use of nets and traps in the fishing industry. Describe the problems associated with the accidental capture of species such as dolphins and sea turtles. How are fishing practices changing to protect these species? With models or graphics, demonstrate to your class the problems and solutions.

Farm Resources

Obtain photographs or pictures of a farm, a ranch, or an orchard. From the photos, identify different resources in the scene. Classify each resource as natural or artificial, limited or unlimited, and renewable or nonrenewable. Also identify waste or potential sources of waste. How many of these wastes could be reduced, recycled, or reused? How would that be done?

Grocery Goes Green

You are the manager of a small grocery store, part of a regional chain. Your district manager has asked you to make your store a "green model" for the district. She isn't sure how you should go

BUSINESS & MARKETING

about this but says, "Look into reduced packaging. Try to encourage recycling, things like that. You'll come up with things, I'm sure." Make a list of things you'll do—action items—to become a green model. Divide your list in two: things you can do that won't cost much and things that require a lot of money to start up. Under each action item, write the potential benefit to the store and to the public. Present your suggestions to the class. As a class, make a list of the best ideas presented. Give them to local store managers to encourage resource conservation.

Animal Rights, Cosmetics, and Furs

Research how animal rights groups are affecting the cosmetic and fur industries. Collect articles and advertisements from manufacturers and animal rights groups. Form two teams and debate the issue.

Traveling Light

Imagine that you work in a store that sells recreational camping supplies. Customers often ask you to help them plan a camping trip so that they won't leave behind a single scrap of waste or have to carry too much along the trail. Devise a low-waste packing list and low-waste guidelines for your customers.

Waste Task Force

You and your classmates represent a cross section of your community's businesses, citizens, city officials, sanitation workers, and other entities that are concerned with the problems caused by waste in your community. Your group has been charged with making plans for improving the waste situation in your community, developing a timeline for their implementation, educating business and industry on their role in the plan, and educating the public to be more responsible in dealing with waste. Divide into groups and let each group tackle some part of this plan. Some activities that groups might undertake are the following:

- Develop a plan to educate your neighborhood or community to stop improperly disposing of hazardous waste. Your campaign should educate people about what can and can't be in their regular garbage pickup. Be aware that a negative message might not be effective. Include alternative disposal for problem items. Set goals and detail strategies. Consider forums such as public meetings, articles, and door-to-door "waste ambassadors" to accomplish your goals. Write the script and plan the graphics for a brief public service ad for television, telling people what not to put in their garbage.

- Start a recycling program for municipal solid waste from households or your school. Decide what recycling efforts would be the most successful way to start a program. You might wish to contact a recycling program that is already established. Plan to phase in those efforts within a year. Develop a presentation to take to the town council for its approval.

- Plan a source reduction festival to raise people's awareness of source reduction. List the kinds of exhibits, demonstrations, and entertainment that the festival could feature.

FAMILY & CONSUMER SCIENCE

Energy Savings at Home

Research ways of making your home more energy-efficient. Use books or magazines on building and maintenance; talk to a representative from the electric company, gas company, or other utility; or use the Internet. Prepare a brochure of ways to save energy at home. Discuss the cost and effectiveness of each measure.

Comparing Sprays

Compare the spray pattern, mist characteristics, convenience, and economy of aerosol products and pump-spray products. Discuss the environmental effects of each type of product. Compare the advantages and disadvantages of these products.

HEALTH CAREERS

Hospital Waste

Interview the waste manager of a local hospital. Get answers to the following questions: How does the hospital separate black-bag and red-bag waste? Does the hospital use reusable or disposable surgical instruments, food service containers, diapers, bed pads, and pillows? What determines the hospital's waste policies? What materials are recycled? Write up your findings in your logbook.

Artificial Body Parts

Using the library or the Internet, research the use of petroleum products in artificial body parts. Some parts to consider are artificial limbs, heart valves, and replacement joints. Try to come up with resources that could replace petroleum products in each artificial body part. Make a chart of possible replacement materials.

INDUSTRIAL TECHNOLOGY

Hazmat Technicians

Investigate the occupation of a hazardous materials (hazmat) technician. Your teacher will suggest sources. Find out what such a technician does and the educational requirements. Make a one-page fact sheet on the career.

Steel Can Recycling

Visit a recycling center or use the Internet or other resources to find out about the magnetic separation process that is used to separate steel cans from other recyclables. Explain the technology as simply and briefly as possible in your logbook, using diagrams.

Desalination of Water

Research the various methods for the desalination (removal of salts) from water. What is involved? What are the costs? What countries use desalination for their source of fresh water? Prepare a report in your logbook and share the information with your class.

CHAPTER 24 SUMMARY

- Both living and nonliving natural resources are essential to our survival.

- Natural resources must be used wisely and managed properly.

- Unlimited resources will probably be available as long as they are needed. Air and sunlight are considered to be unlimited.

- Limited resources, such as fossil fuels, will no longer be available some day. Petroleum is nonrenewable because of the vast time that is required to make it. Nuclear fuels are more efficient than fossil fuels and more controversial because of their dangerous wastes.

- Trees and animals are considered renewable because of their ability to reproduce.

- Some natural resources are part of biogeochemical cycles. Water moves from the atmosphere to the ground through plants and animals and back to the atmosphere.

- Pollution is a form of degradation.

- Recycling extracts useful materials from waste.

- Source reduction can involve decreasing the use of a resource. It can also mean dismantling waste into parts that can be recycled.

- Reusing a resource returns it to use without much processing.

- Restoration returns populations of living resources to healthy levels.

- Some modern landfills recycle gas and water that is collected from decomposing trash.

- Incineration and composting are two alternatives to using landfills.

CHAPTER 24 ASSESSMENT

Concept Review

See TRB, Chapter 24 for answers.

1 Explain how electrical power plants convert the energy of fossil fuels to electricity.

2 Identify the products of fossil fuel burning that degrade natural resources.

3 Does organic matter represent living or nonliving resources? Give a reason for your answer.

4 How does a drop of ocean water become a drop of water in your community's reservoir?

5 Discuss the relationship between clear-cutting forest and degrading of natural resources.

6 Solar energy can be converted directly to electrical energy in solar panels. Identify the natural resource that converts solar energy to chemical energy.

7 Explain why oxygen is an important natural resource.

8 Explain how water quality might be affected by a landfill located on the shore of a lake.

9 A wildlife manager provides high-quality habitat for a threatened population of deer. Which of the four Rs of resource conservation is involved?

10 Give an example of a nonconsumptive use of plants.

11 In Europe, auto makers are trying to make their cars over 80% reusable by the year 2000. Describe which natural resources this will help to conserve.

12 A used car is disassembled, and new uses are found for some of its parts. Which of the four Rs of resource conservation is involved?

13 List industries that rely in some way on animal products as a natural resource.

14 List several ways in which your community can reduce waste.

15 Under what circumstances will a population of living organisms be a renewable resource?

16 What are some useful ways to classify the Earth's natural resources?

17 Predict when alternative energy sources will become more widely used.

18 Evaluate ways in which people can reduce pollution.

Think and Discuss

19 Discuss the pros and cons of developing nuclear power plants to provide electricity.

20 Debate the merits of levying heavy taxes on gasoline and discouraging the importing of foreign oil.

21 Why are there water shortages even though the amount of water on the Earth is nearly constant?

22 A dairy farmer calls in a veterinarian to surgically remove embryos from a prized milk cow and transplant the embryos into the uterus of several other cows. Decide whether this improves the quality or quantity of a natural resource.

In the Workplace

23 You have been asked to serve on the waste minimization team for your office. What might you suggest to reduce, reuse, or recycle typical office wastes?

24 You work in a factory that produces polymers. One of the by-products that must be disposed of is a hazardous material. You find out that a nearby factory uses in its manufacturing process the by-product that you produce. You work out a deal with the other factory's plant manager to send it your waste product, which will be used in its process. Who benefits?

25 You are on the design team for a new line of skiwear. You'll advertise it as being environmentally friendly. Which of the following fills would you pick for your skiwear and why? Virgin polymers (made from petroleum and never used in any other product), recycled polymers (made from plastic drink bottles), or down (gathered from geese).

26 A large landfill has been sealed up with clay. The city asks you to oversee the restoration of the land as a park and wildlife preserve. What steps will you take to develop and implement a plan?

Investigations

The following questions refer to Investigation 24A:

27 You have three plates labeled A, B, and C. Each contains an equal amount of a different type of crushed rock (with a pH of 7.0–7.5). You add an equal volume of sulfuric acid solution (pH 3.0) to each plate. The results are in the table below. Which of the samples has the best buffering capacity?

Plate	pH
A	3.6
B	6.5
C	5.7

28 A fisheries manager wants to reclaim a small fishing pond that has a very low pH as a result of a single year of heavy acid rain. He can buy a truckload of lime to add to the pond. What information does he need in order to know how much lime to buy?

The following questions refer to Investigation 24B:

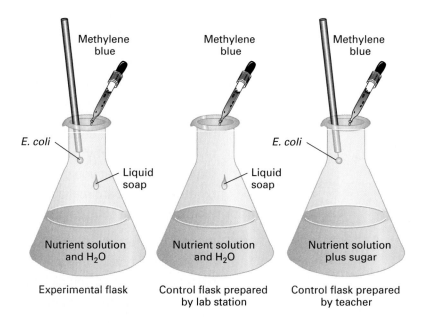

Experimental flask | Control flask prepared by lab station | Control flask prepared by teacher

29 In the figure above, methylene blue (added to the flask) turns colorless if microorganisms break down organic material. Which flask will still be blue after inoculation?

30 What purpose did the control flask prepared by the teacher have in the experiment?

APPENDIX A: SAFETY FIRST

Labs and activities are the most exciting part of science, but they involve potentially harmful materials and activities. The three items in a lab setting that can be the most dangerous are the chemicals, glassware, and any heat source. The first safety rule to remember is *tell your teacher about any spill, accident, or fire.*

The most important safety factor in a laboratory setting is YOU! By following these safety rules, you can protect yourself and others from harm.

SAFETY RULES

- Safety goggles are required at all times in the lab unless your teacher specifies otherwise.
- Notify your teacher of any allergies you may have.
- Never put your face directly over anything that contains chemicals and inhale deeply.
- Be smart—do not eat or store any edible materials in a lab setting.
- Do not apply makeup in the lab.
- Never use your mouth to pipette anything.
- To protect your clothing, lab coats or safety aprons can be worn. Personal protective equipment such as gloves or dust masks may be used at times.
- To protect your feet, you should wear shoes at all times, and the shoes should cover your feet—no sandals or open-toed shoes.
- Know what you are going to do. Read procedures, know the materials and equipment involved in the experiment, and know any hazards associated with each item.
- Do not play around in the lab.
- Never do any unauthorized experiments.
- Wash your hands with hot water and soap after each laboratory or activity, even if gloves were used.
- Your teacher will provide you with specific safety precautions on certain activities and investigations.

APPENDIX B: KEEPING YOUR LOGBOOK

Your *CORD Biology* logbook is a key element to your success in this course. It is in your logbook that you will record everything you do for this course. Keeping a written record might seem like busywork to you now, but the skills that you will be developing are skills that will serve you well beyond this course.

Keeping written records is critical for scientists. During the course of their research, they record all of the details of every experiment that they try. This allows them to better interpret their data and to share their work with others. Detailed records are also required for inventors to patent their inventions. Medical professionals must keep detailed records on each of their patients. Environmental technicians keep records of air and water quality. Bioreactor operators keep records of flow rates, temperatures, pressures, and so forth at different steps in the process. Virtually any job you can name requires that some sort of records be kept.

Your logbook can be any notebook, but it is preferable that it be a bound notebook because these are sturdier and are more typical of those used in business and industry. Throughout the course, as you complete various Activities, Learning Links, Investigations, and Career Applications, you will be directed to keep your work in your logbook. (You should use a different notebook for your class notes because your teacher will periodically take up your logbook to check your work and you might not get it back for a couple of days.) Whenever you record information in your logbook, put the date in the left margin. Also, put a title on your entry, such as "Activity 4-1: Looking at Chromosomes," or "Investigation 4B." Accurately dating and labeling your entries will make it easier for you to refer back to specific notes at a later time when you are directed to do so. Also, you will better be able to share information with your teacher or classmates if you can turn quickly to the material in question.

When you make your entries in your log, write as clearly and concisely as possible. If you are describing a procedure that you followed, write with enough detail that someone who was not working with you could follow your notes and repeat the procedure from your notes. If you worked with others on a particular item, be sure to include their names in your notes. If you are researching a topic for a Learning Link or Career Application, be sure to include the source of information with your notes. For example, write down the Internet addresses of sites that had helpful information. If you get information from someone in the community, include the person's name, where he or she works, and a phone number, address, or e-mail address where they can be reached. Then if a question comes up, as you prepare your presentation and report to the class, you can go back and find missing information.

Sometimes as you are working, you might find information on some other topic that is of interest to you, or you might have a question about something you read. Make a note of it in your logbook so that you can follow up on it at a later time. Mark it with asterisks or some other symbol so that it is clear to someone else that the information is not part of the current entry. You should make your logbook work for you, but it must be able to communicate your work to someone else.

APPENDIX C: SCIENTIFIC METHOD

When scientists study a problem, such as effects of caffeinated soda, they follow a process called the scientific method. A scientist would not watch a television program on caffeinated soda and consider himself or herself an expert. A scientist would not take the word of one user about what caffeinated soda does to the human body. Nor would he or she listen to friends about the evils of caffeine and draw conclusions. Instead, a scientist would approach the subject carefully and follow several guidelines.

OBSERVATION

First the scientist must observe what he or she will study. Observation is the gathering of information about the world around us through the use of some or all of the five senses— sight, hearing, smell, taste, and touch. If the scientist is studying the effects of caffeinated soda on the body, he or she would collect data on people who drink caffeinated soda. The senses of sight and hearing can be used to observe individuals drinking caffeinated soda. The scientist would also find out what other researchers had observed about drinkers of caffeinated soda. The scientist can gather valuable information from studying the works of others and from personally observing caffeinated soda drinkers.

HYPOTHESIS

After carefully examining a problem, a scientist will develop a hypothesis. The hypothesis is the scientist's explanation of the observed data. Before developing a hypothesis on caffeinated soda drinkers' study habits, a scientist would observe drinkers of caffeinated soda while they studied on several occasions. He or she would notice any changes in energy level after a person drinks a caffeinated soda. One hypothesis that the scientist might make after this observation is that "caffeinated soda helps students study for longer periods of time."

EXPERIMENTATION

A scientist does not end his or her investigation with a hypothesis. The hypothesis must then be tested in a controlled study. In the case of caffeinated soda drinkers, a scientist wants to finds out what happens from several drinkers. Much more observation must be done at this stage of the process. The scientist may study how reactions differ with the age of the user, how caffeinated soda affects women versus men, or how reactions differ between an experimental group (caffeinated soda drinkers) and a control group (noncaffeinated soda drinkers).

ANALYZING THE RESULTS

After a scientist gathers what he or she thinks is enough information, the results of the experimentation must be analyzed. When the scientist analyzes data, he or she is attempting to find out if the information collected is reliable. So the scientist will ask questions about the data he or she has collected. Does the data include enough sources? Does the data include all age groups, both male and female, and different ethnic groups? Do the results seem to be supported by other respected researchers in the field? Is there information that doesn't support the hypothesis?

DRAWING CONCLUSIONS

After all of the information has been carefully gathered and analyzed, the scientist forms conclusions. A conclusion must be based solely on the information gathered. Each conclusion must be backed up with supporting evidence from experimentation. If the hypothesis is that "caffeinated soda helps students study for longer periods of time," the research must show a definite pattern of such behavior by observed users before the study can conclude that the hypothesis was confirmed. Medical evidence might be included, along with supporting research from other scientists. If the original hypothesis is not supported by the collected data, the scientist must decide whether or not the method used to gather data was valid before deciding that the hypothesis was wrong. If other factors were unaccounted for in the experiment, the scientist might redesign the experiment.

THEORY

If a hypothesis can withstand extensive testing, it may become a theory. A theory is a prediction that is supported by a large amount of experimental data. The scientist must be able to reproduce the data collected to support the original hypothesis. The scientist repeats these experiments to ensure the correct results were obtained. Then other scientists test the original hypothesis using their own experiments; these experiments are also repeated and tested against other experiments. Even after a hypothesis is accepted as a theory, it is still not regarded as an unchangeable fact. Theories are tested over and over again.

APPENDIX D: PERIODIC TABLE OF THE ELEMENTS

Group

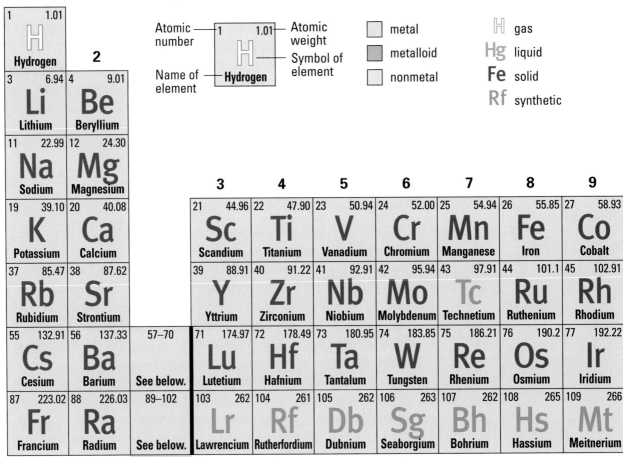

18

					2 4.00 **He** Helium

13 **14** **15** **16** **17**

5 10.81 **B** Boron	6 12.01 **C** Carbon	7 14.01 **N** Nitrogen	8 16.00 **O** Oxygen	9 19.00 **F** Fluorine	10 20.18 **Ne** Neon
13 26.98 **Al** Aluminum	14 28.09 **Si** Silicon	15 30.97 **P** Phosphorus	16 32.07 **S** Sulfur	17 35.45 **Cl** Chlorine	18 39.95 **Ar** Argon

10 **11** **12**

28 58.70 **Ni** Nickel	29 63.55 **Cu** Copper	30 65.39 **Zn** Zinc	31 69.72 **Ga** Gallium	32 72.59 **Ge** Germanium	33 74.92 **As** Arsenic	34 78.96 **Se** Selenium	35 79.90 **Br** Bromine	36 83.80 **Kr** Krypton
46 106.4 **Pd** Palladium	47 107.87 **Ag** Silver	48 112.41 **Cd** Cadmium	49 114.82 **In** Indium	50 118.71 **Sn** Tin	51 121.75 **Sb** Antimony	52 127.60 **Te** Tellurium	53 126.90 **I** Iodine	54 131.29 **Xe** Xenon
78 195.08 **Pt** Platinum	79 196.97 **Au** Gold	80 200.59 **Hg** Mercury	81 204.38 **Tl** Thallium	82 207.2 **Pb** Lead	83 208.98 **Bi** Bismuth	84 (209) **Po** Polonium	85 (210) **At** Astatine	86 (222) **Rn** Radon
110 269 **Uun**	111 272 **Uuu**	112 277 **Uub**						

66 162.50 **Dy** Dysprosium	67 164.93 **Ho** Holmium	68 167.26 **Er** Erbium	69 168.93 **Tm** Thulium	70 173.04 **Yb** Ytterbium
98 251.08 **Cf** Californium	99 252.08 **Es** Einsteinium	100 257.10 **Fm** Fermium	101 258.10 **Md** Mendelevium	102 259.10 **No** Nobelium

APPENDIX E: CLASSIFICATION OF ORGANISMS

The classification of species is always changing as new information is discovered. The questions of whether the Archaea should be treated as a Kingdom separate from Monera and how to distinguish species of asexually reproducing organisms, for example, make classification of bacteria and many protists difficult. Sections, an extra level of classification between kingdoms and phyla or divisions and subphyla have been added to some especially large or diverse groups. The system shown here is a compromise among several proposals and will certainly change over the next few years as more evidence accumulates. Some small phyla, classes, and orders are not listed; a complete classification of all species would require much more space than is available here. The numbers in parentheses following the names of some groups indicate estimated numbers of species.

Kingdom Monera

Section Archaea (100)
  Division Crenarcheota (includes many species that live best under extreme high temperatures and acid conditions, as well as others that live at lower temperatures. Many are chemoautotrophs)
  Division Euryarcheota (includes methane-producing anaerobes and species that require very high salt concentrations or temperatures)
Section Eubacteria (includes very diverse bacteria, including spore-forming bacilli, spirochetes, photosynthetic cyanobacteria, and most other types; 3000)

Kingdom Protista

Phylum Mastigophora ("animal-like" heterotrophic protists with flagella; 5000)
Division Rhodophyta (multicellular red algae; 4000)
Division Euglenophyta (green single-celled autotrophs with flagella and chloroplasts but without cell walls; 800)
Division Chytridiomycota (golden-colored funguslike cells; 1000)
Division Chlorophyta (green algae; 7000)

Section Alveolata (protists with alveoli: sacs of membrane lying just inside the cell membrane)
  Phylum Pyrrophyta (also called dinoflagellates; single-celled marine organisms with two flagella that include phototrophs, free-living heterotrophs, and animal symbionts; 1000)

Phylum Apicomplexa (formerly called Sporozoa; mostly parasites of animals; 3600)
Phylum Ciliata (heterotrophs with cilia, including *Paramecium;* 6000)

Section Stramenopilata
Division Phaeophyta (brown algae, including kelp; 1500)
Division Chrysophyta (golden-brown algae, yellow-green algae, diatoms; 12,000–15,000)
Division Labyrinthulomycota ("slime nets" or net slime molds)
Division Oomycota (formerly classified as fungi; water molds and plant parasites such as white rusts and downy mildews; 400)

Section Gymnomycota (slime molds)
Division Acrasiomycota (cellular slime molds; 26)
Division Myxomycota (true slime molds; 400)

Kingdom Plantae

Division Bryophyta (liverworts, hornworts, mosses; 16,000–23,000)
Division Psilophyta (whisk ferns; 10–15)
Division Lycophyta (club mosses; 1500)
Division Sphenophyta (horsetails; 15–25)
Division Pterophyta (ferns; 10,000–12,000)
Division Coniferophyta (pines, firs, spruce, other conifers; 500–550)
Division Anthophyta (angiosperms; 235,000–300,000)
Class Dicotyledonae (maples, oaks, dandelions, most vegetables and ornamental flowers; 225,000–250,000)
Class Monocotyledonae (grasses, palms, tulips, lilies; 50,000–70,000)

Kingdom Fungi

Division Zygomycota (600–1000)
Division Ascomycota (sac fungi, including yeasts, cup fungi, and most of the fungi in lichens; 30,000–50,000)
Division Basidiomycota (mushrooms, puffballs, other club fungi; 25,000–50,000)
Division Deuteromycota (asexual fungi; 25,000–40,000)

Kingdom Animalia

Phylum Porifera (sponges; 5000)

Section Radiata (radially symmetrical animals with a single opening through which food is taken in and digestive waste is expelled)
Phylum Cnidaria (*Hydra,* jellyfish, sea anemones, corals, and other animals with stinging cells; 10,000–12,000)
Phylum Ctenophora (comb jellies: jellyfishlike animals without stinging cells; 90–150)

Section Protostomia (animals in which the opening of the blastula-stage embryo becomes the mouth and the anus forms later)
Phylum Platyhelminthes (planaria, tapeworms, liver flukes, and other flatworms; 10,000)
Phylum Nemertea (proboscis worms or ribbon worms; 650)
Phylum Acanthocephala (spiny-headed worms; 500)

Phylum Rotifera (rotifers; 1700)
Phylum Nemata (nematodes; 12,000)
Phylum Bryozoa ("moss animals" that use cilia to sweep food particles in seawater into their mouths; 4000)
Phylum Brachiopoda (lamp shells that resemble mollusks; 300)
Phylum Mollusca (mollusks)
    Class Polyplacophora (chitons, with flattened bodies and eight shell plates; 600)
    Class Gastropoda (snails, slugs, sea slugs, limpets, abalones; 25,000)
    Class Bivalvia (also known as Pelecypoda; clams, oysters, scallops, and others with two shells; 7500)
    Class Cephalopoda (octopus, squid, cuttlefish, chambered nautilus; 600)
Phylum Annelida
    Class Polychaeta (sandworms, tubeworms, and other aquatic worms with external gills that also assist motility as simple legs or fins; 8000)
    Class Hirudinea (leeches; 500)
    Class Oligochaeta (earthworms and related worms without external gills; 3100)
Phylum Tardigrada (water bears; 300)
Phylum Arthropoda (animals with jointed exoskeletons; over 2,000,000)
  Subphylum Chelicerata (arthropods with two main body segments and the anterior pair of limbs modified as pincers or fangs)
    Class Merostomata (horseshoe crabs; 4)
    Class Pycnogonida (sea spiders; 1000)
    Class Arachnida (spiders, ticks, scorpions, mites, and others with eight legs; 100,000–700,000)
  Subphylum Uniramia (also known as Mandibulata; arthropods with antennae and with the anterior pair of limbs modified as jaws)
    Class Crustacea (arthropods with two pairs of antennae and with the anterior three pairs of appendages modified as mouthparts; lobsters, crabs, shrimp, pillbugs, water fleas, crayfish, barnacles; 26,000–40,000)
    Class Chilopoda (centipedes; 2500)
    Class Diplopoda (millipedes; two pairs of legs per body segment; 20,000–60,000)
    Class Insecta (insects; arthropods with three body segments, six legs, one pair of antennae, and in most cases two pairs of wings; 1,000,000–2,000,000)
      Order Thysanura (bristletails, silverfish, firebrats; no wings)
      Order Collembola (springtails; no wings)
      Order Ephemeroptera (mayflies)
      Order Odanata (dragonflies, damselflies; 5000)
      Order Plecoptera (stoneflies)
      Order Orthoptera (grasshoppers, crickets, roaches, praying mantis; 30,000)
      Order Dermaptera (earwigs; 1000)
      Order Isoptera (termites; 2000)
      Order Mallophaga (biting lice, bird lice)
      Order Anoplura (sucking lice; 2400)
      Order Hemiptera (true bugs, bed bugs, assassin bugs; 55,000)
      Order Homoptera (aphids, scale insects, leafhoppers)
      Order Coleoptera (beetles, weevils; 500,000–800,000)
      Order Neuroptera (lacewings)
      Order Trichoptera (caddisflies; 7000)
      Order Lepidoptera (moths, butterflies, skippers; 140,000)
      Order Hymenoptera (bees, wasps, ants; 90,000)

Order Diptera (flies, mosquitoes; 80,000)
Order Siphonaptera (fleas; no wings; 1200)

Section Deuterostomia (animals in which the opening of the blastula-stage embryo becomes the anus and the mouth forms later)
Phylum Echinodermata (sessile or sedentary animals with bilaterally symmetrical larvae, radially symmetrical adults, external spines or prickly skins, and a water vascular system that functions in movement; 6000)
Class Asteroidea (sea stars; 1800–2200)
Class Crinoidea (crinoids, sea lilies, sea fans; 60–650)
Class Ophiuroidea (brittle stars)
Class Echinoidea (sea urchins, sand dollars; 860)
Class Holothuroidea (sea cucumbers; 900)
Phylum Chordata (Chordates: animals with notochords, hollow, dorsal nerve cords, pharyngeal slits, and a tail that extends beyond the posterior end of the nerve cord)
Subphylum Urochordata (also called Tunicata; sea squirts; 2000)
Subphylum Cephalochordata (lancelets; 30)
Subphylum Vertebrata (animals with backbones)
Class Agnatha (jawless fish such as lampreys, hagfish, and slime hags; 50)
Class Chondrichthyes (fish with cartilage skeletons, including sharks, skates, and rays; 800)
Class Osteichthyes (fish with skeletons of bone; 30,000)
Class Amphibia (semiaquatic vertebrates that may undergo metamorphosis from a gill-bearing larval stage to an air-breathing adult form; frogs, toads, newts, salamanders; 3100)
Class Reptilia (turtles, snakes, lizards, crocodiles; 7000)
Class Aves (birds; animals with feathers; 8600)
Class Mammalia (vertebrates with hair and mammary glands; kangaroos, shrews, bats, mice, seals, whales, deer, cats, humans; 4100)
Order Monotremata (egg-laying mammals; platypus, spiny anteater or echidna)
Order Marsupialia (also called Metatheria; mammals with a maternal pouch in which older embryos complete development; kangaroos, opossums, koalas)
Order Insectivora (small insect-eating mammals; shrews, moles)
Order Chiroptera (bats)
Order Primata (primates: omnivorous mammals with thumbs that enable grasping, forward-facing eyes, and a large cerebrum; lemurs, monkeys, apes, humans)
Order Edentata (sloths, anteaters, armadillos)
Order Lagomorpha (mammals with long, chisel-like front teeth and long hind legs adapted for jumping; rabbits, hares)
Order Rodentia (mammals with continuously growing chisel-like front teeth; mice, rats, squirrels, beavers, porcupines)
Order Cetacea (dolphins, porpoises, whales)
Order Carnivora (predators with sharp canine teeth and flatter molars; dogs, cats, weasels, bears, seals, walrus, raccoons)
Order Proboscidea (elephants)
Order Sirenia (manatees, dugongs)
Order Perissodactyla (herbivorous mammals with hooves and an odd number of toes on each foot; horses, rhinoceroses, tapirs)
Order Artiodactyla (herbivorous mammals with hooves and an even number of toes on each foot; deer, sheep, goats, pigs, cattle, giraffes, camels, hippopotamuses)

GLOSSARY

A

abscisic acid A plant hormone that inhibits plant growth and promotes dormancy.

acid A substance that increases the concentration of hydrogen ions in a solution, making it more acidic.

actin A thin protein filament in a muscle cell.

action potential The electrical pulse that travels along a neuron, caused by the entry of sodium ions through the cell membrane.

active site The location on an enzyme that binds to a specific type of reactant.

active transport Use of cellular energy to transport substances in a direction opposite to that in which diffusion would carry them.

adapt To adjust to changes in the environment in a way that increases the chance of survival.

adaptation The adjustment in structure or behavior by which a species becomes better suited to survive in an environment.

adaptive radiation The splitting of a population into many species with different specializations that reduce competition between them.

adenosine triphosphate (ATP) A substance that supplies large amounts of energy to cells.

adrenal gland A gland that releases the hormones adrenaline and cortisol.

aerobe An organism that uses oxygen in respiration.

algal bloom An overgrowth of algae, usually due to warm temperatures and/or an oversupply of nutrients, often from runoff containing nitrogen from fertilizers.

allele One of several forms of the same gene.

altruism A form of behavior in which individuals act for the benefit of others.

alveoli Microscopic air sacs in lungs where oxygen is absorbed and CO_2 is released.

ameboid movement Movement of a eukaryotic cell along surfaces by the flow of cytoplasm into temporary extensions of the cell.

anabolic steroids Synthetic hormones that promote muscle growth.

anaerobe An organism that lives in an oxygen-deprived environment.

analogous Unrelated but having the same function.

anaphase The stage of mitosis during which the chromatids come apart and move to opposite ends of the cell.

angiosperm A flowering plant.

annual A plant that lives for a year or less.

anterior The front, or head, of an organism.

anthropologist A scientist who studies the origins of humans and the development of their societies.

antibodies Animal proteins that bind to specific foreign cells and substances, marking them for destruction by the cells of the immune system.

anticodon A triplet of nucleotide bases in tRNA that binds to a complementary codon in mRNA.

antigens Foreign molecules that stimulate an immune response.

anus The posterior opening through which an organism releases the waste that remains after absorbing the nutrients from food.

apical meristem Plant tissue that supplies cells for growth, located in the tips of roots and the buds of shoots.

apicomplexan A member of a phylum of mostly parasitic protist species.

appendage Limb.

appendicular skeleton The bones of the limbs, shoulder, and pelvic girdle.

aquaculture Cultivation of economically important aquatic organisms.

aquifer Underground water supply.

arrhythmia A change or irregularity in the rhythm or force of a heartbeat.

arteriole A small blood vessel that branches from an artery to the capillaries.

artery A blood vessel that carries blood away from the heart.

artificial selection Selectively breeding certain members of a population to produce or maintain desired traits.

atom The smallest unit of an element.

atrium One of the top two chambers of the heart that receives either deoxygenated blood from the body (right atrium) or oxygenated blood from the lungs (left atrium).

autoimmune disease A disease such as diabetes, arthritis, and lupus that occurs when the body's immune system attacks the body itself.

autotroph An organism that uses nonbiological sources of energy and materials to produce its own carbohydrates, lipids, and amino acids.

auxin A plant hormone produced in an actively growing part of the plant and transported to another part of the plant where it produces a growth effect.

axial skeleton The skeleton of the trunk and head.

axon The threadlike extension of a neuron that conducts an action potential along its length.

bacillus The rodlike shape of many types of bacteria; one of the three major shapes of bacteria.

bacteriophage A virus that infects and destroys bacteria.

base A substance that, when added to a solution, increases the solution's concentration of hydroxyl ions, making it more alkaline or more basic.

bilateral symmetry A body plan in which the left and right sides are nearly mirror images.

biochemical pathway A series of chemical reactions in which a product of one reaction becomes the reactant in the next reaction.

biogeochemical cycle A cycle that describes the movement of a key nutrient, such as nitrogen or oxygen, through the environment.

biomass The total dried weight of the organisms in a population.

biome A community of organisms that is adapted to specific climate and soil conditions.

biosphere The parts of the Earth where organisms live.

B lymphocyte A white blood cell that matures in the bone marrow.

body plan The general structure of an organism.

brain stem The part of the brain that controls heart rate, breathing rate, and the diameter of blood vessels.

bronchus (plural bronchi) Either of the two primary divisions of the trachea that lead respectively into the right and the left lung.

bursa (plural bursae) Fluid-filled sac that provides cushioning between bones and tendons.

C

callus A disorganized mass of plant cells.

Calvin cycle The second stage of photosynthesis.

cambium A meristem that produces xylem cells on its inner face and phloem cells on its outer face.

capillary The smallest blood vessels, where blood exchanges substances with the surrounding tissues.

capsule A protective layer around a bacterial cell wall.

cardiovascular system A network of vessels through which a heart pumps blood.

carnivore An organism that eats meat.

carpel The female sex organ of a flower that includes the stigma, ovary, and style.

carrier A heterozygote with a normal phenotype and a recessive disease-causing allele.

carrying capacity The maximum population size for a particular species that an environment can support.

catalyze To increase the rate of a chemical reaction without being consumed or created by the reaction.

cell cycle The phases of growth and division in eukaryotic cells.

cell division The process by which a cell divides to become two new cells.

cell-mediated immunity Immune system response involving interactions between white blood cells.

cell membrane A very thin boundary of lipid and protein that separates a cell's contents from its surroundings.

cell plate A layer of vesicles that appear across the middle of a dividing plant cell and eventually fuse with the cell membrane.

cell respiration The process of breaking down sugars to release energy.

cell wall The rigid part of some types of cells that surrounds the cell membrane.

central nervous system The part of the vertebrate nervous system consisting of the brain and spinal cord.

centriole One of a pair of cylindrical bundles of protein fibers found in the cytoplasm of most eukaryotic cells other than plants, that may help to organize the separation of chromosomes when the cell divides.

centromere A protein structure found in chromosomes that joins together each pair of chromatids.

cerebellum The part of the brain, lying below the cerebrum at the rear of the brain, that is responsible for muscle coordination and maintaining posture and balance.

cerebrum The part of the brain responsible for perception, thought, and voluntary movement.

cervix The hollow, narrow, neck area of the uterus leading to the vagina.

chemical bond The force that binds atoms to each other when they share or exchange electrons.

chemical energy Energy produced or consumed by chemical reactions.

chemical reaction A change that occurs when a substance decomposes or combines with other substances to form a new substance.

chemoautotroph An organism that obtains energy by oxidizing inorganic substances.

chitin A tough, nitrogen-containing carbohydrate that forms much of the exoskeletons of arthropods and the cell walls of some fungi.

chlorophyll A pigment found in plants and cyanobacteria that absorbs sunlight during photosynthesis.

chloroplast An organelle found only in plants and photosynthetic protists that absorbs sunlight and uses it to drive the synthesis of sugars from carbon dioxide and water.

chromatid One of two identical strands of a duplicated chromosome.

chromosome A long molecule of DNA that encodes instructions for manufacturing a cell's proteins.

chyme The mass of partly digested food expelled by the stomach into the small intestine.

cilia Short fibers on the surface of some eukaryotic cells that help to move the cell or something in the surrounding environment.

ciliate A protist that moves by beating cilia which are arranged in rows along its surface.

circadian rhythms Physiological and behavioral characteristics that follow a daily pattern.

class A subdivision of a phylum.

climax community A community that is at the last stage of succession.

clone A group of genetically identical organisms produced by asexual reproduction.

club fungi Members of the fungal division Basidiomycota, consisting mostly of species that form mushrooms or puffballs.

coacervate An artificial, selectively permeable, cell-like structure composed of proteins, carbohydrates, and lipids.

coccus The round shape of many types of bacteria; one of the three major shapes of bacteria.

codominance Condition in which both alleles of a gene are expressed.

codon A sequence of three bases in messenger RNA that specifies a certain amino acid.

coevolution The evolution of two species in response to one another.

colony A group of organisms of the same species that live in close association.

communicable disease A disease in which a pathogen is transmitted directly or indirectly from one host to another.

community All the interacting species within an area.

complement A protein of the immune system that causes invading cells to burst and die.

complementary Relating to the precise pairing of bases between the two strands of DNA or between one strand of DNA and one strand of RNA.

complex carbohydrate A polymer composed of many linked molecules of simple sugars.

compost A mixture of decaying organic matter used as fertilizer.

composting The mixing of organic matter with soil to encourage the release of nutrients from the organic matter.

compound A substance consisting of atoms of two or more elements bonded together.

conditioning A form of learning in which a behavior is associated with a stimulus, usually as a result of training.

congenital Present at birth.

contractile vacuole An organelle in ciliates that pumps excess water out of the organism.

cornea The transparent outer layer of the eye.

corpus luteum The part of the egg follicle left behind after ovulation.

cotyledons The seed leaves of an angiosperm embryo.

crossing over The exchange of corresponding sections of homologous chromosomes during cell division, especially during meiosis.

cuspids Sharp, pointed teeth.

cyclosporine A drug that suppresses the cell-mediated response by cytotoxic T cells without shutting down the humoral response.

cyst A protective capsule that forms around a dormant microorganism.

cytokinin One of a group of plant hormones that stimulate cell division.

cytoplasm The fluid content of a cell.

cytoplasmic streaming A form of active transport that causes nutrient-rich fluid to flow through cytoplasm toward the growing tips of hyphae in fungal cells, or within certain elongated plant cells.

cytoskeleton An internal framework of protein fibers in eukaryotic cells.

defibrillator An electronic device that uses an electrical shock to correct irregular heart rhythms.

degrade To lower in quality.

dendrite Small branching threads of a neuron that are stimulated by the axons of other neurons.

density-dependent factors Factors that affect population growth and depend on population density.

density-independent factors Factors that affect population growth and do not depend on population density.

deoxyribonucleic acid (DNA) Organic molecules that make up the chromosomes of the cells and contain genetic information in the form of a chemical code.

dermal tissue The tissue that forms the outer layer of a plant.

dermis The middle layer of the skin of vertebrates.

diabetes mellitus A disease often caused when the pancreas fails to produce insulin, thereby preventing sugar from getting to the body's cells and depriving them of energy.

diastole The stage of the cardiac cycle in which the heart muscle is relaxed, allowing the ventricles to fill with blood.

differentiate To become specialized.

diffusion The random movement of molecules from a region of high concentration to regions of lower concentration.

digest To break down food into a form that an organism can use.

dihybrid A genetic cross between organisms that are heterozygous for two genes.

diploid Having two sets of chromosomes.

dominant Describes an allele that is expressed whether or not another (recessive) allele of the same gene is also present.

dormant Inactive.

dorsal Upper or back section.

double fertilization A process in which two sperm cells unite with two cells in the embryo sac to form the zygote and the endosperm.

dysentery An intestinal disease characterized by symptoms such as bloody diarrhea, nausea, vomiting, and sometimes liver damage.

ecosystem A community and its nonliving environment, including soil, water, and climate.

ectotherm An animal whose temperature depends on its environment.

electron A particle of an atom having a negative electrical charge.

element One of the 92 basic types of matter; an element cannot be reduced to simpler substances.

embryo A small mass of unspecialized cells, descended from a single cell, that grows and develops into an adult plant or animal.

endocrine system A system of glands that secrete hormones directly into the bloodstream.

endocytosis The process by which cells take in substances by surrounding them and engulfing them in vesicles composed of part of their cell membranes.

endodermis A single cell layer surrounding the center of a plant's root that transfers nutrient ions to the interior of the root by active transport.

endoplasmic reticulum A set of membrane-enclosed channels that carry substances throughout eukaryotic cells.

endoskeleton An internal skeleton made of bone.

endospore A thick internal wall produced by bacteria when exposed to harsh conditions.

endotherm An animal that generates its own body heat through its own metabolism.

energy The ability to do work.

entomologist A scientist who studies insects.

enzyme A protein molecule that speeds the rate of chemical reactions in organisms: a type of catalyst.

epidemic A rapid and extensive increase in the occurrence of a disease above its usual level.

epidermis 1. The surface layer of vertebrate skin. 2. The outer layer of a plant.

epididymis A coiled tube about 6m long in the back of each testis.

epiglottis A flap of cartilage that closes the trachea so that food won't enter the lungs.

epithelial cells Cells that form the covering of the body and the lining of organs and cavities.

epithelium Animal tissue that forms the covering or lining of all body surfaces.

estrogen The female sex hormone that stimulates the thin uterine lining to start growing blood vessels and glands that could nourish a fertilized egg.

estrus A time during which some female animals are receptive to mating and produce ova that can be fertilized.

ethylene A plant hormone responsible for fruit ripening, growth inhibition, and aging.

eukaryote An organism whose cells contain DNA within a nucleus and membrane-bound organelles.

eutrophication An increase in available nutrient levels in an environment that leads to an increase in various populations.

evolution Change in a species over time.

exocrine gland A gland that has ducts that carry secretions away from the gland.

exocytosis A process in which a vesicle fuses with the cell membrane causing its contents to move out of the cell.

exon The segment of a gene that contains coding for a protein.

exotics Non-native species that are introduced into an ecosystem.

exponential growth A kind of population growth in which the size of a population doubles regularly within a given length of time.

external digestion A process in which an organism excretes substances that break down its food before the organism absorbs the nutrients.

external fertilization Fertilization that takes place outside the parents' bodies.

facilitated diffusion A process by which organisms speed up the diffusion of substances into or out of cells without using their own energy to actively transport them.

fermentation Partial breakdown of carbohydrates by cells in the absence of oxygen.

fission The process by which prokaryotic cells divide.

flagellum A whiplike appendage that enables some cells to swim.

flower The sexual reproductive structure of a complex plant (angiosperm).

food chain An arrangement of species according to the order in which one uses another as a food source.

food pyramid A graphic representation of the foods needed to maintain a healthy diet.

forage To look for food.

fossils The preserved or mineralized remains or traces of organisms that lived long ago.

founder effect An extreme form of genetic drift that occurs when an isolated population forms from a few individuals.

free nerve endings The dendrites of a sensory neuron, the simplest type of sensory receptor.

fruit A plant organ that contains seeds.

fruiting body A reproductive structure in which spores form.

fungicide A chemical that kills fungi.

G

G1 phase The first growth phase of the cell cycle that occurs during interphase before DNA synthesis begins.

G2 phase The second growth phase of the cell cycle that occurs during interphase after DNA synthesis.

gallbladder A hollow sac attached to the liver where bile is stored.

gamete A haploid reproductive cell produced by meiosis in a sexually reproducing organism (sperm, egg, or pollen).

ganglion A cluster of neurons.

gene The portion of a DNA molecule that codes for a particular protein.

gene pool All the alleles in a population at any one time.

genetic drift Evolution in which a gene pool changes in random ways because of an event such as a natural disaster.

genome All of the DNA within one cell of an organism.

genotype An individual organism's specific combination of alleles.

genus (plural: genera) A group of closely related species.

germinate To sprout.

germination The sprouting of seeds or spores.

giberellins A group of plant growth regulators that stimulate growth.

gills 1. The thin, flat sheets on the lower surface of a mushroom cap that produce spores. 2. Organs through which some water-dwelling animals absorb dissolved oxygen from the water and release CO_2.

gland An organ that secretes a particular substance into the bloodstream or a duct.

global warming The raising of global temperatures due to the greenhouse effect.

glycogen A form of starch found in animal cells.

glycolysis The first stages in the breakdown of sugars in cell respiration.

Golgi body An organelle that consists of a stack of membranous sacs that collects substances from the endoplasmic reticulum and releases them in vesicles that release their contents in exocytosis.

grafting An artificial plant reproduction technique in which a branch or stem from one plant is joined to the roots and stem of another.

gray matter Nerve tissue of the brain and spinal cord that contains nerve-cell bodies as well as nerve fibers and has a brownish gray color.

greenhouse effect The blanketing of the Earth by carbon dioxide and other gases, trapping heat close to the Earth.

ground tissue The tissue that fills a plant and gives it shape.

guard cells A pair of cells that controls the opening and closing of pores leading to the internal tissues of a plant.

gymnosperm A nonflowering seed plant.

habituation A loss of sensitivity to an event or object after repeated exposure.

hair cells Sensory receptors in the inner ear.

half-life How long it takes half of a sample of a radioactive element to decay.

haploid Having only one set of chromosomes.

hemophilia An X-linked recessive disease in which a blood-clotting factor is absent.

herbivore An animal that eats only plants.

hermaphrodite An individual who has both male and female sex organs.

heterotroph An organism that obtains energy by oxidizing carbon compounds produced by other organisms.

heterozygous Having different alleles at corresponding locations on two matching chromosomes.

histamine An inflammation-inducing substance released by certain cells in the presence of antigens.

homeostasis The maintenance of a steady internal state.

homologous 1. Having evolved from the same ancestral structure. 2. Chromosomes carrying the same gene.

homozygous Having identical alleles at corresponding locations on two matching chromosomes.

hormone A chemical regulator that is produced in glands and affects other parts of an animal's body.

humoral immunity Immune system response in which antibody molecules in the blood bind to specific antigens.

hydrophilic Attracted to water.

hydrophobic Repelled by water.

hydroponics Growing plants without soil in solutions of water and fertilizer.

hydrostatic Relating to the pressure of water or another liquid.

hypertonic A solution with a higher concentration of solutes than the cytoplasm of cells in contact with it.

hypha (plural: hyphae) One of the long, thin strands of cells that make up most fungi.

hypothalamus A gland at the base of the brain that sends messages to other endocrine glands and functions in maintaining constant internal body conditions.

hypotonic A solution with a lower concentration of solutes than the cytoplasm of cells in contact with it.

immune system An organ system that protects animals from parasites, pathogens, and foreign substances.

imperfect fungi Members of the fungal division Deuteromycota consisting of molds that have no known sexual stages.

incisors Flat, slicing teeth usually found in the front of the mouth.

incomplete dominance Inheritance in which an active allele does not entirely compensate for an inactive allele.

infertility The inability to reproduce.

inflammation A local response to injury, including increased blood flow, redness, pain, and tissue swelling.

interferon A protein that vertebrates' cells secrete when they become infected by viruses, bacteria, or foreign cells; it makes other cells prepare defenses against the infective agent by producing proteins that slow the invaders' reproduction.

internal fertilization Fertilization in which the male deposits sperm into the female's reproductive tract, fertilizing the ovum inside the female's body.

internode The stem sections between the nodes of a plant.

interphase The stage in the eukaryotic cell cycle between cell divisions, when the cell grows and produces more ribosomes, mitochondria, and other organelles.

intron The segment of a gene that does not contain coding for proteins.

invertebrate An animal that lacks a backbone.

ion An atom or molecule that has gained or lost one or more electrons and therefore has an electrical charge.

iris The colored layer of the eye behind the cornea that regulates the amount of light admitted to the eye.

isotonic A solution with the same concentration of solutes as the cytoplasm of cells in contact with it.

karyotype An orderly display of an organism's chromosomes.

kidneys Excretory organs that remove wastes from the blood or other body fluids.

kinesiologist One who studies the anatomy and mechanics of human movement.

lactation The making and release of milk.

large intestine The last section of the digestive tract through which digested food passes before the unabsorbed waste is excreted.

law of independent assortment Mendel's law stating that the inheritance of "factors" (alleles) that control one trait is not related to the inheritance of factors that control other traits.

leachates Water and dissolved or suspended materials that have moved downward through the soil.

lens The part of the eye that focuses light.

leukocyte White blood cell.

lichen A symbiotic mixture of fungi and green algae or cyanobacteria.

ligaments Bands of connective tissue that hold bones together at joints.

light reactions The first stage of photosynthesis in which the chloroplast absorbs light energy and produces ATP and NADPH.

lignin A hard, plastic-like substance that strengthens the walls of some plant cells.

limited resource A natural resource that is expected to be entirely exhausted someday.

limiting factor The resource needed for population growth that is in shortest supply.

lipase An enzyme that breaks fats into smaller molecules.

liver A complex organ in the body of which one role is to secrete bile for the digestion of fats in the small intestine.

long-day plant A plant that flowers only after exposure to short nights.

lungs Internal organs through which larger land animals absorb oxygen from the environment.

lymph Colorless fluid that circulates in the vessels of the lymphatic system that is derived from blood plasma and contains white blood cells.

lymphatic system An open circulatory system that gathers excess fluid from the body and drains it into the veins of the cardiovascular system.

lymphocyte A white blood cell that forms specific antibodies that bind to specific antigens.

lysogeny A method of viral reproduction in which the virus integrates its DNA within the host's chromosomes and remains dormant; each time the host cell divides, it replicates the viral DNA with its own.

lysosome A small vesicle inside a cell that contains enzymes that help break down proteins and other substances.

M phase The part of the eukaryotic cell cycle when mitosis occurs.

major histocompatibility complex (MHC) The genes that code for the proteins on cell surfaces that allow vertebrates to recognize their own cells.

medulla A part of the brain stem that acts as a pathway for information between the cerebrum and motor nerve fibers.

meiosis The process by which diploid cells divide to produce haploid gametes.

memory cells The B cells in the immune system that are activated during a second infection by the same antigen-bearing agent.

menopause The end of a woman's reproductive years.

menstrual cycle Monthly series of hormonal changes leading to maturation of the ovum and uterine preparation for possible pregnancy.

menstruate To shed the uterine lining in nonpregnant females.

meristem A specialized zone in a plant that continually supplies new cells to the plant.

messenger RNA An RNA copy of DNA that serves as the guide for protein synthesis.

metabolism The chemical reactions carried out by an organism.

metamorphosis A process by which larvae of many animal species develop into very different adults.

metaphase The stage in mitosis during which chromosomes line up across the center of the cell.

microsphere Tiny droplets of proteinlike substances that exhibit some of the properties of life, including a selectively permeable boundary like a cell membrane.

mitochondrion The eukaryotic organelle where aerobic cell respiration occurs.

mitosis Division of a eukaryotic cell, resulting in two genetically identical daughter cells.

molar A back tooth having flattened surfaces adapted to grinding.

molecule Two or more atoms joined by chemical bonding.

molt To shed part or all of an outer covering.

monoculture Large areas in which the only crop is a single species.

monohybrid A genetic cross between organisms that are heterozygous for one gene.

monomer A subunit of a polymer.

motile Able to move around.

motor neuron A neuron that stimulates a muscle.

mutagen Any agent that is capable of producing mutations.

mutation A change in a cell's DNA sequence.

mutualism A symbiosis that benefits both partners.

mycorrhiza The mutualistic relationship of soil fungi and the roots of most plants.

myelin A lipid coating that insulates the axons in a bundle from one another.

myosin A thick protein filament in muscle cells.

N

natural resource A substance that exists naturally on the Earth and that humans can put to use.

natural selection The process by which the organisms that are best suited to an environment tend to survive and reproduce while less well-adapted organisms tend to die out.

negative feedback A control mechanism that helps to maintain homeostasis, in which overproduction of a substance inhibits the production of the same substance.

nephron One of the microscopic filtering units in the kidney.

nerve net A nervous system in which nerve cells are loosely organized in a netlike pattern and action potentials travel in both directions along a nerve cell.

neuron A nerve cell—the basic unit of the nervous system.

neurotransmitter A chemical substance that relays nerve impulses across a synapse.

neutral pH The midpoint of the pH scale (7) between the acidic and basic ranges.

niche A species' relationships with its living and nonliving environment.

node A place on a plant stem from which leaves grow.

nonrandom mating Reproductive behavior in which individuals are more likely to mate with other individuals of a particular phenotype.

nonrenewable resource A resource that can't replace itself or be replaced by other processes.

notochord A stiff dorsal rod that helps organize development of the chordate embryo.

nucleic acid A polymer composed of nucleotides.

nucleoid The part of a prokaryotic cell that contains the chromosome; also known as the nuclear region.

nucleotide A monomer of DNA or RNA, consisting of a sugar molecule, a phosphate group, and a nitrogen-containing organic base.

nucleus 1. The center of an atom consisting of protons and sometimes neutrons. 2. The part of a eukaryotic cell that contains the chromosomes.

O

omnivore An animal that eats both plants and animals.

oocyte The female gametes that mature to form the ova or egg cells.

organ A group of tissues that carry out a certain function.

organelle A specialized part of a eukaryotic cell.

orthopedist A doctor who is trained to preserve and restore the human musculoskeletal system.

osmosis Diffusion of water through a membrane.

osteocyte A cell that is responsible for bone growth and shape.

osteon A cylinder of layered bone matrix and cells.

ovary An organ in which ova are produced.

ovule A hollow, egg-producing chamber in a seed plant.

oxidation Loss of electrons during a chemical reaction.

oxidized Condition of having lost electrons during a chemical reaction.

paleontologist A scientist who studies fossils.

pancreas A long, flattened gland that produces digestive juices and insulin.

parasitism A symbiotic relationship in which one organism lives in or on another, harming the host.

parthenogenesis A form of reproduction in which offspring are produced from an unfertilized egg.

pathologist A scientist who studies disease.

pedigree A chart of an individual's ancestors that is used to determine genetic history.

pellicle A rigid protein layer inside the cell membrane of some protists.

peptide bond A chemical bond between the nitrogen of one amino acid and the carboxyl carbon of another amino acid, formed by the elimination of water.

peptidoglycan A polymer consisting of polysaccharide and peptide chains.

peristalsis Waves of muscle contraction along a tubular organ.

perennial A plant that lives through several years.

permafrost A permanently frozen layer of soil just below the surface in tundra biomes.

petal A circle of flower parts located just inside the sepal.

pH A scale for describing the concentration of hydrogen ions in a solution.

pharyngeal slit The slits in a chordate embryo that develop into the gill slits in fish.

pharynx A cavity behind the mouth that carries air to the trachea and food toward the esophagus.

phenotype The inherited traits that an individual expresses.

pheromone Hormones that affect the behavior or metabolism of other animals when released into the environment.

phloem Columns of living cells that actively transport substances such as sugars and amino acids throughout the plant body.

photoautotroph An organism that obtains its energy from photosynthesis.

photoperiodism The control of plant flowering by control of night length.

photoreceptor A cell that detects the presence of light.

photorespiration A series of enzyme-catalyzed reactions that interfere with photosynthesis by combining a five-carbon sugar molecule with a molecule of O_2 to produce a three-carbon sugar molecule and two molecules of CO_2.

photosynthesis The use of light energy by organisms to produce carbon compounds.

phylum (plural: phyla) A subdivision of a kingdom.

phytoplankton Algae and cyanobacteria that live near the ocean surface and are consumed by animals and heterotrophic protists.

pineal gland A small endocrine gland that is located in the brain near the hypothalamus and secretes the hormone melatonin.

pioneer species The first organism to survive in a new environment.

pituitary gland An endocrine gland attached to the hypothalamus that controls the activity of other endocrine glands.

placenta Mass of small blood vessels and associated tissues across which materials are exchanged between embryo and mother.

plant growth regulators Substances produced in plants that regulate their growth and development.

plasma The fluid component of blood that consists of a water solution of various proteins, salts, gases, and other substances.

plasma cells The B cells in the immune system that are released in the bloodstream during the primary response to an antigen.

plasmid A small circular piece of DNA found in a bacterium in addition to the cell's chromosome.

plasmodium A colorful slime mold mass that contains many nuclei but is not divided into individual cells.

pneumonia A lung infection.

polar bond The bond between two atoms that attract electrons unequally.

polar molecule A molecule that has one or more polar bonds.

pollen Grains that contain the male gametes of a seed-producing plant.

pollen tube A plant structure that will begin to grow once a pollen grain lands on the stigma. It carries the sperm cells to the egg nucleus.

polymer A large molecule consisting of many similar subunits (monomers).

population A group of individuals of the same species that live in one area and breed with others in the group.

population density The number of individuals per unit of space.

population distribution The pattern in which organisms in a population tend to spread out evenly or unevenly in the area they occupy.

posterior Rear.

predator A meat-eating organism that kills its own food.

preformation A theory that each organism contains within it all its future descendants in miniature form; this theory was popular in the seventeenth and eighteenth centuries.

primary consumers An organism that eats the primary producers in a food web.

primary growth A plant's first, main root.

primary immune response The production of antibodies associated with the first exposure to an antigen.

primary root The embryonic plant root that lengthens, breaks through the seed coat, and begins to absorb water from the soil, turning downward as it grows longer, no matter which end of the seed is up.

primordial First, or primitive.

product A substance that results from a chemical reaction.

progesterone A female sex hormone secreted after ovulation during the menstrual cycle and pregnancy.

prokaryote A type of cell that has a simple structure and does not have the nucleus or the other organelles of eukaryotic cells.

prophase The first stage of mitosis during which the nuclear membrane breaks up and the chromosomes separate from one another.

prostaglandin A hormone-like substance that is produced by many vertebrate cell types.

protein A polymer composed of amino acids.

proton A particle in the nucleus of an atom having a positive electrical charge.

pseudopodium A fingerlike protrusion of cytoplasm that functions in ameboid movement.

pupa The dormant, intermediate stage between larva and adult in most types of insects.

pupil An opening in the iris that lets light into the eye.

R

radial symmetry A body plan in which similar parts are arranged around a central axis.

radiant energy Energy such as light that is transferred by radiation.

reactant A substance that takes part in a chemical reaction.

recycling Returning a used product to a raw material that can be used to make a new product.

reduction To gain electrons.

reflex arc A short series of neurons that transmits a signal directly and involuntarily from a sensory neuron to a muscle or gland.

regenerate To grow back missing parts.

replication The process of making a copy, especially of DNA.

recessive Describes an allele that is not expressed if a dominant allele of the same gene is present.

reciprocal altruism Behavior that benefits another in hopes that there will be future benefit for the person exhibiting the behavior.

reduced Condition of having gained electrons during a chemical reaction.

renewable resource A natural resource that is capable of reproducing by itself.

reproductive barrier Any physical, behavioral, or other obstacle that prevents members of two populations from breeding with one another.

resting potential The difference in voltage between the outside and inside of a neuron when it is not transmitting an action potential.

restriction enzyme An enzyme that cuts DNA molecules at specific base sequences.

retina The tissue layer at the back of the eye that contains the photoreceptor cells.

retrovirus A type of RNA virus that contains an enzyme that transcribes its RNA into DNA once it is inside the host.

reverse transcriptase The enzyme in a retrovirus that transcribes its RNA into DNA inside the host cell.

rhizomes Plant roots that spread horizontally and send up new shoots.

ribosome The structure in cells where protein molecules are synthesized.

root cap Tough tissue that protects a plant's root tip as it pushes through soil.

root hairs A delicate extension of a plant's epidermal cells that helps to increase the surface area for absorption of water and minerals.

rumen One compartment of a herbivore's stomach that acts like a large fermentation vat with many microorganisms.

ruminant A herbivorous animal that has a stomach that is divided into several compartments, one of which is the rumen, from which food is returned to the mouth for further chewing.

S

sac fungi Members of the fungal division Ascomycota that produce sexual spores in a sac consisting of cell walls; sac fungi include the yeasts and cup fungi.

saliva The secretion from salivary glands in the mouth that contain enzymes that break down starch.

scrotum The external sac that contains the testes in males.

secondary consumers An animal located in the third or higher level of a food web that consumes the tissue of other animals rather than of plants.

secondary growth Plant growth that produces the thick stems (or trunks) and roots of trees.

secondary immune response Activation of the memory cells with exposure to an antigen for which the body has already produced antibodies.

secondary roots A branched root system formed by the primary root of a plant.

seedling A young plant grown from seed.

segregation Separation.

selectively permeable Allowing some substances to pass through, but not others.

semen A combination of sperm cells and fluid that protects and nourishes the sperm.

seminiferous tubules Small tubules that make up the testes.

sensory neuron A neuron that transmits a stimulus from the environment.

sensory receptor A cell that is specialized to respond to a stimulus by producing action potentials.

sepal The part of a flower that encloses and protects the bud.

sequencing Determining the exact order of nucleotides in DNA.

sessile Permanently attached to one spot.

short-day plant A plant that flowers only after exposure to long nights.

small intestine A digestive organ in which chyme is exposed to digestive enzymes and juices after passing through the stomach.

social hierarchy An organized group of animals of the same species based on dominance relationships within the group.

sociobiology A theory based on the idea that all animal behavior has a significant genetic component and can be explained in terms of adaptation.

solute A substance that is dissolved in another substance.

solvent A substance in which another substance is dissolved.

source reduction The reduction of the amount and/or toxicity of the waste that enters the waste stream.

spawning The synchronized release of gametes into a water environment.

speciation The process of forming new species.

species A group of organisms that share a set of inherited characteristics; members of a sexually reproducing species are capable of producing fertile offspring with each other, but not with members of other species.

species diversity The number of species in a community and how common each species is.

S phase The phase of the eukaryotic cell cycle during interphase when DNA is replicated.

spirillum A rigid, spiral-shaped bacterium.

spirochete A slender, flexible, spiral-shaped bacterium.

spore A tough, dry structure containing a single fungal or bacterial cell.

stamen The part of a flower that produces pollen.

stolon Horizontal plant stems that send down roots to start new plants.

stomata Microscopic pores in plant leaves.

succession The development of an area by groups of species that succeed one another in time.

surface waters Water that is visible on the Earth's surface such as oceans, lakes, rivers, and streams.

surfactant A protein that keeps the alveolar sacs from collapsing and staying shut.

suture A fibrous joint of the skull plates that appears as the connective tissue in the fontanelle ossifies.

symbiosis A close relationship between organisms of two different species, from which one or both of the organisms benefit.

synapse A narrow gap between neurons across which neurotransmitters carry neural signals.

syndrome A set of symptoms that, together, indicate the presence of a particular disease or abnormality.

synovial fluid A viscous fluid that lubricates the cartilage layers of a joint.

Cole/Corbis. 281: ©Photo Researchers, Inc.; Lester V. Bergman/Corbis; Lester V. Bergman/Corbis; Lester V. Bergman/Corbis. 286: Lester V. Bergman/ Corbis; Runk/Schoenberger from Grant Heilman; Lester V. Bergman/Corbis. 287: Runk/Schoenberger from Grant Heilman. 288: Science Pictures Limited/Corbis. 289: Lester V. Bergman/Corbis; Robert Pickett/Corbis. 291: Jeffrey L. Rothan/Corbis; Corbis/Joseph Sohm; ChromoSohm Inc.; Eric and David Hosking/Corbis. 292: Courtesy of CPI/Photo by S. Crawford. 295: Anthony Bannister, ABPL/Corbis. 297: Grant Heilman Photography; Science Photo Library/Photo Researchers, Inc. 298: Science Photo Library/Photo Researchers, Inc.; Science Photo Library/Photo Researchers, Inc. 299: Lester V. Bergman/Corbis; Lester V. Bergman/Corbis; Christi Carter/Corbis. 300: Science Pictures Limited/Corbis. 301: Science Photo Library/Photo Researchers, Inc.; Lester V. Bergman/Corbis. 305: Grant Heilman Photography. 306: Owen Franken/Corbis.

Chapter 9
314: Jamie Aarron; Papilio/Corbis. 315: Warren Uzzle/Photo Researchers, Inc. 319: George McCarthy/Corbis. 320: Runk/Schoenberger from Grant Heilman. 321: Becky Luigart-Stayner/Corbis. 327: ©Brian Hawkes/Photo Researchers, Inc.; Michael Fogden and Patricia Fogden/Corbis. 329: Microfield Scientific Limited/Science Photo Library/Photo Researchers, Inc. 330: Runk/ Schoenberger from Grant Heilman; Pat O'Hara/Corbis. 333: Charles O'Rear/Corbis. 334: Paul A. Souders/Corbis. 335: Sally A. Morgan, Ecoscene/Corbis. 336: Adam Woolfitt/Corbis; Lester V. Bergman/Corbis. ©1999, PhotoDisc, Inc. 338: Peter Reynolds, The Frank Lane Picture Agency/Corbis; Runk/Schoenberger from Grant Heilman; Charles O'Rear/Corbis; Patrick Johns/Corbis.

Chapter 10
346: Stephen Frink/Corbis. 347: Owen Franken/Corbis. 348: ©1999, PhotoDisc, Inc.; Brandon D. Cole/Corbis; Douglas P. Wilson; Frank Lane Picture Agency/Corbis; ©1999, PhotoDisc, Inc. 350: ©Biophoto Associates/Science Source/Photo Researchers, Inc.; Science Photo Library/Photo Researchers, Inc.; Brandon D. Cole/Corbis. 351: ©1999, PhotoDisc, Inc.; ©1999, PhotoDisc, Inc.; Hulton-Deutsch Collection/Corbis; Brandon D. Cole/Corbis. 353: Kevin Schafer/Corbis. 355: Lester V. Bergman/Corbis; ©Don Fawcett/Photo Researchers, Inc.; Secchi-Lecaque/CNRI/Science Photo Library/Photo Researchers, Inc. 358; Prof. P. Motta/Dept. of Anatomy/ University "La Sapienza", Rome/Science Photo Library/Photo Researchers, Inc. 359: Lester V. Bergman/Corbis. 360: ©David Phillips/Science Source/Photo Researchers, Inc. 361: Stuart Westmorland/Corbis; Runk/Schoenberger from Grant Heilman. 362: Paul A. Souders/Corbis. 364: ©Robert Talbot. 365: Stuart Westmorland/Corbis; ©1999, PhotoDisc, Inc. 370: Stephen Frink/Corbis; John Rutger/Corbis; Runk/Schoenberger from Grant Heilman; Runk/Schoenberger from Grant Heilman; Runk/Schoenberger from Grant Heilman; Stephen Frink/Corbis; Anthony Bannister; ABPL/Corbis; George McCarthy/Corbis; Brandon D. Cole/Corbis; ©1995, PhotoDisc, Inc. 374: Stephen Frink/Corbis; Douglas P. Wilson; Frank Lane Picture Agency/Corbis. 375: Dorling Kindersley Limited, London/Corbis; Stephen Frink/Corbis; Grant Heilman. 376: Stuart Westmorland/Corbis. 378: Brandon D. Cole/Corbis; Brandon D. Cole/Corbis; ©1999, PhotoDisc, Inc.; Kevin Schafer/Corbis; Corbis; ©1999, PhotoDisc, Inc.; ©1999, PhotoDisc, Inc.

Chapter 11
388: Science Photo Library/Photo Researchers, Inc. 389: Photo Courtesy of Industrial Light and Magic/Archive Photos. 391: Sally A. Morgan, Ecoscene/ Corbis; George McCarthy/Corbis. 392: Corbis. 396: Peter Neill/Corbis. 398: SPL/Photo Researchers, Inc. 400: Corbis-Bettman; Anthony Bannister; ABPL/Corbis. 404: Lester V. Bergman/Corbis. 405: Dennis Jones/Corbis. 407: Science Photo Library/Photo Researchers, Inc. 411: ©1999, PhotoDisc, Inc. 414: Science Photo Library/Photo Researchers, Inc. 415: ©1999, PhotoDisc, Inc. 416: Stanford Athletic Dept./Photo Researchers, Inc.

Chapter 12
427: ©1999, PhotoDisc, Inc. 428: ©1999, PhotoDisc, Inc. 429: AP Wide World; Corbis; Corbis. 430: Peter Arnold; ©Manfred Kage/Peter Arnold; ©1999, PhotoDisc, Inc. 435: ©Matt Meadows/Peter Arnold. 443: ©1994 I. Steinmark/ Custom Medical Stock Photo. 451: ©1998 Custom Medical Stock Photo.

Chapter 13
462: CNRI/Science Photo Library/Photo Researchers, Inc. 436: ©Coco McCoy/Rainbow. 465: Corbis; Peter Arnold; ©Ed Reschke/Peter Arnold; Lester V. Bergman/Corbis. 471: James Stevenson/Science Photo Library/Photo Researchers, Inc. 474: Corbis-Bettman. 477: ©Coco McCoy/Rainbow; ©Darviz M. Pour/Photo Researchers, Inc. 482: Corbis. 487: Corbis-Bettman; Corbis-Bettman.

Chapter 14
498: Lester V. Bergman/Corbis. 499: Corbis/Kevin R. Morris. 500: Corbis/Richard T. Nowitz. 501: Corbis/Kevin R. Morris. 502: ©Robert Talbot. 510: TempSport/Corbis. 517: Corbis. 520: David H. Wells/Corbis. 523: Corbis-Bettman; SPL/Photo Resarchers, Inc. 524: Corbis/Alison Wright. 527: ©1999, PhotoDisc, Inc. 528: ©1999, PhotoDisc, Inc. 529: Breast Screening Unit, Kings College Hospital, London/Science Photo Library/Photo Researchers, Inc.

Chapter 15
538: Anthony Bannister, ABPL/Corbis. 539: Owen Franken/Corbis. 541: ©S. Camazine/K. Vischer/Photo Researchers, Inc.; Lester V. Bergman/Corbis. 543: Corbis/Stephen Frank. 544: ©1999, PhotoDisc, Inc. 549: Lester V. Bergman/Corbis; Oliver Meckes/Photo Researchers, Inc. 561: Corbis/Kit Kittle. 562: Lester V. Bergman/Corbis.

Chapter 16
572: ©David M. Phillips/Photo Researchers, Inc. 573: Charles O'Rear/Corbis. 574: Corbis/Ted Spiegel. 575: Lester V. Bergman/Corbis. 576: Lester V. Bergman/Corbis. 577: Corbis/George Lepp. 581: ©Biophoto Associates/Science Source/Photo Researchers, Inc. 582: David H. Wells/Corbis. 590: Science Pictures Limited/Corbis; ©1999, PhotoDisc, Inc.; ©1999, PhotoDisc, Inc. 591: ©SIU 1996/Photo Researchers, Inc.; Corbis/Lester V. Bergman. 593: ©1999, PhotoDisc, Inc.; ©1999, PhotoDisc, Inc. 594: Corbis/David H. Wells; Science Photo Library/Photo Researchers, Inc.; Science Photo Library/Photo Researchers, Inc.; Science Photo Library/Photo Researchers, Inc.; Science Photo Library/Photo Researchers, Inc. 596: Christine Covey/Corbis. 597: ©1999, PhotoDisc, Inc.

Chapter 17
606: ©Phil Farnes/Photo Researchers, Inc. 607: Charles O'Rear/Corbis. 610: Corbis/Anthony Bannister; Corbis/The Purcell Team; ©Stephen Dalton/Photo Researchers, Inc.; Corbis. 611: Corbis/Sally A. Morgan/Ecoscene; ©1997 Margaret Miller/ Photo Researchers, Inc. 612: Corbis/Vittoriano Rastelli; Corbis/Buddy Mays. 614: ©Ken Brate/ Photo Researchers, Inc.; ©1991 David and Hayes Norris/ Photo Researchers, Inc.; ©1991 David and Hayes Norris/Photo Researchers, Inc.; Anthony Bannister/ABPL/Corbis; Owen Franken/Corbis; Eric and David Hoskins/Corbis. 616: Corbis-Bettman. 618: Corbis/Joe McDonald. 619: Aaron Haupt/Photo Researchers, Inc. 620: Lester V. Bergman/Corbis; Corbis-Bettman; Science Source/Photo Researchers, Inc. 621: Corbis/Nik Wheeler. 623: Science Source/Photo Researchers, Inc. 624: ©1999, PhotoDisc, Inc. 625: ©Gregory G. Dimijian/Photo Researchers, Inc.; ©Clem Haasner/Corbis. 626: Corbis/Karl Ammann.

Chapter 18
636: Julia Lewis/Corbis. 637: ©Michael Boys/Corbis. 640: Corbis/Michelle Garrett. 641: Kevin Schafer/Corbis; ©Scott Camazone/Corbis; Owen Franken/Corbis; Joseph Sohm, ChromoSohm, Inc./Corbis. 342: Vince Streano/Corbis. 645: Phillip Gould/Corbis; ©Walter H. Hodge/Peter Arnold; David Lees/©Corbis. 646: Andrew Syred/Science Photo Library/Photo Researchers, Inc. 648; Kevin Schafer/Corbis. 650: Runk/Schoenberger for Grant Heilman; Runk/Schoenberger for Grant Heilman; Adam Woolfitt/Corbis; Runk/Schoenberger for Grant Heilman; Larry Lefever for Grant Heilman; Alfred Pasieka/Peter Arnold. 651: Corbis/Adam Woolfitt. 652: Roger Tidman/Corbis.

Chapter 19
662: Richard Hamilton Smith/Corbis. 663: Michael S. Yamashita/Corbis. 664; Ed Young/Corbis. 666: Runk/Schoenberger for Grant Heilman; Runk/ Schoenberger for Grant Heilman. 667: Coco McCoy/Rainbow. 668: Owen Franken/Corbis; Vince Streano/Corbis; Lester V. Bergman/Corbis; Runk/Schoenberger for Grant Heilman; Runk/Schoenberger for Grant Heilman. 669: Grant Heilman for Grant Heilman; Corbis/Science Source; Science Photo Library/Photo Researchers, Inc. ©1996, PhotoDisc, Inc. 672: Kevin Schafer/ Photo Researchers, Inc. 674: Runk/Schoenberger for Grant Heilman. 676: Runk/Schoenberger for Grant Heilman. 677: ©1995, PhotoDisc, Inc.; Ed Young, Corbis. 679: Katherine Karnow/Corbis; Neil Rabinowitz/Corbis.

Chapter 20
690: Corbis/Paul Jordan. 691: Vince Streano/Corbis. 694: Dr. Jeremy Burgess/ Photo Researchers, Inc.; Michael Long/Corbis. 696: Andrew Syred/Photo Researchers, Inc. 698: Andrew Syred/Photo Researchers, Inc. 699: Owen Franken/Corbis. 704: Corbis-Bettman. 705: Holt Studios/Willem Harnick/ Photo Researchers, Inc.; Gary Birch/Corbis.

Chapter 21
714: Michelle Garrett/Corbis. 715: Pablo Corral V/Corbis. 717: Tony Arruza/Corbis; Kevin Schafer/Corbis; Gary W. Carter/Corbis. 718: David Hoskins/Corbis. 719: Ted Spiegel/Corbis; Ted Spiegel/Corbis. 721: David H. Wells/Corbis. 722: UPI/Corbis-Bettman; SPL/Photo Researchers, Inc. 725: Oliver Meckes/ Photo Researchers, Inc.; Steve Raymer/Corbis; Stephanie Maze/Corbis. 726: Jeremy Horne/Corbis.727: Richard Hamilton Smith/Corbis; Richard Hamilton Smith/Corbis.

Chapter 22
736: Tony Wilson/Corbis. 737: Sally Morgan/Corbis. 740: Corbis-Bettman; Livia Goodsell/Corbis; Lester V. Bergman/Corbis. 742: Corbis-Bettman; ©1999, PhotoDisc, Inc. 743: Michael McCoy/Photo Researchers, Inc. 749: ©Robert Talbot. 751: Tony Wilson/Corbis. 752: Hugh Clark/Corbis. 753: Corbis/Paul Jordan. 754: Kevin Schafer/Corbis; Vince Streano/Corbis. 755: UPI/Corbis-Bettman.

Chapter 23
764: Raymond Gehman/Corbis. 765: Johnathan Blair/Corbis. 767: Joseph Sohm, ChromoSohm Inc./Corbis. 768: Michael Boys/Corbis. 777: Raymond Gehman/Corbis; Richard Hamilton Smith/Corbis. 778: Stephanie Maze/Corbis; Sally Morgan/Corbis. 779: Phillip Gould/Corbis. 783: James Marshall/Corbis. 785: W.G. Baxter/Corbis. 786: Francesco Munzo/Corbis; Andrew Brown/ Corbis. 787: David Meunch/Corbis. 788: Wolfgang Kaehler/Corbis. 789: Joel W. Rogers/Corbis; Stuart Westmorland/Corbis.

Chapter 24
798: Brandon D. Cramer/Corbis. 799: Georg Gerstez/ Photo Researchers, Inc. 801: Vince Streano/Corbis; Paul A. Souders/Corbis. 802: Tim Wheeler/Corbis. 803: Roger Ressmeyer/Corbis; Sally Morgan/Corbis. 810: Bob Krist/Corbis. 812: SLP/Photo Researchers, Inc. 814: Gary Braasch/Corbis; James A. Amor/Corbis. 815: UPI/Corbis-Bettman; Vince Streano/Corbis. 816: Corbis-Bettman; Joel Rogers/Corbis. 818: Sally Morgan/Corbis.

systole The stage of the cardiac cycle in which the heart muscle contracts, pumping the blood from the ventricles to the body.

T

taxis A movement in response to a stimulus in the environment.

telophase The last stage of mitosis, during which a new nuclear membrane forms around the cluster of chromosomes at each pole and the chromosomes are grouped into new nuclei.

tendon A band of connective tissue that connects a muscle to a bone.

terrestrial Relating to land.

tertiary consumers Carnivores that feed on secondary consumers.

testes The male sex organ in which sperm is produced.

testosterone A male sex hormone.

thallus A branching plant or algal body made of filaments of cells.

theory An explanatory hypothesis that has been supported by many observations.

thermoregulation The process of maintaining optimal body temperature.

thylakoid A membrane sac within chloroplasts and cyanobacteria where the light reactions of photosynthesis take place.

thyroid gland A large gland lying at the base of the neck that produces the hormone thyroxin that controls the metabolic rate of the body.

tidal volume The volume of air an animal inhales and exhales with each breath; in humans it averages about 500 mL.

tinnitus Ringing in one or both ears.

tissue Layers or masses of differentiated cells that are suited for particular tasks.

T lymphocyte A white blood cell that matures in the thymus gland.

top consumers The carnivore species that stands atop the food chain.

transcription The process of copying the DNA code into RNA.

translation Biological synthesis of protein according to the information encoded in the nucleotide sequence of mRNA.

transverse fission A process in which an organism reproduces by splitting in half and regenerates the missing parts.

trophic level Any of the various levels found within the food web.

trophic structure The pattern of transfer of energy and nutrients from one species to another in a community.

trypanosomes A group of parasitic zooflagellates that cause diseases such as African sleeping sickness and Chagas' disease.

trypanosomiasis African sleeping sickness; infection by trypanosomes that are transmitted by the tsetse fly and that attack the central nervous system.

turgor Outward pressure of a cell against its cell wall that results from swelling of the cell as it absorbs water.

U

uterus A muscular organ in female animals and many other mammals that holds the developing offspring.

V

vacuole A large vesicle in a plant cell that contains food, water, or wastes.

vagina The tubular passageway that connects the uterus to the outside of the body.

vascular tissue Tissue that transports food, water, and minerals throughout vascular plants.

vas deferens Sperm duct.

vector An organism that transmits a parasite from one host to another.

vein A blood vessel that carries blood to the heart.

ventral Relating to the belly or underside of an organism.

ventricle One of the lower two heart chambers that receives blood from the atria and pumps it through valves either to the lungs (right ventricle) or to the body (left ventricle).

venule A small blood vessel that carries blood from a vein to capillaries.

vertebra (plural vertebrae) One of the bones that make up the vertebral column (backbone) that encloses and protects the spinal cord.

vertebrates Animals with backbones.

vesicle A cell organelle consisting of a tiny fluid-filled sac of lipid membrane.

vestigial Describes body structures that have lost most or all of their function during evolution.

villi Fingerlike projections in the small intestine that increase its absorption area.

virulence A pathogen's ability to cause disease.

vital capacity The maximum volume of air that can be inhaled and exhaled during forced breathing.

weathering The natural processes by which rocks disintegrate.

white matter Nerve tissue that is mostly made up of myelinated nerve fibers, giving it a whitish color, and underlying the gray matter of the brain and spinal chord.

work Using energy to bring about change.

xylem A tissue in plants that carries water and minerals from roots to shoots.

zero population growth The point at which a population's death rate and birthrate are equal, making the growth rate zero.

zooflagellates Animal-like protists with flagella.

zoonosis A disease that can be transmitted from animals to humans.

zooplankton Microscopic organisms, including protists and animals such as small anthropods, that float near the ocean's surface and serve as food for fish.

zygote The diploid cell formed by the fusion of male and female gametes during fertilization.

ACKNOWLEDGMENTS

CORD would like to thank the following people for their contributions.

Dr. Gerald A. Soffen, Director of Biology Programs, NASA Headquarters

Dr. Eric Knudsen, Fisheries Biologist, USGS - Alaska Biological Science Center

Catherine L. Wicklund, M.S., C.G.C., Genetic Counselor, University of Texas Health Science Center at Houston, Department of Obstetrics, Gynecology, and Reproductive Sciences

Cindy Atwell, RDH, BS, Registered Dental Hygienist

Dr. Jack Vance, General Dentistry

Jerry McMillion, Water Quality Coordinator, City of Waco Utility Services Department

Dr. Ira A. Levine, Ph.D., President and CEO, PhycoGen, Inc.

Jay Mertz, Owner, Rabbit Hill Farm

D. Lynn Glass, LAT, ATC, Clinic Coordinator, HEALTHSOUTH Sports Medicine and Rehabilitation

Mr. Dale A. Ester, Arizona Chapter President, American Association of Kidney Patients

Joseph Grubic, Curator of Mammals and Birds, Cameron Park Zoo

Dave Force, Vice President for Animal Training, SeaWorld of Texas

Hollis Malone, Manager of Horticulture, Opryland USA

Lance Vining, Manager, Asplundh

Pat LoPachin, Production Manager, Color Spot-Waco Facility

CREDITS

systole The stage of the cardiac cycle in which the heart muscle contracts, pumping the blood from the ventricles to the body.

taxis A movement in response to a stimulus in the environment.

telophase The last stage of mitosis, during which a new nuclear membrane forms around the cluster of chromosomes at each pole and the chromosomes are grouped into new nuclei.

tendon A band of connective tissue that connects a muscle to a bone.

terrestrial Relating to land.

tertiary consumers Carnivores that feed on secondary consumers.

testes The male sex organ in which sperm is produced.

testosterone A male sex hormone.

thallus A branching plant or algal body made of filaments of cells.

theory An explanatory hypothesis that has been supported by many observations.

thermoregulation The process of maintaining optimal body temperature.

thylakoid A membrane sac within chloroplasts and cyanobacteria where the light reactions of photosynthesis take place.

thyroid gland A large gland lying at the base of the neck that produces the hormone thyroxin that controls the metabolic rate of the body.

tidal volume The volume of air an animal inhales and exhales with each breath; in humans it averages about 500 mL.

tinnitus Ringing in one or both ears.

tissue Layers or masses of differentiated cells that are suited for particular tasks.

T lymphocyte A white blood cell that matures in the thymus gland.

top consumers The carnivore species that stands atop the food chain.

transcription The process of copying the DNA code into RNA.

translation Biological synthesis of protein according to the information encoded in the nucleotide sequence of mRNA.

transverse fission A process in which an organism reproduces by splitting in half and regenerates the missing parts.

trophic level Any of the various levels found within the food web.

trophic structure The pattern of transfer of energy and nutrients from one species to another in a community.

trypanosomes A group of parasitic zooflagellates that cause diseases such as African sleeping sickness and Chagas' disease.

trypanosomiasis African sleeping sickness; infection by trypanosomes that are transmitted by the tsetse fly and that attack the central nervous system.

turgor Outward pressure of a cell against its cell wall that results from swelling of the cell as it absorbs water.

uterus A muscular organ in female animals and many other mammals that holds the developing offspring.

vacuole A large vesicle in a plant cell that contains food, water, or wastes.

vagina The tubular passageway that connects the uterus to the outside of the body.

vascular tissue Tissue that transports food, water, and minerals throughout vascular plants.

vas deferens Sperm duct.

vector An organism that transmits a parasite from one host to another.

vein A blood vessel that carries blood to the heart.

ventral Relating to the belly or underside of an organism.

ventricle One of the lower two heart chambers that receives blood from the atria and pumps it through valves either to the lungs (right ventricle) or to the body (left ventricle).

venule A small blood vessel that carries blood from a vein to capillaries.

vertebra (plural vertebrae) One of the bones that make up the vertebral column (backbone) that encloses and protects the spinal cord.

vertebrates Animals with backbones.

vesicle A cell organelle consisting of a tiny fluid-filled sac of lipid membrane.

vestigial Describes body structures that have lost most or all of their function during evolution.

villi Fingerlike projections in the small intestine that increase its absorption area.

virulence A pathogen's ability to cause disease.

vital capacity The maximum volume of air that can be inhaled and exhaled during forced breathing.

weathering The natural processes by which rocks disintegrate.

white matter Nerve tissue that is mostly made up of myelinated nerve fibers, giving it a whitish color, and underlying the gray matter of the brain and spinal chord.

work Using energy to bring about change.

xylem A tissue in plants that carries water and minerals from roots to shoots.

zero population growth The point at which a population's death rate and birthrate are equal, making the growth rate zero.

zooflagellates Animal-like protists with flagella.

zoonosis A disease that can be transmitted from animals to humans.

zooplankton Microscopic organisms, including protists and animals such as small anthropods, that float near the ocean's surface and serve as food for fish.

zygote The diploid cell formed by the fusion of male and female gametes during fertilization.

ACKNOWLEDGMENTS

CORD would like to thank the following people for their contributions.

Dr. Gerald A. Soffen, Director of Biology Programs, NASA Headquarters

Dr. Eric Knudsen, Fisheries Biologist, USGS - Alaska Biological Science Center

Catherine L. Wicklund, M.S., C.G.C., Genetic Counselor, University of Texas Health Science Center at Houston, Department of Obstetrics, Gynecology, and Reproductive Sciences

Cindy Atwell, RDH, BS, Registered Dental Hygienist

Dr. Jack Vance, General Dentistry

Jerry McMillion, Water Quality Coordinator, City of Waco Utility Services Department

Dr. Ira A. Levine, Ph.D., President and CEO, PhycoGen, Inc.

Jay Mertz, Owner, Rabbit Hill Farm

D. Lynn Glass, LAT, ATC, Clinic Coordinator, HEALTHSOUTH Sports Medicine and Rehabilitation

Mr. Dale A. Ester, Arizona Chapter President, American Association of Kidney Patients

Joseph Grubic, Curator of Mammals and Birds, Cameron Park Zoo

Dave Force, Vice President for Animal Training, SeaWorld of Texas

Hollis Malone, Manager of Horticulture, Opryland USA

Lance Vining, Manager, Asplundh

Pat LoPachin, Production Manager, Color Spot-Waco Facility

CREDITS

Cole/Corbis. 281: ©Photo Researchers, Inc.; Lester V. Bergman/Corbis; Lester V. Bergman/Corbis; Lester V. Bergman/Corbis. 286: Lester V. Bergman/Corbis; Runk/Schoenberger from Grant Heilman; Lester V. Bergman/Corbis. 287: Runk/Schoenberger from Grant Heilman. 288: Science Pictures Limited/Corbis. 289: Lester V. Bergman/Corbis; Robert Pickett/Corbis. 291: Jeffrey L. Rothan/Corbis; Corbis/Joseph Sohm; ChromoSohm Inc.; Eric and David Hosking/Corbis. 292: Courtesy of CPI/Photo by S. Crawford. 295: Anthony Bannister, ABPL/Corbis. 297: Grant Heilman Photography; Science Photo Library/Photo Researchers, Inc. 298: Science Photo Library/Photo Researchers, Inc.; Science Photo Library/Photo Researchers, Inc. 299: Lester V. Bergman/Corbis; Lester V. Bergman/Corbis; Christi Carter/Corbis. 300: Science Pictures Limited/Corbis. 301: Science Photo Library/Photo Researchers, Inc.; Lester V. Bergman/Corbis. 305: Grant Heilman Photography. 306: Owen Franken/Corbis.

Chapter 9
314: Jamie Aarron; Papilio/Corbis. 315: Warren Uzzle/Photo Researchers, Inc. 319: George McCarthy/Corbis. 320: Runk/Schoenberger from Grant Heilman. 321: Becky Luigart-Stayner/Corbis. 327: ©Brian Hawkes/Photo Researchers, Inc.; Michael Fogden and Patricia Fogden/Corbis. 329: Microfield Scientific Limited/Science Photo Library/Photo Researchers, Inc. 330: Runk/Schoenberger from Grant Heilman; Pat O'Hara/Corbis. 333: Charles O'Rear/Corbis. 334: Paul A. Souders/Corbis. 335: Sally A. Morgan, Ecoscene/Corbis. 336: Adam Woolfitt/Corbis; Lester V. Bergman/Corbis. ©1999, PhotoDisc, Inc. 338: Peter Reynolds, The Frank Lane Picture Agency/Corbis; Runk/Schoenberger from Grant Heilman; Charles O'Rear/Corbis; Patrick Johns/Corbis.

Chapter 10
346: Stephen Frink/Corbis. 347: Owen Franken/Corbis. 348: ©1999, PhotoDisc, Inc.; Brandon D. Cole/Corbis; Douglas P. Wilson; Frank Lane Picture Agency/Corbis; ©1999, PhotoDisc, Inc. 350: ©Biophoto Associates/Science Source/Photo Researchers, Inc.; Science Photo Library/Photo Researchers, Inc.; Brandon D. Cole/Corbis. 351: ©1999, PhotoDisc, Inc.; ©1999, PhotoDisc, Inc.; Hulton-Deutsch Collection/Corbis; Brandon D. Cole/Corbis. 353: Kevin Schafer/Corbis. 355: Lester V. Bergman/Corbis; ©Don Fawcett/Photo Researchers, Inc.; Secchi-Lecaque/CNRI/Science Photo Library/Photo Researchers, Inc. 358; Prof. P. Motta/Dept. of Anatomy/ University "La Sapienza", Rome/Science Photo Library/Photo Researchers, Inc. 359: Lester V. Bergman/Corbis. 360: ©David Phillips/Science Source/Photo Researchers, Inc. 361: Stuart Westmorland/Corbis; Runk/Schoenberger from Grant Heilman. 362: Paul A. Souders/Corbis. 364: ©Robert Talbot. 365: Stuart Westmorland/Corbis; ©1999, PhotoDisc, Inc. 370: Stephen Frink/Corbis; John Rutger/Corbis; Runk/Schoenberger from Grant Heilman; Runk/Schoenberger from Grant Heilman; Stephen Frink/Corbis; Anthony Bannister; ABPL/Corbis; George McCarthy/Corbis; Brandon D. Cole/Corbis; ©1995, PhotoDisc, Inc. 374: Stephen Frink/Corbis; Douglas P. Wilson; Frank Lane Picture Agency/Corbis. 375: Dorling Kindersley Limited, London; Stephen Frink/Corbis; Grant Heilman. 376: Stuart Westmorland/Corbis. 378: Brandon D. Cole/Corbis; Brandon D. Cole/Corbis; ©1999, PhotoDisc, Inc.; Kevin Schafer/Corbis; Corbis; ©1999, PhotoDisc, Inc.; ©1999, PhotoDisc, Inc.

Chapter 11
388: Science Photo Library/Photo Researchers, Inc. 389: Photo Courtesy of Industrial Light and Magic/Archive Photos. 391: Sally A. Morgan, Ecoscene/Corbis; George McCarthy/Corbis. 392: Peter Neill/Corbis. 398: SPL/Photo Researchers, Inc. 400: Corbis-Bettman; Anthony Bannister; ABPL/Corbis. 404: Lester V. Bergman/Corbis. 405: Dennis Jones/Corbis. 407: Science Photo Library/Photo Researchers, Inc. 411: ©1999, PhotoDisc, Inc. 414: Science Photo Library/Photo Researchers, Inc. 415: ©1999, PhotoDisc, Inc. 416: Stanford Athletic Dept./Photo Researchers, Inc.

Chapter 12
427: ©1999, PhotoDisc, Inc. 428: ©1999, PhotoDisc, Inc. 429: AP Wide World; Corbis; Corbis. 430: Peter Arnold; ©Manfred Kage/Peter Arnold; ©1999, PhotoDisc, Inc. 435: ©Matt Meadows/Peter Arnold. 443: ©1994 I. Steinmark/Custom Medical Stock Photo. 451: ©1998 Custom Medical Stock Photo.

Chapter 13
462: CNRI/Science Photo Library/Photo Researchers, Inc. 436: ©Coco McCoy/Rainbow. 465: Corbis; Peter Arnold; ©Ed Reschke/Peter Arnold; Lester V. Bergman/Corbis. 471: James Stevenson/Science Photo Library/Photo Researchers, Inc. 474: Corbis-Bettman. 477: ©Coco McCoy/Rainbow; ©Darviz M. Pour/Photo Researchers, Inc. 482: Corbis. 487: Corbis-Bettman; Corbis-Bettman.

Chapter 14
498: Lester V. Bergman/Corbis. 499: Corbis/Kevin R. Morris. 500: Corbis/Richard T. Nowitz. 501: Corbis/Kevin R. Morris. 502: ©Robert Talbot. 510: TempSport/Corbis. 517: Corbis. 520: David H. Wells/Corbis. 523: Corbis-Bettman; SPL/Photo Resarchers, Inc. 524: Corbis/Alison Wright. 527: ©1999, PhotoDisc, Inc. 528: ©1999, PhotoDisc, Inc. 529: Breast Screening Unit, Kings College Hospital, London/Science Photo Library/Photo Researchers, Inc.

Chapter 15
538: Anthony Bannister, ABPL/Corbis. 539: Owen Franken/Corbis. 541: ©S. Camazine/K. Vischer/Photo Researchers, Inc.; Lester V. Bergman/Corbis. 543: Corbis/Stephen Frank. 544: ©1999, PhotoDisc, Inc. 549: Lester V. Bergman/Corbis; Oliver Meckes/Photo Researchers, Inc. 561: Corbis/Kit Kittle. 562: Lester V. Bergman/Corbis.

Chapter 16
572: ©David M. Phillips/Photo Researchers, Inc. 573: Charles O'Rear/Corbis. 574: Corbis/Ted Spiegel. 575: Lester V. Bergman/Corbis. 576: Lester V. Bergman/Corbis. 577: Corbis/George Lepp. 581: ©Biophoto Associates/Science Source/Photo Researchers, Inc. 582: David H. Wells/Corbis. 590: Science Pictures Limited/Corbis; ©1999, PhotoDisc, Inc.; ©1999, PhotoDisc, Inc. 591: ©SIU 1996/Photo Researchers, Inc.; Corbis/Lester V. Bergman. 593: ©1999, PhotoDisc, Inc.; ©1999, PhotoDisc, Inc. 594: Corbis/David H. Wells; Science Photo Library/Photo Researchers, Inc.; Science Photo Library/Photo Researchers, Inc. 596: Christine Covey/Corbis. 597: ©1999, PhotoDisc, Inc.

Chapter 17
606: ©Phil Farnes/Photo Researchers, Inc. 607: Charles O'Rear/Corbis. 610: Corbis/Anthony Bannister; Corbis/The Purcell Team; ©Stephen Dalton/Photo Researchers, Inc.; Corbis. 611: Corbis/Sally A. Morgan/Ecoscene; ©1997 Margaret Miller/ Photo Researchers, Inc. 612: Corbis/Vittoriano Rastelli; Corbis/Buddy Mays. 614: ©Ken Brate/ Photo Researchers, Inc.; ©1991 David and Hayes Norris/ Photo Researchers, Inc.; ©1991 David and Hayes Norris/Photo Researchers, Inc.; Anthony Bannister/ABPL/Corbis; Owen Franken/Corbis; Eric and David Hoskins/Corbis. 616: Corbis-Bettman. 618: Corbis/Joe McDonald. 619: Aaron Haupt/Photo Researchers, Inc. 620: Lester V. Bergman/Corbis; Corbis-Bettman; Science Source/Photo Researchers, Inc. 621: Corbis/Nik Wheeler. 623: Science Source/Photo Researchers, Inc. 624: ©1999, PhotoDisc, Inc. 625: ©Gregory G. Dimijian/Photo Researchers, Inc.; ©Clem Haasner/Corbis. 626: Corbis/Karl Ammann.

Chapter 18
636: Julia Lewis/Corbis. 637: ©Michael Boys/Corbis. 640: Corbis/Michelle Garrett. 641: Kevin Schafer/Corbis; ©Scott Camazone/Corbis; Owen Franken/Corbis; Joseph Sohm, ChromoSohm, Inc./Corbis. 342: Vince Streano/Corbis. 645: Phillip Gould/Corbis; ©Walter H. Hodge/Peter Arnold; David Lees/©Corbis. 646: Andrew Syred/Science Photo Library/Photo Researchers, Inc. 648: Kevin Schafer/Corbis. 650: Runk/Schoenberger for Grant Heilman; Runk/Schoenberger for Grant Heilman; Adam Woolfitt/Corbis; Runk/Schoenberger for Grant Heilman; Larry Lefever for Grant Heilman; Alfred Pasieka/Peter Arnold. 651: Corbis/Adam Woolfitt. 652: Roger Tidman/Corbis.

Chapter 19
662: Richard Hamilton Smith/Corbis. 663: Michael S. Yamashita/Corbis. 664; Ed Young/Corbis. 666: Runk/Schoenberger for Grant Heilman; Runk/Schoenberger for Grant Heilman. 667: Coco McCoy/Rainbow. 668: Owen Franken/Corbis; Vince Streano/Corbis; Lester V. Bergman/Corbis; Runk/Schoenberger for Grant Heilman; Runk/Schoenberger for Grant Heilman. 669: Grant Heilman for Grant Heilman; Corbis/Science Source; Science Photo Library/Photo Researchers, Inc. ©1996, PhotoDisc, Inc. 672: Kevin Schafer/Photo Researchers, Inc. 674: Runk/Schoenberger for Grant Heilman. 676: Runk/Schoenberger for Grant Heilman. 677: ©1995, PhotoDisc, Inc.; Ed Young, Corbis. 679: Katherine Karnow/Corbis; Neil Rabinowitz/Corbis.

Chapter 20
690: Corbis/Paul Jordan. 691: Vince Streano/Corbis. 694: Dr. Jeremy Burgess/ Photo Researchers, Inc.; Michael Long/Corbis. 696: Andrew Syred/Photo Researchers, Inc. 698: Andrew Syred/Photo Researchers, Inc. 699: Owen Franken/Corbis. 704: Corbis-Bettman. 705: Holt Studios/Willem Harnick/ Photo Researchers, Inc.; Gary Birch/Corbis.

Chapter 21
714: Michelle Garrett/Corbis. 715: Pablo Corral V/Corbis. 717: Tony Arruza/Corbis; Kevin Schafer/Corbis; Gary W. Carter/Corbis. 718: David Hoskins/Corbis. 719: Ted Spiegel/Corbis; Ted Spiegel/Corbis. 721: David H. Wells/Corbis. 722: UPI/Corbis-Bettman; SPL/Photo Researchers, Inc. 725: Oliver Meckes/ Photo Researchers, Inc.; Steve Raymer/Corbis; Stephanie Maze/Corbis. 726: Jeremy Horne/Corbis. 727: Richard Hamilton Smith/Corbis; Richard Hamilton Smith/Corbis.

Chapter 22
736: Tony Wilson/Corbis. 737: Sally Morgan/Corbis. 740: Corbis-Bettman; Livia Goodsell/Corbis; Lester V. Bergman/Corbis. 742: Corbis-Bettman; ©1999, PhotoDisc, Inc. 743: Michael McCoy/Photo Researchers, Inc. 749: ©Robert Talbot. 751: Tony Wilson/Corbis. 752: Hugh Clark/Corbis. 753: Corbis/Paul Jordan. 754: Kevin Schafer/Corbis; Vince Streano/Corbis. 755: UPI/Corbis-Bettman.

Chapter 23
764: Raymond Gehman/Corbis. 765: Johnathan Blair/Corbis. 767: Joseph Sohm, ChromoSohm Inc./Corbis. 768: Michael Boys/Corbis. 777: Raymond Gehman/Corbis; Richard Hamilton Smith/Corbis. 778: Stephanie Maze/Corbis; Sally Morgan/Corbis. 779: Phillip Gould/Corbis. 783: James Marshall/Corbis. 785: W.G. Baxter/Corbis. 786: Francesco Munzo/Corbis; Andrew Brown/Corbis. 787: David Meunch/Corbis. 788: Wolfgang Kaehler/Corbis. 789: Joel W. Rogers/Corbis; Stuart Westmorland/Corbis.

Chapter 24
798: Brandon D. Cramer/Corbis. 799: Georg Gerstez/ Photo Researchers, Inc. 801: Vince Streano/Corbis; Paul A. Souders/Corbis. 802: Tim Wheeler/Corbis. 803: Roger Ressmeyer/Corbis; Sally Morgan/Corbis. 810: Bob Krist/Corbis. 812: SLP/Photo Researchers, Inc. 814: Gary Braasch/Corbis; James A. Amor/Corbis. 815: UPI/Corbis-Bettman; Vince Streano/Corbis. 816: Corbis-Bettman; Joel Rogers/Corbis. 818: Sally Morgan/Corbis.

Aristotle, 223
arrhythmia, 436
arteries, 431
arterioles, 431
arthropods, 374–376, 394
artificial insemination (AI), 585, 590
artificial organs, 356
artificial skin, 558
aseptic technique, 244
asexual reproduction, 128, 574–577
    artificial, 718–719
    budding in, 575
    characteristics of, 575
    cloning and, 574–575
    fungi and, 320–321, 325–326, 328,
      329
    genotypes and, 129–130
    parthenogenesis as, 576–577
    plants and, 716–720
    protists and, 290
    regeneration as, 576
    slime molds and, 305
assessments, chapter. See review
    questions
Association of Official Seed Analysts
    (AOSA), 665
astrobiology, 7
athlete's foot, 541
athletic trainers, 400–401
atoms, 11
atrium, 433
autoclave, 242
autoimmune diseases, 560–562
autotrophs, 92–102
    chemoautotrophs, 100–101
    defined, 89
    heterotrophs compared to, 89–91
    photosynthesis and, 92–99, 639
auxin, 702–703
avascular necrosis, 414
axial skeleton, 408–409
axons, 501, 503

B

babies:
    early development of, 596–597
    infant mortality and, 741–742
    premature, 17
bacillus, 242
bacteria, 238–248
    antibiotics and, 54, 62–63, 256
    careers related to knowledge of,
      243, 268–270
    cell organization of, 240–242
    competition between fungi and, 337
    complex cells compared to, 63
    diseases caused by, 260–267
    ecological roles of, 246–248
    energy obtained by, 100–101
    environmental tolerance of, 245–246

    genetic engineering and, 255–259
    growth and division of, 68
    helpful, 9, 56, 238
    intestinal, 549
    modern biotechnology and, 254–259
    pathogenic, 541
    products made by, 258
    review questions on, 272–275
    sampling in food, 239
    shapes of, 242
    structure and function of, 54–56,
      239–242, 244–245
    summary points on, 271–272
    used in food processing, 56, 238,
      240
bacteriophages, 251
ball-and-socket joints, 416
base (chemical), 27
Beaumont, William, 474
beekeepers, 721–722
bees, 362, 543
behavior of animals. See animal
    behavior
bilateral symmetry, 352
bile, 478
biochemical pathway, 87
biocontrol agents, 542
biodegradable resources, 817
biogeochemical cycles, 805–806
biological clocks, 618
biological communities, 766–771
biology, 3
biomass, 774–775
biomes, 782–789. See also ecosystems
    aquatic, 788–789
    chaparrals, 786
    coniferous forests, 787–788
    deciduous forests, 787
    deserts, 785
    grasslands, 785–786
    terrestrial, 783–785
    tropical rain forests, 786–787
    tundra, 788
biosphere, 782
Biosphere 2, 782–783
biotechnology, 254–259
    excretory systems and, 362
    genetic engineering and, 255–259
Biotechnology Industry Organization
    (BIO), 32
biotic potential, 741
birth control, 591–592, 755–756
birth process, 596
birthrates, 742, 743, 747, 751
blindness:
    color, 144
    day, 138
blood:
    circulation of, 430
    donating, 428–429
    heart as pump for, 431–432

blood pressure:
    cardiovascular system and, 439–440
    high blood pressure, 361–362
blood type, 143
B lymphocytes, 553
body plans, 352–354
body temperature, 610–611
bones, 391–393, 404–413
    anthropological study of, 405
    appendicular skeleton and, 408, 410
    axial skeleton and, 408–409
    blood production in, 407
    cartilage and, 407–408
    dynamic nature of, 408
    frame size and, 392–393
    injury and disease of, 412–413
    osteons and, 405–406
    vertebrate adaptation and, 410–412
boom-and-bust cycle, 747, 748
Boveri, Theodor, 135
Bowerman, Bill, 813
Boysen-Jensen, Peter, 703
bradycardia, 436
brain. See also nervous systems
    hemispheric disconnection in,
      506–507
    major regions of, 510–511
brain stem, 510
breathing. See respiratory systems
breeding patterns, 581
bronchi, 446
Brown, Robert, 42
brown algae, 291
brown recluse spider, 540–541
bubonic plague, 541
budding:
    in asexual reproduction, 575
    of yeast cells, 43
bursae, 416
business and marketing careers:
    animals, 381–382
    bacteria and viruses, 268
    behavior of animals, 629
    cardiovascular systems, 455–456
    cellular biology, 74
    chemistry of life, 32
    classical genetics, 148–149
    ecosystems, 791–792
    energy metabolism, 112
    evolution and classification of
      species, 230
    fungi, 340–341
    immune systems, 564–565
    musculoskeletal systems, 418
    natural resources, 819–821
    nutrition, 490–491
    plants, 654–655, 683, 708–709,
      730–731
    populations, 757–758
    protists, 308
    reproductive systems, 599

respiratory systems, 455–456
senses, 531

C

calcium levels, 524–526
callus, 719
calories:
daily requirements for, 488
energy needs and, 103
weight loss and, 104–105
Calvin, Melvin, 98, 99
Calvin cycle, 95–97
cambium, 677
cancer cells, 65, 72
capillaries, 431
capillary exchange, 441
capsule, 241
carbohydrate loading, 110–111, 482
carbohydrates, 481–482
carbon cycle, 91
carbon dioxide, 83
cardiac muscles, 401–402
cardiovascular systems, 428–442. *See also* circulatory systems
blood pressure and, 439–440
blood transfusions and, 428–429
capillary exchange in, 441
careers related to, 455–457
circulation of blood in, 430
conduction system of the heart and, 436–437
diffusion process and, 429
heart as pump for, 431–432
history of scientific study of, 438
kidneys and, 451–454
miscellaneous functions of, 434
pulmonary circulation and, 448–449
review questions on, 459–461
rhythms of the heart and, 435, 436
summary points on, 458–459
variety of, 432–433
career applications:
animals, 381–383
bacteria and viruses, 243, 268–270
behavior of animals, 621–622, 628–631
cellular biology, 40–41, 74–76
chemistry of life, 32–33
classical genetics, 123–124, 148–149
ecosystems, 780, 790–793
endocrine systems, 531–535
energy metabolism, 112–114
evolution and classification of species, 229–232
fungi, 340–342
immune systems, 564–567
molecular genetics, 182–184
musculoskeletal systems, 400–401, 405, 418–420

natural resources, 808–809, 819–822
nervous systems, 512–513, 531–535
nutrition, 482, 490–493
plants, 638–639, 643, 648–649, 654–657, 664–665, 682–685, 701–702, 708–710, 716, 730–733
populations, 739, 757–760
protection and defense, 564–567
protists, 307–309
reproductive systems, 585, 598–601
carnivores, 353, 774
carpels, 680
carrier, 126
carrying capacity, 746–747, 748
cartilage, 407–408
catalyze, 15
cell cycle, 68–69
cell division, 66–67
cell-mediated immunity, 552
cell membrane, 45–52
diffusion of substances through, 48–50
healthy skin and, 46
structure and function of, 46–47
transport of larger particles through, 51–52
cell metabolism, 81–117
autotrophs and, 92–102
careers related to, 112–114
chemical energy and, 84–85
environmental interaction and, 83–84, 88–91
fermentation process and, 107–110
heterotrophs and, 103–111
oxidation and, 85–87
photosynthesis and, 92–99
review questions on, 115–117
sugar metabolism and, 105–106
summary points on, 114–115
cell plate, 71
cell respiration, 105, 107
cells, 39–79. *See also* animal cells; plant cells
bacterial, 54–56, 63
careers related to, 40–41, 74–76
cloning from, 44
complex, 57–62, 63
controlling division of, 72
environmental interactions of, 45–52
growth and division of, 66–72
identifying, 53
life cycle in, 65, 68–71, 72
making models of, 63
nerve, 500–505
review questions on, 77–79
stages of mitosis in, 69–71
summary points on, 76–77
theory of, 41–42, 43
types of, 53–64
cell walls, 49, 241

census, 749
Centers for Disease Control (CDC), 296
central nervous system (CNS), 509–511
centrifuge, 430
centrioles, 61
centromere, 69
cerebellum, 510
cerebrum, 510
cervix, 588
chaparrals, 786
Chargaff, Erwin, 162
Chase, Martha, 252
chemical bond, 11–12
chemical digestion, 471–473
chemical energy, 9–10, 84–85
chemical reactions, 12, 14–15
chemistry of life, 1–37
careers related to, 32–33
characteristics of life and, 2–8
environmental changes and, 24–31
life cycle and, 20–23
organic structure and, 9–19
origin-of-life theories and, 214–219
primordial soup and, 216
review questions on, 35–37
summary points on, 34–35
chemoautotrophs, 100–101, 247
child bearing:
birth process and, 596
birthrates and, 742, 743, 747, 751
contraception and, 591–592
family planning and, 755–756
fetal testing and, 595
infant development and, 596–597
pregnancy and, 592–595
China, population growth in, 751
chitin, 318
Chlamydomonas, 281, 288, 290
chlorophyll, 93
chloroplasts, 58, 93
chordates, 377–379
chromatids, 69
chromosome painting, 121
chromosomes, 55, 120–127
hereditary diseases and, 120–122
meiosis process and, 130–133
sex, 143–144
chyme, 472, 474
cilia, 61, 548–549
ciliates, 300–301
circadian rhythms, 528, 617
circulatory systems, 359–360, 428–442. *See also* cardiovascular systems
blood circulation in, 430
cardiac muscles and, 401–402
diffusion and, 429
miscellaneous functions of, 434
open vs. closed systems, 432–433
cities, ecosystems in, 779–780
class, 225